PLANT CAST PRECAST AND PRESTRESSED CONCRETE

A DESIGN GUIDE

by
WILLIAM R. PHILLIPS
and
DAVID A. SHEPPARD

Sponsored by
The Prestressed Concrete Manufacturers Association of California

FIRST EDITION FIRST PRINTING 1977
SECOND EDITION FIRST PRINTING 1980
SECOND EDITION SECOND PRINTING 1980

Library of Congress Catalog Card Number: 80-84400
ISBN 0-937040-17-7

Cover photograph:

Site #1A – State Office Building – Sacramento

This 4-level office building contains 230,000 sq. ft. of prestressed concrete double tees in the floors and roof. Support elements are exposed aggregate precast concrete bents 12 ft. wide by 50 ft. high. These 2-column, 4-beam bents are attached to prestressed concrete pile foundations and are part of the lateral load resisting system. Pictured is an 8 ft. wide by 24 in. deep prestressed concrete double tee being erected.

Printed in U.S.A.

PRESTRESSED CONCRETE MANUFACTURERS ASSOCIATION OF CALIFORNIA

The Prestressed Concrete Manufacturers Association of California, Inc. (PCMAC) is a nonprofit organization formed twenty years ago to further the causes of specifier and student education, and to encourage standardization and the increased use of precast and prestressed concrete in California.

The past work of various PCMAC committees has resulted in industry guidelines for prestressed concrete bridge applications, improved contract proposal provisions, connection details, transportation and erection procedures, project specifications, product liability, and information in such traditional problem areas as differential camber, tolerances, waterproofing, and exterior finishes. PCMAC-sponsored research has been conducted in many specific fields of application, such as liquid storage tanks, wall panel connectors, and prestressed concrete piling subjected to extreme seismic loading induced by layered soil movements. PCMAC developed an inspector's manual that is used by personnel involved in performing in-plant inspection of precast and prestressed concrete production.

More recently, attention has been directed to the uniqueness of the California marketing situation because of the seismic design requirements. Because of this, PCMAC has become involved with planning areas of required research and testing in line with expanded code acceptance of the use of precast and prestressed concrete elements in resisting lateral seismic forces.

In 1976, the project to develop an educational text for precast and prestressed concrete design was conceived to help fill the void in this area of the educational development of most of the region's designers and specifiers. This text is the result of the efforts and funding of PCMAC producer and associate member companies.

The primary objective of this Design Guide is to assist in teaching the design and application of precast and prestressed concrete, and to serve as a valuable reference document for designers engaged in building design. The Design Guide complements technical handbooks published by the Prestressed Concrete Institute by bridging the gap between theory and practice. Care has been taken to have the material presented be as accurate as possible; however, the Prestressed Concrete Manufacturers Association of California, Inc., cannot accept responsibility for incorrect engineering designs resulting from misuse of, or errors and omissions in, the text material.

This book is distributed by the Prestressed Concrete Institute through an agreement with the Prestressed Concrete Manufacturers Association of California, Inc. PCI's Technical Activities Committee reviewed the book and recommended that it receive wide distribution to engineers, both in design offices and in prestressing plants, and be used as a practical resource document in all engineering schools and colleges.

Although the book relates to practice by the California producers and is heavily oriented to severe seismic loadings under the provisions of the Uniform Building Code, the engineering assumptions, logic and solutions are clearly presented and easily applicable to any other practices and code requirements. Further, suggested solutions to many of the design problems concerning production and handling of product are not available in other published documents. The Technical Activities Committee agrees with the authors that the book complements and supplements current PCI technical books and reports, especially the PCI Design Handbook.

It is for these reasons that this book is made available by the Prestressed Concrete Institute with the conviction that its wide distribution will greatly extend needed knowledge in precast, prestressed concrete design. However, the Prestressed Concrete Institute cannot accept responsibility for incorrect engineering designs resulting from misuse of, or errors and omissions in, the material contained in the book.

201 N. Wells Street/Chicago, Illinois 60606
Telephone: 312/346-4071

PREFACE

This design guide on plant cast precast and prestressed concrete is the result of a project conceived and funded by the Prestressed Concrete Manufacturers Association of California (PCMAC).

The first edition was developed by California Polytechnic State University at San Luis Obispo, in collaboration with certain key industry personnel representing PCMAC, and was published in 1977. Since it was desired to have the text serve equally for practicing design professionals, as well as for students, an extensive technical edit of the first edition was required to achieve this end. The purpose of the text remains as originally conceived in 1976 - to introduce the student, architect, or engineer to precast concrete and its potential in the ever increasing search for efficiency and form, and to eliminate some of the current mystery and misconceptions surrounding its design and use. We feel that the second edition is suitable for use at the undergraduate level in universities as well as in evening courses or design seminars for practicing designers.

The result is a practical and thorough text covering the items one must know in order to design plant cast precast and prestressed concrete in a "how to do it" fashion. In Parts A & B, a good general discussion of the mechanics and advantages of precast concrete production and use is given, followed by presentation of related design aspects of the material, as well as detailed design procedures and the present state of the art of overall structural design with precast and prestressed concrete as permitted by the 1979 Edition of the Uniform Building Code. Part C covers the design of components, with an emphasis on production and erection considerations that must be understood for efficient design. Finally, detailed guide specifications are presented which are comprehensive in their coverage, as opposed to the partially adequate documents available previously. Most important is the discussion preceding the guide specifications pointing out the high points of concern to student and designer; good specifications are a facet of design that is often overlooked.

Overall, the text tells the student what he really needs to know to design prestressed concrete. What is known prior to beginning the design. What preliminary design aids are required to save wasted time with unnecessary trial designs. What stresses are critical and why (and which ones can usually be ignored). And finally, what are inherent pitfalls we should warn the student about in the design of a particular component (camber, deflection, connections, creep effects, loss of prestress, effects of extreme seismic disturbance, temporary handling stresses, erection considerations, tolerances, etc.).

This is the real significance of the text. The various existing design manuals and design texts presuppose a knowledge of the essential practical aspects of plant cast precasting which for the most part exists at neither the student level nor at the level of the practicing professional. This book bridges the gap between theory and practice.

As a prerequisite to using this text, it is mandatory that the student or designer be familiar with the principles of basic reinforced concrete design. For this reason, the elements of ultimate strength theory related to flexure and shear design are not covered. However, they are treated in example form in the component design sections (Part C) and in Chapter 3 of the Second Edition of the PCI Design Handbook.

Certain individuals made essential contributions to the generation of the text:

CALCULATIONS - Craig Oka, who literally burned the midnight oil in fulfilling his commitment prior to departing for study in Copenhagen.

SKETCHES AND DETAILS - Barry Isakson, lent intelligence and imagination as well as a good hand in fulfilling a demanding assignment.

TYPING - Rita Deller, patient and reliable, trying to decipher illegible scrawl while learning about the precast business, page by page.

SPECIAL INFORMATION - on the intricacies of transportation was provided by Al Deller (no relation to Rita) of Progressive Transportation Company in Irwindale.

HISTORICAL INFORMATION - about some past projects shown pictorially provided by Gordon McWilliams, of Stepstone Inc. in Gardena.

TECHNICAL INFORMATION - about the Site 1A State Office Building shown on the cover furnished by Karl Stricker, Senior Structural Engineer, Office of the State Architect, Sacramento.

In closing, we would like to dedicate this book to Lloyd Compton, President of Kabo-Karr Corporation in Visalia, without whose vision and prescience the text would not have been developed.

William R. Phillips
San Luis Obispo

David A. Sheppard
San Francisco

February 1980

TABLE OF CONTENTS

BENEFICIAL STANDARD LIFE INSURANCE COMPANY BUILDING - LOS ANGELES

Sierra white granite aggregate was revealed by deep sand-blasting in the plant. These handsome exterior "T" shaped units also support vertical floor loads for this building, constructed in 1966.(Photo by Eugene Phillips)

PRECAST CONCRETE - A MATERIAL AND A METHOD

Precast concrete, what is it? Basically, precast concrete is defined as concrete which is cast in some location other than its final position in the finished structure. Precast concrete elements are reinforced either with mild steel reinforcing bars or with prestressing strands. When prestressing is employed for the production of precast concrete members, pretensioning is the method usually used, where the strands are tensioned prior to pouring the concrete in long lines in the precasting operation. Prestressing is discussed in greater detail on pp. 145-150.

Precast concrete is both a construction method and a construction material. Precast concrete is produced under rigid quality control conditions in a precasting plant. The concrete strengths used range from 4000 psi to 6000 psi with the higher strengths being preferred to ensure durability and high cycle production rates in the plant. The forms used are of better quality than those normally used for cast in place concrete; hence, truer shapes and better finishes are obtained. Cast in place concrete requires a greater quantity of formwork and results in minimal form re-use of up to a maximum of ten times. For precast concrete, finished wood and fiberglass forms may be used up to 50 times with minor rework; concrete and metal forms have practically unlimited service lives. With precast concrete, the Architect is offered greater freedom in design with a lower total form and product cost, given sufficient project size and repetition of units.

By placing the forms on a vibrating table or with the use of external vibrators placed on the forms, a higher degree of consolidation is obtained which, along with lower water cement ratio and higher cement contents results in higher strengths and truer surfaces and corners.

Repetitious casting of elements also lends itself to pretensioning, saving reinforcing and ensuring crack control during handling and under service loads.

In using precasting, there is a greater control of the surface texture allowing achievement of many textures which are not easily obtainable with cast in place concrete.

As a material, precast concrete may be used either nonstructurally or structurally.

As a method of construction, precast concrete can greatly reduce total project construction time since the units or components are cast and stockpiled while other phases of the building process are performed, thereby telescoping the total required construction time. Due to steam or controlled curing methods, precast concrete can be cast in the afternoon, cured at elevated ambient temperatures seven to ten hours during the night, the units removed from the forms the next morning and the forms readied for reuse the same day. With steam or controlled curing the concrete can reach two-thirds of its strength in 10 to 14 hours, whereas cast in place concrete forms can usually not be removed in less than 7 days without reshoring.

In using precast concrete, continuous uninterrupted erection of components is possible, quickly forming the structural frame and enclosing the building. Precast concrete also lends itself to fast track construction methods. This total savings in time equates to lower interest paid on the construction loan, earlier occupancy and a quicker return on the owner's investment.

Precast concrete is a unit system of construction wherein a combination of a number of identical or similar components are assembled to produce the total building or structure. The precast components may be prestressed piling, single or multi story precast columns, prestressed concrete girders, single tee, double tee, hollow core plank or solid slab floor or roof members, and wall panels consisting of solid units, single or double tee members or insulated sandwich units. The wall units may be either load bearing or non load bearing. Other types of precast components are curtain wall cladding panels, sun shades, complete cell units such as hotel rooms or apartment units of one or more rooms, and architectural precast concrete units serving as fire protective covering for steel columns and spandrel beams or as forms for poured in place concrete. The selection of other types of precast units is limited only by the imagination of the architect and the mold maker's skill.

At this point, it should be pointed out that a project should be of a certain minimum size in order to spread out certain fixed costs such as plant set-up, mold costs and erection mobilization costs over a sufficiently broad base to make the use of precast concrete economically viable. Minimum project sizes are those that would generate 10,000 square feet of architectural precast concrete panels, 15,000 square feet of prestressed concrete deck members, or 1000 lineal feet of standard prestressed or precast concrete components, such as girders, columns or piling.

The structural system is usually a shear wall and diaphragm system; however a ductile moment resisting frame is possible through the use of precast shell units and poured in place concrete. The actual system used depends on the functional requirements of that particular project.

The total building may consist of precast concrete or a combination of precast concrete and either structural steel or poured in place concrete. The components can be structural prestressed, architectural precast, or a combination of both. The potential of the architectural expression possible is unlimited.

The feasibility of the systems building concept results from a large number of identical parts leading to industrialized production, combined with a guaranteed long-term market for the system. The economic advantage is fully realized only if the planning rules of modular coordination with many like units are observed. The component size limit is usually controlled by transportation and erection considerations.

Coarse aggregates used may be either lightweight or normal weight subject to local availability and cost. Where an exposed aggregate surface is desired, the same aggregate can be used throughout the section or the section composed of two mixes with the more expensive special aggregate used in that portion to be exposed, and a standard structural concrete mix used for the backup.

Both 40 and 60 ksi yield strength reinforcing bars are used. 40 ksi yield steel is usually used for #3 and #4 bars and 60 ksi yield steel is used for #5 bars and larger. Prestressing strand of 270 ksi specified ultimate strength is normally used. Strand is available in 3/8, 7/16 and 1/2 inch diameters. Where a smaller strand is used, say 1/4 or 5/16, then grade 250 ksi is used. Welded wire fabric is commonly used to reinforce both structural and architectural precast concrete elements. Cement may be natural grey, white or colored with an additive. Colored concrete may also be achieved by addition of predetermined amounts of color additives when the concrete is batched.

A number of visual presentations will be used in the initial sections to illustrate applications and practical aspects of precast and prestressed concrete manufacture and installation. Following these are sections which discuss design considerations which help in understanding the uniqueness of precast concrete technology.

TYPICAL FRAMING WITH STANDARD COMPONENTS

Certain standard precast and prestressed concrete structural shapes are currently made by the prestressed concrete manufacturers in California. These components are used together to form structures which fall into two general categories:

1. Frame structures

2. Shear wall structures

There are various combinations of the above systems which form sub-categories that allow the architect to optimize the function of the building, and at the same time minimize the architectural constraints imposed by the structural lateral force resisting system of the building. The Uniform Building Code currently permits the use of precast concrete elements to resist seismic lateral forces providing the design and detailing used satisfy the Code requirements for cast in place concrete. To date, most structures that have been built have used the precast elements as "pin ended" non lateral load resisting elements, with the lateral forces being resisted by other methods of construction, such as masonry or poured in place concrete shear walls. The use of the vertical precast or prestressed concrete elements to resist lateral forces has been limited to those situations where the load transfer is direct from the precast concrete element to the supporting foundation or the resisting soil medium, such as for double tee wall panels in industrial building construction and for bearing or sheet piling.

In this section, we shall first see how the basic building blocks of the prestressed concrete industry are put together in buildings, and then we shall briefly survey seismic design of precast prestressed concrete buildings as related to current design concepts used for cast in place concrete, and indicated areas of required research.

On the following two pages are shown the basic standard structural shapes which we use to build structures in precast concrete. Whenever possible the designer should attempt to use the standard shapes and sizes available in the region of the jobsite. This will reduce the product cost due to the savings in mold costs. Just about any structural configuration may be made using these shapes in combination with one another. In addition to normal deck systems consisting of tees or slabs, "spread" deck systems are often used, which feature channels, or tees spread apart with hollow core plank or solid slabs spanning transversely between the stemmed members. Alternate cross sections should be permitted in bidding in order that particular manufacturers can bid their most competitive sections.

STANDARD STRUCTURAL SHAPES

	Approximate Size Ranges		
	Width	Depth	Span
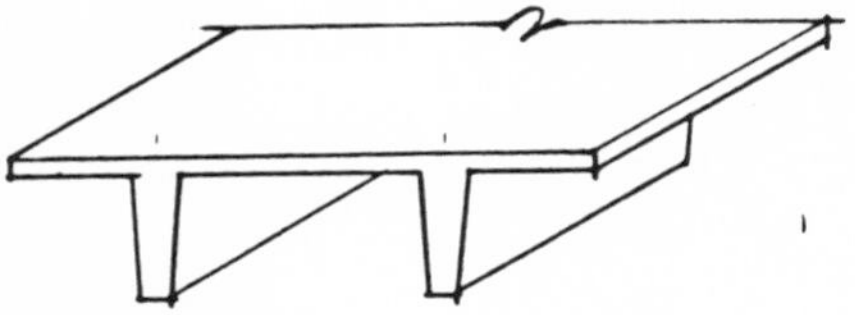 DOUBLE TEE	4' - 12'	10" - 41"	30' - 90'
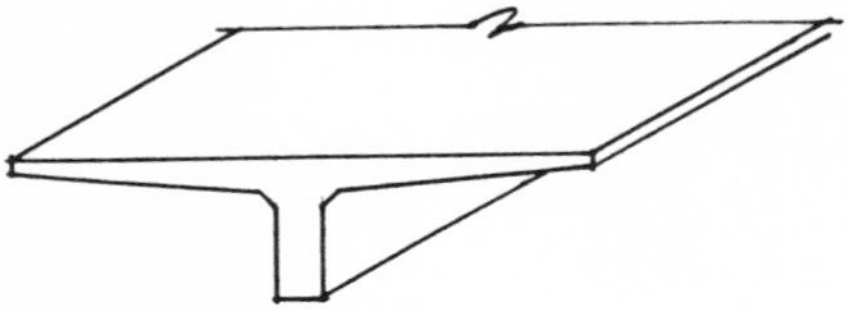 SINGLE TEE	6' - 12'	16" - 48"	30' - 110'
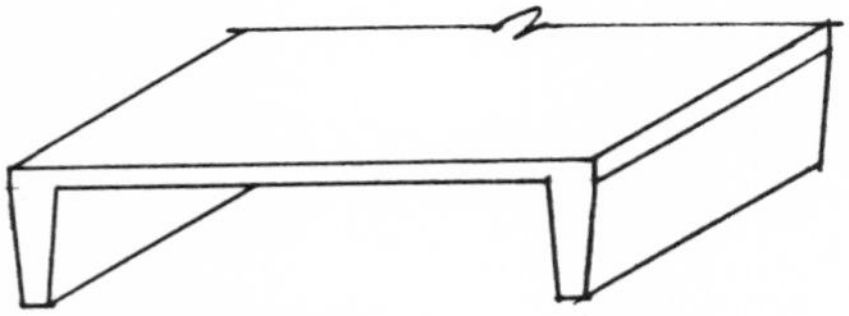 CHANNEL SLAB	6' - 12'	24" - 42"	40' - 90'
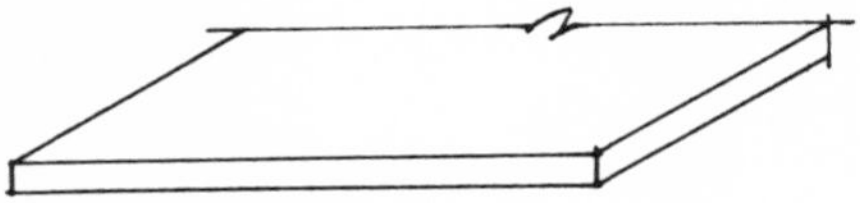 FLAT SLAB	8' - 12'	3" - 6"	14' - 22' (35' with shoring)
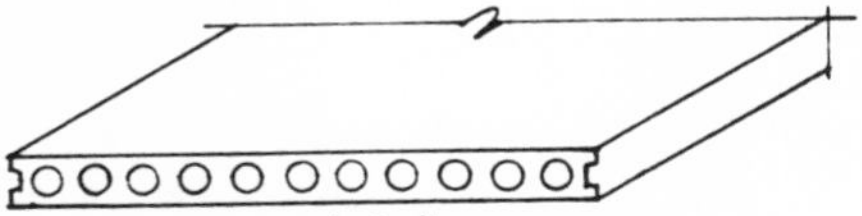 HOLLOW CORE PLANK	40" - 8'	6" - 12"	16' - 42'
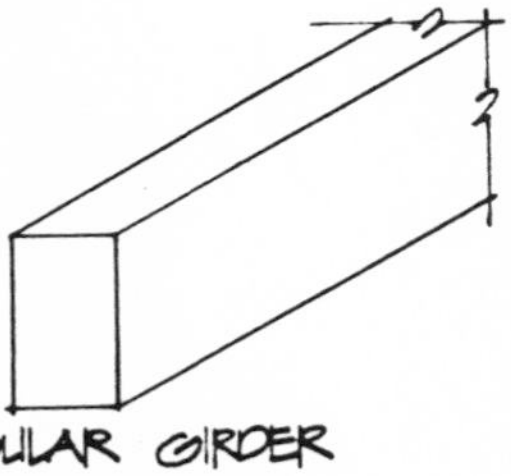 RECTANGULAR GIRDER	12" - 36"	18" - 48"	24' - 70'

STANDARD STRUCTURAL SHAPES

	Approximate Size Ranges		
	Width	Depth	Span
INVERTED TEE GIRDER	12" - 24"	18" - 48"	24' - 48'
LEDGER BEAM	12" - 30"	18" - 48"	24' - 48'
COLUMN	10" - 24"	12" - 24"	
BEARING PILE	12" - 24"	12" - 24"	
SHEET PILE	4' - 8'	10" - 16"	

On pp. 35 - 36, the selection of a module is discussed. The hypothetical building on the following page illustrates the concept of using standard components on an eight foot module. For the two story structure, only every other interior column is taken up to the roof, since the girders supporting the relatively light roof loads can span a much greater distance as compared to the girders supporting the heavy floor loads. The exterior wall panels are eight feet wide in both directions, making it mandatory that the eight foot module be held in the diretion of deck member span as well as in the modular width direction of the deck member. When the exterior elements enclosing the building in the deck member span direction are non-modular, then the module in the deck member span direction is one foot. If the wall panels on the ends of the building are designed as non-load bearing cladding panels, then ledger beams are required to support the ends of the double tees and carry the loads to the columns. This situation also allows the panels to be non-modular in width, if required. Load bearing panels would normally be designed in a modular width, with corbels on the inside face to support the double tee stems.

On the tee floor and roof, a structural topping is usually used to form the horizontal diaphragm to transmit wind and seismic lateral forces to vertical/lateral load resisting elements in the structure, in this case the exterior wall panels. The prestressed concrete deck elements may also be used without a topping, in which case the diaphragm may be formed by interconnecting units together with flange weld plates, or by tying the units together with poured in place concrete closures and shear friction reinforcing. The interior girders may be either inverted tee girders, with the tees resting on the ledges of the girder, or may be rectangular with the tees sitting on top. The columns are usually made in one piece, with corbels to support girders framing in at intermediate levels. The wall panels are supported on grade beams spanning between spread footings, or on continuous strip footings. Where prestressed concrete piling are used, the footing is called a pile cap. The wall panels are interconnected by flange weld plates, and are also connected to the supporting footing by welded or grouted connections, as well as being connected to the poured in place floor slab. Single or double tees may also be used as wall panels with openings for fenestration placed between the tee stems in the flanges.

If spread deck systems are used, then the module is dependent upon the standard width of the infill product spanning between the spread apart supporting channels, single tees or double tees. Several types of supporting elements are used with spread channels, among which are double columns (one at each channel stem), tree columns (tree arms support channels) or solid masonry or precast wall columns, designed to take lateral forces as well as the floor vertical loads.

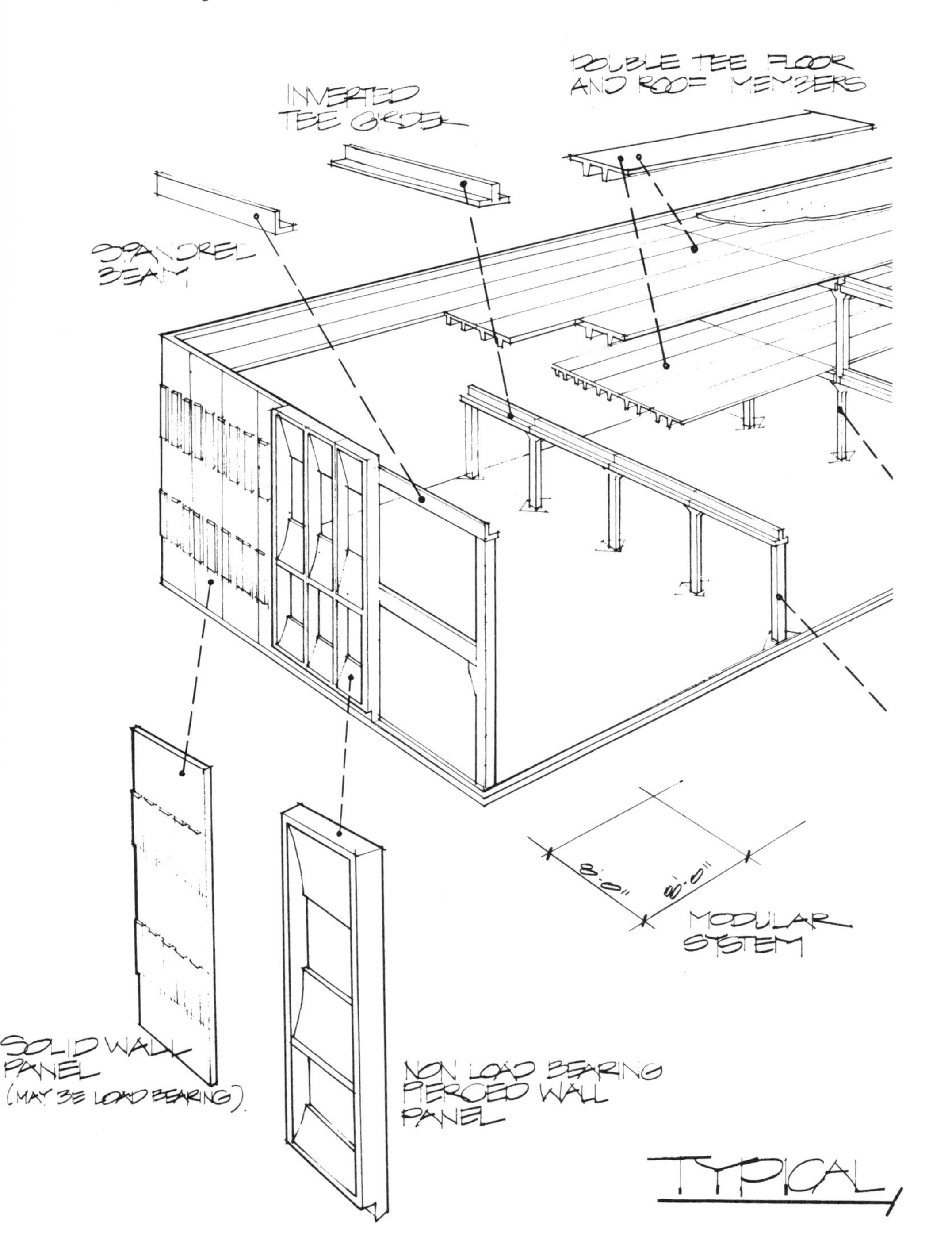
DOUBLE TEE FLOOR
AND ROOF MEMBERS
INVERTED
TEE GIRDER
SPANDREL
BEAM
8'-0"
8'-0"
MODULAR
SYSTEM
SOLID WALL
PANEL
(MAY BE LOAD BEARING).
NON LOAD BEARING
PIERCED WALL
PANEL
TYPICAL

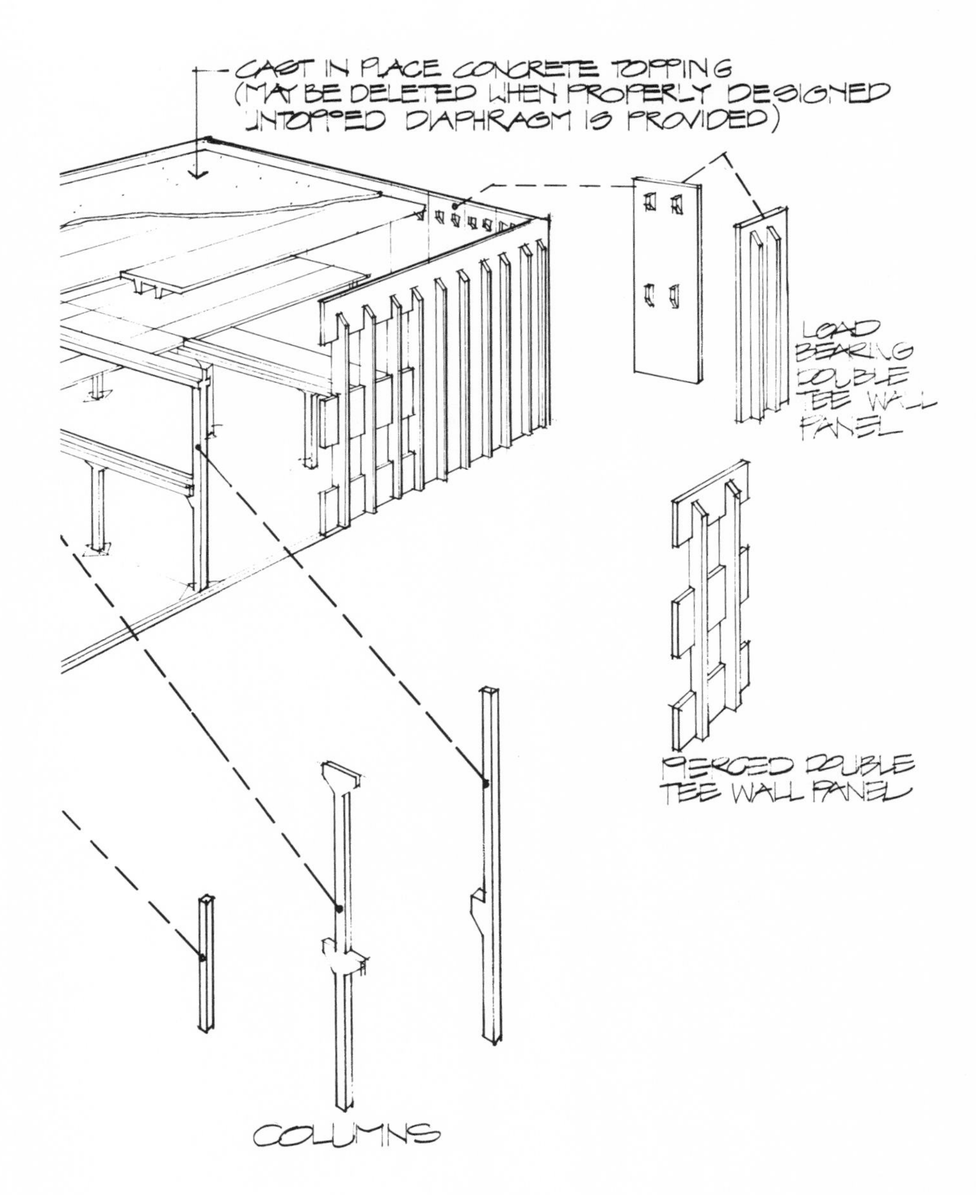

FRAMING W/ STANDARD SHAPES

As was stated at the beginning of this section, the Uniform Building Code requires that precast concrete elements or systems may provide lateral force resistance, provided that the resulting construction and confinement details conform to those required for cast in place concrete. In addition, the connections are required to be monolithic near points of maximum stress, capable of development to cause the building to act as a single, integral unit, and detailed to eliminate brittle type failures. Precast concrete construction has the potential to be used to provide part or all of the building structure using 3 basic design philosophies:

1. The precast and prestressed concrete elements function primarily as forms and/or "pin ended" structural elements and reduce or eliminate the need for temporary shoring and forming. The precast concrete units remain as an integral part of the final construction, and contain reinforcing to carry vertical loads (or perpendicular wind or seismic forces on panel elements) to independent lateral load resisting systems through diaphragms or properly designed connectors. The precast and prestressed concrete elements do not form a part of the lateral load resisting system in the building.

2. The precast and prestressed concrete elements form a structural system unto itself, by supporting vertical loads as well as serving to form the lateral load resisting system in transmitting these forces to the foundation. The individual precast concrete pieces are integrated into a total structure by various types of mechanical or grouted connection devices. Cast in place concrete is not normally a part of the system except where used to simplify certain connections or ties between elements.

3. Precast and prestressed concrete elements are used in various combinations of the two previous concepts taking advantage of the positive aspects of both systems. For example, precast concrete shear walls with mechanical connectors may be combined with cast in place concrete ductile moment resisting frames to provide the required total damping and energy absorption to resist cyclic seismic loading.

Unfortunately, very little test data exists to substantiate the performance of connections under cyclic loading such as is induced by earthquake ground motions. On the following pages, we shall comment upon the seismic design of the indicated structure types as related to the current design concepts used for cast in place concrete. We shall also indicate potential areas of application for prestressed concrete in using design philosophies "2" and "3" above, and will allude to indicated required research.

BASIC STRUCTURAL CONFIGURATIONS

PERIMETER SHEAR WALLS

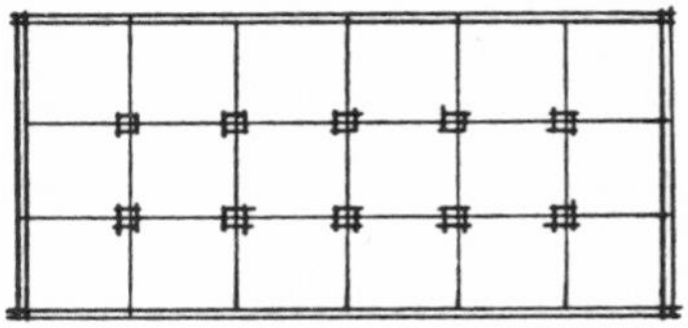

INTERIOR CORE SHEAR WALLS

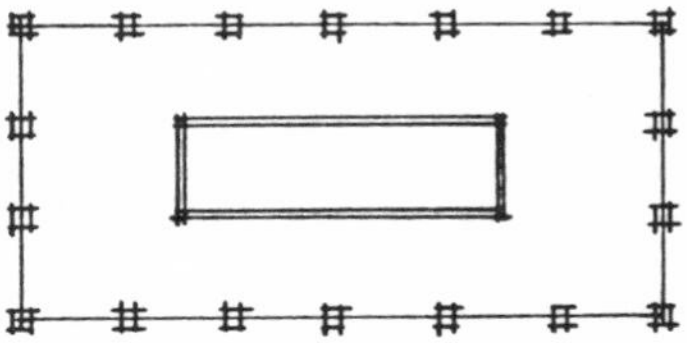

DUAL WALL-FRAME

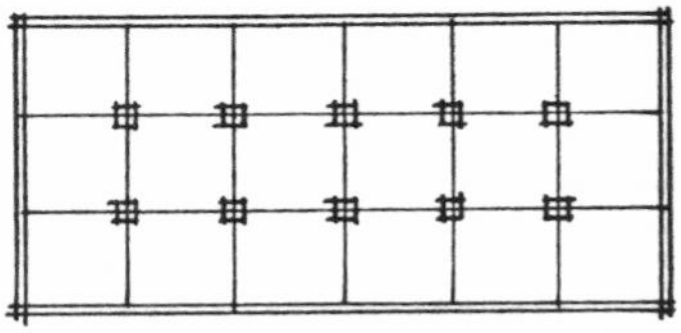

UNIFORM DUCTILE FRAME

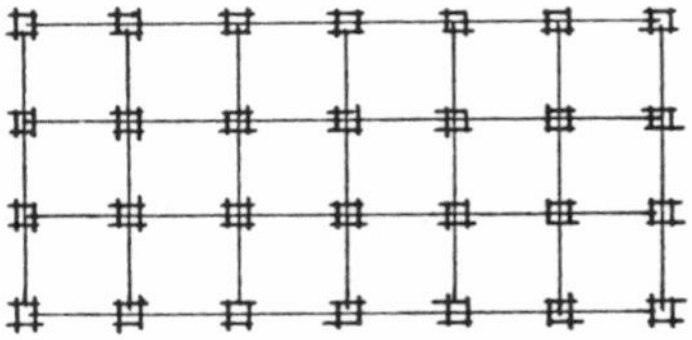

PERIMETER DUCTILE FRAME

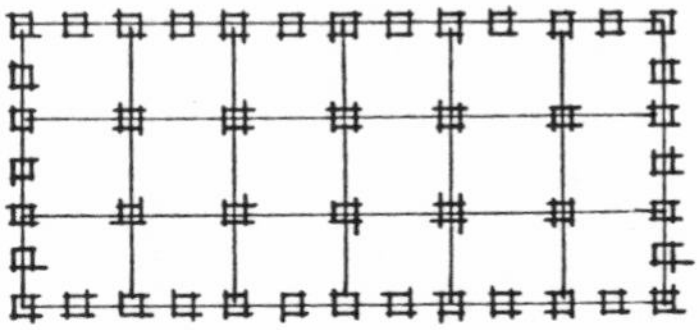

COMBINED WALLS & FRAMES

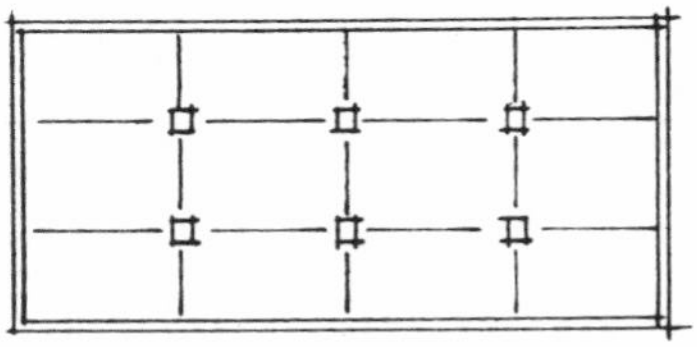

CATEGORIES OF PRECAST CONCRETE BUILDING FRAMING SYSTEMS

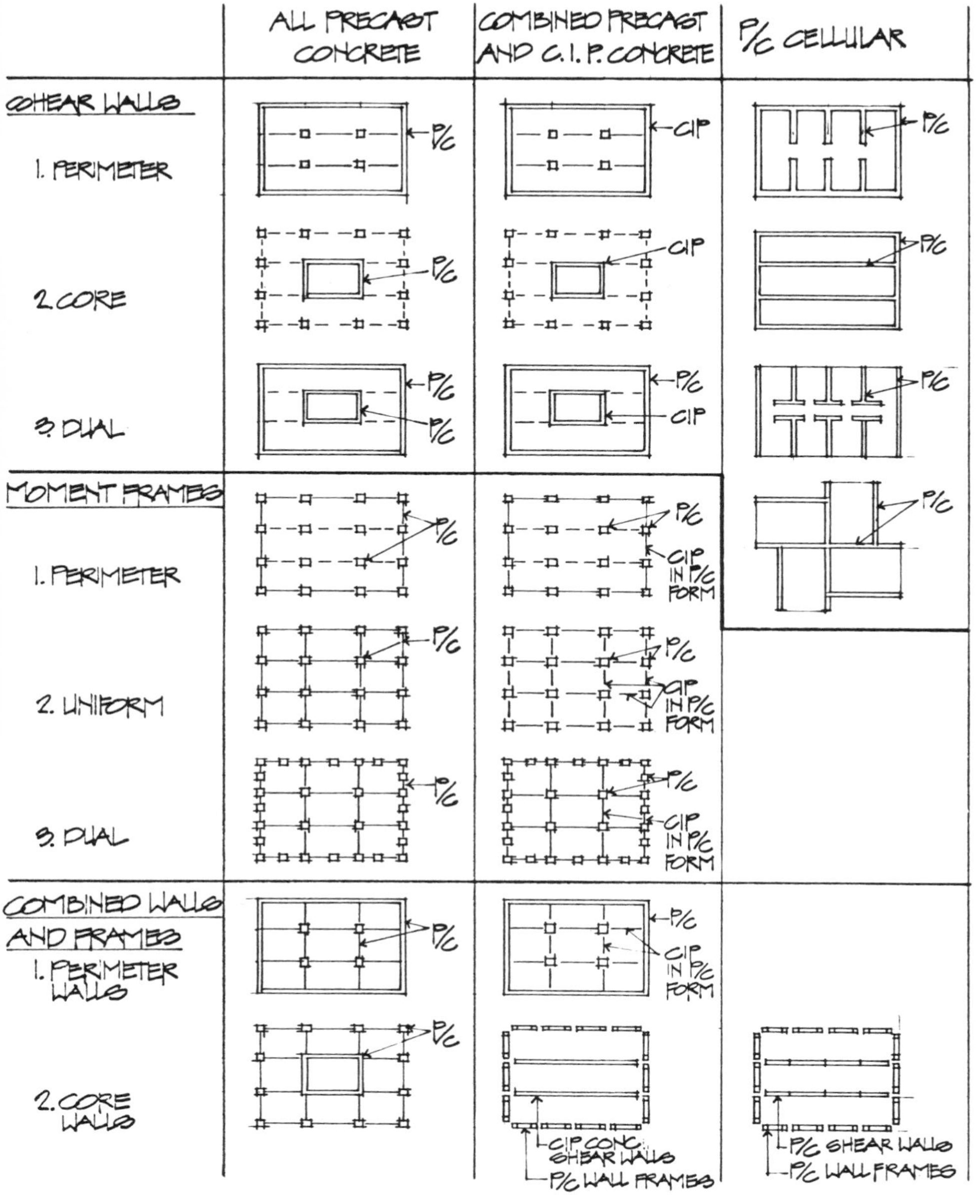

	ALL PRECAST CONCRETE	COMBINED PRECAST AND C.I.P. CONCRETE	P/C CELLULAR
SHEAR WALLS			
1. PERIMETER	P/C	CIP	P/C
2. CORE	P/C	CIP	P/C
3. DUAL	P/C P/C	P/C CIP	P/C
MOMENT FRAMES			
1. PERIMETER	P/C	P/C CIP IN P/C FORM	P/C
2. UNIFORM	P/C	P/C CIP IN P/C FORM	
3. DUAL	P/C	P/C CIP IN P/C FORM	
COMBINED WALLS AND FRAMES			
1. PERIMETER WALLS	P/C	P/C CIP IN P/C FORM	
2. CORE WALLS	P/C	CIP CONC. SHEAR WALLS P/C WALL FRAMES	P/C SHEAR WALLS P/C WALL FRAMES

SHEAR WALL SYSTEMS

1. PERIMETER LOCATION

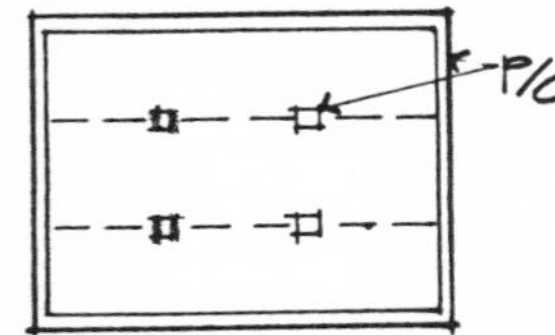

COMBINED PRECAST AND CAST IN PLACE

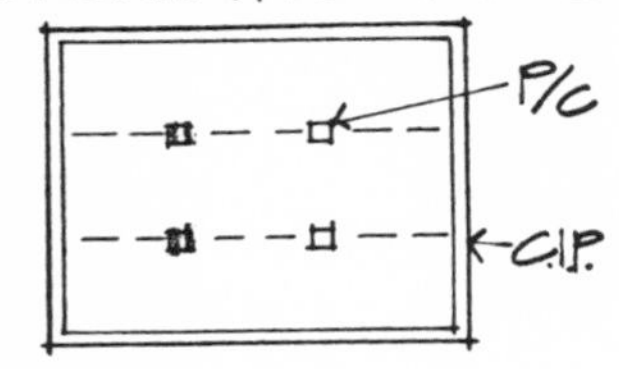

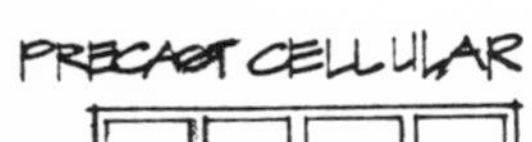

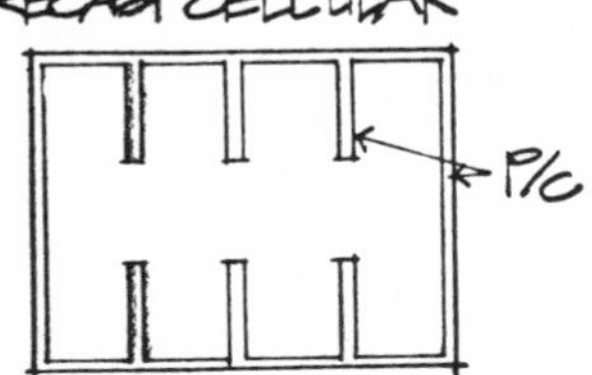

SEISMIC PERFORMANCE COMMENTARY

Permitted Under Uniform Building Code

- Horizontal and vertical connections are provided by cast in place concrete closures. Horizontal joints, where provided, may use a combination of grouting, dry packing, and/or cast in place concrete to provide a monolithic connection. Vertical joints may require shear castellations, horizontal loop steel and vertical field placed reinforcing.
- Vertical connections, such as weld plates, may be used to transfer shear provided they are designed to perform elastically under design earthquake loading.
- Boundary elements are required in walls with aspect ratio $h/d > 4$.
- Openings in walls should be carefully located to avoid local frame action in panels.
- A continuous horizontal strut or diaphragm is required to transmit shear; a positive connection is required between wall and supporting foundation.

Areas requiring Further Research

- Coupled wall concept, utilizing intentionally designed yield hinges in coupling links to provide system ductility.

SHEAR WALL SYSTEMS

2. CORE LOCATION

ALL PRECAST

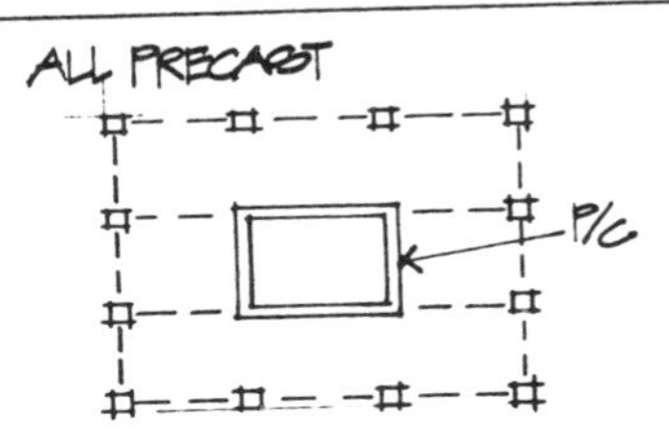

COMBINED PRECAST AND CAST IN PLACE

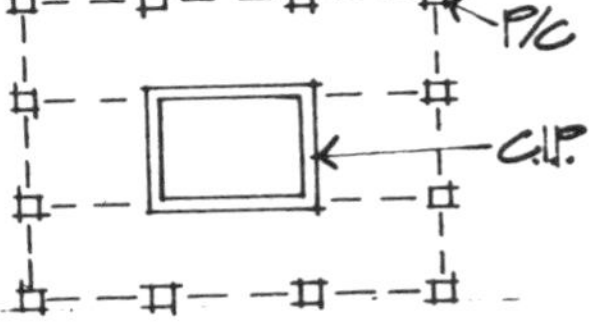

PRECAST CELLULAR

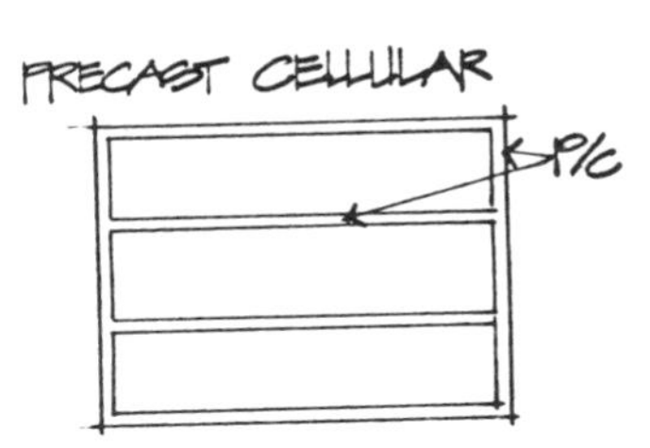

SEISMIC PERFORMANCE COMMENTARY

Permitted Under Uniform Building Code

(same as for perimeter location)

Areas Requiring Further Research

(same as for perimeter location)

- Precast beam and column elements support vertical loads only.
- Interior core walls should be coupled and analyzed for torsion. Caution should be exercised in designing taller structures with only core walls providing lateral rigidity.

SHEAR WALL SYSTEMS

3. DUAL LOCATION

SEISMIC PERFORMANCE COMMENTARY

(Same comments as for previous shear wall systems)

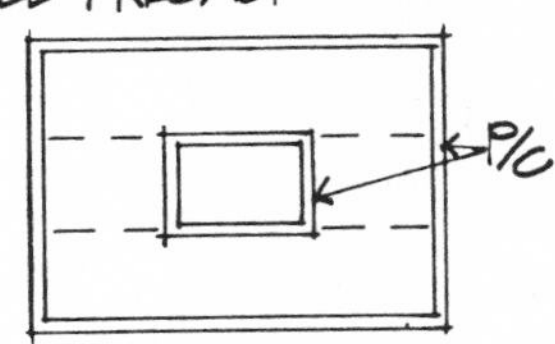

COMBINED PRECAST AND CAST IN PLACE

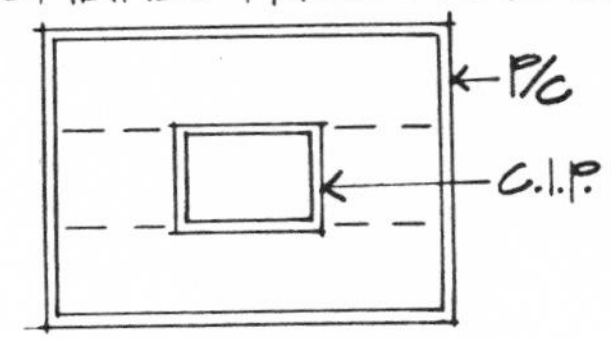

PRECAST CELLULAR

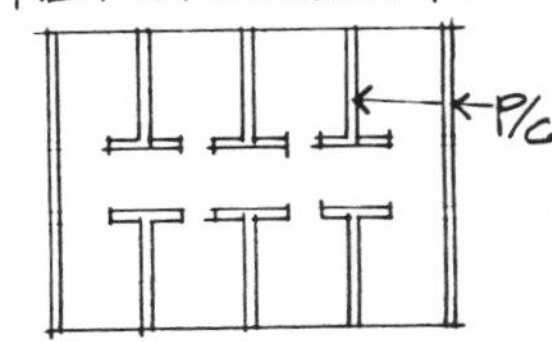

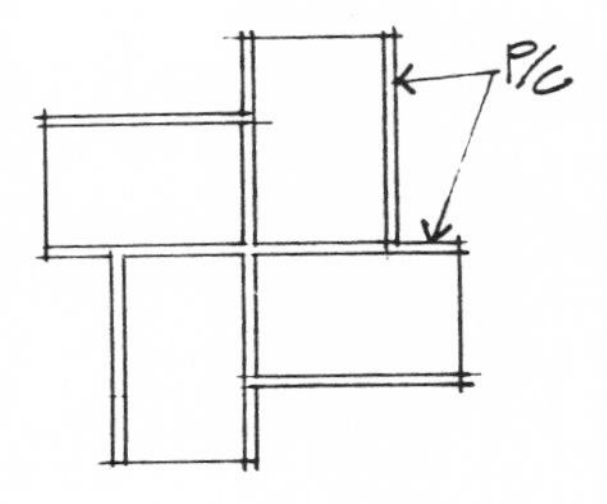

MOMENT FRAME SYSTEMS

1. PERIMETER LOCATION

ALL PRECAST

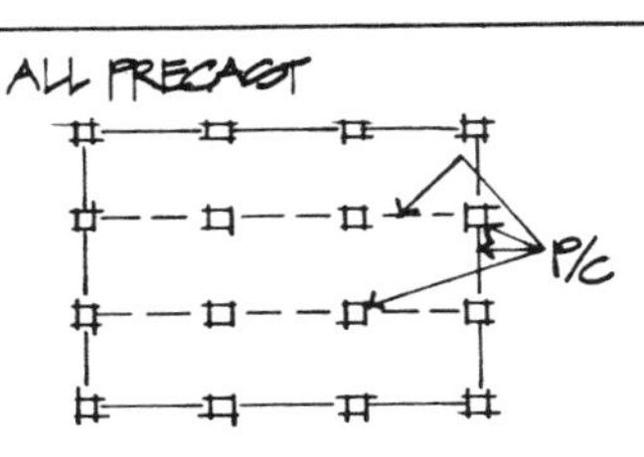

COMBINED PRECAST AND CAST IN PLACE

P/C

C/P IN P/C FORMS

2. UNIFORM LOCATION

ALL PRECAST

P/C

COMBINED PRECAST AND CAST IN PLACE

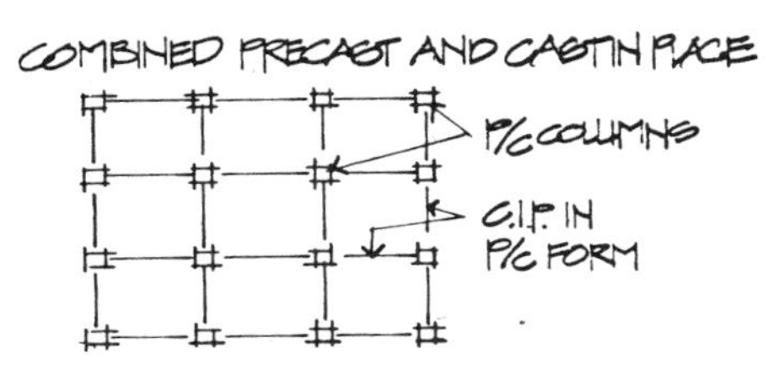

SEISMIC PERFORMANCE COMMENTARY

Permitted Under Uniform Building Code

. All precast or combined precast concrete systems use detailing conforming to the ductile moment resisting frame provisions of the Code. The most frequently used method employs single unit or "tree" columns with extended reinforcing spliced to field placed longitudinal bars at non critical locations. "U" shaped spandrels may incroporate required confinement reinforcing as well as containing closure pours forming a complete monolithic structure.

. For perimeter frames, interior beams and columns support vertical loads only.

. For uniform location, all elements provide lateral rigidity.

Areas Requiring Further Research

. Great potential exists for forming ductile moment resisting frames by post tensioning of precast concrete beam and column segments. This type of construction can achieve great energy absorption with the prestressed concrete system remaining in the elastic cracked range.

(continued)

MOMENT FRAME SYSTEMS

3. DUAL LOCATION

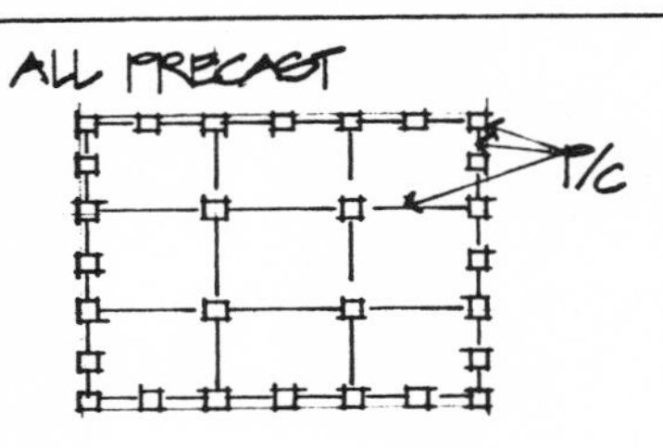

COMBINED PRECAST AND CAST IN PLACE

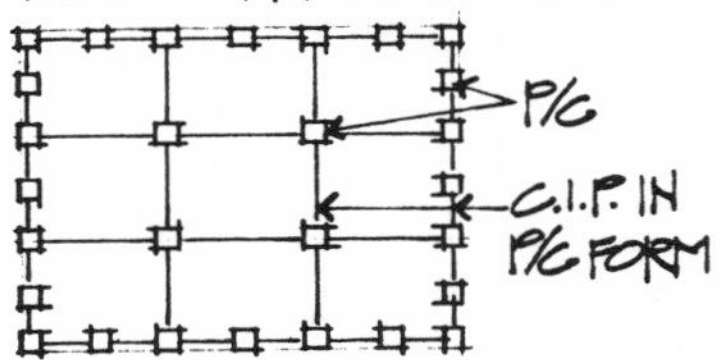

Required Research (cont.)

Much of this research has been performed by Professor Robert Park and others at the University of Canterbury in New Zealand. The status of this work has progressed to the extent that ductile moment resisting frame provisions for prestressed concrete are being drafted for the New Zealand Code. Detailed connection and specific system testing remain to be performed.

COMBINED WALL & FRAME SYSTEMS

1. PERIMETER WALLS AND INTERIOR FRAMES

SEISMIC PERFORMANCE COMMENTARY

(Comments made previously regarding walls and frames apply)

PRECAST

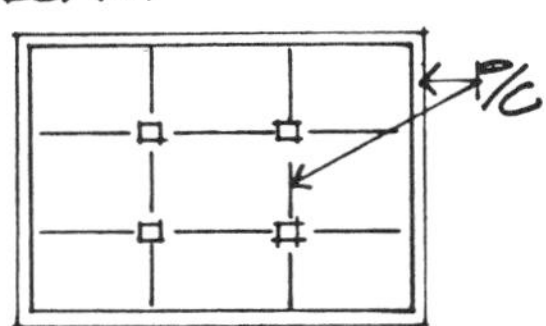

COMBINED PRECAST AND CAST IN PLACE

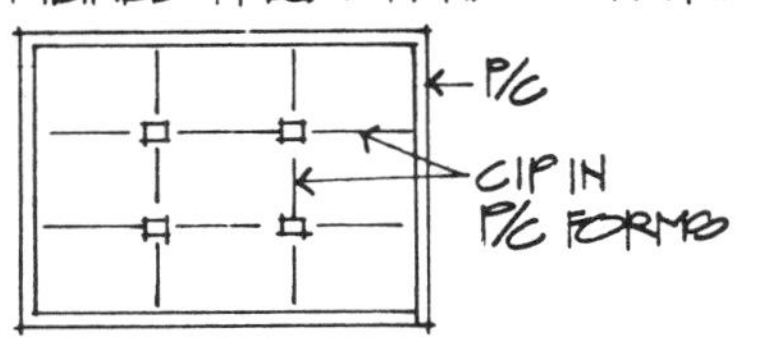

2. CORE WALLS AND EXTERIOR FRAMES

PRECAST

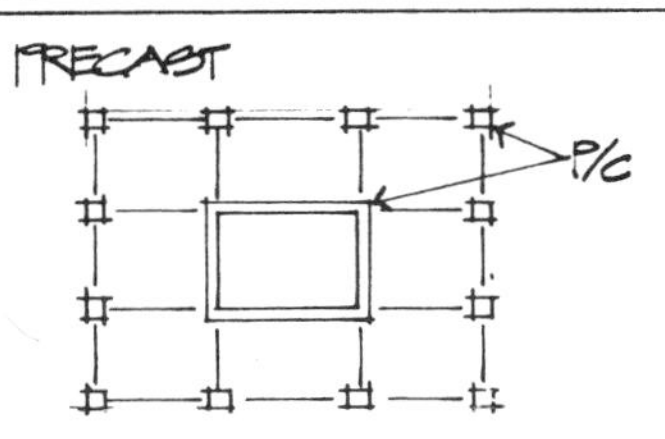

COMBINED PRECAST AND CAST IN PLACE

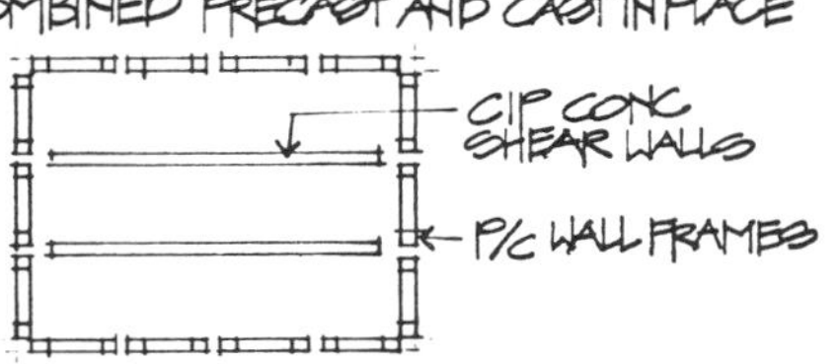

Precast concrete wall frames are usually proportioned such that the resultant beam stiffness is less than the frame column stiffness. If the beam element is a deep stiff section, such as a spandrel, then the column element should be designed to perform elastically under the design earthquake loading.

PRECAST CELLULAR

P/C SHEAR WALLS

P/C WALL FRAMES

UNION BANK - OCEANGATE BUILDING - LONG BEACH

A perimeter moment frame serves to provide lateral stability for this 16 story building, shown under construction in 1975. The precast concrete columns were spliced vertically at certain intervals, and contained reinforcing extending out into "U" shaped precast spandrels at each floor. Field reinforcing bar splices were made, and additional confinement reinforcing was placed prior to making cast in place concrete closures. The completed construction conformed to 1976 Uniform Building Code requirements for cast in place concrete ductile moment resisting frames. The floor system consists of 12 foot wide by 36" deep single tees with 12" thick stems, which are framed into the exterior columns and an intermediate vertical load carrying frame. The space between the spread apart single tees is spanned with 8" deep hollow core plank. The single tees were shored at the flange tips as well as below the stems until the cast in place concrete topping and closures were poured and cured. (See pp. 99-144 for a more extensive pictorial coverage of various types of total precast concrete buildings that can be constructed using current Uniform Building Code Provisions.)

PRECAST PLANT OPERATION

PLANT ORGANIZATION

Precast and prestressed concrete plants are organized generally along the lines shown in the chart below. Very large organizations would have more specialized positions within the various categories shown, but basically, an efficient flow of work can be effected with this organization.

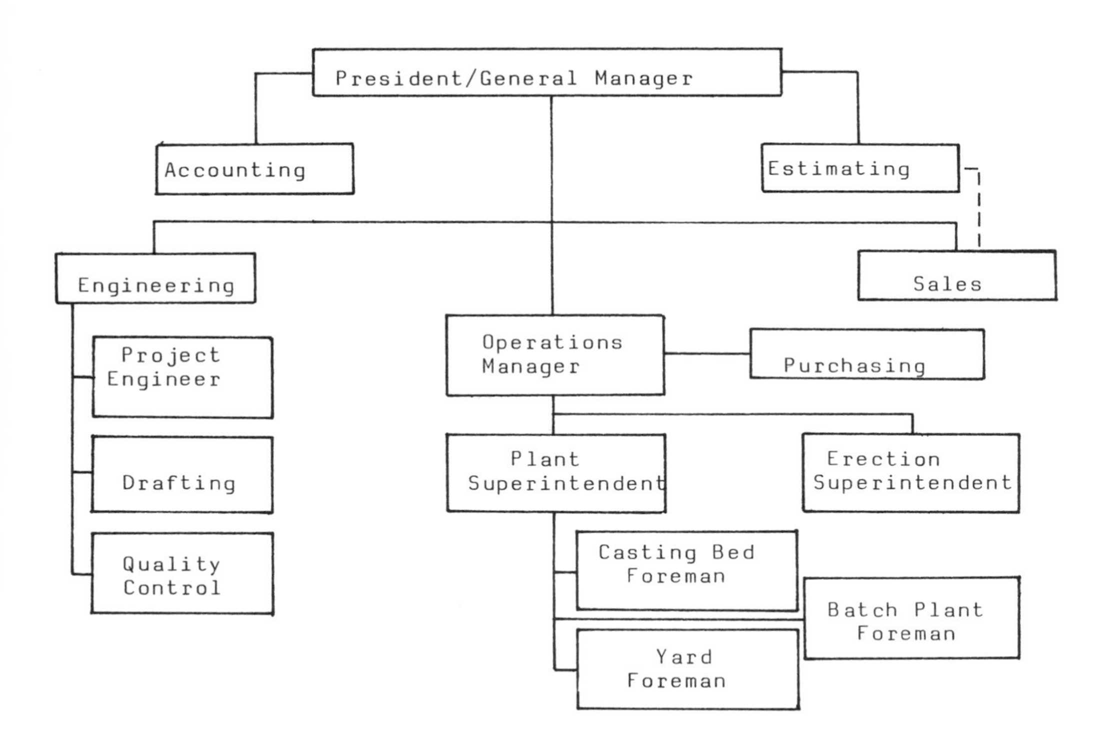

The president or general manager, in addition to overseeing the entire operation, maintains close fiscal control over the company. Daily feedback on estimated versus actual costs is given to accounting from the plant and field. Any significant variance from programmed costs warrants his immediate attention. Similar rigid controls are maintained over the estimating function. All estimates and quotations over a certain dollar amount require the president's review and approval. Sales has limited authority in reviewing and approving quotations. Pricing decisions should only be made by the president.

The Sales Department keeps current on work that is coming up for bid. Sales also maintains a projections file so that longer range forecasts may be made. This information is a result of data gained from contacts with architects, engineers, and general contractors. Sales furnishes budget prices on work being analyzed by designers in preliminary planning stages. There are 2 principle ways in which Sales obtains contracts for the precasting firm. The first is where he prepares a subcontract proposal for a list of general contractors bidding plans and specifications for projects designed in the traditional architect/engineer approach. The second method, which is gaining in popularity especially in parking structures, housing and other types of structures where the bulk of the project is precast, is the design/construct contract with an owner, developer, or construction manager. The design/construct approach requires the precaster to organize a team of required design skills and construction services that do not exist in his organization, and places a heavy reliance on his sales and engineering personnel to properly evaluate the project for design variables and risks which may not be apparent from the outline drawings and information made availabe by the owner. In either process, sales has the responsibility of defining the extent of the work provided by the precaster in the contract agreement, and coordinating with the other departments in the company as required.

The Estimating Department prepares total cost information on the project based upon the material take-off, construction schedule, and existing and projected labor and material costs in the plant and in the field. Written quotations are obtained for major items bought from others, such as miscellaneous iron, hauling, and erection equipment. Actual historical data on plant labor is used in estimating labor on future projects, where possible. When all the project costs are totalled, the President, Sales Manager, and Estimating Manager confer and decide on a suitable mark-up, or gross margin to be assigned to the project, which includes overhead, general and administrative costs, and profit. The total plant and field costs, added to the gross margin, equal the selling price for the project.

If the precaster is the successful bidder, a written subcontract or contract agreement is signed with the contractor or owner as applicable. Then the Sales Department releases information to Accounting, Engineering, and Operations instructing them to proceed with the project. Engineering will perform its product design, and proceed to prepare drawings based upon approval of the product design by the Architect and the Structural Engineer. For architectural precast concrete projects, additional approvals are required; one for shape, which permits mold fabrication to proceed, and one for design, which allows reinforcing and embedded items to be ordered. Finish approval is usually handled by the use of a mock-up, which is provided for in the specifications. Operations prepares casting schedules while engineering approvals are being obtained. Project coordination in a small to medium sized company is usually handled by Operations; in large precast concrete companies, a special Project Coordinator or Project Manager position is usually established to make sure that the project proceeds in a timely manner.

The Engineering Department usually has the responsibility for quality control. Each plant should have its own inspection and testing facilities as a part of an established quality control program. This program is conducted as outlined in the PCI Manuals for quality control (PCI MNL-116 and 117). These in-plant quality control programs are periodically certified by either the PCI Plant Certification Program, or by the ICBO Plant Certification Program as a part of the ICBO fabricator approval registration system.

The Purchasing Agent, in addition to buying materials required for fabricating the precast elements, is also responsible for contracting for outside services, such as hauling and erection, where these capabilities do not exist within the precaster's organization.

The production of the units proceeds with the fabrication of molds where required, fabrication of reinforcing cages, placing and stressing of prestressing strands, placing embedded items and inserts, placing and vibrating concrete, curing concrete (usually overnight), stripping the precast concrete units, performing any required finishing and patching, and placing units into storage. Throughout the production process, ongoing quality control checks are performed at certain critical stages, such as after bed layout, during stressing, prior to casting (cylinder making, slump, and air content checks) and prior to release of prestress. Quality control also performs an after casting check to assure dimensional correctness, satisfactory placement of inserts and blockouts, etc. At that time, the QC seal of approval is placed on the unit, along with the piece mark and date cast. Units are then taken to storage where they are stored by shipping load if and when possible. This is done when the erection sequence is known prior to shipping.

The Erection Superintendent has a very important position. He must ascertain that the site is indeed ready to receive precast concrete elements. He coordinates with the General Contractor to assure proper access and availability of lines and grades. He checks interfacing structures and supporting bearings provided by others to assure they are satisfactory. He checks locations of anchor bolts, inserts, and embedded items required to erect the precast units to assure that the assumed tolerances on the drawings are realistic. He continues to maintain close control over the erection to make sure that required coordination is given to allow quick erection of the units. He makes sure that required erection bracing is installed as shown on the bracing drawings to assure the safety of jobsite personnel. Finally, he sees that all required cleanup is performed and punch list items taken care of to fulfill requirements for payment of any retainage owed the precaster by the general contractor.

MERIDIAN HILL PARK - WASHINGTON D.C.

John J. Earley pioneered the development of exposed aggregate concrete as an architectural finish that was the forerunner of plant finishing techniques still used to this day. This railing element and the walls in the foreground were fabricated in situ in about 1920. The control over mix proportions with low water-cement ratio, and the retardation of surface matrix (using sugar syrup coating on the forms) resulted in exposed natural aggregates to achieve the desired color, excellent weathering characteristics, and durability. Inspection of this 60 year old project in December of 1979 revealed the concrete in an as-new condition, supporting the statement that properly proportioned concrete in a properly designed structure has an infinite life, even in severe extremes of temperature and weathering such as are experienced in the Northeastern USA.

HISTORY OF PRECAST PRESTRESSED CONCRETE

The earliest history of concrete dates back to the period of the Roman Empire. Cement consisting of slaked lime and pozzolan (volcanic ash containing silica) was mixed with gravel, broken tile and brick to form a concrete. This material was placed between or on forms to create domes, vaults and walls. The majority of the concrete was mixed and cast in place. As the art was perfected, concrete was used to create sculptures, fountains, and decorative cast stone. Marble or cut stone was the preferred material for facing walls and for sculptured surfaces; however in some areas concrete was used. Slaked lime was discovered prior to the Roman Empire.

The weakness of masonry and concrete in tension was recognized at an early age. In Egypt as early as 2000 B.C., metal was used to help tie stone together. Due to the problems involved in producing the metal, it was not used to a great extent. During the Roman period, iron chains were used with concrete to resist the thrust in domes. Iron straps were also used in some architectural type structures.

Between the fall of the Roman Empire and the 1700s, the production of concrete was not widely spread. This was due to the lack of availability of pozzolan. In England in 1756, John Smeaton rediscovered the art of making hydraulic cement using natural cement rock. In 1881 Canvass White discovered natural cement rock in Madison County, New York which he used in the construction of the Erie Canal. Portland Cement was invented in 1824 in England and first produced in the United States in 1871. In 1850 Joseph Monier of France developed reinforced concrete, the art of combining metal with concrete. He used this new material to cast garden pots, tubs, tanks and sculptures. This was probably when precast reinforced concrete was first used.

The date of the earliest development of the concept of prestressed concrete is unknown. In 1886 P. H. Jackson, a San Francisco structural engineer obtained a patent on a system of tightening steel rods through voided precast concrete blocks to form slabs. In 1888, C. E. W. Doehring of Germany secured a patent for reinforced concrete with metal tensioned prior to loading. Neither of these two methods were successful because the prestressing was lost through creep and shrinkage, and relaxation of the low strength steels used. In 1908 C. R. Steiner of the United States developed a method of retightening the reinforcing bar after shrinkage and creep had occurred. In 1925 R. E. Dill of the United States used high strength steel bars, coated to prevent the concrete from bonding to the steel, which were post tensioned. Due to the high cost of the high strength steel, the method was not economically feasible.

The greatest impetus to the development of a practical system of prestressing came from the work of E. Freyssinet of France who started using high strength steel wires for prestressing in 1928. With a yield point of approximately 180,000 psi, the wires were tensioned to around 150,000 psi, leaving an effective prestressing force of over 125,000 psi after losses. Freyssinet worked with both pretensioned and post-tensioned systems, but a German, E. Hoyer, is generally credited with developing the first practical system of pre-tensioning. In the late 1930s and early 1940s several practical end-anchorage systems for post-tensioned work were perfected, the most widely accepted being those of Freyssinet and G. Magnel, a Belgian professor.

Based principally on the work of these two men, France and Belgium led in the development of prestressed concrete as a structural material following World War II, but England, Germany, Holland, and Switzerland quickly followed suit and made important contributions.

The development of prestressing in the United States was along somewhat different lines. Instead of linear prestressing, circular prestressing took the lead, primarily in the work of the Preload Company which developed special wire-winding machines. Between 1935 and 1953, over 700 circular tanks were constructed by this system in this country and abroad. The first major linear prestressed structure in the United States was the Walnut Lane Bridge in Philadelphia, PA. Begun in 1949, it is a 160 ft. post-tensioned highway span utilizing the Magnel system of end anchorage. The bridge was designed by Gustave Magnel using I girders which were cast on a false work and post-tensioned. The first pretensioned beams in this country were cast in 1951 for a 24-ft. high bridge near Hershey, PA. The first prestressed railway bridge in the U.S. was constructed on the Burlington in 1954.

The Arroyo Seco foot bridge near Pasadena was the first prestressed bridge to be built in California. Built in 1950, it used a button ended rod and anchorage system. By 1951 the concept of prestressing was well accepted. In 1954 the Prestressed Concrete Institute was founded in the United States. The first widely used text book on prestressed concrete was written by T. Y. Lin during his Fulbright Fellowship while studying under Gustave Magnel. The book was published in 1954. By 1957 there were over 60 prestressed bridges contracted for in California and as a result the Bridge Division of Public Roads published a "Criteria for Prestressed Concrete Bridges". In 1956 the 24 mile Lake Pontchartrain Bridge near New Orleans was completed. It used pretensioned and precast element.

In the United States, after the success of the Walnut Lane Bridge in 1949, several small precast concrete plants began experimenting and producing pretensioned prestressed concrete components for building structures. These were hollow slabs, flat slabs, channel slabs, beams, single tees, double tees and pilings. The first high rise prestressed concrete building, the Diamond Head Apartment house was built in 1956-57 in Hawaii. Designed by Alfred Yee, this building was 14 stories tall and consisted of prestressed I beams and cast in place slabs. In Germany, Ulich Finsterwalder designed many creative bridges using prestressed concrete. A good example is the Mangfall Bridge at Dorching. It is a two level cantilever bridge built in 1957.

The more rapid growth in Europe can be attributed to the major rebuilding task facing most of the European countries following the devastation of World War II, coupled with critical material shortages. Structural steel in particular was scarce and expensive, while labor costs were still quite low.

In this country progress was slower, but has since developed rapidly as the physical capabilities were proven in Europe. Rising steel costs, material shortages during the Korean conflict, the expanded highway construction program and the development of mass production methods to minimize labor costs have all been factors leading to the expansion of the use of plant cast prestressed and precast concrete.

Throughout the United States from 1901 thru 1940 there was much experimenting with precast concrete. Most of these experiments were for building fascias and consisted of wall panels cast on tilt beds at the jobsite or in a plant and trucked to the job site. A good example exists in several storefronts constructed between 1906 and 1912 in Des Moines, Iowa. The panels were cast on a tilt platform at the job site. In several portions of the panels, hollow walls were produced by first casting a 2" layer of concrete and then a 2" layer of sand followed by a second 2" layer of concrete. The two concrete layers were tied together with reinforcing. As the wall was tilted up, the sand was washed out. In 1906, a railroad bridge was also precast and lifted by a railroad crane to its final position. During this period, plant cast precast concrete was also used to produce decorative elements for buildings, such as capitals for columns, railings, etc. With the advent of prestressing and high strength concrete, the technology developed as we know it today; a fabrication method in which concrete strengths of 6000 psi are common as compared to 2000 and 3000 psi prior to 1950, using quality control and efficiency in both design, production and erection. For a more detailed review of the early history of the precast and prestressed concrete industry in North America, the interested reader should refer to the excellent series of articles entitled <u>Reflections on the Beginnings of Prestressed Concrete in America</u> (Parts 1 through 9, May-June 1978 through May-June 1980, PCI Journals).

THE FUTURE OF PRECAST CONCRETE

We are on the threshold of a new era in modern construction methodology - the era of prefabrication. After several false starts, precast concrete will finally come into its own, principally due to the incipient disappearance of fossil fuels and the role this will play in the resultant required restructuring of our society. The beginnings of this new era can already be seen in Europe and parts of South America. Two basic trends will highlight the advanced development of plant cast precast concrete technology:

1. There will occur a development of standard precast concrete shapes, similar in concept to the format which exists in the structural steel industry today. This standardization of both structural and architectural forms in precast will facilitate modular design, and will be a natural complement to the establishment of the design-build concept as the predominant building method used in constructing total precast and prestressed concrete structures.

2. With the increasing scarcity of precious fossil fuels, the automobile will gradually fade from the scene as the predominant mode of transportation. This will force a drastic restructuring of our entire way of living as we know it today. Precast and prestressed concrete will be the prevailing structural material used in building the new society and living environment of the 21st century; housing and living groups will reform in vertical clusters containing essential services for day-to-day living, surrounded by open areas and parks, and interconnected by monorails and people movers. Feeder lines serviced by electric powered vehicles will move people individually and in groups to the mass transit systems. The freeways will serve as the rights of way for the mass transportation systems. Rail transit will re-assert itself as the predominant mode of long distance conveyance for goods and people. The use of fossil fuels will be restricted to major industrial production , essential air transportation, and national defense. At the same time, this will cause traditional on-site construction methods to no longer be economically feasible. Plant cast precast concrete systems building will be used to provide energy efficient, permanent structures, with components being installed with electric hoisting equipment. Solar energy systems will interface with precast concrete sandwich panel construction both above and below grade. Prestressed concrete, with its most efficient ratio of supported load versus weight of steel reinforcing, will be utilized to the maximum. Hybrid materials will be used to replace mild steel reinforced concrete and structural steel as building materials due to the increasing cost and scarcity of fossil fuels for steel production. Glass fiber reinforced

concrete, polymer concrete, and other composites exhibiting high tensile strength will be required. Other manufacturing processes not currently employed in the precast industry will be used in production of these materials, such as extrusion molding, sintering, ceramic processes, automation of all sorts for conveying, vibrating, removal of excess moisture to facilitate early stripping, and computerized cutting, storing and handling methods.

The beginnings of the coming change are already apparent. Some of these are evident in California today, such as:

Monorail structures (Disneyland)
Prestressed concrete railroad ties (B.A.R.T.)
Mass transportation structures (B.A.R.T.)
People movers (Los Angeles -planned)
Solar housing (Savell System)
Systems Building (Forest City Dillon)
Glass fiber reinforced concrete cladding panels for buildings.

In Russia, all prestressed precast concrete shapes are standardized in a National products manual used by architects and engineers throughout the Soviet Union. In Russia, polymer impregnated concrete is used extensively for power poles; studies in this country on glass-fly ash composites show that power poles can be produced at one fourth the cost of comparable wood poles. Cities in Europe are already re-arranging living modes to adapt to an automobile-less society, and will not suffer the travail that will accompany our adjustment, which to date has been delayed for political and other reasons.

The architect of the future will face an ever increasing challenge to design for the greatest structural and construction efficiencies. Economics point toward the increasing use of precast concrete with all its advantages. In many ways, new opportunities are created for designers in creating different architectural forms, as well as totally precast seismic resistant structures, combining the advantages of precasting and pretensioning with post-tensioning.

ADVANTAGES OF PLANT CAST PRECAST CONCRETE

High quality precast and prestressed concrete components are plant manufactured under ideal quality control conditions while foundation and site work proceeds at the same time, allowing delivery and erection from truck to structure on precise and predetermined construction schedules. On larger projects, this "telescoping" of critical path functions results in significantly reduced construction time, reducing high on-site labor costs and interim financing charges, and allows earlier initial occupancy and use of the completed structure. Other advantages of plant cast precast and prestressed concrete are outlined below:

- SHALLOW CONSTRUCTION DEPTH Prestressed concrete deck members achieve long spans in a minimum of construction depth, thereby reducing the overall building height and space requiring cooling and heating.

- HIGH LOAD CAPACITY Prestressed concrete possesses the structural strength and rigidity essential to accommodate heavy loads, such as those resulting from heavy manufacturing equipment.

- DURABILITY Precast concrete is exceptionally resistant to weathering, abrasion, impact, corrosion, and the general ravages of time.

- LONG SPANS Fewer supporting columns or walls result in more useable floor space, allowing greater flexibility in interior design, efficiency and economy.

- FLEXIBILITY FOR EXPANSION Precast concrete components can easily be designed to facilitate future horizontal or vertical expansion. Thus, with the design of precast concrete buildings being on a modular system, necessary additions can be made with low removal cost.

- LONG ECONOMIC LIFE Precast concrete buildings give added years of service with a minimum of repairs and maintenance.

- LOW MAINTENANCE Dense precast and prestressed concrete components factory cast in smooth steel, concrete, or fiberglass forms exhibit smooth surfaces which resist moisture penetration, fungus and corrosion. High density concrete reduces the size and quantity of surface voids, thereby resisting the accumulation of dirt and dust. Prestressing controls crack propagation, thus assuring product integrity. The clean maintenace free properties of precast concrete result in operating cost savings for the owner, and have advantages for specific occupancies, such as food processing plants and electronics manufacturing facilities.

. MOLDABILITY INTO DESIRED SHAPES AND FORMS Precast concrete, being a plastic material in the liquid state may be cast in complex shapes and have intricate surface textures. With optimum reuse of forms, 3 dimensional sculptural designs are possible which cannot be achieved with vertically formed site cast concrete or "tilt-up" site cast precast.

. ATTRACTIVE APPEARANCE With architectural treatment of surfaces, the pattern, texture, and color variations of precast and prestressed concrete are practically unlimited. Its inherent beauty enhances the image of an occupant by suggesting good taste, stability, and permanence. This increases saleability and rentability, keeping occupancy high.

. READY AVAILABILITY Daily cyclic casting of standard structural shapes assures that mass produced structural components can be furnished to satisfy even "fast track" construction schedules.

. ECONOMY In addition to low first cost advantages, precast concrete construction results in other cost savings. Expensive on-site labor costs are reduced. Higher strength, lighter prestressed concrete elements result in savings in foundation costs. By quickly enclosing the building during favorable weather conditions, the interior finish work of other trades may proceed during inclement weather or in winter. Faster construction with factory cast precast concrete elements results in reduced general contractor overhead.

. QUALITY CONTROL Plant cast precast concrete components are fabricated under optimum conditions of forming, fabrication and placement of reinforcing cages and embedded items, vibration of low slump concrete and curing not achievable at the jobsite. Other special manufacturing techniques are made possible in the factory. In-plant quality control programs maintain these high standards of manufacturing.

. FIRE RESISTANCE Precast concrete, being non-combustible, is an excellent material to prevent the spread of fire within a building or between buildings. This inherent benefit in concrete construction assures occupant safety and results in low fire insurance premiums. (See Section 16 for an expanded discussion on Fire Resistance.)

. LOW NOISE TRANSMISSION One method of reducing sound transmission is by increasing the mass of the barrier or intervening elements, such as walls and floors. The density of precast concrete provides excellent sound reduction properties. It is excellent for occupancies such as multi-story apartments, condominiums, hotels, motels, music studios, auditoriums, and schools. Good sound control results in lower tenant turnover. (See Section 15 for an expanded discussion on Sound Transmission.)

- ENERGY CONSERVATION Precast concrete, in addition to being able to incorporate thermal insulation cast in during its manufacture, also contributes additional savings due to the inherent thermal lag it develops in modifying the rate at which heat or cold moves through the material. (See Section 14 for an expanded discussion of Energy Conservation.)

- CONTROL OF CREEP AND SHRINKAGE Since precast and prestressed concrete elements are usually kept in storage after casting for 30 to 60 days prior to delivery to the jobsite, a significant portion - 50% or more - of the long term creep and shrinkage movements may occur before the components are incorporated into the structure. This reduces the amount of long term movement required to be recognized in the building design when compared to cast in place and post tensioned concrete construction. Plant cast precast concrete elements use high strength concrete with a lower water cement ratio which minimizes the amount of potential creep movement.

- ELIMINATION OF FORMWORK Precast and prestressed concrete elements may be erected without falsework, maintaining traffic over travelled ways. Precast elements are quickly erected in buildings providing an instant working platform for other trades. Precast concrete elements may also serve as forms for cast in place concrete, such as in column covers or spandrel units which also serve as the finished exterior for the building.

- SPEED OF CONSTRUCTION Precast concrete construction reduces construction time by reducing on site forming and casting for cast in place concrete structures, and elimination of time consuming infill construction required for steel framed buildings with metal decking, exterior enclosures, hung ceilings and fireproofing operations. Other trades also have instant access to the erected prestressed concrete deck members which form a stable secure floor. This savings in time afforded by plant cast precast construction results in significant interest savings in the builders construction financing, as well as reducing the general contractor's overhead. Interim financing charges and loan fees for construction loans are normally 3 to 6% over the prime rate, which at the time of this editing was 15½%. Earlier occupancy also brings earlier returns on the initial investment. An example of savings in interest for a 4 story office building is shown below. The building has an area of 500,000 square feet and an estimated construction cost of $35.00 per square foot. For this illustration, the construction cost is assumed to be the same for structural steel as for precast concrete. Assume the construction loan interest rate is 20%. Use 15% as the mortgage rate.

PROJECT BUDGET

ITEM	Structural Steel 16 months	Precast Concrete 12 months
Estimated Construction Cost	$17,500,000	$17,500,000
Development Cost		
Land	2,000,000	2,000,000
Construction Insurance	100,000	100,000
Architect's Fee (5%)	875,000	875,000
Business Management (1%) (including accounting and financing arrangements)	175,000	175,000
Initial Leasing Expense	100,000	100,000
Misc. & Contingency	450,000	450,000
Subtotal	21,200,000	21,200,000
Financing Costs		
Interest on Construction Loan 75% of subtotal @20% Interest Assuming Loan Average Balance of 50%		
Steel Building		
0.75 x 21,200,000 x 0.20 x 0.50 x 16/12	2,120,000	
Precast Building		
0.75 x 21,200,000 x 0.20 x 0.50 x 12/12		1,590,000
Interest Expended on Equity 25% of Subtotal @ 15% Interest. Assume equity was expended at beginning of project		
Steel Building		
0.25 x 21,200,000 x 0.15 x 16/12	1,060,000	
Precast Building		
0.25 x 21,200,000 x 0.15 x 12/12		795,000
Total Projected Cost	24,380,000	23,585,000
Net Savings		795,000

This savings due to a four month reduction in construction time amounts to 4½% of the total construction cost.

ADMINISTRATION BUILDING - CIVIC CENTER, INGLEWOOD

Note how the modular regularity of the exterior precast concrete panels obviously coincide with the structural module selected within.(Photo by Eugene Phillips)

SELECTION OF A MODULE

To achieve the advantages of industrialization a large number of like units must be used in the building design. This requires that a module be established. The module selected must be convenient for the precast manufacturing operation as related to product and casting bed widths, in-plant handling, transportation to the job site and erection. The module must also work with the intended use of the building. In that regard, the architect may want to consult with local precasters at an early state in the design process.

The module selected is subject to limitations depending upon whether the direction being considered is the dimension of the deck member span or the dimension across the width of the individual deck member units. In the direction of member span, a fairly wide latitude is available subject to the load carrying capacity of the deck member being considered. Shipping and handling may limit the optimum length to less than 80 feet. Therefore, unless wall panel modules exist which would dictate this dimension, the module in the direction of floor span may vary in increments of one foot.

In the direction of deck member widths, a larger module is normally chosen. Hollow core plank are made in 4', 8' and 40 inch widths. (3 - 40 inch planks equal 10 feet.) Solid slabs are made in widths of 8' or 12'. Double tees and single tees are made in widths of 8', 10' or 12'. Member widths may be made up to 12' and still be shipped without requiring escorts. Considering all these variables, the most common module selected is 8 feet, which leads to optimum bay widths of from 24 feet to 80 feet, varying in multiples of 8 feet. Of course, depth limitations on supporting girder sizes will usually limit the bay width to between 24 feet and 40 feet. In any event, the module selected in the direction across the width of the member should <u>always</u> be a module of 2 feet. The floor to floor height need not be a module.

<u>In summary the following are suggested modules for use in project planning and layout</u>:

- In the direction of member span (unless dictated by wall panel dimensions): 1 foot.

- In the direction transverse to member span: 8 feet.

TRANSPORTATION

Precast concrete manufacturers usually include in their contracts not only the production, but also the shipping and erection of the units. Some of the manufacturers have their own equipment for transporting the units to the job site. Others will subcontract out the transportation. The units are normally shipped by truck. Consequently, the routing selected as well as trucking regulations have a definite effect upon the unit size and weight.

Precast concrete units are usually supported on two points to avoid undue strain caused by the flexibility of the truck bed in route to the job site. As such, the units must also be designed for two point support. Often the same two points are used in stripping and plant handling. For erection, a two point system is used if the unit is a double tee, inverted tee, "L" beam, rectangular beam or hollow core slab. However, if the unit is a wall panel which must be rotated, then one or more of these two points and a top insert can be used for unloading. Where the size of the precast unit is such that more than two supports are required, then a rocker system is used.

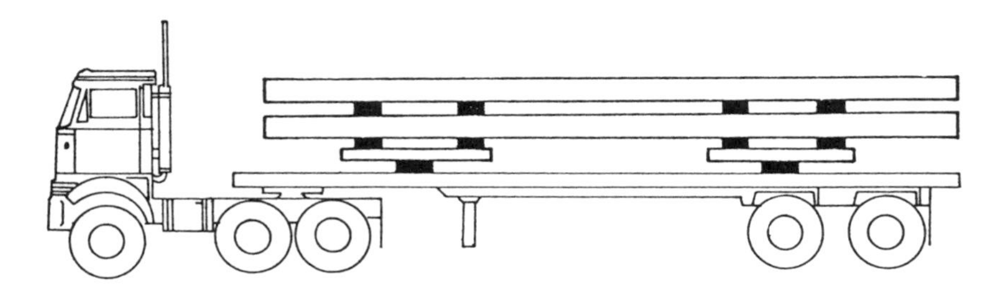

ROCKER SYSTEM

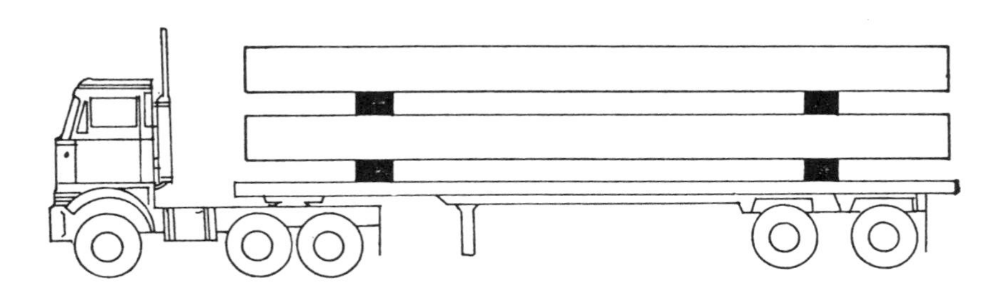

TYPICAL TWO POINT SUPPORT

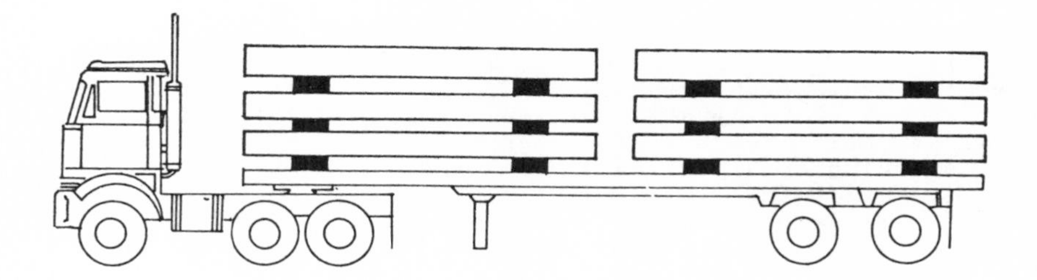

WALL PANELS LAID FLAT

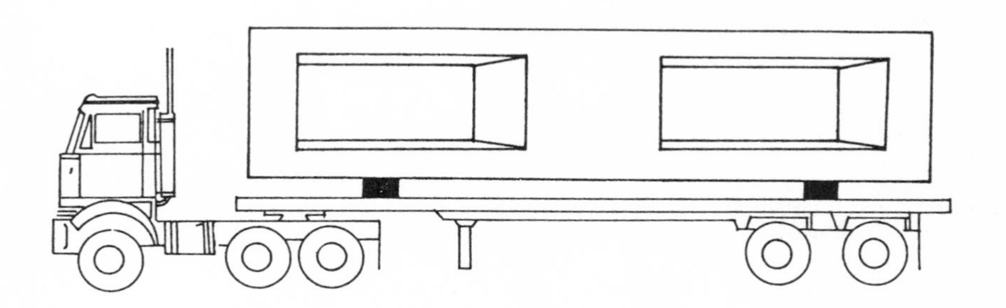

WALL PANELS SUPPORTED EDGEWISE

The interstate highway system, the California freeway system and most major highways and bridges are designed for HS20-44 loading. However, secondary road systems may restrict the maximum weight and height due to bridge limitations. Thus it is important to check out the route if heavy or tall loads are to be transported. The California vehicle code pertaining to the weights allowable without special permissions are complex, but the following information will serve as a guide. Excerpts from the "Vehicle Code Size and Weight Law Summary", June 1976 are given here. Unless special permission is obtained, the following limitations are mandatory:

Width   Maximum width of a vehicle or load is 96 inches.

Vehicle Height   No vehicle shall exceed a height of 13 ft-6 in. measured from the surface on which the vehicle stands except that a mast or boom of a fork lift may extend up to 14 ft.

Load Height   No load on a vehicle may exceed a height of 14 ft. measured from the surface on which the vehicle stands.

Vehicle Length (1) No vehicle shall exceed a length of 40 ft. except a vehicle in combination.
(2) Length of Combination of Vehicles. No combination of vehicles shall exceed 65 ft. when coupled together except that the following combinations may not exceed 60 ft.:

(a) Truck tractor and semi trailer.

(b) Truck tractor, auxiliary dolly and semi trailer.

(3) Extension Devices. Any extension or device including any adjustable axle added to the front or rear of a vehicle used to increase the carrying capacity of a vehicle, shall be included in measuring the length of the vehicle. A draw bar shall be included in measuring the overall length of a combination of vehicles.

Load Length (a) Front Projections. No load may extend more than 3 feet in front of the front bumper or foremost part of the front tires if there is no front bumper.
(b) Rear Projections. The load upon any motor vehicle alone or independent load upon a trailer or semi trailer shall not extend to the rear beyond the last point of support for a distance greater than that equal to two thirds of the wheel base of the vehicle carrying the load, except that the wheel base of a semi trailer shall be considered as the distance between the rear most axle of the semi trailer.

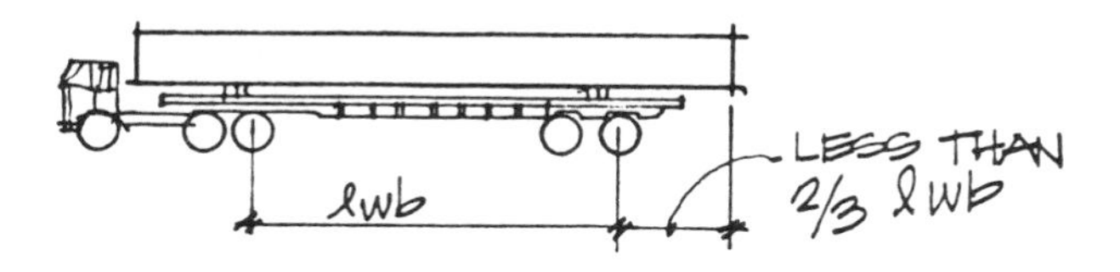

(c) Combinations of Vehicles. The load upon any combination of vehicles shall not exceed 75 feet measured from the front extremity of the vehicle or load to the rear extremity of the last vehicle or load.

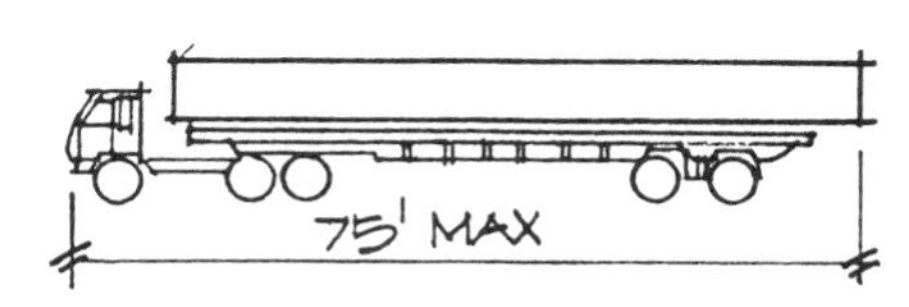

TRANSPORT VEHICLE		COMPONENT l	COMPONENT w	COMPONENT h	VEHICLE l	VEHICLE w	VEHICLE h	MAX TOTAL WT. TONS ‡	SPECIAL PROVISIONS
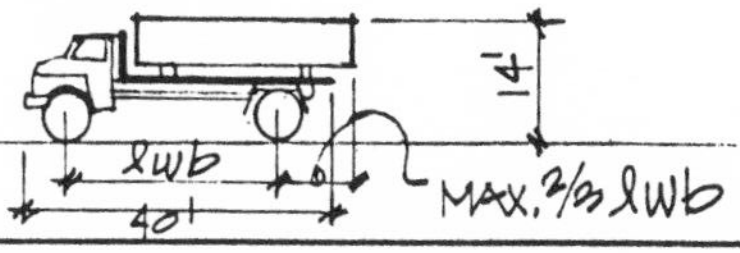 14'; lwb; 40'; MAX. 2/3 lwb	TRUCK	*	8'-4"	14'	40'	8'	13'-6"	23.25	NONE
	OVER WIDTH OR LENGHT								PERMISSION
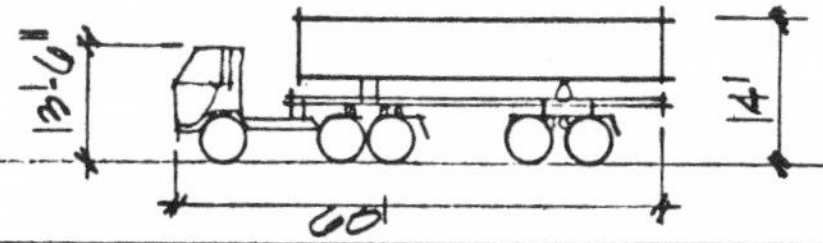 13'-6"; 14'; 60'	SEMI NO OVERHANG	*	8'-4"	14'	60'	8'	13'-6"	40	
	OVERWIDTH								PERMISSION
 MAX. 75'; lwb; 60'; MAX 2/3 lwb	SEMI W/ OVERHANG	*	8'-4"	14'	60'	8'	13'-6"	40	
	OVERWIDTH								PERMISSION
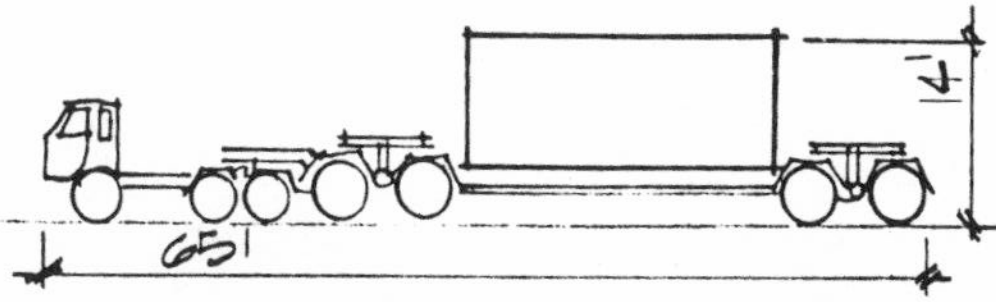 14'; 65'	LOW LOADER	*	8'-4"	14'	65'	8'	13'-6"	40	
	OVERWIDTH								PERMISSION
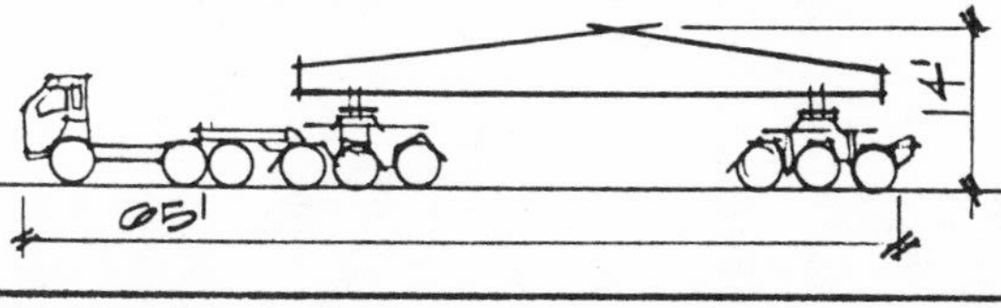 14'; 65'	WITH DRAWBAR	*	8'-4"	14'	65'	8'	13'-6"	40	
	SPECIAL TRANSPORTS			14'		13'-6"			PERMISSION NEEDED

* LENGHT VARIES SUBJECT TO THE VEHICLE AND ITS WHEELBASE
‡ VERIFY WEIGHT W/ TABLE AND DEPT. OF MOTOR VEHICLES.

<u>Axle Gross Weight</u> (a) The gross weight imposed upon the highway by the wheels on any one axle of a vehicle shall not exceed 20,000 pounds and the gross weight upon any one wheel, or wheels, supporting one end of an axle, and resting upon the roadway, shall not exceed 10,500 pounds, except that the gross weight imposed upon the highway by the wheels on any front steering axle of a motor vehicle shall not exceed 12,500 pounds.
(b) The following vehicles are exempt from the front axle weight limits specified in this section:

(1) Trucks transporting vehicles.

(2) Dump trucks.

(3) Cranes.

(4) Transit mix concrete or cement trucks and trucks that mix concrete or cement at or adjacent to a jobsite.

(5) Motor vehicles that are not commercial vehicles.

(6) Truck or truck-tractor with a front axle at least four feet to the rear of the foremost part of the truck or truck-tractor, not including the front bumper.

(7) Trucks equipped with a fifth wheel when towing a semi-trailer.

<u>Ratio of Weight to Length</u> The total gross weight in pounds imposed on the highway by any group of two or more consecutive axles shall not exceed that given for the respective distance in the following table.

<u>Gross Weight to Length Table</u>

Distance in feet between the extremes of any group of 2 or more consecutive axles	2 axles	3 axles	4 axles	5 axles	6 axles
4	34,000	34,000	34,000	34,000	34,000
5	34,000	34,000	34,000	34,000	34,000
6	34,000	34,000	34,000	34,000	34,000
7	34,000	34,000	34,000	34,000	34,000
8	34,000	34,000	34,000	34,000	34,000
9	39,000	42,500	42,500	42,500	42,500
10	40,000	43,500	43,500	43,500	43,500
11	40,000	44,000	44,000	44,000	44,000

	2 axles	3 axles	4 axles	5 axles	6 axles
12	40,000	45,000	50,000	50,000	50,000
13	40,000	45,500	50,500	50,500	50,500
14	40,000	46,500	51,500	51,500	51,500
15	40,000	47,000	52,000	52,000	52,000
16	40,000	48,000	52,500	52,500	52,500
17	40,000	48,500	53,500	53,500	53,500
18	40,000	49,500	54,000	54,000	54,000
19	40,000	50,000	54,500	54,500	54,500
20	40,000	51,000	55,500	55,500	55,500
21	40,000	51,500	56,000	56,000	56,000
22	40,000	52,500	56,500	56,500	56,500
23	40,000	53,000	57,500	57,500	57,500
24	40,000	54,000	58,000	58,000	58,000
25	40,000	54,500	58,500	58,500	58,500
26	40,000	55,500	59,500	59,500	59,500
27	40,000	56,000	60,000	60,000	60,000
28	40,000	57,000	60,500	60,500	60,500
29	40,000	57,500	61,500	61,500	61,500
30	40,000	58,500	62,000	62,000	62,000
31	40,000	59,000	62,500	62,500	62,500
32	40,000	60,000	63,500	63,500	63,500
33	40,000	60,000	64,000	64,000	64,000
34	40,000	60,000	64,500	64,500	64,500
35	40,000	60,000	65,500	65,500	65,500
36	40,000	60,000	66,000	66,000	66,000
37	40,000	60,000	66,500	66,500	66,500
38	40,000	60,000	67,500	67,500	67,500
39	40,000	60,000	68,000	68,000	68,000
40	40,000	60,000	68,500	70,000	70,000
41	40,000	60,000	69,500	72,000	72,000
42	40,000	60,000	70,000	73,280	73,280
43	40,000	60,000	70,500	73,280	73,280
44	40,000	60,000	71,500	73,280	73,280
45	40,000	60,000	72,000	76,000	80,000
46	40,000	60,000	72,500	76,500	80,000
47	40,000	60,000	73,500	77,500	80,000
48	40,000	60,000	74,000	78,000	80,000
49	40,000	60,000	74,500	78,500	80,000
50	40,000	60,000	75,500	79,000	80,000
51	40,000	60,000	76,000	80,000	80,000
52	40,000	60,000	76,500	80,000	80,000
53	40,000	60,000	77,500	80,000	80,000
54	40,000	60,000	78,000	80,000	80,000
55	40,000	60,000	78,500	80,000	80,000
56	40,000	60,000	79,500	80,000	80,000
57	40,000	60,000	80,000	80,000	80,000
58	40,000	60,000	80,000	80,000	80,000
59	40,000	60,000	80,000	80,000	80,000
60	40,000	60,000	80,000	80,000	80,000

In addition to the weights specificed above, two consecutive sets of tandem axles may carry a gross weight of 34,000 pounds each if the overall distance between the first and last axles of such consecutive sets of tandem axles is 36 feet or more. The gross weight of each set of tandem axles shall not exceed 34,000 pounds and the gross weight of the two consecutive sets of tandem axles shall not exceed 68,000 pounds.

The above weights are gross. To determine the payload, the weight of the transporting unit must be deducted. Once the maximum payload is determined the number of units to be included within the payload can then be determined. If the weight exceeds that in the table, then a special permit is required. If the shipping route includes minor roads with bridges then the highway department should be consulted as to the load capacity of the bridges.

Overweight and Oversize Permits   Permits for oversize and overweight vehicles and loads may be issued under certain conditions by the California Department of Transportation with respect to highways under its jurisdiction or by local authorities with respect to highways under their jurisdiction.

Some of the increases allowed with permits are as follows:

1. On State Roads (CAL-TRANS)
   - Increased load width from 8' to 12' - no escort required.
   - Load width from 12' to 14' - front escort required. (Absolute maximum width: 14'-0)
   - Overweight loads - no escorts required.
   - Over 100' in length (load + truck) - front escort required.

2. On City and County Roads, Local Jurisdictions May Grant Permits Subject to the Following Limitations
   - Increased load width from 8' to 14' - no escorts required.
   - Load width from 14' to 15' - front and rear escorts required.
   - Load width from 15' to 20' - front and rear escorts required; loads may only be moved at night between the hours of 1 to 6 a.m.
   - Over 100' in length (load + truck) - front and rear escorts required.
   - Overweight loads - no escorts required.

Permit loads for overweight loads are limited by bridge classification. CAL-TRANS uses a color code system to rate the capacity of bridges. The lighter classification is "green", and the heavier classification is "purple." On the next page, some of the axle loads for various transportation configurations are given.

WEIGHT CHART (CAL-TRANS)

	Color Class	Tr. Ax. Width	Maximum Gross	Axles 2 and 3	Axles 4 and 5	Axles 4, 5 & 6	Axles 6 and 7
1. 3 Axle Tractor & 2 Axle Semitrailer	Green	8'	NA	40,200	40,200		
	Purple		NA	46,300	46,300		
2. 3 Axle Tractor & 2 Axle Semitrailer (16 tires on 4 & 5)	Green	8'	NA	40,200	46,000		
	Purple		NA	46,300	53,100		
3. 3 Axle Tractor & 2 Axle Semitrailer (16 tires on 4 & 5)	Green	Minimum		40,200	49,700		
	Purple	10'	NA	46,300	57,400		
4. 3 Axle Tractor & 3 Axle Semitrailer	Green	8'		40,200		43,200	
	Purple			46,300		49,900	
5. 3 Axle Tractor & 2 Axle Aux. Trailer & 2 Axle Semitrailer (16 tires 4 & 5-6 & 7)	Green	8'	124,000	40,200	46,000		46,000
	Purple		142,000	46,300	53,100		53,100
6. 3 Axle Tractor & 2 Axle Aux. Trailer & 2 Axle Semitrailer (16 tires 4 & 5-6 & 7)	Green	Minimum	130,000	40,200	49,700		49,700
	Purple	10'	152,000	46,300	57,400		57,400

Component Loadings	GREEN	PURPLE
Single Axle	24,000	27,000
Tandem Axles	40,200	46,300
3 Axle Group	43,200	49,900
Trunnion Tandem Axles (8' or over but less than 10' out-to-out width on TIRES).	46,000	53,100
Trunnion Tandem Axles (10' Minimum out-to-out width on TIRES).	49,700	57,400

Five or seven axle trunnion tandem units may be allowed bonus weights in excess of those listed here if certain conditions are met. The California Division of Highways' Bridge Dept. will rate a certain unit upon presentation of a drawing containing the following information:

a. Axle spacing, hub-to-hub, between each axle
b. Tires per axle.
c. Out-to-out width of tires on axle.
d. Tare weight for steering, drive and trailer axles. (connected in train)
e. Kingpin location referenced to an axle.
f. License and equipment number of tractor, auxiliary and/or semitrailer.

ERECTION

If any one element is the key to a successful plant cast precast project, it has to be erection. Enormous sums of money can be lost by inefficient installation; on the other hand, understanding the problems and solutions involved with putting the precast pieces together in the field can give astute precasters a competitive edge and dramatize the most significant aspect of plant cast precast concrete construction - speed. Proper planning through all phases of the building process is essential for efficient erection to be assured. Preliminary design decisions by the Architect and the Structural Engineer, the attention to detail given by the precast concrete manufacturer at bid time, the manufacturer's product design prior to fabrication, and the coordination between the erector and the general contractor are all vital to the success of the erection phase. Everyone involved with the project should understand the importance of erection and how it is affected by so many factors.

PRELIMINARY DESIGN DECISIONS BY THE A/E

Designers are usually making decisions as to sizes and weights of precast concrete elements early in the project. As a general statement, component weights should be limited to 11 tons or less, for both architectural and structural precast concrete projects, unless precast concrete erection personnel are consulted. Heavier weights of up to 22 tons may be handled provided good access is available immediately adjacent to the erection point. Designers should be aware of the importance of proper access, especially in tight city sites between existing structures or on stepped, spread-out buildings where heavy panels coupled with long reaches may require heavy and expensive erection equipment not available locally. Realistic tolerances between precast concrete elements and other materials are required. For example, construction tolerances for cast in place concrete or structural steel supporting elements must usually be added to fabrication tolerances for the precast concrete elements to determine realistic precast concrete panel clearances, ledge dimensions for cast in place concrete corbels, etc. Other materials, such as window walls, partitions, and glazing should be designed and specified recognizing the existence of tolerances required for proper precast concrete installation. Most importantly, the structural engineer should define connection requirements early in the project for architectural precast concrete cladding panels.

PRECASTER DECISIONS AT BID TIME

The precast concrete manufacturer usually seals his own fate in decisions he makes in estimating at bid time. This is especially true of decisions affecting erection. Care should be taken that the connections designed work, and allow for required erection tolerances. If re-design is required at bid time, make sure the general contractor is aware of the effect of changes to other trades, or else these extras will become back charges to the precast concrete manufacturer's account. Also, the precast concrete manufacturer should determine the construction schedule for the erection of the precast concrete elements and tie this into the subcontract agreement, with escalation provisions for increases in field labor, equipment rental rates, and fuel for delays in the project. Delays of one year or more are not uncommon on large projects, and these delays can have disastrous effects on erection costs due to inflation.

PRODUCT DESIGN

There are specific installation and erection considerations for individual precast and prestressed concrete components that are covered in Part C of this Design Guide.

The shape of member influences the type of hoisting equipment required for erection, the number of men required to set or guide the member in place, the sequence of placement, the time and difficulty of attaining the required vertical and horizontal alignments, the time and difficulty of completing temporary and permanent connections, and the time and difficulty of releasing the member from the lifting device (crane).

It may be stated that inasmuch as each shape poses different erection problems, then the shape of the member influences the design of the structural connections, which should facilitate installation as well as in-place stability.

Vertical member shapes (load-bearing wall panels and columns) tend to require the most care during erection because these shapes often act as "benchmarks" for the members which they support. Erection procedures must be developed to make sure that load-carrying panels and columns are plumb. Connections for these members should be designed to facilitate temporary alignment as well as permanent alignment and stability.

The wide flange and narrow stem of the single tee make that shape a relatively difficult member to erect in that a single tee may fall over on its side unless braced or otherwise stabilized in some manner. Stability is often accomplished by means of a combination of detailing and sequencing. For example, the first single tee in the erection sequence is welded by flange weld plates to another member or members in the structural system; the second tee is connected to the first; and subsequent tees are erected in a similar manner, each connected to the previously erected tee with these erection connections which may also form part of the final connection system for the precast elements.

The remaining structural shapes (double tees, beams and slabs) tend to be relatively easy to erect in that they do not pose any special stability problems and often require no temporary connections during the erection sequence. The erection crew simply places these shapes into position as quickly as possible (without erection connections) until all units are erected; then the crew "goes back" and completes the permanent connections at a later time.

Precast concrete cladding panels are probably the most critical from the standpoint of the inter-relation of product design to erection. The connections should be designed to allow immediate stability and alignment to be obtained, maximizing erection productivity. These connections should accomodate building tolerances as well as product tolerances. The selection of bolted, welded, or grouted connections should be made with erection being always kept in mind. Auxiliary bolted connections are often selected to provide alignment and temporary erection bracing, even though the final connection may be made at some future time.

Certain drawings are essential for both installation and field inspection personnel:

<u>Erection Drawings</u>: Building elevations and/or plans showing member layout with piece marks. It is strongly recommended that the marking procedure used for structural steel be used for precast elements, i.e. - piece mark on top left of piece corresponding to same location on erection drawings (right side for opposite hand members).

<u>Connection Detail Sheets</u>: Cross referenced to erection drawings and Bills of Material.

<u>Erection Hardware Bills</u>: Loose erection materials should be clearly marked by piece mark or better yet, color coded and cross referenced to connection details. This material should be segregated into individual kegs or cans at the jobsite marked with corresponding piece marks and color code.

<u>Bracing Plan</u>: Erection bracing and shoring required for precast concrete elements not stable or built into the structure prior to release from the erection equipment. This plan is prepared by a registered civil engineer in the employ of the precast concrete manufacturer and furnished to the general contractor and field installation and inspection personnel prior to beginning erection.

Anchor Bolt/Insert Plate Location Plan: prepared by the precaster for other trades setting hardware cast in and interfacing with precast concrete elements. Many times, the precaster will also provide setting templates for critical cast-in embedments.

Field Work Sheet: showing alterations required to be performed to existing structures prior to erecting precast.

Product handling, as it relates to removing elements from the truck and rotating into the vertical position, as required ("tripping up") should be shown on the erection drawings. (See pp. 165-176)

COORDINATION WITH THE GENERAL CONTRACTOR PRIOR TO BEGINNING ERECTION

Prior to beginning the erection of precast or prestressed concrete elements, certain coordination is required. First, it should be determined that sufficient access to, around, and inside the site, as applicable, on well-compacted roads is present. Other trades should not interfere with the uninterrupted installation of the precast concrete elements. Good base lines and control points for line and grade should be provided by the general contractor. The precast erector "lays out" the project to make sure that embedded items are cast in correctly, that the building frame, where supporting precast concrete elements, is installed within allowable tolerances, and provides sufficient marks so that the precast concrete elements may be installed without a lot of juggling and re-positioning to correctly space out the precast concrete units. The sequence of erection is re-confirmed with the general contractor since it was first ascertained in order to establish the casting priority for production of the precast concrete elements. Based upon this sequence of erection, the erector makes up load lists for the plant so that the pieces are shipped as they are required for installation on the building. The load list will also indicate the location of individual piecemarks on the truck, so that time is not lost in the field in double handling pieces.

Additional coordination is sometimes required when an owner or general contractor supplied crane is available for erection of precast concrete units. Subcontract agreements in this case should read to the effect that the crane is "fully manned and operated for the exclusive use of the precaster during precast concrete erection."

ERECTION EQUIPMENT

Erection equipment used for installing precast concrete components usually varies with the height of the building.

Tall Buildings - Greater than 16 stories
(Architectural Precast Concrete Cladding Panels)
a. Fixed towercrane

b. Monorail System with Chicago Boom
c. Guy Derrick - Stiffleg Crane

Medium Sized Buildings - 5 to 16 stories
(Architectural and Structural Precast)
a. Portable Towercrane or Fixed Towercrane
b. Crawler Crane: 140 ton - 200 ton
c. Rubber tired truck crane: 125 ton - 140 ton

Low Rise Buildings - Up to 4 stories
(Architectural and Structural Precast)
a. Rubber tired truck crane: 50 ton - 140 ton
b. Hydro to 50 ton (GFRC; light precast)

The vast majority of precast concrete falls into the low rise category erected with truck cranes. Erection capacities of some typical truck cranes found in California are listed here.

Some Typical Rubber-tired Truck Crane Capabilities

50 Ton P & H 100' Boom (16,000 lb. cwt)	
30' Radius	22T
50' Radius	10T
70' Radius	6T

80 Ton American 110' Boom (Full Cwt)	
40' Radius	20T
60' Radius	11T
80' Radius	7T

100 Ton P & H 120' Boom (25,000 Cwt)	
40' Radius	21T
50' Radius	16T
70' Radius	10T
90' Radius	7T

125 Ton American 140' Boom (34,000 Cwt)	
50' Radius	21T
60' Radius	16T
80' Radius	10T
110' Radius	6T

140 Ton Manitowoc 140' Boom (74,000 Cwt)	
65' Radius	21T
80' Radius	16T
100' Radius	11T
120' Radius	8T

Notes:
1. All capacities shown are with outriggers extended and set
2. Cwt = Counterweight
3. Radius is from center pin (center of rotation) to boom tip
4. All loads lifted over side or rear of crane.

Oftentimes it is necessary to temporarily store precast concrete elements on the ground because of occasional problems with loads out of sequence, etc. When done, jobsite storage should be performed in the same manner as done in the plant, using proper supporting dunnage in the proper locations. Care should be taken to protect the precast concrete components from staining from mud, etc.

Erection crews vary in composition, depending upon the extent and variety of work performed at the jobsite. In general, the basic erection crew for either architectural or structural precast consists of a crane operator and oiler, an ironworker foreman and 4 ironworkers, and any additional skills depending upon other work required as a part of the erection, such as welding (welders), grouting (masons), patching (cement finishers), and shoring (laborers). Sometimes, the ironworker foreman also serves as the overall erection supervisor, but it is wiser for the precast concrete manufacturer to have his own employee on the job to provide overall supervision and coordination with the plant and the general contractor.

Certain major decisions affecting the erection are usually made early in project planning, and their impact is seen much later when it can be said that hindsight is better than foresight. One of these involves the interrelation of cast in place concrete work that supports precast concrete elements. In general, cast in place concrete walls, shafts, frames or other supporting elements for deck members should be designed so that it can all be cast before precast erection begins, allowing the precast to be installed in one continuous operation, without expensive "on and off" charges for cranes and resultant loss in productivity as compared with the continuous operation. The erection mode which is most efficient for total precast concrete buildings is by bay, erecting a stack of precast members required to complete a vertical portion of the structure without moving the crane, and then moving to the next crane set-up point where the operation is repeated. Long vertical elements, such as wall panels or columns are "tripped up" into the vertical position using 2 lines. Handling inserts are used also as final connection points to the buildings where possible. "Choking" precast elements (lifting with long cable slings completely around the element without using fixed handling points) is only employed as a last resort, for reasons of safety and potential damage to the precast surfaces. If done, choking should be done with protective dunnage and pads to preclude this scarring.

In closing, personnel safety is the most important concern for the precast erector. Totally precast concrete structures are veritable "houses of cards", until closures are made connecting the elements together and to the lateral force resisting system. Adequate bracing to resist wind loads or seismic forces, as shown on bracing plans prepared by a registered civil engineer is mandatory for all projects of this nature. Cladding panels should be tied into the structure immediately before releasing from the erection equipment. Written procedures should also be furnished erection personnel for this type of erection.

In conclusion, prior planning of every crane move is recommended in order to minimize lost time in the field. Below are shown schematic representations for crane sequencing for light and heavy erection loads.

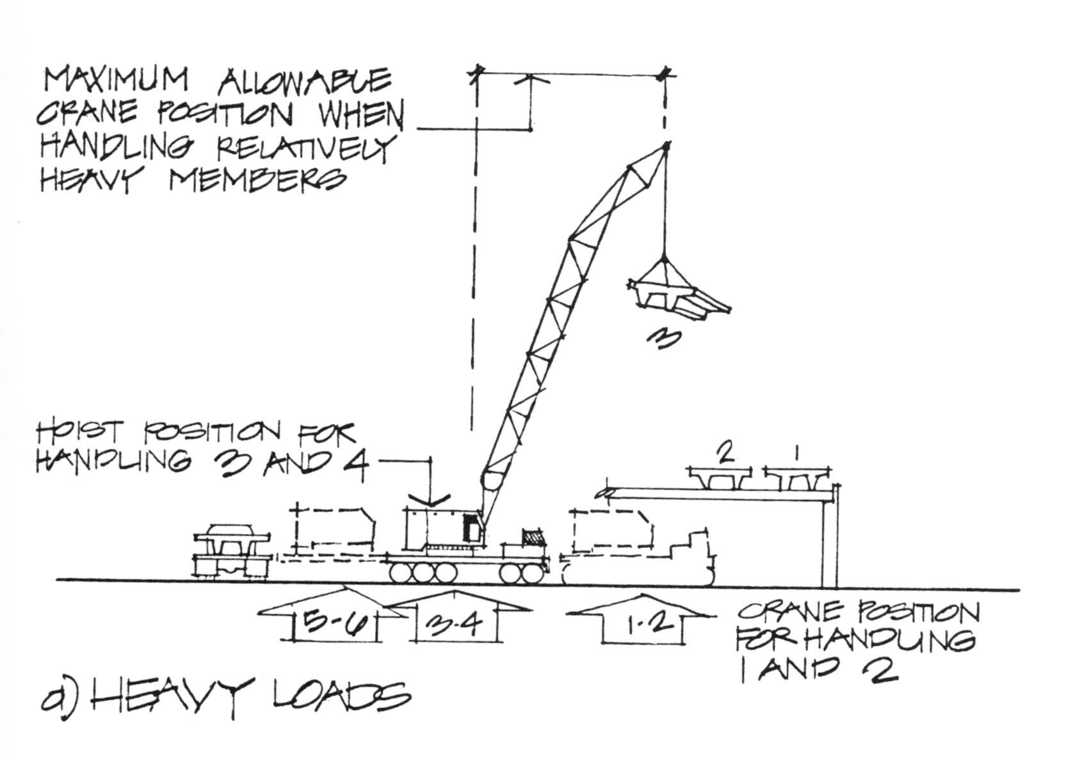

a) HEAVY LOADS

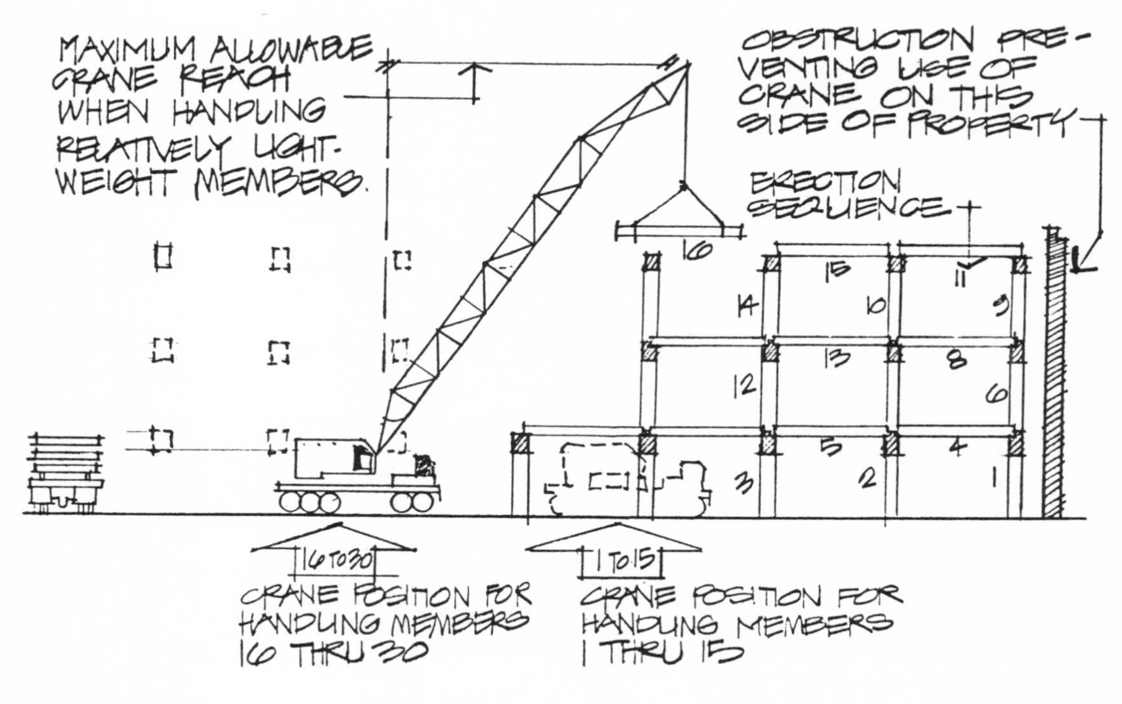

b). LIGHT LOADS

ENERGY CONSERVATION

In the State of California, the State Energy Resources Conservation and Development Commission has adopted energy conservation regulations for buildings. The regulation for non-residential buildings became effective February 6, 1977. The regulations for residential buildings went into effect February 22, 1975. The primary purpose of the regulations is to conserve depleting energy sources such as natural gas, fuel oil and electrical energy. These regulations were adopted in the California Administrative code. The non-residential phase is in Title 21, Chapter 2, Subchapter 4, Article 2, and the residential phase is in Title 25, Chapter 1, Article 5, Section 1094, and Title 24, Part 6, Division T 25, Chapter 1, Subchapter 1, Article 5, Section T 25-1094.

The requirements are in terms of "U" values. "U" value is defined as the transmission of heat flow through a square foot of wall, floor or roof surface, air to air per degree of temperature difference in BTU/hour. The BTU is a British Thermal Unit, a unit of heat measurement - that quantity of heat necessary to raise the temperature of one pound of water one degree fahrenheit. U is equal to the reciprocal of the sum of the resistance to heat flow. U = 1/R

$$U = \frac{1}{Ro + R_a + R_1 + R_2 + R_3 + R_4 + R_i}$$

$R_i = \frac{1}{f_i}$ = Resistance of Inside Air Film

$R_o = \frac{1}{f_i}$ = Resistance of Outside Air Film

$R_a = \frac{1}{C_a}$ = Resistance of Core Air Space

$R_1 = \frac{X_1}{k_1} = \frac{1}{C_1}$ = Resistance of first material - i.e. concrete shell

$R_2 = \frac{x_2}{K_2} = \frac{1}{C_2}$ = Resistance of second material - i.e. core insulation

$R_3 = x_3 = \frac{1}{C_3}$ = Resistance of third material - i.e. concrete wall material

$$R_4 = \frac{x_4}{k_4} = \frac{1}{C_4}$$ = Resistance of fourth material - i.e. finished wallmaterial

$$U_{ow} = \frac{U_{wall}\ A_{wall}\ \underline{MCF} + U_{window}\ A_{window} + U_{door}\ A_{door} + UA_{\substack{other \\ elements}}}{A_{ow}}$$

U_{ow} = The average thermal transmission of the gross wall area

A_{ow} = Gross wall area

MCF = Mass correction factor used for walls or roofs.

The mass correction factor takes into account the thermal inertia properties of concrete. Dense concrete walls and roofs store large quantities of heat (cold) and respond slowly to temperature changes. This is also referred to as the "flywheel effect." Concrete construction possesses an inherent advantage due to its mass inertia, which affects the thermal dynamic response of the building. For example, lightweight wood, metal or glass exteriors cannot store large quantities of heat, whereas heavier, more massive materials in walls, partitions, floors, and roofs store greater amounts of heat and subsequently release it as the demand is increased by a change in the weather and by solar effects. Portland Cement Association Publication EB089.01B (Simplified Thermal Design of Building Envelopes for use with ASHRAE Standard 90-75) contains charts which allow one to determine the MCF to be used to adjust theoretical U values for thermal lag which result in as much as a 35 percent reduction in the U value. This significantly reduces energy consumption in heating and cooling.

If a wall panel is made up of both solid and hollow portions or insulated portions, then the following averaging method is used:

$$U_{panel} = \frac{U_{solid}\ A_{solid} + U_{hollow} + U_{insulated}\ A_{insulated}}{A_{panel}}$$

NOTE: Percentages can be used in lieu of areas in the above equation.

RESIDENTIAL BUILDINGS

The requirements for residential buildings in terms of maximum U values are as follows:

Walls U = 0.08 Walls and spandrels with insulation between framing and the effects of the framing not considered.

Walls U = 0.095	Walls and spandrels with insulation not effecting the framing of the framing system considered.
Walls U = 0.12	Walls weighing between 26 and 40 pounds per square foot and in areas of 3500 or less degree days.
Walls U = 0.16	Walls weighing over 40 pounds per square foot and in areas of 3500 or less degree days.
Ceiling or roofs U = 0.05	With insulation between framing members and the effects of the framing not considered.
Ceiling or roofs U = 0.06	With insulation not penetrated by the framing or the effects of the framing considered.
Floors U = 0.10	In areas with from 3001 to 4500 degree days.
Floors U = 0.08	In areas with over 4500 degree days.

Glazed portions of walls should be a maximum of 20% of the gross floor area for low rise buildings and 40% of the exterior wall area for high rise buildings. A high rise building is defined as containing four or more stories, excluding basements, parking and non-habitable areas. The Uniform Building Code requires a minimum of 10% the floor area of habitable rooms in glazing with one half of that amount openable for ventilation, unless mechanical ventilation is provided. (UBC-79, Sec. 1205)

The regulations also state that the U value of any component of the walls, ceiling, roof system or floor including the glazing may be increased and the U value of other components decreased until the overall heat gain or loss of the total building does not exceed that total resulting from conformance with the stated U values. The building official may approve any alternate design including designs utilizing non-depleting energy systems such as solar or wind, provided he finds that the total system does not use more depleting energy than that given by the stated U values.

The above paragraph allows the architect to increase the glazing area above the maximum if he uses insulating glass (double glazing) subject to the degree days for his area or he may decrease the U value for the walls, roof or floors. The reverse is also true; that is, he may increase the U value of the walls, roof, or floor system if he uses insulating glass without increasing the window area or if he decreases the window area. Thus it now becomes possible for the architect to balance the area and U values of the different materials so as not to exceed the maximum total allowable heat loss or gain.

The zone related to the annual heating degree days would determine if floor and foundation insulation would be necessary as well as if insulating glass would be required.

Zone 1 - 2500 or less heating degree days will not require insulation of foundation walls or walls of heated basements or crawl spaces.

Zone 2 - 2500 to 3000 heating degree days, foundation walls of heated spaces must have a U value of 0.15, but floors need not be insulated.

Zone 3 - 3000 to 4500 heating degree days, floors over unheated basements or crawl spaces must have a U value of 0.10.

Zone 4 - 4500 plus heating degree days, the floor must have a U value of 0.08 and all windows in the building must have insulating or special glazing.

NON RESIDENTIAL BUILDINGS

For non-residential buildings using the standard design concept, the minimum requirements for the thermal design of the exterior envelope are as follows: for a building which is both heated and cooled, the more stringent of either the heating or the cooling requirements shall govern. The U value of any component such as the walls, roof/ceiling or floor may be increased, and the U value of other components decreased until the overall heat gain or loss for the entire building envelope does not exceed the total resulting from the stated U values. The overall U value of the wall, roof or floor portion of the envelope is governed by the degree days of the building location for heating and by the latitude for cooling. The annual degree days for different parts of California are listed in this section. The maximum U_{ow} (U overall for walls), the maximum U_{or} (U overall for roofs) and U_f (U overall for floors) related to heating degree days are indicated in the following graphs. Values that may be used for the Mass Correction Factor (MCF) are given below:

WEIGHT OF WALL CONSTRUCTION	MCF
0-25 lb/sq.ft.	1.00
26-40	0.85
41-80	0.75
81 AND ABOVE	0.65

FOR AREAS WITH MORE THAN 3500 DEGREE DAYS MCF = 1.00

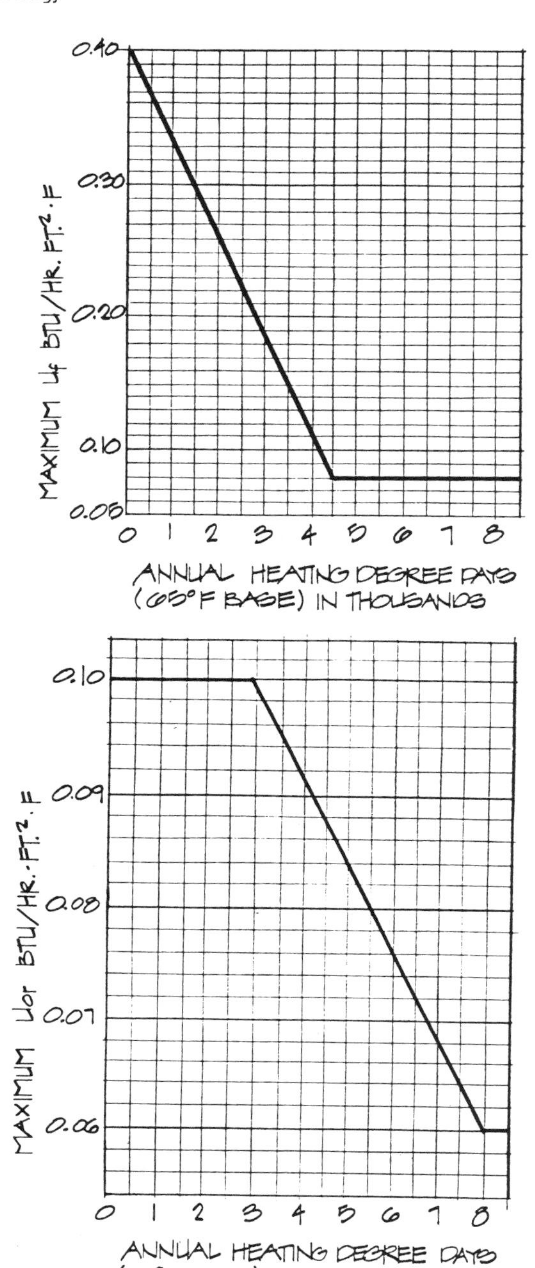
0.40
0.30
0.20
0.10
0.05
MAXIMUM Uf BTU/HR. FT.2 . F
0 1 2 3 4 5 6 7 8
ANNUAL HEATING DEGREE DAYS
(65°F BASE) IN THOUSANDS
0.10
0.09
0.08
0.07
0.06
MAXIMUM Uor BTU/HR. FT.2 . F
0 1 2 3 4 5 6 7 8
ANNUAL HEATING DEGREE DAYS
(65°F BASE) IN THOUSANDS

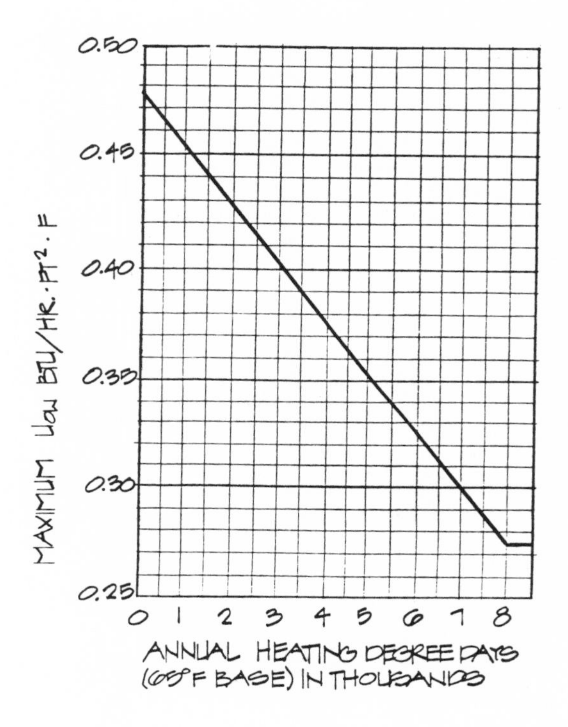

For roof/ceiling areas, U_{or} is the combined thermal transmittance value for the gross roof area, including skylights. The roof assembly is considered as those components of the roof envelope through which heat flows. For buildings which are heated only, skylight areas up to 5% of the gross roof area are exempt from the U_{or} calculations. Skylight area in excess of 5% must be included in the calculations. U_f value of the floor system is for heated spaces over unheated spaces or outdoors.

$$U_{ow} = \frac{U_{wall} A_{wall} MCF + U_{window} A_{window} + U_{door} A_{door}}{A_{ow}}$$

where U_{ow} = the average thermal transmittance of the gross wall area, Btu/hr/ft^2/F

A_{ow} = the external exposed above grade gross wall area of the building that faces heated spaces, ft^2

U_{wall} = the thermal transmittance of all elements of the opaque wall area, adjusted for the effect of framing in the insulated building section, Btu/hr/ft^2/F

A_{wall} = opaque wall area, ft^2

MCF = Mass Correction Factor, value given in the preceding table. This Mass Correction value corrects for thermal storage in the wall system and is related to the mass of the wall and the annual degree days for the area.

U_{window} = the thermal transmittance of the window area, $Btu/hr/ft^2/F$.

A_{window} = window area including sash, ft^2

U_{door} = the thermal transmittance of the door, considered as an assembly, including the frame, Btu/hr ft^2 F

A_{door} = door area including frame, ft^2

Note: Where more than one type of wall, window and/or door is used, the term or terms for the exposure shall be expanded into its sub-elements, as:

$$U_{wall_1} A_{wall_1} MCF_1 + U_{wall_2} A_{wall_2} MCF_2, \text{ etc.}$$

$$U_{or} = \frac{U_{roof} A_{roof} + U_{skylight} A_{skylight}}{A_{or}}$$

Where U_{or} = the average thermal transmittance of the gross roof/ceiling area, $Btu,/hr/ft^2/F$

A_{or} = the external exposed gross roof/ceiling area of the building over heated spaces, ft^2

U_{roof} = the thermal transmittance of all elements of the opaque roof/ceiling area, adjusted for the effect of framing in the insulated building section, $Btu/hr/ft^2/F$

A_{roof} = opaque roof/ceiling area, ft^2

$U_{skylight}$ = the thermal transmittance of the skylight area, $Btu/hr/ft^2/F$

$A_{skylight}$ = skylight area, ft^2

Note: Where more than one type of roof/ceiling and/or skylight is used, the U x A term for that exposure shall be expanded into its sub-elements, as:

$$U_{roof_1} A_{roof_2} + U_{roof_2} A_{roof_2}, \text{ Etc.}$$

COOLING

For cooling, the overall thermal transmittance value of the walls in Btu/hr/ft^2 for the gross exterior wall area including window areas that enclose interior cooled spaces shall not exceed the value shown in the following graph.

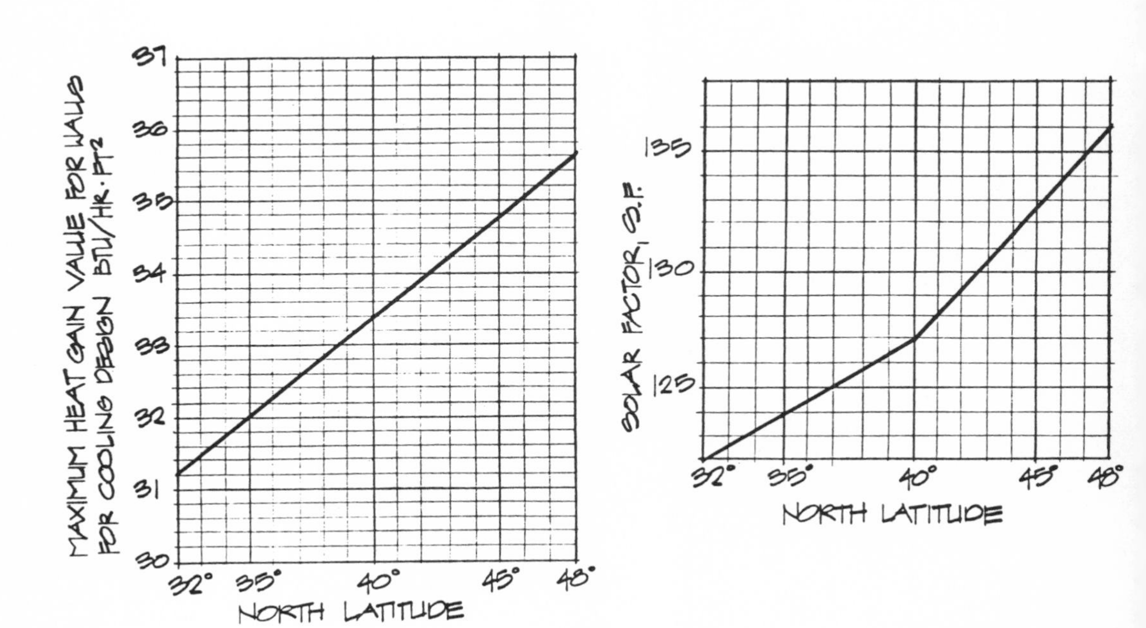

Windows or portions of windows that, because of orientation or fixed exterior shading devices which are never exposed to direct sunlight between April 21 to October 21, shall be considered as having a solar gain factor (SF) of 30 Btu/hr.

For cooling, the overall thermal transmittance value of the roof/ceiling in Btu/hr/ft^2 for the gross area of the exterior roof consisting of opague roof areas and fenestration that enclose interior cooled spaces shall not exceed the value 41 x U_{or} for heating. The following equation is used to determine the acceptable combinations of opague roof and fenestration areas:

$$\text{Btu/hr ft}^2 = \frac{41\ U_r A_r \times A_c M_c + 118 SC_s A_s + TU_x A_x}{A_{or}}$$

(overall thermal transfer value)

where: U_r = The thermal transmittance of opague roof, Btu/hr/ft^2/F.

A_r = Area of opague roof, ft^2.

A_c = Absorptance coefficient.

M_c = Mass coefficient. This factor compensates for the storage effect of walls due to their mass.

SC_s= Shading coefficient of skylights.

A_s = Area of skylights, ft^2.

T = Temperature difference between exterior and interior design conditions, F.

U_s = The thermal transmittance of skylight, Btu/hr/ft^2/F.

Wt, lb/ft^2	Class	M_c
0-15	Light	1.00
16-40	Medium	0.92
41 and above	Heavy	0.84

Surface	Absorptance	A_c
Asphalt, "dark roof"	0.90	1.00
Gravel	0.70	0.79
ASHREA "light roof"	0.45	0.52
Intense white	0.35	0.42

The following equation is used for walls:

$$\text{Btu/hr/ft}^2 = \frac{(U_w \times A_w \times TD_{eq}) + (A_f \times SF \times SC) + (U_f \times A_f \times T)}{A_{ow}}$$

where: U_w = The thermal transmittance of opaque walls and Btu/hr/ft^2/F.

A_w = Area of opaque wall, ft^2.

TD_{eq}= Value given in the table below.

A_f = Area of fenestration, ft^2.

SF = Solar factor, Btu/hr/ft^2. Values are given in the graph on page 14.8.

SC = Shading coefficient of fenestration.

U_f = The thermal transmittance of fenestration, $Btu/hr/ft^2/F$.

T = Temperature difference between exterior and interior design conditions, F.

A_{ow} = Total area of wall opposite cooled spaces, ft^2.

Note: Where more than one type of wall and/or fenestration is used, the respective term(s) shall be expanded into sub-elements, such as $(U_{w1} \times A_{w1} \times TD_{eq1}) + (U_{w2} \times A_{w2} \times TD_{eq2})$, etc.

WEIGHT OF WALL CONSTRUCTION	TDeq
0.25 LB/FT^2	44
26-40	37
41-70	30
70 and above	23

SUMMARY OF THE ENERGY REQUIREMENTS FOR CALIFORNIA

Residential type occupancies:	Max "U" Value
Ceilings or roofs with framing not penetrating the insulation	0.06
Floors in areas of 3001 to 4500 degree days	0.10
*Walls weighing over 40 lb/sq. ft.	0.16
Walls weighing between 26 and 40 lb/sq. ft.	0.12
Window area, maximum (single glazing):	
For low rise building 3 or less stories	20% of the floor area
For high rise buildings	40% of the exterior wall area

If the glass area is reduced or insulative glass is used, the wall "U" value may be adjusted such that the total heat loss based upon the standard is not increased.

* Precast concrete wall panels usually weigh in excess of 40 lb/ft^2. Thus the standard value is 0.16 for residential occupancies.

For Commercial buildings:

(non-residential) the maximum U value is based upon the annual degree days.

	Annual Heating Degree Days	Design Temp. Winter	Design Temp. Summer	U_{ow}	Heating U_{or}	U_{of}	Lat.
San Diego	1439	42	83	.44	-.10	.30	33.02
Los Angeles	2061	42	86	.43	.10	.25	34
Santa Barbara	2290	34	84	.42	.10	.25	34.4
Visalia	2546	32	100	.41	.10	.22	36.4
San Jose	2656	34	88	.41	.10	.22	37.2
San Francisco	3080	42	80	.40	.10	.18	37.7
Sacramento	2782	29	97	.41	.10	.20	38.4

From the above data, it can be seen that the roof system must be insulated to achieve a U value of 0.10 or 0.06. If the floor system is over an open or unheated space it must also be insulated to achieve a U value between .18 and 0.30 or .10 for residential occupancies.

To achieve an average U value between .40 and .44, the wall system must also be partially or totally insulated. That portion to be insulated and to what degree is determined by the quantity and type of glass and the quantity or uninsulated concrete in the precast pierced wall units.

The wall insulation is furnished by one of several methods:
(a) Insulation applied to the inside wall surface with wallboard attached.

(b) Full sandwich wall panels, i.e. two concrete wythes - one structural and one nonstructural.

(c) A partial sandwich panel which is a total structural unit. However, the problem here may be differential temperature volume changes with induced shear stresses in the webs parallel to the flanges.

In using the "U_{ow}" equation, the designer has the option of selecting the quantity and type of glass and varying the thickness and type of insulation in the sandwich panel portion to achieve the required U value for the wall assembly.

- EXAMPLES OF PRECAST CONCRETE HEAT TRANSMISSION VALUES.

1. SOLID WALL NO INSULATION:

5½"

	NORMAL WEIGHT CONCRETE (R)	LIGHT WEIGHT CONCRETE (R)
OUTSIDE SURFACE AIR FILM	.17	.17
5½" CONCRETE	.41	1.04
INSIDE SURFACE AIR FILM	.68	.68
R TOTAL	1.26	1.89
U = 1/R	.79	.529

R PER INCH OF NORMAL WEIGHT CONCRETE IS 0.075
R PER INCH OF LIGHTWEIGHT CONCRETE
OF 100 lb/ft.cub. = 0.24
OF 110 lb/ft.cub = 0.19
LIGHTWEIGHT STRUCTURAL CONCRETE IS USUALLY 110 lb/ft.cub.

2. SOLID WALL WITH INSULATION ONE SIDE:

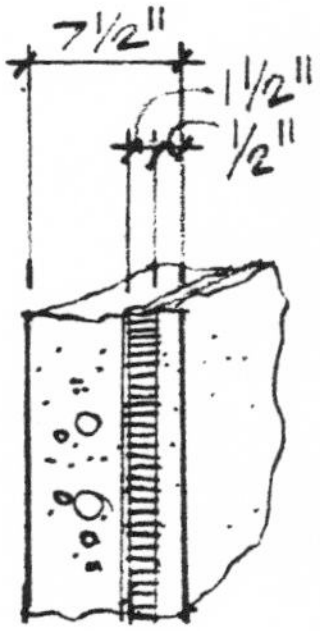

7½" 1½" ½"

	NORMAL WEIGHT CONCRETE (R)	LIGHT WEIGHT CONCRETE
OUTSIDE SURFACE AIR FILM	.17	.17
5½" CONCRETE	.41	1.04
1½" POLYSTYRENE	6.25	6.25
½" GYPSUM BOARD	.45	.45
INSIDE SURFACE AIR FILM	.68	.68
R TOTAL	7.96	8.59
U = 1/R	.126	.116

NOTE: FOR INSULATIONS OTHER THAN THAT NOTED ABOVE CORRECT THE INSULATION R VALUE ACCORDING TO THE FOLLOWING:

THICKNESS	1"	1½"	2"
POLYURETHANE	6.25	9.38	12.5
POLYSTYRENE	4.17	6.26	8.34
FIBERGLASS BOARD	4.00	6.00	8.00
FIBERGLASS BLANKET	3.70	5.55	7.40

3. SANDWICH WALL PANELS:

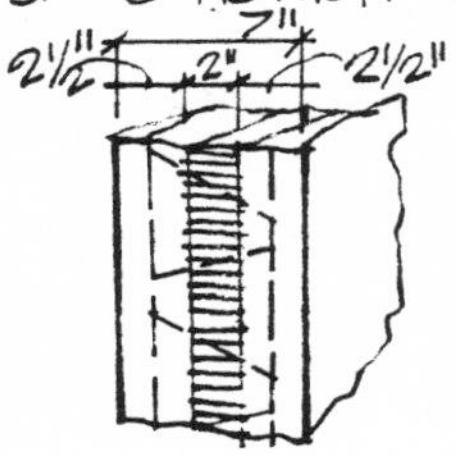

7" 2½" 2" 2½"

	NORMAL WEIGHT CONCRETE (R)	LIGHT WEIGHT CONCRETE
OUTSIDE SURFACE AIR FILM	.17	.17
2½" CONCRETE	.188	.48
2" POLYSTYRENE	8.34	8.34
2½" CONCRETE	.188	.48
INSIDE SURFACE AIR FILM	.68	.68
R TOTAL	9.57	10.15
U = 1/R	.104	.098

4. MARBLE FACED SANDWICH PANEL:

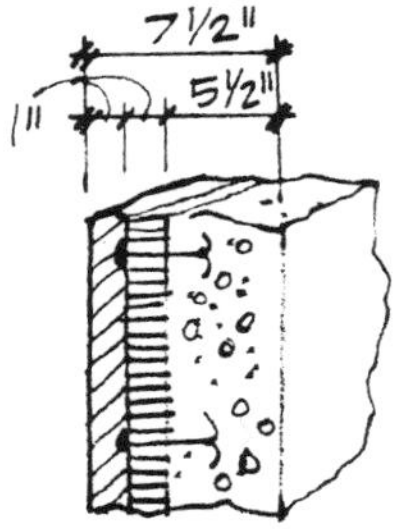

7 1/2" 5 1/2" 1"

	NORMAL WEIGHT CONCRETE	(R)	LIGHT WEIGHT CONCRETE
OUTSIDE SURFACE AIR FILM	.17		.17
1" MARBLE FACING	.07		.07
1" POLYSTYRENE	4.17		4.17
5 1/2" CONCRETE	.41		1.04
INSIDE SURFACE AIR FILM	.68		.68
R TOTAL	5.5		6.13
U = 1/R	.18		.163

5. RIBBED SANDWICH PANEL:

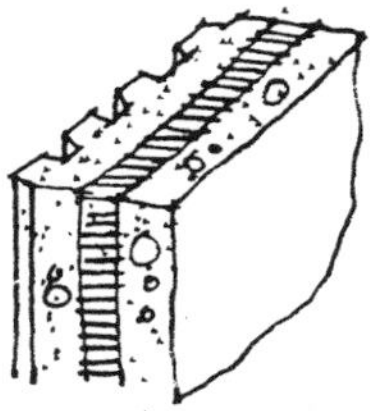

	NORMAL WEIGHT CONCRETE	(R)	LIGHT WEIGHT CONCRETE
OUTSIDE SURFACE AIR FILM.	.17		.17
3" AVE. CONCRETE THICKNESS	.225		.57
1" POLYURETHANE	6.25		6.25
2 1/2" CONCRETE	.188		.48
INSIDE SURFACE AIR FILM	.68		.68
R TOTAL	7.51		8.15
U = 1/R	.133		.123

6. HOLLOW CORE WALL PANEL:

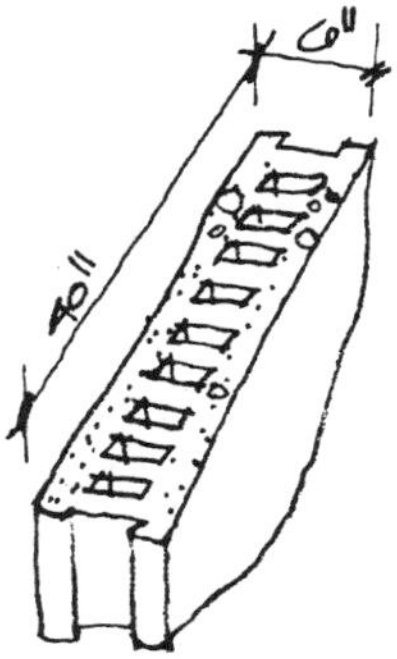

6" 40"

(CELL PORTION OF PANEL EQUALS 59% OF PANEL LENGHT)	NORMAL WEIGHT CONCRETE	(R)	LIGHT WEIGHT CONCRETE
OUTSIDE SURFACE AIR FILM	.17		.17
1" CONCRETE	.08		.19
AIR SPACE 3/4" AND LARGER	.97		.97
1 1/4" CONCRETE	.09		.24
INSIDE SURFACE AIR FILM	.68		.68
R TOTAL	1.99		2.25
U_c = 1/R	.50		.45

(NON CELL PORTION EQUALS 41% OF PANEL LENGHT)		
OUTSIDE SURFACE AIR FILM	.17	.17
6" CONCRETE	.45	1.14
INSIDE SURFACE AIR FILM	.68	.68
R TOTAL	1.3	1.99
U = 1/R	.77	.50

$$U_{ave} = \frac{U_1A_1 + U_2A_2}{A_{ave}} = \frac{.50 \times .59 + .77 \times .41}{1.00} = .61$$

$.45 \times .59 + .50 \times .41 = .47$

NOTE: IF THE CELLS ARE FILLED W/ VERMICULITE R= 3" ave. x 2.08/INCH = 6.24 IN LIEU OF THE .97 FOR THE AIR SPACE AND U_c = .137, .133 AND U_{ave} = .396, .394.

7. FLAT SLAB ROOF SYSTEM:

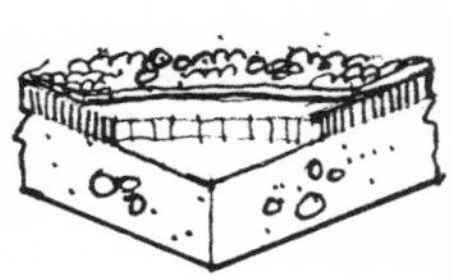

	NORMAL WEIGHT CONCRETE (R)	LIGHT WEIGHT CONCRETE
OUTSIDE SURFACE AIR FILM	.17	.17
3/8" BUILT UP ROOFING	.33	.33
1 1/2" POLYURETHANE	9.38	9.38
5" CONCRETE	.38	.95
INSIDE SURFACE AIR FILM	.61	.61
R TOTAL	10.87	11.44
U = 1/R	.092	.0875

8. DOUBLE OR SINGLE TEES:

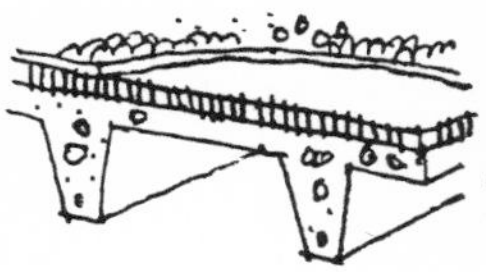

	NORMAL WEIGHT CONCRETE (R)	LIGHT WEIGHT CONCRETE
OUTSIDE SURFACE AIRFILM	.17	.17
3/8" BUILT UP ROOFING	.33	.33
1 1/2" POLYSTYRENE	6.26	6.26
2 1/2" CONCRETE FLANGE	.188	.48
ACOUSTICAL TILE	1.89	1.89
INSIDE SURFACE AIR FILM	.61	.61
R TOTAL	9.448	9.74
U = 1/R	.1058	.1027

9. HOLLOW CORE SLAB 8":

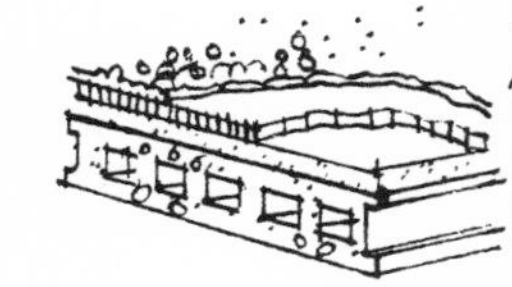

(CELL PORTION OF PANEL EQUALS 56% OF PANEL WIDTH)	NORMAL WEIGHT CONCRETE (R)	LIGHT WEIGHT CONCRETE
OUTSIDE SURFACE AIR FILM	.17	.17
3/8" BUILT UP ROOFING	.33	.33
3" VERMICULITE INSULATING CONCRETE TOPING	3.00	3.00
1 1/4" CONCRETE	.09	.24
AIR SPACE 3/4" AND LARGER	.97	.97
1 1/2" CONCRETE	.11	.28
INSIDE SURFACE AIR FILM	.61	.61
1 1/2" FIBERGLASS ROOF INSULATION	6.00	6.00
R TOTAL	11.28	11.6
U = 1/R	.088	.086

NON CELL PORTION = 44%		
OUTSIDE AIR FILM	.17	.17
3/8" BUILT UP ROOFING	.33	.33
1 1/2" FIBERGLASS	6.00	6.00
3" VERMICULITE CONCRETE	3.00	3.00
8" CONCRETE	.40	1.52
INSIDE AIR FILM	.61	.61
R TOTAL	10.71	11.63
U = 1/R	.093	.084

$$\text{AVE. } U = \frac{U_1 A_1 + U_2 A_2}{A_1 + A_2} = \frac{.088 \times .56 + .093 \times .44}{1.0} = .09 \ \& \ .086$$

EXAMPLE — NON RESIDENTIAL BUILDING

GIVEN THE FOLLOWING OFFICE BLDG. TO BE CONSTRUCTED IN SAN JOSE, CALIFORNIA.

- HEATING DEGREE DAYS FOR SAN JOSE IS 2650.
- THE MAXIMUM VALUE OF U_{ow} FROM THE CHART FOR 2650 DEGREE DAY IS 0.41.
- IF GLASS CONSTITUTES 40% OF THE WALL AREA AND SINGLE GLAZING IS PROPOSED THEN $U_{GLASS} = 1.13 \times 40\% = 0.45$ WHICH IS GREATER THAN 0.41, HENCE NOT PERMITTED.
- IF INSULATING GLASS IS USED AND THE U VALUE IS 0.67 THEN:

$$U_{ow} = \frac{U_{WALL} A_{WALL} MCF + U_{GLASS} A_{GLASS}}{A_{ow}}$$

$$\frac{0.41 \times 1.00 - 0.67 \times .40}{0.60 \times 0.75} = U_{WALL}$$

$U_{WALL} = 0.237$

MCF = 0.75 FOR WALLS BETWEEN 41 TO 80 lb/sq.ft.

IF THE GLASS AREA WERE REDUCED TO 30% OF THE WALL AREA THEN:

$$\frac{0.41 \times 1.00 - 1.13 \times 0.30}{0.70 \times 0.75} = U_{WALL} = 0.135$$

IF INSULATING GLASS IS USED:

$$\frac{0.41 \times 1.00 - 0.67 \times 0.30}{0.70 \times 0.75} = U_{WALL} = 0.398$$

IF GLASS WERE REDUCED TO 25% OF THE WALL AREA THEN:

$$\frac{0.41 \times 1.00 - 1.13 \times 0.25}{0.75 \times 0.75} = U_{WALL} = 0.227$$

THE ABOVE COMBINATIONS OF GLASS AREA AND TYPE OF GLASS WILL GIVE THE MAXIMUM U VALUE PERMITTED FOR THE WALLS, UNLESS THE U VALUE OF THE ROOF IS DECREASED. FROM THE REQUIRED WALL U VALUE THE TYPE OF WALL PANEL INSULATIVE SYSTEM CAN BE SELECTED. IF THE WALL PANEL SYSTEM IS FIRST SELECTED, THEN THE TYPE AND MAXIMUM AREA OF GLASS CAN BE DETERMINED.

THE U VALUE FOR THE ROOF SYSTEM WITHOUT SKYLIGHTS FOR THE DEGREE DAY RATING OF SAN JOSE IS 0.10. FOR FLOORS OVER UNHEATED OR OUT-DOOR AREAS, THE MAXIMUM U VALUE IS 0.21.

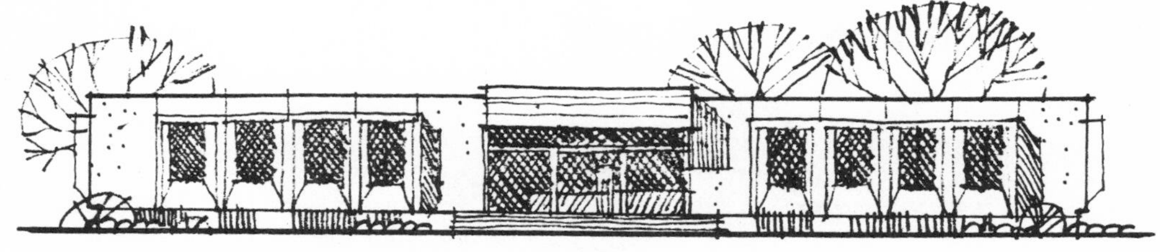

HEATING REQUIREMENTS

VISALIA - ANNUAL DEGREE DAYS VALUE IS 2546, DESIGN TEMPERATURE FOR WINTER - 32°, AND FOR SUMMER - 100°

REQUIREMENTS: $U_{ow} = .41$, $U_{or} = .10$, $U_{of} = .215$. ROOF MUST BE INSULATED, THEREFORE USE DOUBLE TEES W/ 1 1/2" POLYSTYRENE, $U = .10$.

A_f = AREA OF GLASS = 6'x7'x32 = 1344 SQ. FT

A_{wp} = AREA OF PIERCED PANELS LESS GLASS = (8 X 13 X 32) − 1344 = 1984 SQ. FT

A_d = AREA OF GLASS ENTRANCE DOORS = (8 X 13 X 6) = 624 SQ FT

A_{ws} = AREA OF NON-PIERCED PANELS (8 X 13 X 22) = 2288 SQ FT

A_{ow} = AREA TOTAL WALL = 6240 SQ FT

U_{wp} = U OF 5 1/2" CONCRETE LIGHTWEIGHT = .53

U_f = U OF INSULATING GLASS = .67

U_d = U OF GLASS FOR ENTRANCE DOORS = 1.13

MCF = MASS CONCRETE FACTOR FOR WALLS = .95

$$U_{ow} = \frac{U_w A_w MCF + U_f A_f + U_d A_d}{A_{ow}}$$

$$U_{ws}\,MAX = \frac{(U_{ow} + A_{ow}) - (U_{wp} A_{wp} MCF) - (U_f A_f) - (U_d A_d)}{A_{ws}\,MCF}$$

$$= \frac{(.41 \times 6240) - (.53 \times 1984 \times .95) - (.63 \times 1344) - (624 \times 1.13)}{2288 \times .75}$$

$$= .0957$$

SOLID PANELS MUST BE INSULATED BY: (a) INSULATION APPLIED TO THE INSIDE SURFACE, OR BY (b) SANDWICH PRECAST PANEL.

(a) 5 1/2" LIGHT WT. WALLS WITH 1 1/2" POLYURETHANE INSULATION ON THE SURFACE.

$$U = \frac{1}{\Sigma R} = \frac{1}{R_{oa} + R_c + R_i + R_{gb} + R_{ia}} = \frac{1}{.17 + 1.04 + 9.38 + .45 + .68}$$

$$= .085$$

(b) SANDWICH PANEL OF 2 1/2" LIGHT WT. CONCRETE, 1 1/2" POLYURETHANE, AND 2" LIGHT WT. CONCRETE.

$$U = \frac{1}{\Sigma R} = \frac{1}{R_{oa} + R_c + R_f + R_c + R_{ia}} = \frac{1}{.17 + .48 + 9.38 + .48 + .68}$$

$$= .0894$$

COOLING REQUIREMENTS

LATITUDE 36.4°

REQUIREMENTS

WALLS, MAX, BTU/HR/FT^2		= 32.5
SF	SOLAR FACTOR	= 125
		= 30 (FOR NORTH ORIENTATION, SHADE)
SC	SHADING FACTOR	= .95 (DOORS)
		= .90 (INSULATION GLASS)
TD_{eq}	TEMP. DIFFERENTIAL EQUIVALENT	= 30 (IF THE WALL WEIGHS BETWEEN 41-70 lb/SQ FT.)
ΔT		= 22°

$$BTU/HR/FT^2 = \frac{(U_W A_W TD_{eq}) + (A_f S_f SC) + (U_f A_f T)}{A_{ow}}$$

IF THE BUILDING IS PLACED NORTH AND SOUTH AT 12 NOON, THE ALTITUDE ANGLE IS 66.3°, THE AZIMUTH ANGLE IS 0° HENCE THE EAST AND WEST WALLS ARE SHADED. FOR A 2 FT. PROJECTION AT THE TOP OF THE GLASS THERE IS SHADING:

TAN ∡ X OVERHANG = HEIGHT OF SHADED AREA

TAN 66.3° X 2 FT = 4.5 FT

$U_W A_W TD_{eq}$ =

SOLID PANEL	= .089 X 2288 X 30	= 6109
PIERCED PANEL LESS GLASS	= .53 X 1984 X 30	= 31545

$A_f S_f SC$

INSULATING GLASS		
NORTH, EAST, WEST	= 6' X 7' X 24 X 30 X .90	= 27216
SOUTH, SHADED	= 6' X 4.5 X 8 X 30 X .9	= 5832
UNSHADED	= 6' X 25 X 8 X 125 X .9	= 13500
NON INSULATING GLASS		
NORTH	= 24 X 13 X 30 X .95	= 8892
SOUTH, SHADED	= 24 X 8 X 30 X .95	= 5472
UNSHADED	= 24 X 5 X 125 X .95	= 14250

$U_f A_f \Delta T$

INSULATING GLASS	= 0.67 X 1344 X 22	= 19810
DOORS	= 1.13 X 624 X 22	= 15512
		148,138

148138/6240 = 23.7 BTU/SQ. FT

THE ACTUAL 23.7 BTU/SQ. FT. < MAX. ALLOWED 32.5 BTU/SQ. FT., THEREFORE THE WALL SYSTEM IS ACCEPTABLE.

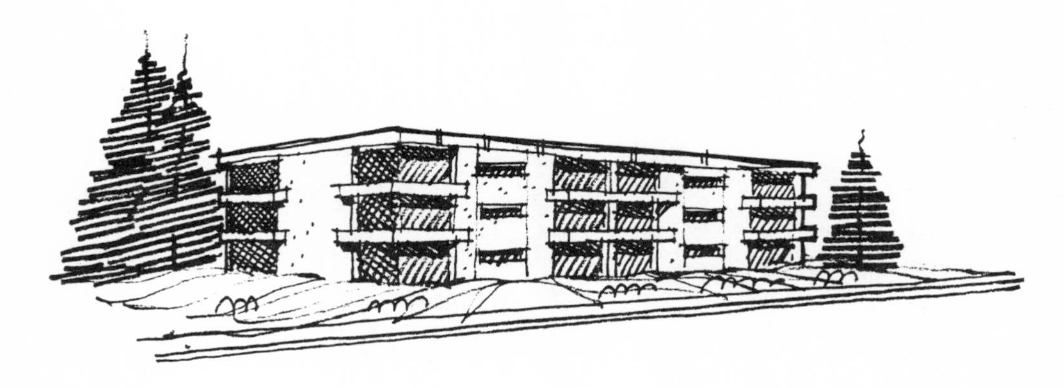

THREE STORY APARTMENT BUILDING
(LOW RISE)

FLOOR AREA = 60' × 120' = 7200 SQ.FT.
TOTAL FLOOR AREA = 3 × 7200 = 21,600 SQ.FT.
PERIMETER = 3 × 10' × 2 (60+120) = 10 800 SQ.FT.
AREA OF GLASS = 3552 SQ.FT

STANDARD REQUIREMENTS FOR HEATING

STANDARD GLASS AT 20% OF FLOOR AREA = .20 × 21,600 × 1.13 = 4881.6
WALL WEIGHT = 5/12 × 110 = 45.8 lb/ft², U_W = .16
= .80 × 10,800 × .16 = 1382.4
MAXIMUM ENVELOPE ALLOWED (U × AREA) = 6264.0

DESIGN SOLUTION № 1

USE INSULATING GLASS, U. = .67 : $A_f \times U_f$ = 3552 × .67 = 2379.8

WALL MAX. U VALUE, $U_W = \dfrac{6264 - 2379.8}{(A_W)\ 10800 - 3552} = .5359$

U VALUE OF 5½" LIGHT WT. CONCRETE WALL IS .529, AND .529 < .5359 THEREFORE WALL INSULATION FOR EXTERIOR WALLS IS NOT NEEDED, ROOF U VALUE MAX. = .06 AND THEREFORE MUST USE INSULATION OVER THE PRECAST CONCRETE ROOF SYSTEM.

DESIGN SOLUTION № 2

USE STANDARD GLASS, U = 1.13 : $A_f \times U_f$ = 3552 × 1.13 = 4013.7

WALL MAX. U VALUE, $U_W = \dfrac{6264.0 - 4013.7}{(A_W)\ 10800 - 3552} = .31$

TO ACHIEVE THE REQUIRED U VALUE, THE WALL MUST BE INSULATED BY: (a) APPLYING INSULATION TO THE INSIDE SURFACE OR (b) USING SANDWICH PANELS. FOR A SANDWICH PANEL W/ A MINIMUM OF 5" OF CONCRETE:

$$R_{\text{INSULATION MIN}} = 1/U - (R_{oa} + R_{conc.} + R_{IN})$$
$$= 1/.31 - (0.17 + 0.96 + .68)$$
$$= 1.41$$

(½" POLYURETHANE) = 3.12
(½" POLYSTYRENE) = 2.08

MULTISTORY APARTMENT BUILDING
(HIGH RISE)

PERIMETER = 8 STORIES × 10' × 2(56+120) = 27200 SQ FT.
GLASS AREA = 9000 SQ. FT (9000/27200 = 33% GLASS
MAX. 40% OF ENVELOP = .40 × 27200 = 10880 SQ. FT.

STANDARD REQUIREMENTS FOR HEATING

STANDARD GLASS = .40 × 27200 × 1.13 = 12294.4
WALL = .60 × 27200 × 0.16 = 2611.2
MAXIMUM ENVELOP ALLOWED (U × AREA) = 14905.6

DESIGN SOLUTION No 1

STANDARD GLASS U = .67 : $A_f \times U_f$ = 9000 × .67 = 6030

WALL MAX U VALUE = $\frac{U_{ow} - U_f}{A_{ow} - A_{glass}} = U_w = \frac{14905.6 - 6030}{27200 - 9000} = .4876$

.487 < .529 FOR A LIGHTWEIGHT 5 1/2" PRECAST WALL PANEL, THEREFORE MUST INSULATE. USE EITHER (a) INSULATION APPLIED TO THE INSIDE SURFACE OR (b) PRECAST SANDWICH PANEL.

DESIGN SOLUTION No 2

IF USING INSULATING GLASS AND NON INSULATED WALL PANELS OF 5 1/2" LT. WT. PRECAST CONCRETE, THE FOLLOWING MAXIMUM GLASS AREA IS ALLOWABLE:

$X A_w U_f + (1-X) A_w U_w = 14905.6$
$X \times 27200 \times .67 + (1-X)(27200 \times .53) = 14905.6$
$X = .128$ OR 12.8% GLASS
MAX. GLASS AREA = 3497 SQ. FT

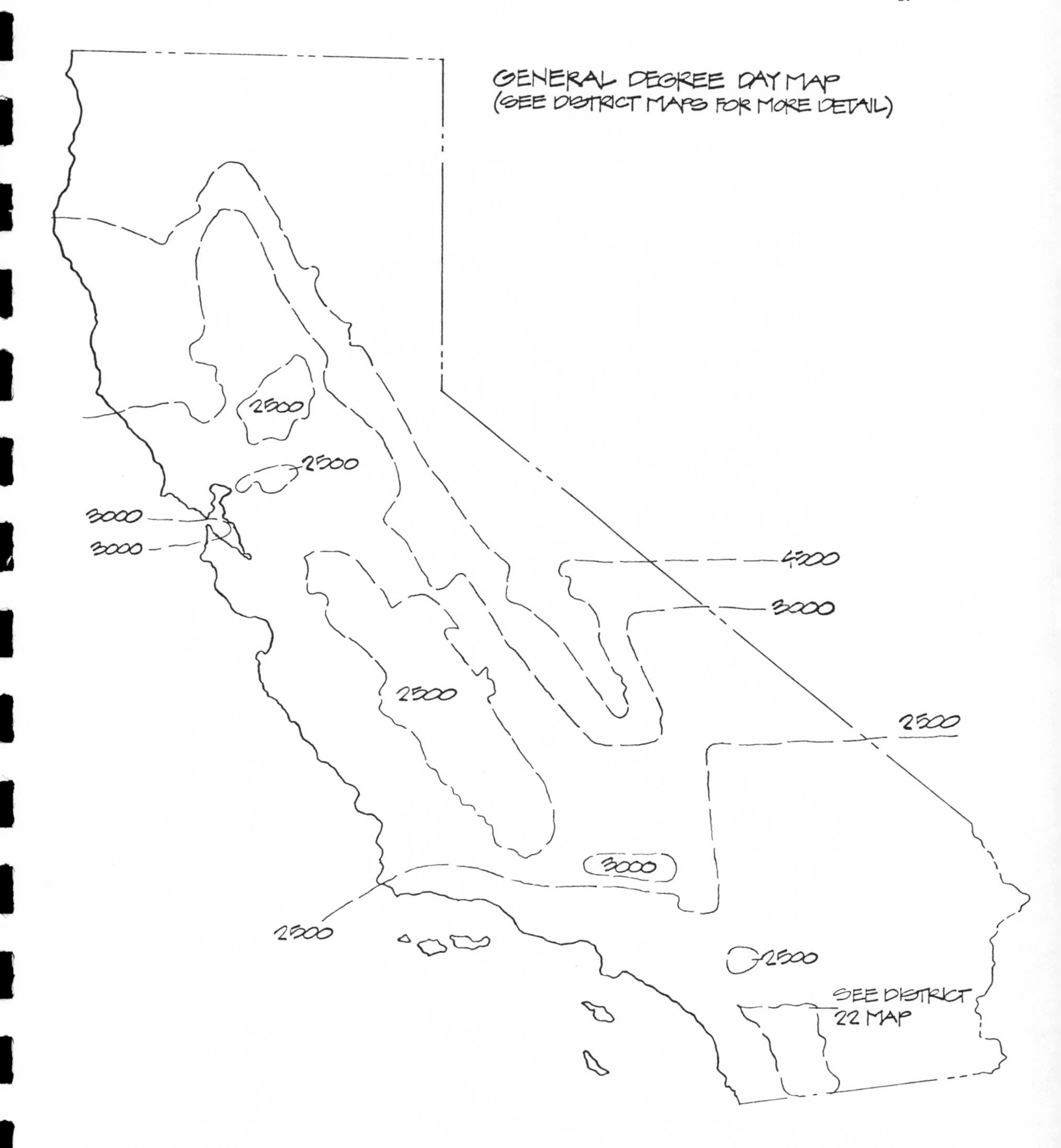
GENERAL DEGREE DAY MAP
(SEE DISTRICT MAPS FOR MORE DETAIL)
2500
2500
3000
3000
4500
3000
2500
2500
3000
2500
2500
SEE DISTRICT
22 MAP

ANNUAL HEATING DEGREE DAYS
FOR
THE STATE OF CALIFORNIA

County	Annual Heating Degree Days	Design Temperature Winter	Design Temperature Summer
Alameda			
Berkeley	2,850	39	84
Fremont	2,906	30	89
Livermore	2,781	28	97
Oakland	2,906	35	85
Alpine			
Markleeville	7,884	8	83
Amador			
Jackson	2,760	31	91
Ione	2,728	28	96
Butte			
Chico	2,795	29	100
Oroville	2,597	30	100
Calaveras			
Calaveras Big Tree	5,736		
Colusa			
Colusa	2,788	30	100
Contra Costa			
Antioch	2,627	30	93
Concord	2,766	32	92
Pittsburg	2,633	32	93
Richmond	2,644	35	85
Del Norte			
Crescent City	4,545	33	69
Elk Valley	5,404		
El Dorado			
Placerville	4,161	25	96
Tahoe Valley	8,198	2	84
Fresno			
Fresno	2,611	29	99
Mendota	2,555	29	100
Glen			
Orland	2,830	30	101
Stony Gorge Res.	3,124	29	99
Willows	2,807	30	100

County	Annual Heating Degree Days	Design Temperature Winter	Summer
Humboldt			
Alderpoint	3,290		
Eureka	4,679	35	65
Scotia	3,954	34	78
Imperial			
Brawley	1,161	34	109
El Centro	1,216	31	109
Imperial	1,060	35	107
Inyo			
Bishop	4,275	16	98
Death Valley	1,205	35	116
Independence	2,995	19	96
Kern			
Bakersfield	2,212	30	101
Delano	2,220	31	103
Inyokern	2,570	23	102
Maricopa	2,165	32	101
Mojave	2,590	25	100
Kings			
Hanford	2,642	28	100
Lemoore	2,960	29	100
Lake			
Lakeport	3,716	25	89
Lassen			
Susanville	6,248	4	89
Los Angeles			
Culver City	1,711		
Edwards A.F.B.	3,123	21	102
Fairmont	3,327	28	94
Long Beach	1,803	38	84
Los Angeles	2,061	42	86
Palmdale	3,088	24	101
Pasadena	1,694	35	93
Pomona	2,166	31	96
San Fernando	1,800	34	97
Madera			
Madera	2,485	30	100
Marin			
Novato	2,815	30	89
San Rafael	2,619	34	90
Mariposa			
Mariposa	3,116	27	96
Yosemite	4,800	18	90

County	Annual Heating Degree Days	Design Temperature Winter	Design Temperature Summer
Mendocino			
Fort Bragg	4,424	34	67
Ukiah	3,030	27	96
Willits	4,160	17	89
Merced			
Los Banos	2,267	28	100
Merced	2,697	29	99
Modoc			
Alturas	6,785	-1	90
Fort Bidwell	6,365		
Monterey			
King City	2,655	25	93
Monterey	2,985	34	82
Salinas	2,959	32	85
Soledad	2,920	29	90
Napa			
Napa	2,690	31	92
St. Helena	2,833	28	95
Nevada			
Nevada City	4,488	20	93
Truckee	8,209	-4	84
Orange			
Huntington Beach	2,361	40	81
Laguna Beach	2,262	37	80
San Juan Capistrano	1,646	39	82
Santa Ana	1,496	33	89
Placer			
Auburn	3,047	31	96
Colfax	3,441	25	89
Roseville	2,899	30	100
Tahoe City	8,162	7	77
Plumas			
Portola	7,055	-1	88
Quincy	5,852	10	93
Riverside			
Beaumont	2,790	28	96
Blythe	1,076	31	109
Corona	1,875	33	95
Elsinore	2,101	30	99
Palm Springs	1,232	32	108
Riverside	2,089	33	98
San Jacinto	2,376		
Sacramento			
Folsom	2,899	30	99
Sacramento	2,782	29	97

County	Annual Heating Degree Days	Design Temperature Winter	Design Temperature Summer
San Benito			
Hollister	2,725	30	91
San Bernardino			
Barstow	2,496	24	102
Daggett	2,203	24	103
Lake Arrowhead	5,200	20	84
Needles	1,072	33	110
Redlands	2,052	34	96
San Bernardino	2,018	32	100
Twentynine Palms	2,006	28	104
San Diego			
Alpine	2,104		
Barrett Dam	2,363		
Bonita	1,857	33	88
Borrego Springs	1,262	28	106
Cabrillo National Monument	1,653		
Campo	3,247		
Chula Vista	2,229	37	78
Cuyamaca	4,469	19	85
El Cajon	1,920	31	95
El Capitan Dam	1,397		
Encinitas	1,952	40	82
Escondido	2,052	33	92
Henshaw Dam	3,652		
Julian Wynola	4,085	19	90
La Mesa	1,492	36	90
Mecca	1,117	30	107
Nellie	4,745		
Oak Grove	3,516	26	95
Oceanside	2,092	38	81
Palomar Mt. Observ.	3,868	23	83
Point Loma	1,860	44	83
Ramona Spaulding	2,223	27	98
San Clemente	1,877	42	78
San Diego	1,439	42	83
Vista	1,760	34	85
Warner Springs	3,470	29	95
San Francisco			
San Francisco	3,080	42	80
San Joaquin			
Lodi	2,785	30	97
Stockton	2,690	30	98
Tracy	2,616	30	98
San Luis Obispo			
Paso Robles	2,890	26	100
San Luis Obispo	2,582	35	90

Counties	Annual Heating Degree Days	Design Temperature Winter	Design Temperature Summer
San Mateo			
Redwood City	2,596	32	86
San Mateo	2,655	36	87
South San Francisco	3,061	36	79
Santa Barbara			
Santa Barbara	2,290	34	84
Santa Maria	2,985	32	82
Santa Clara			
Gilroy	2,808	28	94
Los Gatos	2,794	32	89
Palo Alto	2,869	34	88
Santa Clara	2,566	31	88
San Jose	2,656	34	88
Santa Cruz			
Shasta			
Burney	6,249	5	90
Redding	2,610	31	101
Sierra			
Sierraville	6,953		
Siskiyou			
Fort Jones	5,614		
McCloud	6,007	11	86
Weed	5,870	8	86
Yreka	5,393	13	94
Solano			
Fairfield	2,434	30	95
Vacaville	2,812	29	94
Sonoma			
Cloverdale	2,666	31	96
Healdsburg	2,700	30	94
Petaluma	2,966	29	91
Santa Rosa	2,980	29	93
Stanislaus			
Modesto	2,767	32	98
Oakdale	2,832	28	99
Patterson	2,368	30	100
Sutter			
Yuba City	2,386	31	100
Tehama			
Mineral	7.192	-1	88
Red Bluff	2,688	31	101

County	Annual Heating Degree Days	Design Temperature Winter	Summer
Trinity			
Weaverville	4,935	16	96
Tulare			
Lindsay	2,619	30	100
Porterville	2,563	30	100
Visalia	2,546	32	100
Tuolumne			
Hetch Hetchy	4,797	18	90
Sonora	3,086	29	96
Ventura			
Filmore	2,377	33	90
Oxnard	2,352	35	80
Thousand Oaks	2,425	32	92
Yolo			
Woodland	2,447	30	100
Davis	2,819	30	99
Yuba			
Marysville	2,377	32	100

SOUND BARRIER WALL - SEAL BEACH

6000 L.F. of 18 foot high barrier wall deflect noise from the San Diego Freeway over immediately adjacent residential areas. 12 foot wide prestressed concrete units are supported laterally by prestressed concrete joists anchored in pole-type drilled footings.

SOUND TRANSMISSION

In group R occupancies such as hotels, motels, apartment houses, condominiums, and other non single family dwellings, regulations have been established for sound control by the State of California (California Administrative Code, Title 25, Chapter 1, Subchapter 1, and article 4), H.U.D. (Housing and Urban Development) and the Uniform Building Code, Chapter 35. According to UBC-79, airborne and impact sound insulation must be provided for walls and floor-ceiling assemblies that separate dwelling units from each other and from public spaces (i.e. interior corridors, and service areas). The airborne sound insulation must not fall below the sound transmission class (STC) of 50, or if field tested, 45. All openings for pipes, ducts, electrical outlets, and fixtures are to be sealed and treated. Entrance doors from interior corridors together with their permanent seals must not have an STC rating of less than 26. All floor-ceiling assemblies between separate units or guest rooms shall provide sound impact insulation which will give an impact insulation class (IIC) of 50. Floor coverings may be included in the assembly to obtain the rating but must be retained as a permanent part of the assembly. In addition to the above, the State of California and H.U.D. give the following requirements. For noise insulation from exterior sources, the interior community noise equivalent level (C.N.E.L.) shall not exceed 45 db in any habitable room with all the doors and windows closed. For residential locations having an exterior C.N.E.L. value greater than 60 db, an accoustical analysis is required to show that the structure has been designed to meet the interior C.N.E.L. value of 45. The exterior value shall be determined by local jurisdiction in accordance with the general plan of that city.

The unit of sound measurement is the decible (db) developed by Alexander Graham Bell. It is the log base of sound energy times 10, i.e.:

for an energy of:	10	100	1000
log base:	1	2	3
db:	10	20	30

Sound is measured by special sound level instruments, consisting of a microphone, amplifier and output instrumentation.

The following table is a list of the approximate sound pressure levels of common everyday noises:

Wind and Leaves	10 db
Quiet Conversation	30 db
Refrigerator	30 db
Quiet Radio	40 db
Conversation	50 db
Radio - loud	70 db

Speech	70 db
Car	90 db
Motorcycle	110 db
Rock Band	110 db
Sonic Boom	120 db
Jet Taking Off	130 db

The changes of sound level are:

1-2 db - not noticeable
3 db - just noticeable
5 db - clearly noticeable
10 db - twice as loud
20 db - much louder

The three basic terms used in sound transmission and sound control that the designer should be familiar with are:

<u>SPL (Sound Pressure Level)</u>. The intensity of sound present in a room, or outside environment, measured in decibels (db).

<u>STC (Sound Transmission Class)</u>. The amount of sound reduction given by a wall or floor-ceiling assembly. Basically, the sound pressure level on one side of the barrier, less the STC of the barrier medium, equals the sound pressure level on the other side of the medium. This is also the Sound Transmission Loss (STL) of the medium.

<u>IIC (Impact Insulation Class)</u>. A designation given to a floor assembly to indicate its ability to <u>deaden</u> impact noises going through the floor-ceiling assembly. Insulating mediums or floating floor systems give good values of IIC.

Sound interferes with speech, privacy, concentration, sleep, and community stability. The degree of annoyance depends on the sound pressure level, extent of pure tones, time of day, season, what you are doing and who hears it, and background levels. Annoyance is reduced by a volume reduction in the source, reduction or absorption along the sound's path, and masking (use of cover-up sound).

Sound is reduced by the following methods: (a) Mass - each doubling of mass results in a reduction of 6 db; (b) Discontinuity - use of air spaces of a minimum of 2" - each doubling of the air space results in a reduction of 5 db, and dissimilar materials (materials which vibrate at different frequencies); (c) Distance - 6 db reduction for doubling of distance from a point source and 3 db reduction for doubling the distance from a line source; (d) use of barriers - i.e. trees, earth, mounds, constructed walls, etc.

The way in which the STC of a wall or floor determines accoustical privacy is illustrated below. A good sound barrier should reduce the noise originating in one room to below the background noise level in an adjacent room. The higher the values of STC and IIC the more desirable the material or system.

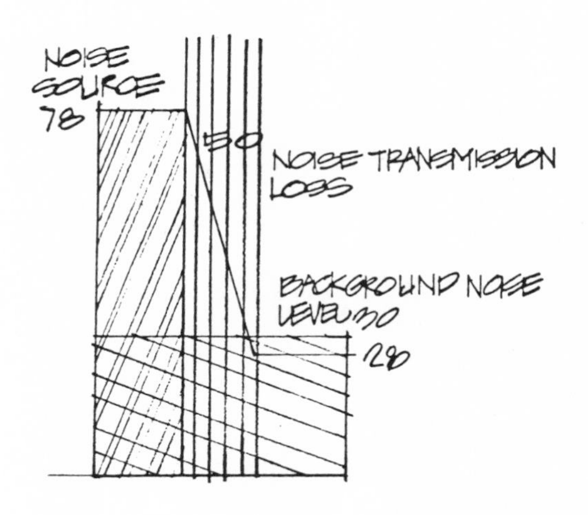

Sound is made up of waves at various frequencies from 0 to 4000 cycles per sec. (hz). The following is an example of sound transmission loss through a 6" block wall:

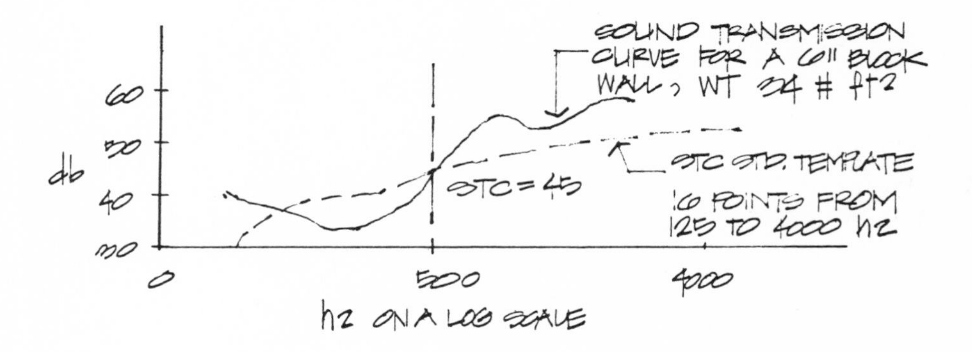

STC = Sound Transmission Class - The point where this standard curve crosses the true material curve at 500 hz, is the STC rating for that material.

The Second Edition of the PCI Design Handbook on pp 6-16 through 6-24 has an excellent summary of acoustical properties of various precast concrete systems and other related materials used in building construction.

SOUND LEAKAGE

Sound leakage occurs from under doors, louvers, wall openings, ducts, and over hung ceiling spaces:

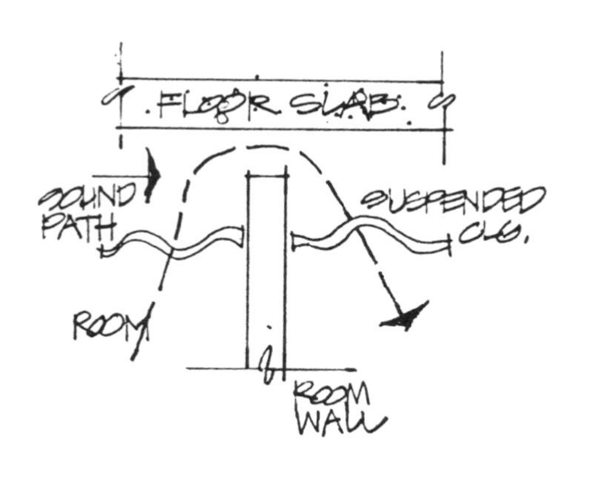

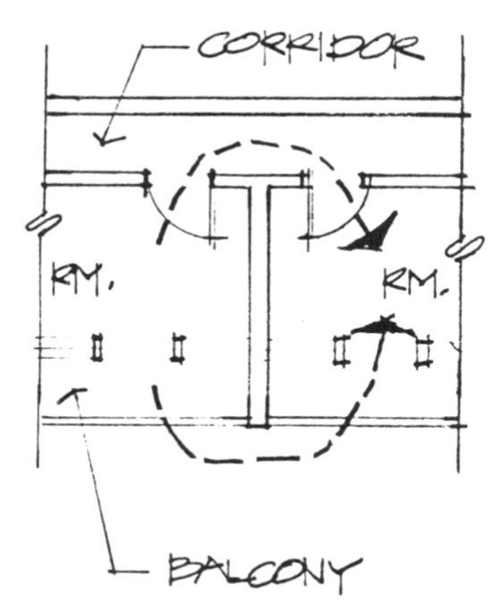

SOUND BARRIER SYSTEMS

Various types of sound barriers used in design are indicated below:

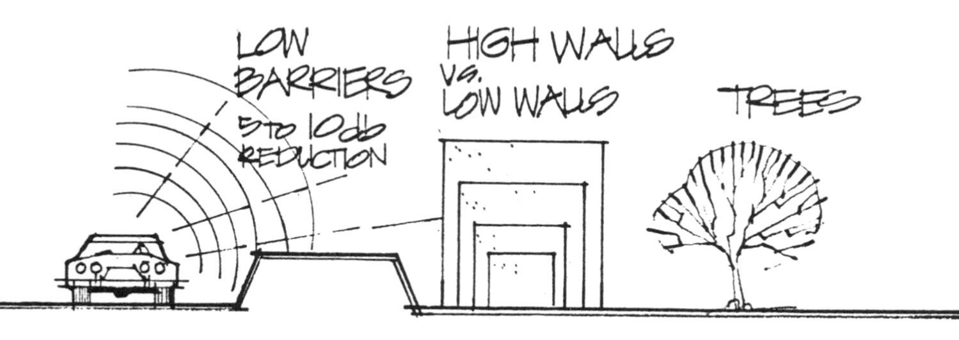

In California, precast and prestressed concrete units have been used to form sound barriers to deflect freeway noise over adjacent residential areas. The effectiveness of these barriers is dramatic, in the complete silence and lack of audible traffic noise one experiences behind the wall. High walls (over 12 feet) are especially suitable to the use of plant cast precast prestressed concrete units.

SOUND BARRIER WALL - SEAL BEACH

This sound barrier wall extends along the 605 Freeway on the north side between Valley View Street and Seal Beach Boulevard. The prestressed concrete solution was the low bid in a series of alternate systems and materials specified for design-construct proposals. Additional savings were realized from the time savings achieved in using precast concrete elements as compared to slow on-site alternate methods of construction. The amount of sound attenuation behind these walls is total, with virtually no noise being heard behind the wall.

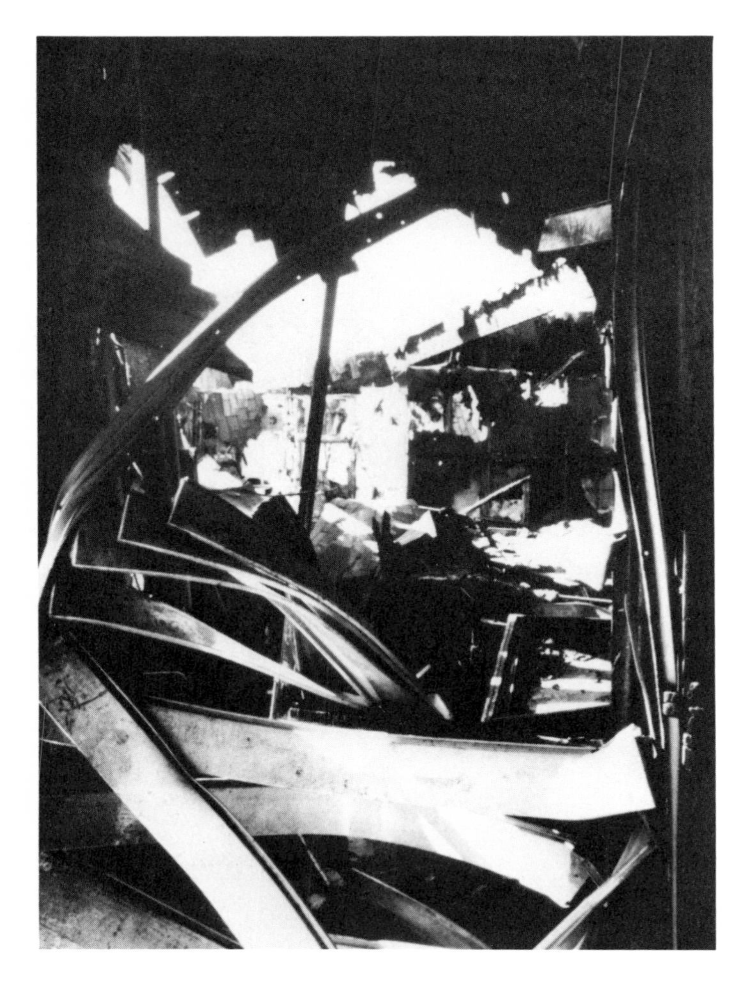

LOS CERROS INTERMEDIATE SCHOOL FIRE - DANVILLE (2 JUN 79)

Over 30 fires occurred in 1979 in California's non-fire proof schools, many of which are sprinklered. In the case above, as is with many others, arson was suspected. Not only did the San Ramon Valley School District have to pay the first $100,000 in cash for repairs, in accordance with the deductable provisions of their insurance policy, but they also spent over $200,000 in additional funds for temporary school quarters for the displaced students. As of this writing (Feb. 80) insurance regulations have blocked reopening of the school in September, and the students may lose even their temporary quarters in distant Walnut Creek. As a result of another school fire (Neil Armstrong Elementary School, San Ramon) the school district in a lawsuit is citing the architects for their failure to design the school to prevent the spread and extent of the fire.

FIRE RESISTANCE AND FIRE PROOFING

The Uniform Building Code requires that the designer consider fire resistance in the design of structures. The amount of fire protection required is determined by various provisions in the Code, which can be somewhat confusing to one unfamiliar with the logic used. Therefore, the procedure for determining the required fire resistance of building components is given here.

1. DETERMINE OCCUPANCY TYPE. Section 501, Table 5-A, and Chapters 6, 7, 8, 9, 10, 11 and 12 of UBC-79 give definitions of the type of occupancy of the structure based upon the type of use envisioned.

2. DETERMINE FIRE ZONE. The intended location of the building will affect required fire resistance of certain portions of the structure.

3. DETERMINE PROXIMITY TO PROPERTY LINES. Increased fire resistive requirements are imposed for exterior walls too close to property lines.

4. DETERMINE TYPE OF CONSTRUCTION. The type of construction (Type I, II, III, IV or V) required for various components will be indicated by the occupancy type, the height of the building, and area per tier. Types of construction are defined in Chapters 17 through 22 of UBC-79. Tables 5-C and 5-D give types of construction required for the areas and heights of structure being considered. See also Section 505 for area requirements for multi-story buildings. Section 709 and Table 7-A give specific requirements for open parking structures.

5. DETERMINE REQUIRED COMPONENT FIRE RATING. Based upon the type of construction required, determine the fire rating required for the building components being designed. Table 17-A of UBC-79 gives a breakdown of type of construction and required fire rating in hours required for the component. Fire ratings for components are as determined by testing with a standard procedure as outlined in ASTM E-119. The component being tested is subjected to a specific fuel load in a standard furnace, which develops a specified temperature and time relationship calibrated to this fuel load. The fire rating of the component is the maximum time that the component can satisfy each of three basic "end point" criteria, which are:

a. support of the design applied load
b. surface and interior integrity preventing the passage of flame through the member
c. for floors or roofs, a limitation on the temperature of the unexposed surface of the deck.

When any of these criteria is exceeded, the test is stopped and the time recorded is the fire rating of the component or system. Shown below is a part of Table 17-A:

Building Element	Non-Combustible				Combustible				
	Type I	Type II			Type III		Type IV	Type V	
	Fire Res.	Fire Res.	1 hr	N	1 hr	N	H.T.	1 hr	N
Exterior Bearing Walls	4	4	1	N	4	4	4	1	N
Exterior Non-Bearing Walls	4	4	1	N	4	4	4	1	N
Interior Bearing Walls	3	2	1	N	1	N	1	1	N
Interior Non-Bearing Walls	1	1	1	N	1	N	1 or HT	1	N
Structural Frame	3	2	1	N	1	N	1 or HT	1	N
Floors	2	2	1	N	1	N	H.T.	1	N
Roofs	2	1	1	N	1	N	H.T.	1	N

N - No fire rating required

H.T. - Heavy Timber

6. DETERMINE THICKNESS AND CONCRETE COVER REQUIREMENTS. Chapter 43 of UBC-79 lists fire resistive standards for materials and components based upon the fire rating requirements determined above. Table 43-A gives required cover over reinforcing and strand for precast and prestressed concrete members. Table 43-B gives required thicknesses of walls and Table 43-C gives required thicknesses of floors and roofs to satisfy fire ratings required. Some of these requirements are listed below:

Component	4 hr.	3 hr.	2 hr.	1 hr.
Solid Concrete Walls, Thickness	6½	6	5	3½
Concrete Floors and roofs, thickness (Normal weight concrete)	6½	5½	4½	3½
Concrete floors and roofs, thickness (Lightweight concrete)	5	4½	4	3
Concrete coverage - pretensioned Members - Grade A concrete*				
Beams and Girders	4	3	2½	1½
Solid Slabs		2	1½	1
Concrete coverage - reinforcing Bars - Grade A Concrete* Columns and Girders:				
12" Wide and Larger	1½	1½	1½	1½
Concrete Joists	1¼	1¼	1	3/4
Floor and Roof Slabs	1	1	3/4	3/4

* For Sand-Lightweight Concrete, cover requirements may be reduced by 25%, but not less than 1½" in girders and columns, nor 3/4" in slabs.

The Uniform Building Code also allows the use of PCI MNL-124 (Design for Fire Resistance of Precast Prestressed Concrete). This publication gives analytical procedures which may be used to determine required cover and concrete thicknesses of prestressed and precast concrete components. Fire ratings of hollow core planks are based upon tests performed by the particular licensing agent of the product under consideration. These tests have been performed in accordance with the requirements of ASTM E-119. Fire resistance ratings of precast and prestressed components and built-up assemblies can also be determined by referring to Fire Resistance Ratings published by the American Insurance Association, New York City, and the Fire Resistance Index, by Underwriters Laboratories in Chicago, Illinois.

ARCHITECTURAL PRECAST CONCRETE CLADDING AS FIRE PROTECTION.

These units may be used to fire proof structural steel members, thus permitting steel framed buildings to be classified as type I and type II construction. The architectural precast concrete units are used to enclose bracing systems, spandrel beams, girders, and columns. These precast concrete enclosures contribute elements of color, texture, and form to help express the design theme of the architect. The entire enclosure can give a sculptured effect to a column or frame. These cladding units are quickly erected, and therefore reduce construction time when compared to on-site methods of providing exterior fireproofing. The joints between cladding units must be protected in order to provide the fire rating required for the individual cladding units (see PCI MNL-124).

FIRE RESISTANCE AND SPRINKLERS

The Uniform Building Code allows increases in maximum allowable heights and floor areas when an automatic system is installed. (Sections 506 and 507) Section 508 allows elimination of fire rating requirements for all one-hour construction when the building is sprinklered. Chapter 33 permits reducing the required number of exits and increasing the distance to exits when sprinklers are used. These reductions in life safety requirements in exchange for automatic sprinklers are known as "trade-offs." The intent of these trade-offs is to make it economically viable for owners to provide for sprinklers in the design of the building. This practice is not new, and has been permitted for many years in our building codes. Recently, there has been an increased push by special interest groups to effect further reductions in life safety requirements for buildings using the trade-off concept. The danger in this philosophy is that total reliance for fire safety and life safety is based upon the 100% reliability of the sprinkler system. Statistics for the past performance of sprinklers indicate that human and mechanical factors render this assumption to be invalid. The Prestressed Concrete Institute, in a recently issued position paper, indicates "strong objection to the concept of trade-offs and endorses the following positive recommendations:

1. BUILDING STRUCTURES MUST BE DESIGNED TO MINIMIZE THE POSSIBILITY OF FAILURE WHEN SUBJECT TO FIRE.

2. CODES SHOULD BE REORIENTED TOWARD REDUCING FIRE HAZARDS, SMOKE-GENERATING MATERIALS, AND COMBUSTIBLE FINISHES.

3. COMPARTMENTATION IS A PROVEN METHOD OF PROVIDING LIFE SAFETY FOR BUILDING OCCUPANTS AND SHOULD BE PROVIDED TO CONTAIN A FIRE WITHIN A LIMITED AREA.

4. SPRINKLERS SHOULD BE REQUIRED IN HAZARDOUS AREAS, PARTICULARLY WHERE COMBUSTIBLE CONTENTS EXIST. HOWEVER, THE STRUCTURAL INTEGRITY OR LIFE SAFETY ASPECTS OF A BUILDING MUST NOT BE IMPAIRED.

In summary, sprinklers are not totally reliable. Safe designs should take into account the possibility of sprinkler failure rather than the hope of sprinkler performance. Use of proven noncombustible materials, compartmentation, smoke detection, and selective use of sprinkler systems offers the best life safety system."

Compartmentation is the "breaking up" of the floor area into smaller subdivisions that contain the spread of fire, such as is obtained with fire resistive partitions between living units in apartments, etc. This aspect of construction has contained fires within relatively small areas until fire fighting equipment and personnel have arrived. Chapter 38 in UBC-79 requires sprinklers in practically all types of occupancies, is independent of type of construction, and does not recognize the tremendous life safety aspects of compartmentation. The insurance industry, however, recognizes the advantage of fire resistant construction; recent rate comparisons indicate a 4 to 1 ratio in favor of concrete for structure and contents. Extended coverage rates for non-fire proofed steel structures are 9 times those for a precast concrete structure.

PCI MNL-124, Design for Fire Resistance of Precast Prestressed Concrete provides an analytical method of evaluating the fire endurance of structures made of precast and prestressed concrete. This manual, mentioned on the previous page, brings together information from many sources and permits the engineer to design for fire endurance without referring directly to standard tests. The manual is a positive contribution for designers, building officials, and insurance underwriters - anyone concerned with fire safety of buildings.

OYSTER POINT MARINA BREAKWATER - SO. SAN FRANCISCO

Adequate cover, coupled with proper concrete mix proportioning and plant casting quality control techniques assure that these prestressed concrete sheet piling units will have longevity, even when subjected to the heavy saline environment of San Francisco Bay, and the great potential for corrosion in the splash zone above low water.

MINIMUM CONCRETE COVERAGE FOR REINFORCING

Minimum clear thicknesses of concrete (cover) are provided over reinforcing bars or prestressing strands in precast concrete members to provide sufficient protection from the elements, and to satisfy the applicable requirements of fire protection. The greater of these two basic requirements governs which amount of cover is required.

1. FIRE PROTECTION: The Uniform Building Code uses minimum cover criteria as outlined in Section 4303 and gives required cover dimensions in Table 43A as one means to satisfy required fire rating requirements for precast and prestressed concrete components in structures.

2. PROTECTION FROM THE ELEMENTS: In an alkaline concrete environment, the reinforcing develops an isolating iron oxide film that prevents corrosion and makes additional protection unnecessary, such as by galvanizing or by providing cathodic protection. Minimum cement contents, low water cement ratios, coupled with compliance with the cover requirements listed in UBC-79 will suffice to provide plant cast precast concrete elements that will last indefinitely. Section 2607(o)3 gives the requirements for plant cast precast elements; Section 2607(o)4 gives criteria for prestressed concrete members. For components that will be subjected to severe ocean exposure, minimum cover should be increased to 2½" over main reinforcing and prestress strands, and 1½" over secondary reinforcing (Prof. Ben C. Gerwick, Jr. in ACI Publication SP47-14). In this regard it should be reiterated that increased cover is just one of the durability requirements accompanied by proper water cement ratio, minimum cement content, proper aggregates, proper tri-calcium aluminate (C3A) content in the cement along with maximum alkali content, and in some cases, extended moist curing.

DESIGN LOADS

The design of precast concrete buildings is similar to the design of buildings in steel, wood, or cast-in-place concrete. The exceptions are that precast units must also be designed for handling and special consideration must be given to joinery. These loads are identified and a discussion of the general nature of the loads is presented in this section. The manner in which the precast components are designed to resist these loads is covered more fully in Part C. The design of total precast and prestressed concrete structures to resist seismic lateral forces, to the extent permitted by the Uniform Building Code, is covered on pages 5 and 99. In general, prestressed and precast concrete components and/or structures are subjected to 4 major categories of loading.

1. Vertical
   a. Dead Loads
   b. Live Loads

2. Lateral
   a. Wind Loads
   b. Seismic Loads

3. Volumetric Change Induced Loads
   a. Shrinkage
   b. Creep
   c. Temperature

4. Handling
   a. Stripping
   b. Shipping
   c. Erection

DEAD LOADS

Dead loads consist of the weight of the structural system plus all permanently attached materials and equipment. In the design of a system, assumptions are often made for the dead load of the member or members being designed. If the assumption is in error by more than 10% then a correction is necessary. The designer will often use the manufacturer's product information tables for determination of member dead load prior to performing product design. Some typical dead loads of floors, ceilings, roofs and walls are given in tabular form on page 8-2 of the Second Edition of the PCI Design Handbook.

LIVE LOADS

Live loads are the result of temporary, transitory weights of people, vehicles, snow, or moving equipment. Governing building codes recommend minimum live loads. The architect or engineer has the responsibility of defining the final design live loads to be used. These may or may not be the same as minimum code recommendations. The final values are based upon experience depending upon the use or occupancy of the building. Table 23-A of the 1979 Edition of the Uniform Building Code gives some recommended minimum live loads to be used in designing buildings. If the tributary area supported by a member is large, a reduction is allowed due to the small probability that loading would occur over the entire area at any one time. Section 2306 of UBC-79 bases this reduction on the amount of area supported over 150 square feet. The amount of live load reduction is limited to 40% for members receiving load from one level only and 60% for other members. Live loads exceeding 100 psf may not be reduced.

WIND LOADS

Wind produces dynamic forces with varying magnitudes due to gusts, altitude, direction, building shape, and shadow or funneling effects caused by surrounding structures. In the design of lower structures, wind loads are treated as static forces whose magnitude depends upon building location and component height above the ground. In UBC-79, wind loads are determined by geographical location in wind pressure zones shown in Figure 4, Chapter 23, and by height above the ground as shown in Table 23-F. Section 2311 gives other requirements for wind design. At the time of this edit, the Structural Engineers Association of California is in the process of developing a revision to the Code which reflects the philosophy used in ANSI A58.1-72.

SEISMIC LOADS

Seismic ground motions occur in random patterns in both the vertical and horizontal directions. They are very complex and nonpredictable. Some of the variables are: focal point depth, type of fault action, wave propagation through the different rock and soil strata, fault location, magnitude and probability. In any earthquake there are usually foreshocks, main shock and aftershocks. No two have ever been recorded to have occurred at the same focal point nor of the same acceleration pattern. High rise buildings are designed to withstand both known earthquake patterns and man made patterns. They are designed by the dynamic concept and compared with the static concept. Acceptable limits are set for story drift.

Low rise buildings and precast concrete components used in buildings are designed according to the static concept considering also story drift limits. The static concept is referred to in the Uniform Building Code as the Equivalent Lateral Force Concept (ELF). Section 2312 of UBC-79 details the requirements of this procedure. Pages 99 to 144 of this book go into greater detail on the basic concepts of seismic design theory.

Seismic activity causes the application of another type of loading on the members and their connections to one another, diaphragm elements, or lateral stiffening elements: repetitive loading. Repetitive loads are cyclical loads which are applied, removed, and reapplied, and in the case of seismic cyclic loads, are characterized by reversals in direction of application. Many structural configurations using precast concrete which appear to offer promise in resisting cyclic seismic loads must have their performance verified under the influence of repetitive (cyclic) loading.

FORCES INDUCED FROM VOLUMETRIC CHANGES

Volume change movements resulting from shrinkage, creep, and temperature if restrained, can induce significant forces in precast concrete component connections and interfacing materials in the structure. The significance of these three variables varies with the type of precast member being considered. In general, all three variables significantly affect precast, prestressed members, while only temperature changes significantly affect non-prestressed members; and the effect of each variable increases as the length (or size) of the member increases.

Shrinkage The degree of precast concrete shrinkage (reduction of concrete volume) brought about by the evaporation of water used in the concrete mix is normally less than that of cast-in-place concrete of the same strength because low water cement ratios are specified for precast concrete mixes. On the other hand, some precast units often tend to be relatively long, therefore the total shrinkage (total change in dimension) of a single unit is sufficiently large to warrant close attention during the design process. Short (approximately 30 feet or less), non-prestressed wall panels constructed with low-slump concrete are usually not affected much by shrinkage. However, many wall panels are 40 feet or more in length and are usually prestressed and fabricated with higher slump concrete. Such long, prestressed wall panels are significantly affected by shrinkage.

All precast, prestressed beams must be carefully designed to account for the effects of concrete volume change brought about by shrinkage. Shrinkage acts to shorten the length of the member. As the member shortens, two important things happen: 1) the force in the prestressed tendons decreases, and 2) the member to which the beam is attached may be subjected to additional forces induced by the shortening movement. The effects of loss-of-prestress-force are considered when selecting the initial prestress force to be applied to the member. The effects of possible transfer or redistribution of forces through differential movement of structural members are considered during the design of the connection details.

Column sections are similar to wall panel sections in that the effects of shrinkage are highly dependent upon the factors of slump, length, the presence or absence of prestressing, and the magnitude of the prestressing force.

<u>Creep</u> Creep is the long term deformation of a material under sustained load. Precast-prestressed members have a greater tendency to creep than do members which are not prestressed. Creep tends to reduce the initial prestress forces placed on the member and reduce the length of the member at the centroid of the prestress force. The effects of creep must be carefully considered in order to prevent the development of excessive stresses and deflections due to loss of prestress within the prestressed member and to prevent the transference of forces caused by the reduction in length of the prestressed member (i.e., connection details must be designed so that no unconsidered force is developed by a change in length of the prestressed member).

The effect of creep on precast-non-prestressed members is the same as that for cast-in-place concrete of equivalent strength and consistency. However, the high-strength and low-slump consistency of concrete used in precast operations is generally more resistant to the effects of creep than that concrete normally used in cast-in-place operations.

The effect of creep on wall panels, beams and columns is similar to that of shrinkage except that (1) creep continues over a longer period of time, (2) lightweight concrete creeps more than normal concrete and (3) creep on load bearing wall panels and columns can be significant even when the panels or columns are not prestressed.

<u>Temperature</u> The effects of temperature on precast-non-prestressed structural members are treated in the same manner as cast-in-place members which use deformed bars as the primary reinforcing element.

When prestressed tendons are used as the primary reinforcing elements, the designer must pay special attention to the effect of temperature changes because (1) such members tend to be relatively long; therefore, the total dimensional change (induced by the temperature change) is represented as one relatively large change at one location (as contrasted with a series of relatively small changes spread out over many locations) and (2) changes in temperature often have a significant effect on the force developed by prestressing strands embedded in long members. (An increase in temperature causes a significant lessening of the pre-stress force on a strand which is not yet embedded in a member. However, once the member is cast, and the strand is bonded to the concrete, an increase in temperature usually results in a negligible increase in strand tension. Temperature changes must be carefully considered when designing connections for all structural members (walls, beams and columns) to make sure that no unexpected or "unaccounted for" force develops within the structural system. Of special significance is the effect of temperature-induced forces on long thin spandrel elements and their connections, as outlined on pages 182 through 185.

Estimates of magnitudes of expected movements resulting from shrinkage, creep, and temperature strains for various precast and prestressed concrete components and structures can be made by use of design information given on pages 4-10 through 4-18 of the Second Edition of the PCI Design Handbook. Designers should consider these movements and forces in the preliminary design stages of projects, especially in composite structures consisting of precast/prestressed concrete components with poured in place concrete closures or post-tensioned slabs. Location and arrangement of lateral stiffening elements is critical in light of long term creep and shrinkage movements to preclude severe distress and cracking. Closer spacing for contraction joints coupled with temporary gaps in post-tensioned slab construction to allow initial creep and shrinkage movements to occur are indicated as being essential practical considerations in this type of construction.

HANDLING LOADS (DEAD LOADS, SUCTION, MECHANICAL BONDING AND IMPACT)

The term "handling" is included here as a loading classification to emphasize the fact that there exists a general set or collection of loads which are extremely important variables affecting the design of precast members, but which are not even relevant to the design of most cast-in-place members.

Handling loads include those induced by the production operations of stripping, storing, transporting and erecting. During each of these operations, the position of the member may vary; the combination of applied loads (dead, suction, mechanical bonding, impact) may vary; the location and characteristics of the member supports may vary; and the strength of the concrete may be significantly different at the time of each operation.

a. Dead Loads (Member Self Weight) A discussion of dead loads is reintroduced at this time to emphasize that it is a major production factor as well as an in-place factor to be considered. The larger the dead load of the member, the greater the work and time involved to move that member. Usually a precast member is moved in the following manner: it is first lifted from the casting bed and placed on a storage pad, then lifted from the storage pad and placed on a trailer for transportation to the site and finally lifted from the trailer to its final resting place. Thus a minimum of three lifting operations is normally involved. When the dead load of the member is reduced, the size of the equipment needed to move the member is reduced and the time to move the member is usually also reduced. However, it should be noted that it is generally less costly to handle one large member than to handle two small members which are equivalent in weight to the large one.

b. Suction Suction forces are developed between concrete and the form in which it is placed. These forces are relatively insignificant for cast-in-place concrete and the stresses developed by the form-stripping operation are usually ignored in the structural analysis of the member. In contrast to this, the suction forces on precast wall panels are significantly large and must be given special consideration during the form-stripping operation. Suction is insignificant in precast-prestressed beam design because when the tensile forces are transferred from the prestressed strands to the concrete in which the strands are embedded, the concrete beam shortens in the form bed; this shortening releases or "breaks" the suction between form and concrete. Thus, there are no suctional forces to resist when removing precast, pretensioned beams from their forms.

c. Mechanical Bonding Mechanical bonding forces are often developed between the concrete and the form in which it is placed. Forces developed by mechanical bonding during form-stripping are similar to those created by suction in that they are relatively insignificant for cast-in-place concrete but often tend to be significant during the stripping operation for all types of precast units. Careful attention must be paid to the architectural detailing of the shape of wall panels to minimize the amount of mechanical bond that develops at panel edges, faces and openings.

The sides of precast beams are never formed as parallel lines. The form sides are always canted slightly in order to assure that no mechanical bond will develop between the form and the concrete. (Recommended minimal slopes or draft are discussed more fully on pp 205-208.) If it were possible to maintain two perfectly parallel sides, it would be unnecessary to provide draft in the form sides. However, there is sometimes some slight bowing of the form sides. The sides are sloped in order to make sure that even after the bowing occurs, the concrete may be lifted "straight up" without encountering any mechanical keys formed by excessive deflections of the form sides.

The amount of draft depends on the design of the formwork itself and how resistant it is to deflecting under the weight of the concrete and the forces generated by the vibrating effort. It is important to give careful consideration to the effects of mechanical bonding regardless of whether the concrete is precast or cast-in-place. Poorly detailed forms for plant-manufactured precast concrete can result in a reduction in the number of times that a form can be re-used, excessive stripping costs, and undesirable damage to the concrete finish.

d. Impact Impact loads are placed on all precast members when they are transported on trucks. The producer attempts to support the member at only two transverse points on the truck trailer (preferably above the trailer axles). Supporting the member at these points tends to minimize the amount of bouncing and thus minimizes the impact forces placed on the member.

A detailed summary of handling forces, impact loads, and design procedures is given on pages 165 through 176 (Product Handling).

SITE #1A - STATE OFFICE BUILDING - SACRAMENTO

8 foot wide by 24" deep prestressed concrete double tees span between precast concrete frames that take wind and seismic lateral forces to the foundations, as well as providing support for vertical loads. The lateral analysis for the building was performed using the Equivalent Lateral Force design procedure outlined in the Uniform Building Code.

SEISMIC DESIGN AND STRUCTURAL SYSTEMS

Three types of structural systems are used in buildings to resist lateral forces resulting from wind or seismic disturbances. These systems are:

1. Shear Wall Systems
2. Moment Frame Systems
3. Combined Wall and Frame Systems

In Section 4, it was shown that the present state of the art in precast concrete construction, and the current level of acceptability in the Uniform Building Code and the Recommended Lateral Force Requirements of the Structural Engineers Association of California (referred to as the "Blue Book") restricts the use of precast concrete in zones of high seismicity to shear wall systems and systems employing details conforming to the ductile moment resisting frame design provisions of UBC-79 for cast-in-place concrete. In this section, we shall briefly discuss an empirical dynamic theory of the seismic response of buildings in resisting seismic forces, the current Equivalent Lateral Force (ELF) design provisions of the Uniform Building Code, and present a procedure which may be used for the rational analysis of shear wall buildings in resisting lateral forces. Finally, some design examples are given to further demonstrate the use of the principles involved in shear wall design with precast concrete elements.

EQUIVALENT LATERAL FORCE DESIGN

The Uniform Building Code provisions attempt to reduce complex dynamic relationships resulting from seismic ground motion effects on buildings into a loading situation resulting from equivalent static lateral forces. In general, the actual induced lateral forces from a severe earthquake are much greater than the forces resulting from the use of the Equivalent Lateral Force procedure. A frequently used method of determining actual seismic forces on buildings is by performing a dynamic analysis based upon a response spectrum for a severe earthquake. The 1940 El Centro quake is commonly used as a model. This earthquake is the most severe quake for which we have accurate recorded data, and it is estimated that the likelihood of its occurrence would be once in 50 years. The maximum recorded ground acceleration for the El Centro quake was 0.33g.

The graph of base shear versus building period also shows graphically the large difference between actual seismic forces and those resulting from the ELF procedure. This difference in energy demand must therefore be satisfied by inelastic action of the structure. In the design of stiff shear wall buildings, the Code assigns severe penalties in the form of higher K factors and greater ultimate load factors. In ductile frame structures, where inelastic action can take place by yielding of the structure, or by formation of plastic hinges to absorb energy, the ELF penalties are less severe. The required energy absorp-

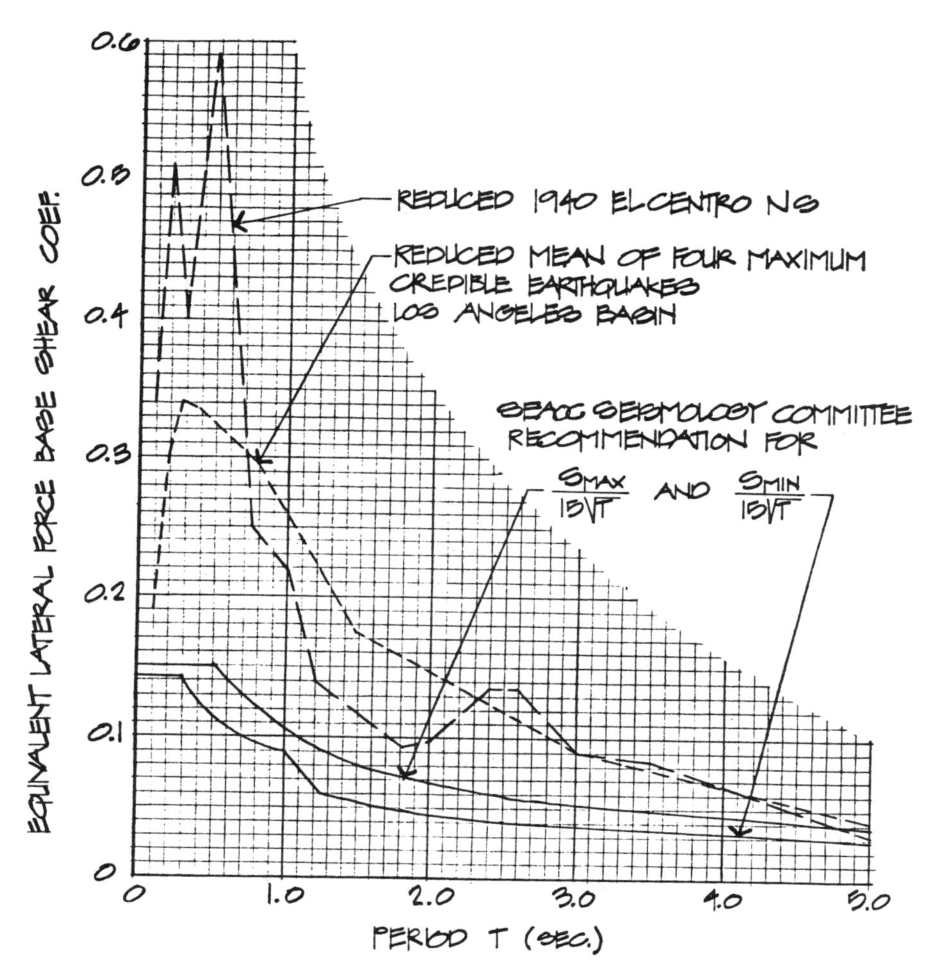

BASE SHEAR COEFFICIENT

tion in the structure is provided by lateral translation of the structure. Ductility, or ductility demand is defined as the ratio of the inelastic movement of the structure to the elastic movement. Due to practical considerations the Code requires that these plastic hinges be made to form in the girder elements of building frames rather than in the columns. In other words, the stiffness of the columns in a ductile moment resisting frame should be greater than the girder stiffness. Therefore, the use of the Code provisions requires that we provide for the required ductility by inelastic action in the girder elements, while the columns usually respond elastically, with the result that the seismic disturbance will not cause the collapse of the building, provided drift limitations are also satisfied. This combination of elastic and inelastic deformation in the girders, in satisfying ductility demand or lateral ductility is called Elasto-Plastic Response.

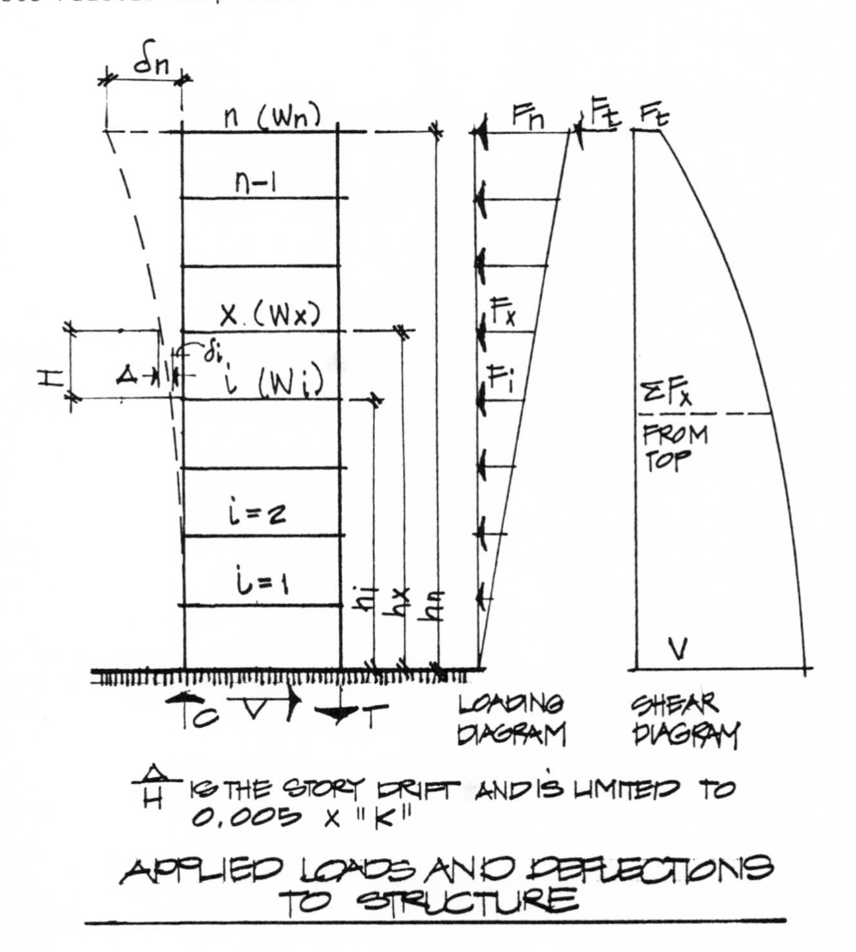

APPLIED LOADS AND DEFLECTIONS TO STRUCTURE

SEAOC

D) MINIMUM EARTHQUAKE FORCES FOR STRUCTURES.

$$V = ZIKCSW \qquad (1\text{-}1)$$

$$C = \frac{1}{15\sqrt{T}} \qquad (1\text{-}2)$$

$$T = 2\pi\sqrt{\frac{\sum_{i=1}^{n} W_i d_i^2}{g\left[\left(\sum_{i=1}^{n-1} F_i d_i\right) + (F_t + F_n) d_n\right]}} \qquad (1\text{-}3)$$

$$T = \frac{0.05 h_n}{\sqrt{D}} \qquad (1\text{-}3A)$$

$$T = 0.10 N \qquad (1\text{-}3B)$$

$$S = 1.0 + \frac{T}{T_s} - 0.5\left(\frac{T}{T_s}\right)^2; \ \frac{T}{T_s} \leq 1 \qquad (1\text{-}4)$$

$$S = 1.2 + 0.6\frac{T}{T_s} - 0.3\left(\frac{T}{T_s}\right)^2; \ \frac{T}{T_s} > 1 \qquad (1\text{-}4A)$$

$$V = F_t + \sum_{i=1}^{n} F_i \qquad (1\text{-}5)$$

$$F_t = 0.07\,TV \qquad (1\text{-}6)$$

$$F_x = \frac{(V - F_t)\, W_x h_x}{\sum_{i=1}^{n} W_i h_i} \qquad (1\text{-}7)$$

SEISMIC RESISTANT DESIGN CONCEPTS

The design of structures in the United States for lateral seismic loads generally follows the provisions of the 1979 Uniform Building Code. The basis for this code is the "Recommended Lateral Force Requirements and Commentary" of the Structural Engineers Association of California. In these documents, criteria and guidelines are established which have as their primary object that a minor earthquake should cause little or no damage, and that a major earthquake should not cause collapse of the structure. As a corollary, the building should behave elastically under expected frequent earthquake or normal wind loadings and should be capable of inelastic, cyclic behavior under infrequent strong earthquakes or abnormal wind loadings.

Since code values for seismic forces are factored downward from expected or recorded earthquake motions, it is incumbent that the designer think not in terms of code values, but realistic forces which may be three to four times these code minimums (See Base Shear Coefficient Graph on page 101. Thus, for concrete structures, cracking of members can be expected, with an increase in building period and resulting increased drift and deformation. Rocking about the foundations can also increase the building deformation. These conditions result in increased energy absorption, and dictate that ductility must be preserved in the member and between members if failure is to be prevented.

The seismic loads that are applied to a building are influenced considerably by the "K" coefficient in the lateral force formula system; a value of 0.67 for ductile type frames to 1.33 for bearing wall buildings without frames and carrying both the vertical and lateral loads. It is possible for the designer to incorporate shear walls into the structural system such that the "K" value can be 0.67, 0.80, 1.0 or 1.33 - the value depending on whether the framing is ductile, vertical load carrying or bearing wall. When the designer is using K = 1.0, he must assure himself that should the walls be heavily damaged during an earthquake that the framing will remain intact to carry the vertical loads.

The SEAOC Recommended Lateral Force Requirements and Commentary note that "the minimum design forces prescribed by the SEAOC Recommendations are not to be implied as the actual forces to be expected during an earthquake. The actual motions generated by an earthquake may be expected to be significantly greater than the motions used to generate the prescribed minimum design forces. The justification for permitting lower values for design are manyfold and include: increased strength beyond working stress levels, damping contributed by all the building elements, an increase in ductility by the ability of members to yield beyond elastic limits, and other redundant contributions."

Most of the current seismic codes in use in the United States are based on these Recommendations, including the 1979 UBC which is the reference for this section. On this basis, the ductility, damping, and drift considerations for a precast structure or the individual elements and their connections, must conform to the same requirements and limitations as for cast-in-place concrete. As an alternative, precast concrete construction could be considered as a separate type of structural system, with recognition given to its own particular response characteristics and behavior. To accomplish this will require extensive research and testing, which is already being carried out in many parts of the world at this time.

The SEAOC Commentary (pg. 20-c), in discussing structures other than buildings which do not have significant damping and do not have elements which are capable of yielding or which may fail by jeopardizing the safety of the structure, suggests that a minimum K value of 2.0 should be used. It is therefore recommended that structures in areas of high seismicity have a degree of ductility to eliminate this high design load.

SEISMIC PERFORMANCE CHARACTERISTICS

Since seismic response of a structure is the action of resisting inertial forces generated by the mass of the elements within the structure as the ground beneath it accelerates and decelerates in a random pattern from earthquake motions, both horizontally and vertically, the structural design must incorporate a shear resistance which is capable of transferring these loads with smooth continuity from the top to the foundations. This capacity is developed by designing various vertical elements within the structure for continuity and stiffness, based on the concept that in order to achieve the best solution, abrupt changes should be avoided. The basic vertical assemblies used for this are rigid frames, braced frames, shear walls, and various combinations of these.

These inertial loads must be accumulated at each floor level and transferred to the vertical shear elements. In addition, for various reasons, some of these vertical elements may be discontinuous, or have minor changes in stiffness at some level. Since the participation of each element depends on its relative stiffness to the total system stiffness and on its position within the structure, any changes in load participation must be redistributed through the floor system. Therefore, it becomes necessary to have a relatively rigid horizontal beam or diaphragm element to function as this transfer member to the vertical elements. The logical "members" are the floors and roofs, which must then be designed for these horizontal forces in addition to their normal vertical load functions.

A concept basic to designing for seismic forces is continuity and providing for a smooth transfer of forces without abrupt changes. This points to a critical part of precast construction, the connections. The cyclic nature of forces to be transferred is not always duplicated by the usual static test loading procedures and thus many of the current design data for connections may not be valid for seismic design, unless arbitrarily low elastic design values are used, which in many cases will not appreciably affect the overall construction cost. Thus, many "non-ductile" precast concrete building systems, in lower height ranges, may be designed using this philosophy.

Terminology

Basic design concepts for earthquake resistive design include consideration of the following terminology:

1. Elastic design is the basic design concept used to establish the physical configuration of structure or detail, and is based on stresses factored below the yield point of the material.

2. Ductility is a measure of the inelastic strength of a material, and is usually defined as the ratio of the maximum deformation prior to ultimate failure to the deformation at initial yield. Thus a brittle material which fractures at the yield point has a ductility factor of 1.0. Our concern for ductility is based on the recognition that we usually cannot realistically design elastically for 100% resistance to potential earthquake forces, except in low level structures. In the first place we don't know the actual maximum magnitude of forces which may be induced into a structure, although an apparent range of forces is identifiable. Even if we did know with certainty, it is usually not economically feasible to design to remain within the elastic range of material stresses. Therefore, the present seismic design concept generally used is based on elastic design criteria for some "expected" level of earthquake excitation, rather than the maximum possible level. The reserve strength of the total structure is thus expected to withstand various degrees of short-term overstress or inelastic action which will prevent structural collapse under the so called "maximum credible earthquake." This does imply various degrees of structural damage, although some buildings in the San Fernando 1971 earthquake appeared to have experienced localized excursions of inelastic action with no visible signs of permanent damage.

3. <u>Damping</u> is a measure of the ability of a member or a structure to absorb the energy generated by the application of an external repetitive type loading. This energy absorption reduces the oscillations resulting from the loading and is expressed as a percentage of critical damping, which is the minimum viscous damping that will allow a displaced system to return to its initial position without oscillation.

4. <u>Drift</u> is the unit lateral displacement of a structure. Our concern with drift is based on the comfort of building occupants, minimizing of secondary stresses in a structure due to eccentricities resulting from the offsetting or vertical load carrying members (the P-Δ effect), and the effects on non-structural elements from lateral displacement between floors. These non-structural elements include elevators, partitions, windows, exterior cladding, vertical pipes, ducts and shafts.

Three references are available which give a good overview on seismic design as related to construction with precast and prestressed concrete elements. They are:

- <u>AMERICAN PRACTICE IN SEISMIC DESIGN</u>
  by Alfred L. Parme
  (PCI Journal Vol. 17-#4 July/Aug 1972)
- <u>STATE OF THE ART REPORT ON SEISMIC RESISTANCE OF PRESTRESSED AND PRECAST CONCRETE STRUCTURES</u>
  by Neil M. Hawkins
  (PCI Journal Nov/Dec 1977 & Jan/Feb 1978)
- <u>SEISMIC DESIGN CRITERIA FOR MULTI STORY PRECAST PRESTRESSED BUILDINGS</u>
  by Sigmunc Freeman
  (PCI Journal May/Jun 1979)

SEISMIC FORCES

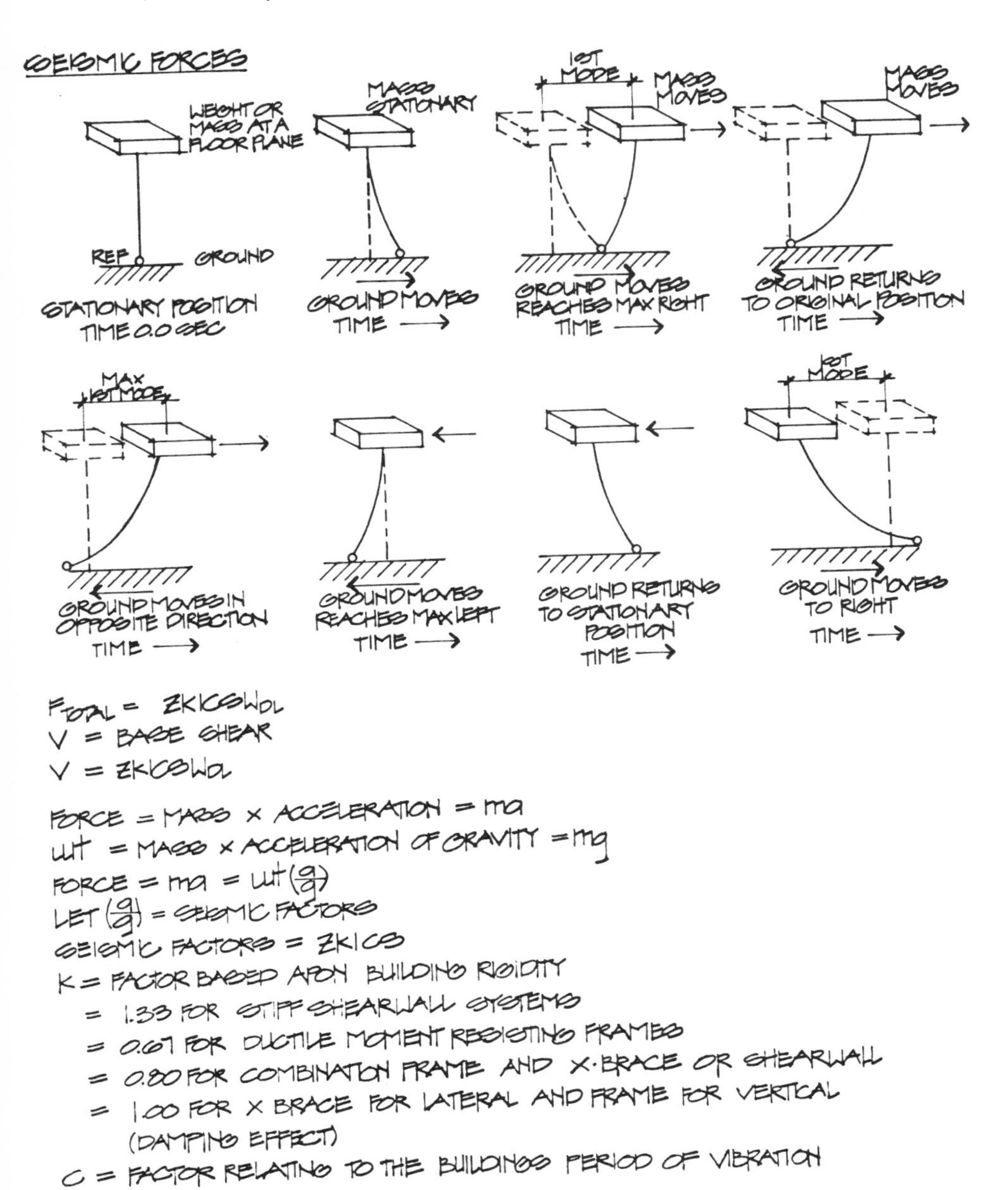

$F_{TOTAL} = ZKICSW_{DL}$

V = BASE SHEAR

$V = ZKICSW_{DL}$

FORCE = MASS × ACCELERATION = ma

WT = MASS × ACCELERATION OF GRAVITY = mg

FORCE = ma = WT$\left(\frac{a}{g}\right)$

LET $\left(\frac{a}{g}\right)$ = SEISMIC FACTORS

SEISMIC FACTORS = ZKICS

K = FACTOR BASED APON BUILDING RIGIDITY
- = 1.33 FOR STIFF SHEARWALL SYSTEMS
- = 0.67 FOR DUCTILE MOMENT RESISTING FRAMES
- = 0.80 FOR COMBINATION FRAME AND X-BRACE OR SHEARWALL
- = 1.00 FOR X BRACE FOR LATERAL AND FRAME FOR VERTICAL (DAMPING EFFECT)

C = FACTOR RELATING TO THE BUILDINGS PERIOD OF VIBRATION

S = SOIL INTERACTION FACTOR

Z = SEISMIC ZONE FACTOR RELATED TO PROBABILITY AND INTENSITY

I = IMPORTANCE OF THE BUILDING TO REMAIN OPERATIVE AFTER THE EARTHQUAKE

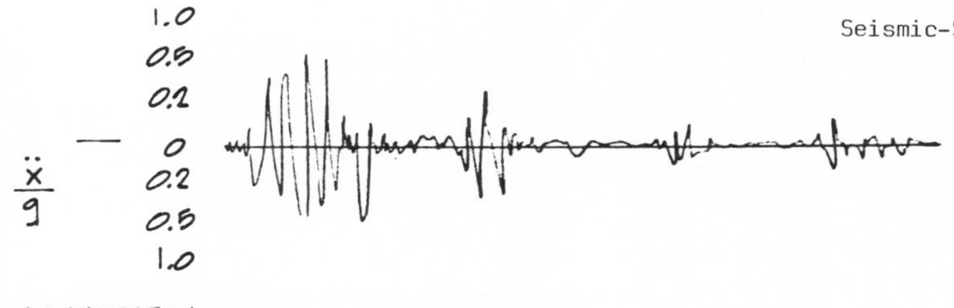

$\frac{\text{ACCELERATION}}{\text{GRAVITY}}$ TIME IN SECONDS

ACCELERATION MEASUREMENT DURING AN EARTHQUAKE

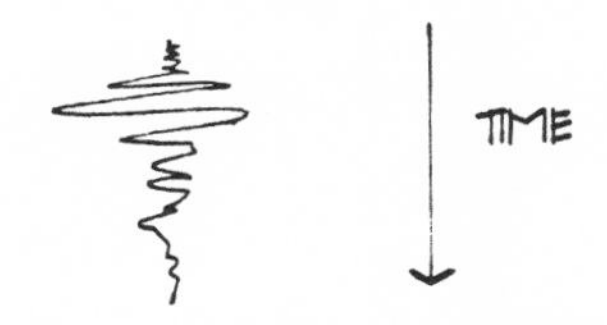

GROUND MOVEMENT (DISPLACEMENT)

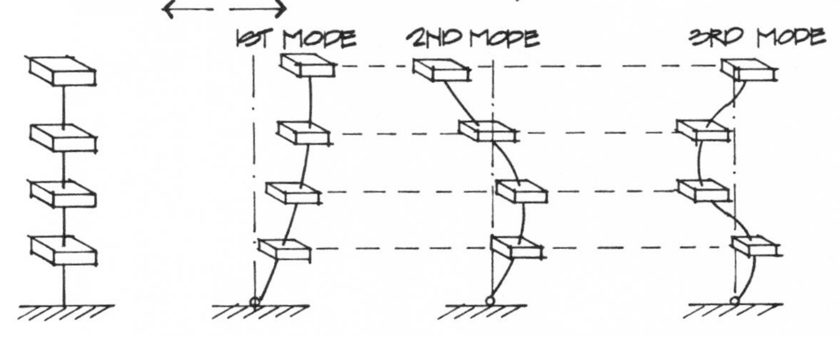

TYPICAL MODE SHAPES DURING SEISMIC DISTURBANCE

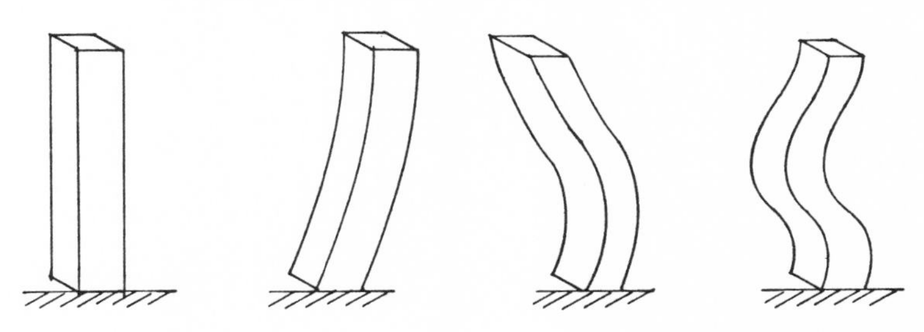

FOR SYMMETRICAL BUILDINGS

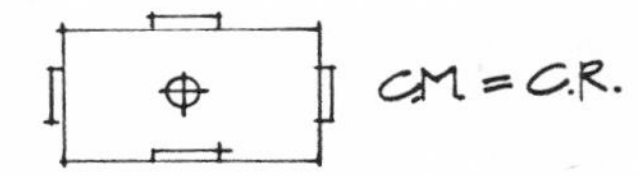

C.M. = CENTER OF MASS OF THE BUILDING

C.R. = CENTER OF RIGIDITY OF ALL STIFFENING ELEMENTS

FOR NON SYMMETRICAL BUILDINGS (WHERE THE CENTER OF MASS IS NOT IN THE SAME LOCATION AS THE CENTER OF RIGIDITY) A TORSIONAL EFFECT WILL OCCUR.

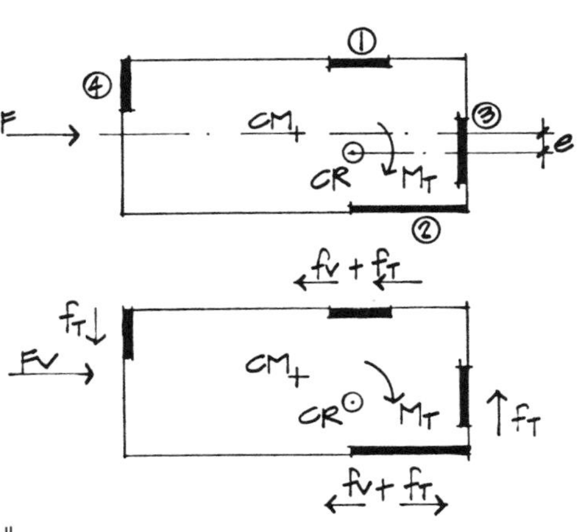

THE TORSIONAL MOMENT = Fe

THE CODE (UBC) REQUIRES A MINIMUM e OF .05 l_{MAX} OF THE BUILDING.

$$f_{V_1} = F_V \frac{r_1}{\Sigma r_1 + r_2}$$

$$f_{V_2} = F_V \frac{r_2}{\Sigma r_1 + r_2}$$

"d" IS THE DISTANCE MEASURED FROM THE CENTROID OF THE WALL ON A LINE NORMAL TO THE THE WALL PASSING THROUGH THE CENTROID OF THE WALL TO A POINT ON A LINE PASSING THROUGH THE CENTER OF RIGIDITY OF THE BUILDING.

IE:

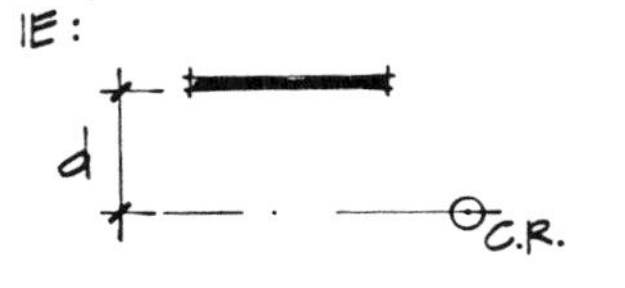

$$f_{T_1} = M_T \frac{r_1 d_1}{\Sigma_i^j r d^2} \quad , j=4 \; i=1$$

$$f_{T_2} = M_T \frac{r_2 d_2}{\Sigma_1^4 r d^2}$$

$$f_{T_3} = M_T \frac{r_3 d_3}{\Sigma_1^4 r d^2}$$

$$f_{T_4} = M_T \frac{r_4 d_4}{\Sigma_1^4 r d^2}$$

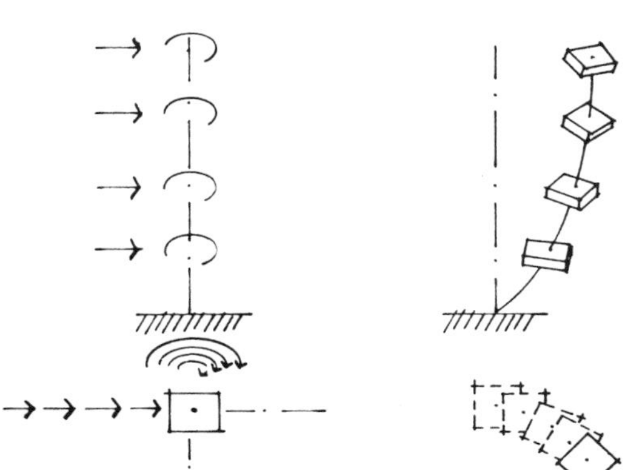

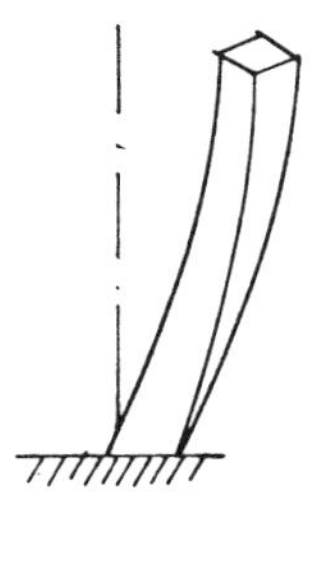

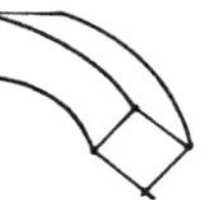

ELEVATION

PLAN

FIRST MODE LATERAL AND TORSION

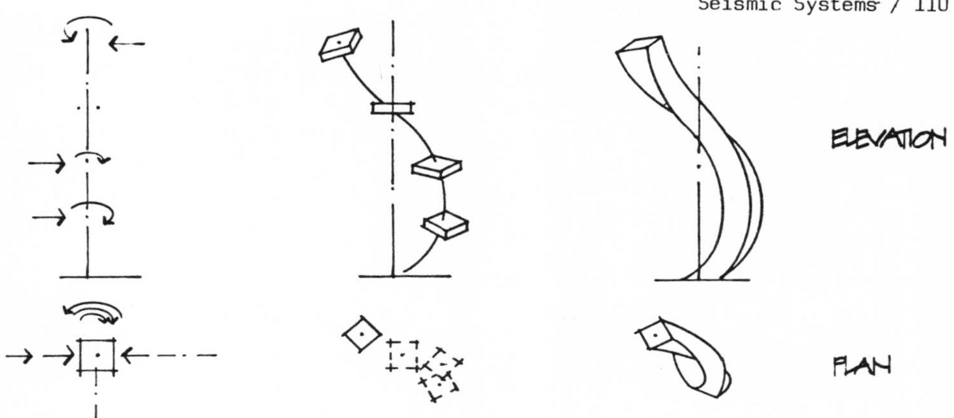

SECOND MODE LATERAL AND TORSION

FOR CONCRETE FLOOR SYSTEMS, THE FLOOR AND ROOF DIAPHRAGMS ARE CONSIDERED RIGID. AS SUCH, CONCRETE, PRECAST CONCRETE, OR MASONRY WALLS ARE DESIGNED FOR FORCES PROPORTIONAL TO THEIR RELATIVE STIFFNESSES.

IF THE WALL IS CAST INTEGRAL WITH THE FLOOR SLABS THEN THE WALL IS ASSUMED TO BE FIXED TOP AND BOTTOM. IF THE WALL IS PRECAST CONCRETE OR IF THE FLOOR-ROOF SYSTEM IS FLEXIBLE SUCH AS WOOD, OR METAL DECKING WITHOUT A CONCRETE TOPPING THEN THE WALL IS DESIGNED AS A CANTILEVER.

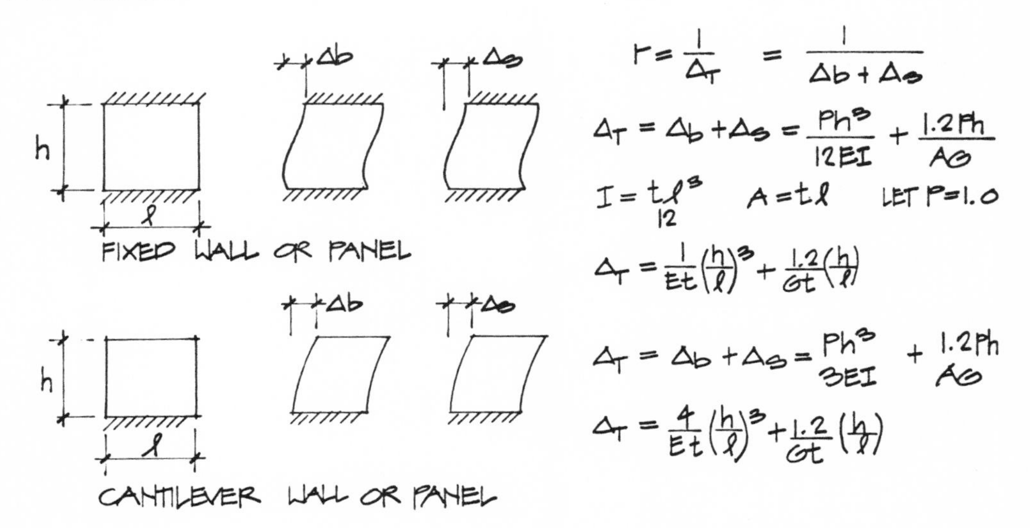

$$r = \frac{1}{\Delta_T} = \frac{1}{\Delta b + \Delta s}$$

$$\Delta_T = \Delta b + \Delta s = \frac{Ph^3}{12EI} + \frac{1.2Ph}{AG}$$

$$I = \frac{t\ell^3}{12} \qquad A = t\ell \qquad \text{LET } P = 1.0$$

$$\Delta_T = \frac{1}{Et}\left(\frac{h}{\ell}\right)^3 + \frac{1.2}{Gt}\left(\frac{h}{\ell}\right)$$

FIXED WALL OR PANEL

$$\Delta_T = \Delta b + \Delta s = \frac{Ph^3}{3EI} + \frac{1.2Ph}{AG}$$

$$\Delta_T = \frac{4}{Et}\left(\frac{h}{\ell}\right)^3 + \frac{1.2}{Gt}\left(\frac{h}{\ell}\right)$$

CANTILEVER WALL OR PANEL

FOR STATIC ANALYSIS WE ASSUME THE FIRST MODE IN FIRST ONE DIRECTION THEN IN A DIRECTION 90° TO THE FIRST: WE CALCULATE THE SEISMIC FORCE USING THE EQUATION $V = ZKICSW_{DL}$. FOR A ONE STORY BUILDING THE FORCE IS APPLIED AT THE ROOF PLANE. FOR A TWO STORY BUILDING THE FORCE IS DIVIDED EQUALLY BETWEEN THE ROOF AND SECOND STORY PLANES. FOR A MULTI-STORY THE FORCE IS PROPORTIONED TO THE ROOF AND EACH FLOOR BY THE EQUATIONS:

$$V = F_t + \sum_{i=1}^{n} F_i$$

$F_T = 0.07TV$ $\quad$ F_T MAX $= 0.25V$

$F_T = 0$ WHEN T IS 0.7 SECONDS OR LESS

$$F_x = (V - F_T)\frac{w_x h_x}{\sum_{i=1}^{n} w_i h_i}$$

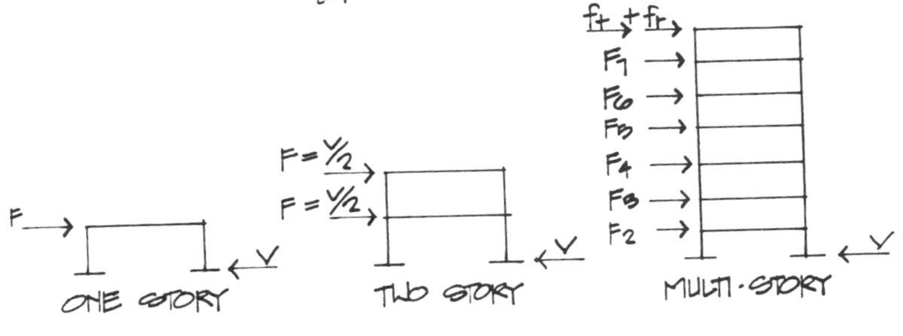

$M_{EXT} = 1/8 \cdot W/L \cdot L^2 = \frac{WL}{8}$

$M_{INT} = T \cdot d$

$\therefore Td = \frac{WL}{8}$

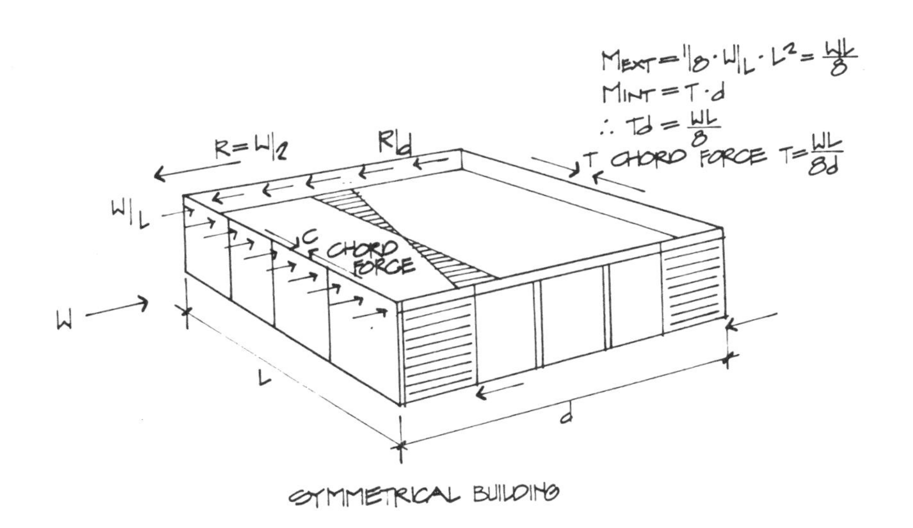

SYMMETRICAL BUILDING

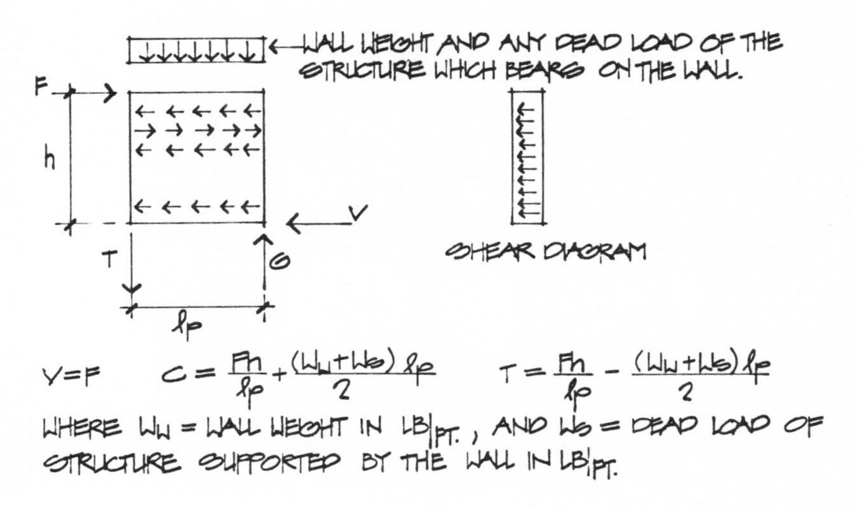

WHEN THE CENTER OF MASS AND CENTER OF RIGIDITY COINCIDE:

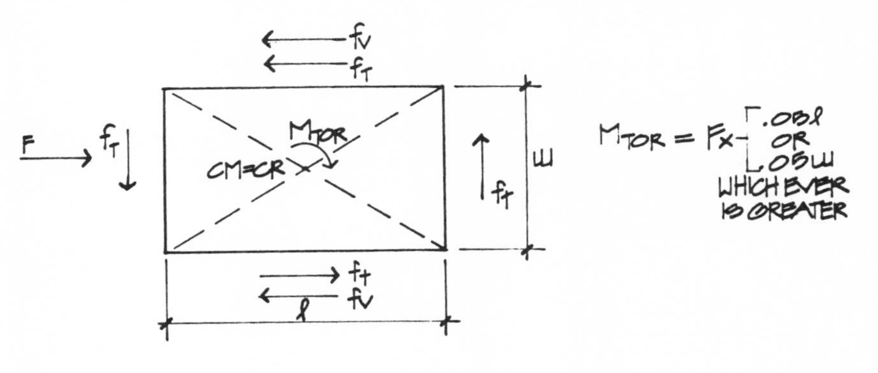

WHEN THE CENTER OF MASS AND CENTER OF RIGIDITY DIFFER:

fv
fT
F
MTOR
CM
CR
e
ft
MTOR = Fe
CM = CENTER OF MASS
CR = CENTER OF RIGIDITY

THE FOLLOWING SHOWS HOW LATERAL FORCES ARE PROPORTIONED TO SHEAR WALLS

$R_{vx} = F\frac{r_x}{\Sigma r}, \quad F_{Tx} = M_{TOR}\cdot\frac{r_x d_x}{\Sigma r d^2}$

d_x - IS THE DISTANCE FROM THE CENTER OF RIGIDITY TO THE RESISTING ELEMENT.

THE RIGIDITY OF THE WALL PANEL IS DETERMED FROM THE FOLLOWING CONCEPT:

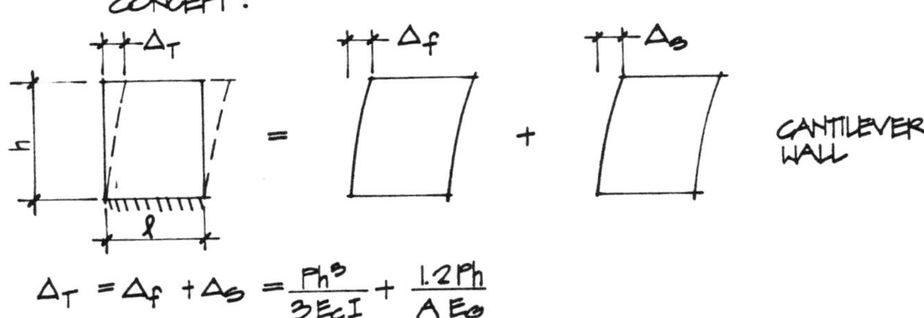

$$\Delta_T = \Delta_f + \Delta_s = \frac{Ph^3}{3E_cI} + \frac{1.2Ph}{AE_s}$$

E_c = MODULUS OF ELASTICITY $= 33w^{1.5}\sqrt{f'_c}$

FOR 5 KSI CONCRETE $E_c = 33\times 144^{1.5}\sqrt{5000} = 4.03\times 10^6$

E_g = SHEAR MODULUS $0.4E_c$

I = MOMENT OF INERTIA IN THE DIRECTION OF BENDING FOR PANELS OF UNIFORM THICKNESS $I = \frac{t\ell^2}{12}$

$\frac{E_c}{E_g} = \frac{E_c}{0.4E_c} = 2.5$, ASSUME $P=1, t=1, E_g=1$

THEN $\Delta = \frac{h^3}{3\times 2.5\times\frac{\ell^3}{12}} + 1.2\frac{h}{\ell} = 1.6\left(\frac{h}{\ell}\right)^3 + 1.2\frac{h}{\ell}$

FIXED WALL OR PIER

$$\Delta_T = \Delta_f + \Delta_s = \frac{Ph^3}{12E_cI} + \frac{1.2Ph}{AE_g} = \frac{h^3}{2.5\ell^3} + 1.2\frac{h}{\ell} = 0.4\left(\frac{h}{\ell}\right)^3 + 1.2\frac{h}{\ell}$$

r = RIGIDITY OR STIFFNESS OF THE WALL PANEL $= \frac{1}{\Delta_T}$

FOR A PIERCED WALL PANEL THE FOLLOWING RELATIONSHIP IS USED:

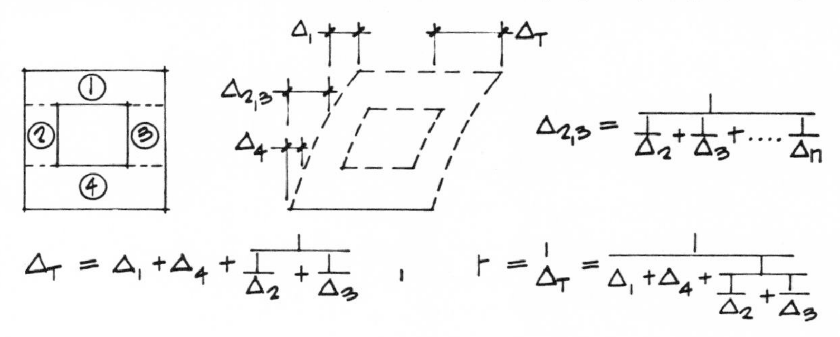

$$\Delta_T = \Delta_1 + \Delta_4 + \frac{1}{\frac{1}{\Delta_2}+\frac{1}{\Delta_3}} \quad , \quad r = \frac{1}{\Delta_T} = \frac{1}{\Delta_1 + \Delta_4 + \frac{1}{\frac{1}{\Delta_2}+\frac{1}{\Delta_3}}}$$

FOR WALLS WITH UNIFORM WIDTH PIERS THE DISTRIBUTION OF THE SHEAR FORCE WITH-IN THAT WALL IS AS FOLLOWS:

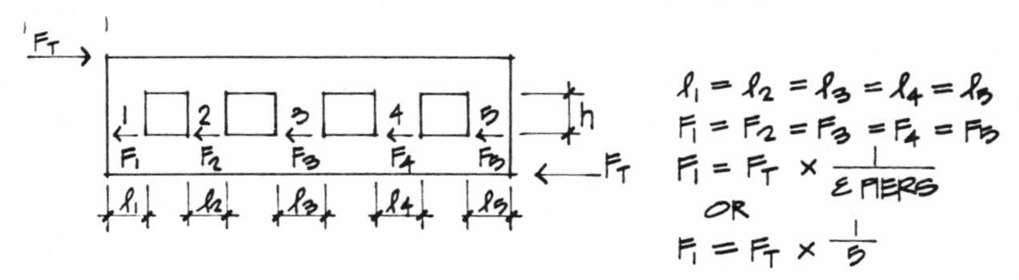

$$l_1 = l_2 = l_3 = l_4 = l_5$$
$$F_1 = F_2 = F_3 = F_4 = F_5$$
$$F_1 = F_T \times \frac{1}{\Sigma \text{ PIERS}}$$
OR
$$F_1 = F_T \times \frac{1}{5}$$

FOR WALLS WITH VARIABLE WIDTH PIERS THE DISTRIBUTION WITH-IN THAT WALL IS: $F_1 = F_T \frac{r_1}{\Sigma r}$

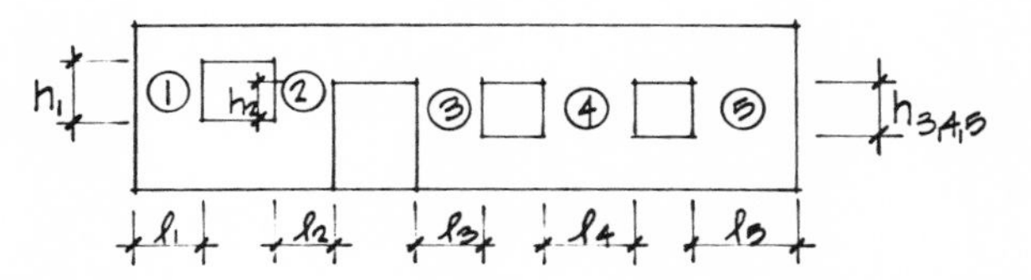

TO ASSIST IN DETERMINING THE SHEAR STRESS IN THE PIER THE FOLLOWING FORMAT IS USED:

WALL: ____________________ F_T = ____________________

PIER	h	l	$\frac{h}{l}$	$0.4 \frac{h}{l}$	$+ 1.2 \frac{h}{l}$	$= \Delta$	$r = \frac{1}{\Delta}$	$\frac{r}{\Sigma r}$	$\times F_T = F_P$	$V_u = \frac{F_P u}{\phi t(l-.17)12}$
						Σr				

SECTION 2627(a) OF THE UNIFORM BUILDING CODE 1979 EDITION FOR "EARTHQUAKE RESISTING SHEAR WALLS" REQUIRES THE FOLLOWING "U" VALUES:

$U = 1.4(D+L) + 1.4E \qquad (27\text{-}1)$

$U = 0.9D + 1.4E \qquad (27\text{-}2)$

"PROVIDED FURTHER THAT 2.0E SHALL BE USED IN FORMULAS (27-1) AND (27-2) IN CALCULATING SHEAR AND DIAGONAL TENSION STRESSES IN SHEAR WALLS OF BUILDINGS OTHER THAN THOSE COMPLYING WITH REQUIREMENTS FOR BUILDINGS WITH K = 0.67

SINCE K = 0.67 RELATES ONLY TO "DUCTILE MOMENT RESISTING SPACE FRAMES WHICH RESIST THE TOTAL REQUIRED LATERAL FORCE THEN:

$U = 1.4(D+L) + 2E$

$U = 0.9D + 2E$

SINCE DEAD LOAD DOES NOT APPLY FOR HORIZONTAL SHEAR THEN $U = 2E$

$V_u = \frac{V_u}{\phi bd} = \frac{F_p U}{\phi t(l' - .17)12}$, $\phi = .85$ $.17 = 2''$ (ASSUMED DISTANCE FROM STEEL TO EDGE OF WALL)

V_c allowable $= 2\sqrt{f'_c}$

$V_u \leq V_c$

TO DETERMINE THE CENTER OF MASS THE FOLLOWING PROCEDURE IS USED:

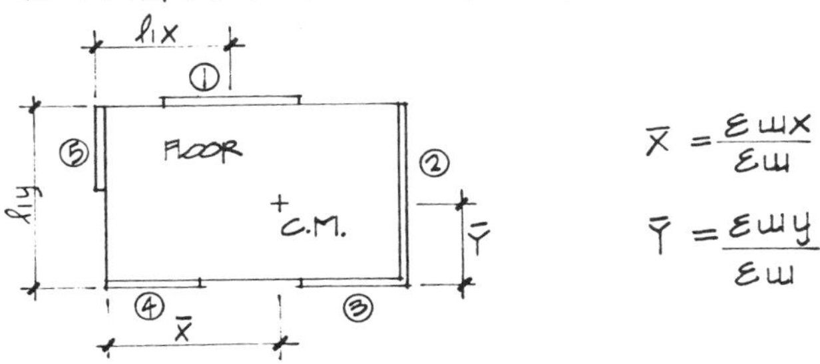

$\bar{X} = \frac{\Sigma wx}{\Sigma w}$

$\bar{Y} = \frac{\Sigma wy}{\Sigma w}$

ELEMENT	AREA	w / SQ.FT.	w	DISTANCE FROM ZERO REFERENCE X	Y	wx	wy
FLOOR SYSTEM							
WALL 1							
2							
3							
4							
5							
		Σ			Σ		

TO DETERMINE THE CENTER OF RIGIDITY THE FOLLOWING PROCEDURE IS USED:

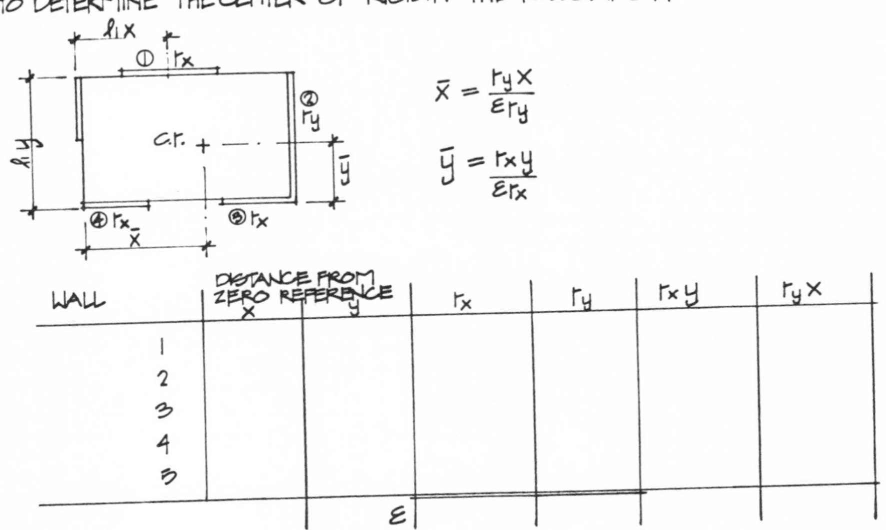

$$\bar{x} = \frac{r_y x}{\Sigma r_y}$$

$$\bar{y} = \frac{r_x y}{\Sigma r_x}$$

WALL	DISTANCE FROM ZERO REFERENCE x	y	r_x	r_y	$r_x y$	$r_y x$
1						
2						
3						
4						
5						
		Σ				

TO DETERMINE THE FORCES RESISTED BY THE WALLS INCLUDING TORSION THE FOLLOWING PROCEDURE IS USED:

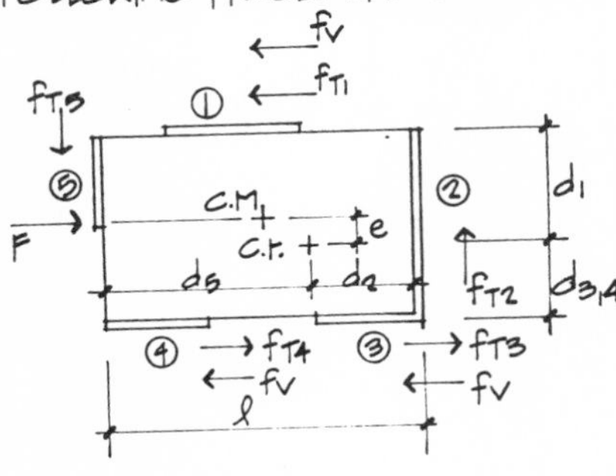

$$M_{TOR} = Fe, \quad e_{min.} = 0.05\ell$$

$$F_{T1} = M_{TOR} \frac{r_1 d_1}{\Sigma r d^2}$$

WALL	r_x	r_y	d	rd	rd^2	$F_V = \frac{r_x}{\Sigma r_x} \times F$	$F_{TOR} = M_{TOR} \frac{r_1 d_1}{\Sigma r d^2}$	$F_S = F_V + F_{TOR}$
1								
2								
3								
4								
5								

$$V_u = \frac{F_S\, U}{\phi\, t(\ell' - .17')\, 12''\,|_1} = \frac{2\, F_S}{.85\, t(\ell' - .17')\, 12''\,|_1}$$

$$V_c = 2\sqrt{f'_c} \qquad V_u \leq V_c \text{ UNLESS SHEAR REINFORCING IS USED}$$

On the following fourteen pages, 2 example problems are presented which demonstrate the design principles discussed in the preceeding portion for single story buildings. The equivalent lateral force procedure of the Uniform Building Code is used to satisfy seismic design requirements.

DESIGN EXAMPLE 1 - TYPICAL OFFICE BUILDING

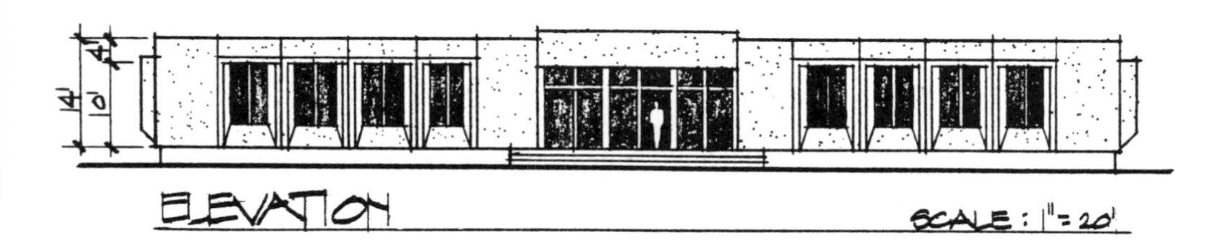

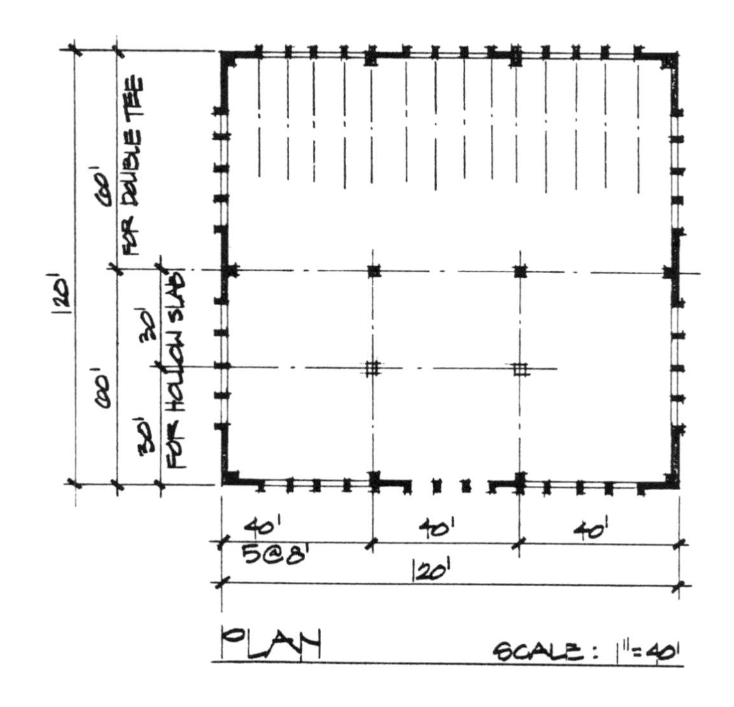

ANALYSIS OF THE EXAMPLE BUILDING FOR SEISMIC FORCES

$V = ZIKCSW$
$V = 1 \times 1 \times 1.33 \times 0.14 W$
$V = 0.1862 W$

Z - FOR ZONE IV = 1
I - FOR AN OFFICE BUILDING = 1
K - FOR A BUILDING WITH SHEAR WALLS AND WITHOUT A VERTICAL MOMENT RESISTING FRAME = 1.33
C - FACTOR BASED UPON THE PERIOD OF THE BUILDING
S - SOIL INTERACTION FACTOR
CS - EQUALS 0.14 MAX
W - DEAD WEIGHT OF THE BUILDING

WEIGHT OF THE BUILDING

ROOF SYSTEM

ROOFING	$6\ LB/FT^2$
INSULATION	$1\ LB/FT^2$
CEILING SYSTEM	$5\ LB/FT^2$
DBL. TEE SEMI-LIGHT WT.	$50\ LB/FT^2$
	$62\ LB/FT^2 \times 120^2/1000 = 892.8$ KIPS

INVERTED TEE GIRDER ASSUME $800\ LB/FT \times 120/1000 =$ 96.0 KIPS

LEDGER BEAMS 2 @ $600\ LB/FT \times 120/1000$ = 144.0 KIPS

WALL PANELS ASSUME AVE. 5" THICK

$$\frac{5'}{12''} \times 144\ LB/FT^2 \times \frac{14}{2} \times \frac{2\ WALLS \times 120'}{1000} = 100.8\ KIPS$$

1233.6 KIPS

$V = 0.1862 W = 0.1862 \times 1233.6 = 229.7$ KIPS

SINCE THE BUILDING IS SYMMETRICAL THE CENTER OF MASS AND THE CENTER OF RIGIDITY BOTH OCCUR AT THE SAME POINT AND IN THE CENTER OF THE BUILDING. THE UNIFORM BUILDING CODE REQUIRES THE BUILDING TO BE DESIGNED FOR A MIN. TORSIONAL MOMENT BASED UPON A MIN. "e" DISTANCE OF 5% OF THE MAX. BUILDING DIMENSION AT THAT LEVEL.

$.05 \times 120' = 6' = e$ $\qquad M_{TOR} = 6 \times 229.7 = 1378.2$ KIP-FT

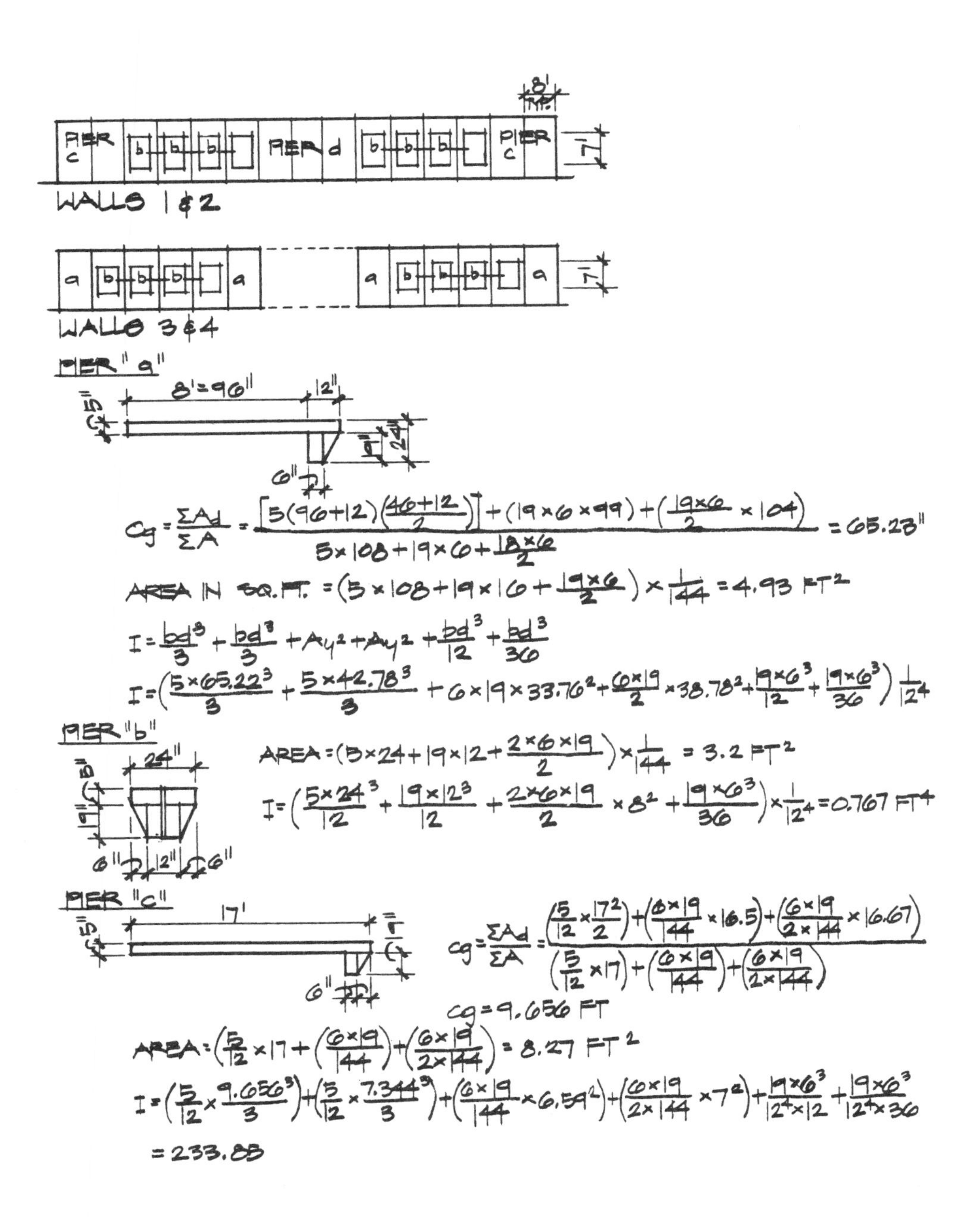
8'
PIER C
PIER d
WALLS 1 & 2
WALLS 3 & 4
PIER "a"
8'=96"
12"
6"
Cg = ΣAd/ΣA = [5(96+12)((96+12)/2)] + (19×6×99) + ((19×6)/2 × 104) / 5×108 + 19×6 + (18×6)/2 = 65.23"
AREA IN SQ.FT. = (5×108 + 19×16 + (19×6)/2) × 1/144 = 4.93 FT²
I = bd³/3 + bd³/3 + Ay² + Ay² + bd³/12 + bd³/36
I = (5×65.22³/3 + 5×42.78³/3 + 6×19×33.76² + (6×19)/2 × 38.78² + 19×6³/12 + 19×6³/36) 1/12⁴
PIER "b"
24"
6" 12" 6"
AREA = (5×24 + 19×12 + (2×6×19)/2) × 1/144 = 3.2 FT²
I = (5×24³/12 + 19×12³/12 + (2×6×19)/2 × 8² + 19×6³/36) × 1/12⁴ = 0.767 FT⁴
PIER "c"
17'
6"
cg = ΣAd/ΣA = (5/12 × 17²/2) + (6×19/144 × 16.5) + (6×19/(2×144) × 16.67) / (5/12 × 17) + (6×19/144) + (6×19/(2×144))
cg = 9.656 FT
AREA = (5/12 × 17 + (6×19/144) + (6×19/(2×144))) = 8.27 FT²
I = (5/12 × 9.656³/3) + (5/12 × 7.344³/3) + (6×19/144 × 6.59²) + (6×19/(2×144) × 7²) + 19×6³/(12⁴×12) + 19×6³/(12⁴×36)
= 233.83

PIER "d"

26'

5"

19" 6"

$$AREA = \left(\frac{5}{12} \times 26\right) + \left(\frac{2 \times 19 \times 6}{144}\right) + \left(\frac{2 \times 19 \times 6}{2 \times 144}\right) = 13.20 \text{ FT}^2$$

$$I = \left(\frac{5 \times 26^3}{12 \times 12}\right) + \left(\frac{6 \times 19}{144} \times 12.25^2\right) + \left(\frac{16 \times 19}{144} \times 12.25^2\right) + \left(2 \times \frac{6 \times 19}{2 \times 144} \times 12.67^2\right) + \left(\frac{19 \times 6^3}{12} \times \frac{1}{12^4} \times 2\right) + \left(\frac{2 \times 19 \times 6^3}{36 \times 12^4}\right) = 975$$

$$\Delta_T = \Delta_f + \Delta_s = \frac{Ph^3}{12 E_c I} + \frac{1.2 Ph}{A E_s}$$

LET $P = 1$, $E_c = 1$ $E_s = 0.4$ $E_c = 0.4$ $h = 7$

THEN $\Delta_T = \dfrac{h^3}{12 I} + \dfrac{3h}{A}$

WALL 1 AND 2

PIER	h	h^3	I	A	$\frac{h^3}{12I}$ +	$\frac{3h}{A}$ =	Δ	$R = \frac{1}{\Delta}$	$\frac{R}{\Sigma R}$
c	7	343	233.8	8.27	0.122	2.54	2.66	.375	.249
b	7	343	0.77	3.2	37.1	6.56	43.68	.02289	.0152
b	7	343	0.77	3.2	37.1	6.56	43.68	.02289	.0152
b	7	343	0.77	3.2	37.1	6.56	43.68	.02289	.0152
d	7	343	975	13.2	0.0293	1.59	1.62	.6172	.4102
b	7	343	0.77	3.2	37.1	6.56	43.68	.02289	.0152
b	7	343	0.77	3.2	37.1	6.56	43.68	.02289	.0152
b	7	343	0.77	3.2	37.1	6.56	43.68	.02289	.0152
a	7	343	233.8	8.27	0.122	2.54	2.66	.375	.249
								ΣR = 1.5045	1.00

WALL 3 AND 4

PIER	h	h^3	I	A	$\frac{h^3}{12I}$ +	$\frac{3h}{A}$ =	Δ	$R = \frac{1}{\Delta}$	$\frac{R}{\Sigma R}$
a	7	343	39	4.94	.733	4.25	4.89	.20	.213
b	7	343	.77	3.2	37.1	6.56	43.68	.023	.0245
b	7	343	.77	3.2	37.1	6.56	43.68	.023	.0245
b	7	343	.77	3.2	37.1	6.56	43.68	.023	.0245
a	7	343	39	4.94	.733	4.25	4.89	.20	.213
a	7	343	39	4.94	.733	4.25	4.89	.20	.213
b	7	343	.77	3.2	37.1	6.56	43.68	.023	.0245
b	7	343	.77	3.2	37.1	6.56	43.68	.023	.0245
b	7	343	.77	3.2	37.1	6.56	43.68	.023	.0245
a	7	343	39	4.94	.733	4.25	4.89	.20	.213
								ΣR = 0.939	1.00

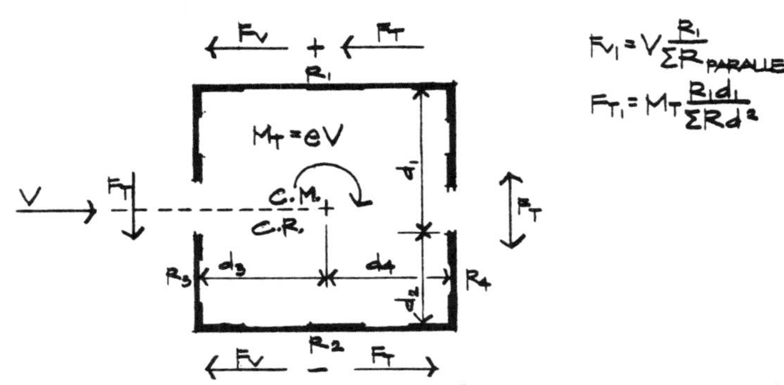

$$F_{V_1} = V\frac{R_1}{\Sigma R_{PARALLEL}}$$

$$F_{T_1} = M_T\frac{R_1 d_1}{\Sigma R d^2}$$

- IF THE WALLS ARE CAST IN PLACE OR PRECAST AS ONE WALL AND THE WINDOWS WERE OF THE SAME HEIGHT THE RIGIDITY OF THE WALL IS THE SUM OF THE RIGIDITIES OF THE PIERS.
- IF THE PRECAST PANELS ARE NOT CONNECTED ALONG THEIR VERTICAL EDGES WITH WELD PLATES OR SHEAR KEYS THEN EACH PANEL IS CONSIDERED AS AN INDEPENDENT WALL WITH ITS OWN RIGIDITY.
- IF THE PANELS ARE CONNECTED ALONG THEIR VERTICAL EDGES WITH WELD PLATES OR SHEAR KEYS THE CONDITION WOULD FALL BETWEEN THE ABOVE TWO, ACTING COMPOSITELY NEAR THAT OF THE FIRST. THE USE OF GROUTED SHEAR KEYS IS PERFERRED DUE TO THE EFFECTS OF VOLUME CHANGE UNDER TENSION OF THE WELD PLATES UNDER CONCENTRATED FORCES.

THIS EXAMPLE WILL ASSUME THE USE OF GROUTED SHEAR KEYS AND THAT THE WALL ONCE ASSEMBLED WILL ACT SIMILAR TO A CAST IN PLACE WALL. SINCE THE WINDOWS ARE OF THE SAME HEIGHT AND THE WALLS ARE OF THE SAME HEIGHT THE RIGIDITY OF A WALL WILL BE BASED UPON THE SUM OF THE RIGIDIES OF THE PIERS.

FROM THE FOLLOWING: $R_1 = R_2 = 1.5$, $R_3 = R_4 = .939$, $V = 229.7^K$, $M_T = 1378.2$ FT-K, $d = 60$ FT

FOR WALL 1

$$F_{V_1} = V\frac{R_1}{\Sigma R_{PARALLEL}} = 229.7 \times \frac{1.5}{2 \times 1.5} = 114.6^K$$

$$F_{T_1} = \frac{R_1 d_1}{\Sigma R d^2} = 1378.2 \times \frac{1.5 \times 60}{(2 \times 1.5 \times 60^2) + (2 \times .939 \times 60^2)} = 7.06^K$$

$$121.66^K$$

FOR WALL 2

$$F_{V_1} = V\frac{R_1}{\Sigma R_{PARALLEL}} = 229.7 \times \frac{.939}{2 \times .939} = 114.6^K$$

$$F_T = M_T\frac{R_3 d_3}{\Sigma R d^2} = 1378.2 \times \frac{.939 \times 60}{(2 \times 1.5 \times 60^2) + (2 \times .939 \times 60^2)} = 4.42^K$$

$$119.02^K$$

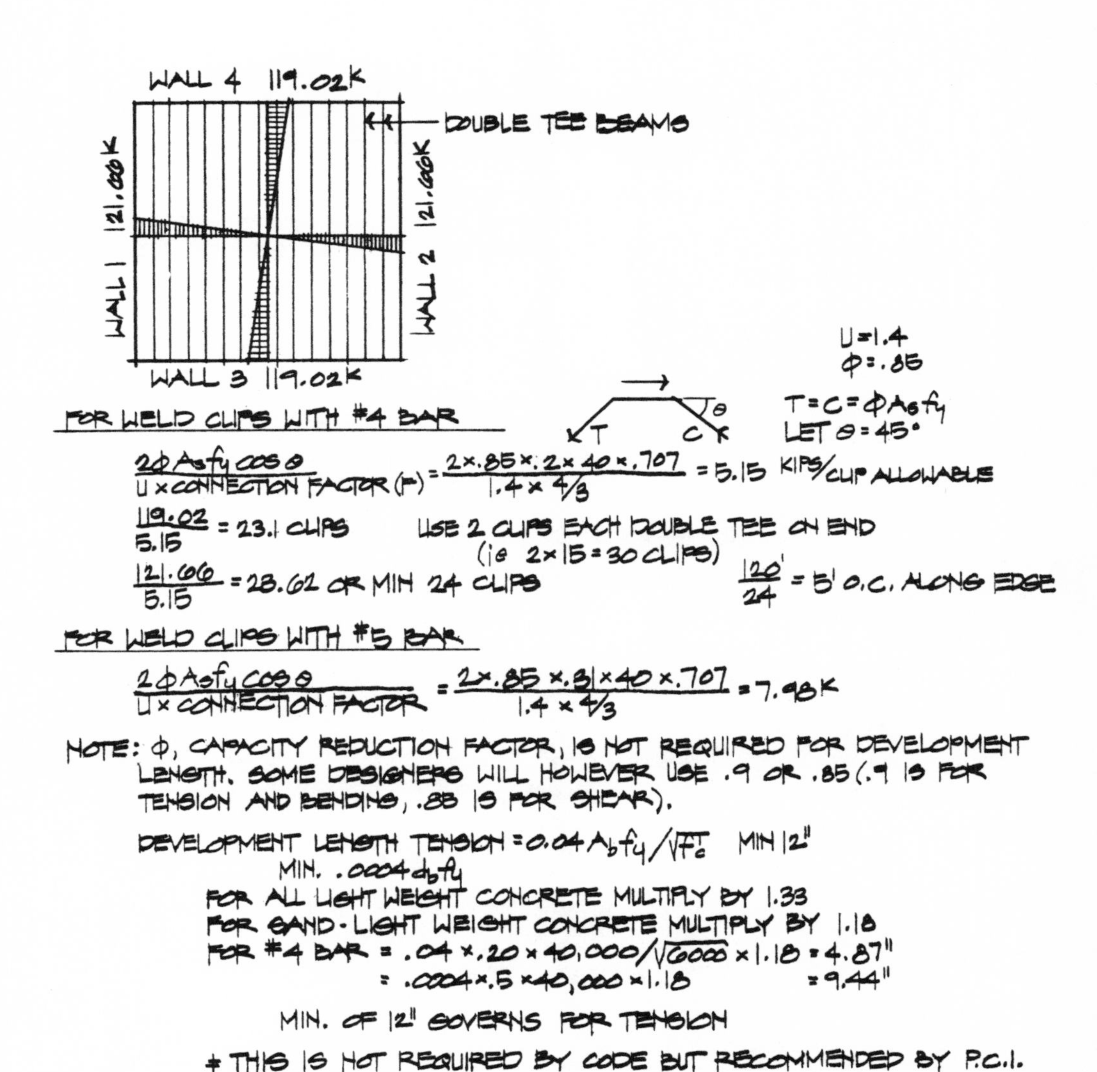

WALL 4 119.02K

WALL 1 121.66K

WALL 2 121.66K

WALL 3 119.02K

DOUBLE TEE BEAMS

$U = 1.4$

$\phi = .85$

$T = C = \phi A_s f_y$

LET $\theta = 45°$

FOR WELD CLIPS WITH #4 BAR

$$\frac{2\phi A_s f_y \cos\theta}{U \times \text{CONNECTION FACTOR (F)}} = \frac{2 \times .85 \times .2 \times 40 \times .707}{1.4 \times 4/3} = 5.15 \text{ KIPS/CLIP ALLOWABLE}$$

$\frac{119.02}{5.15} = 23.1$ CLIPS   USE 2 CLIPS EACH DOUBLE TEE ON END (ie $2 \times 15 = 30$ CLIPS)

$\frac{121.66}{5.15} = 23.62$ OR MIN 24 CLIPS $\frac{120'}{24} = 5'$ O.C. ALONG EDGE

FOR WELD CLIPS WITH #5 BAR

$$\frac{2\phi A_s f_y \cos\theta}{U \times \text{CONNECTION FACTOR}} = \frac{2 \times .85 \times .31 \times 40 \times .707}{1.4 \times 4/3} = 7.98K$$

NOTE: ϕ, CAPACITY REDUCTION FACTOR, IS NOT REQUIRED FOR DEVELOPMENT LENGTH. SOME DESIGNERS WILL HOWEVER USE .9 OR .85 (.9 IS FOR TENSION AND BENDING, .85 IS FOR SHEAR).

DEVELOPMENT LENGTH TENSION $= 0.04 A_b f_y / \sqrt{f'_c}$ MIN 12"

MIN. $.0004 d_b f_y$

FOR ALL LIGHT WEIGHT CONCRETE MULTIPLY BY 1.33

FOR SAND-LIGHT WEIGHT CONCRETE MULTIPLY BY 1.18

FOR #4 BAR $= .04 \times .20 \times 40{,}000 / \sqrt{6000} \times 1.18 = 4.87"$

$= .0004 \times .5 \times 40{,}000 \times 1.18 = 9.44"$

MIN. OF 12" GOVERNS FOR TENSION

* THIS IS NOT REQUIRED BY CODE BUT RECOMMENDED BY P.C.I.

FOR COMPRESSION, DEVELOPMENT LENGTH IS $\dfrac{0.02\, d_b f_y}{\sqrt{f'_c}}$

MIN = $0.0003 f_y d_b$ OR MIN. OF 8"

FOR #4 BARS $\dfrac{.02 \times .5 \times 40{,}000}{\sqrt{6000}} = 5.16''$

$.0003 \times 40{,}000 \times .5 = 6''$

MIN = 8", 8" GOVERNS

TENSION OF 12" > 8" FOR COMPRESSION HENCE 12" LEG REQUIRED FOR THE INSERT AS A MINIMUM

WELDS, TEN CAPACITY $T_w = \phi \times 25 \text{ KSI } l_w t_w$

t_w = THICKNESS OF WELD AT THE THROAT

$\phi = 0.70$ E 70 XX ELECTRODE

FOR #4 INSERT ⊓ 1/4" WELD ROD TO ROD $\dfrac{2 A_s f_y \times .707}{\phi\, 25 \text{ KSI } t_w}$ = LENGTH OF WELD (MIN.)

$\dfrac{2 \times .20 \times 40 \times .707}{.7 \times 25 \times .25} = 2.6''$ OR 3"

MIN. LENGTH = 4 × WELD SIZE, $4 \times 1/4 = 1''$ MIN.

FOR f_y 60 REINF., $\dfrac{2 \times .2 \times 60 \times .707}{.7 \times 25 \times .25} = 3.88$ OR 4"

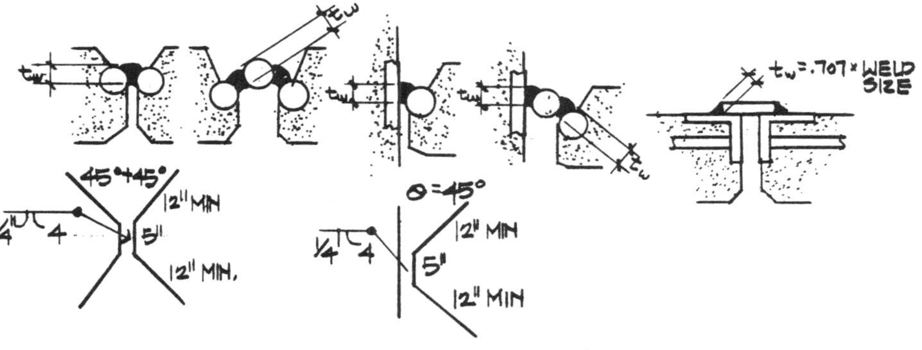

SHEAR IN CONCRETE FLANGE OF THE DOUBLE TEE, 2" FLANGE

$V_u = \dfrac{(f_v + f_t) U}{l} = \dfrac{121.66 \text{ KIP} \times 1.4}{120'} = 1.42 \text{ K/FT}$

$v_u = \dfrac{V_u}{\phi b_w t} = \dfrac{1.42 \text{ K/FT}}{.85 \times 12''/\text{FT} \times 2''} = 0.69$ KSI OR 69 PSI < 131.66 ∴ O.K.

v_c ALLOWABLE WITHOUT SHEAR REINF. = $2\sqrt{f'_c} = 2\sqrt{6000} = 154.9$ PSI

FOR SAND-LIGHT WEIGHT CONCRETE USE $.85 \times 2\sqrt{f'_c} = 131.66$

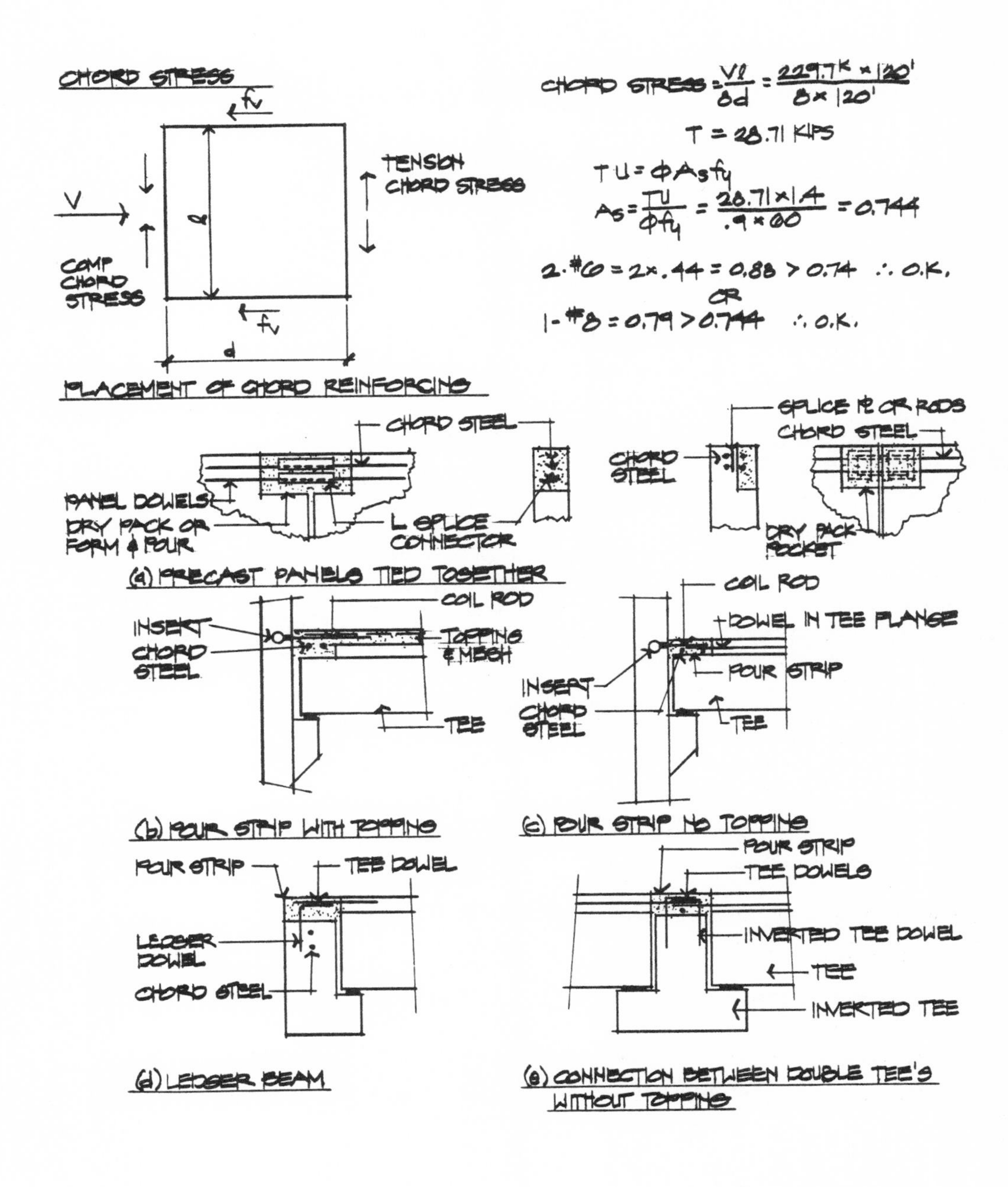
CHORD STRESS
fv
V
d
TENSION CHORD STRESS
COMP CHORD STRESS
fv
d
CHORD STRESS = Vl/8d = 229.7K × 120' / 8 × 120'
T = 28.71 KIPS
TU = φAsfy
As = TU/φfy = 28.71 × 1.4 / .9 × 60 = 0.744
2-#6 = 2× .44 = 0.88 > 0.74 ∴ O.K.
OR
1-#8 = 0.79 > 0.744 ∴ O.K.
PLACEMENT OF CHORD REINFORCING
CHORD STEEL
PANEL DOWELS
DRY PACK OR FORM & POUR
L SPLICE CONNECTOR
SPLICE PL OR RODS
CHORD STEEL
CHORD STEEL
DRY PACK POCKET
(a) PRECAST PANELS TIED TOGETHER
COIL ROD
INSERT
CHORD STEEL
TOPPING & MESH
TEE
COIL ROD
DOWEL IN TEE FLANGE
POUR STRIP
INSERT
CHORD STEEL
TEE
(b) POUR STRIP WITH TOPPING
(c) POUR STRIP NO TOPPING
POUR STRIP
TEE DOWEL
LEDGER DOWEL
CHORD STEEL
POUR STRIP
TEE DOWELS
INVERTED TEE DOWEL
TEE
INVERTED TEE
(d) LEDGER BEAM
(e) CONNECTION BETWEEN DOUBLE TEE'S WITHOUT TOPPING

EXAMPLE PROBLEM

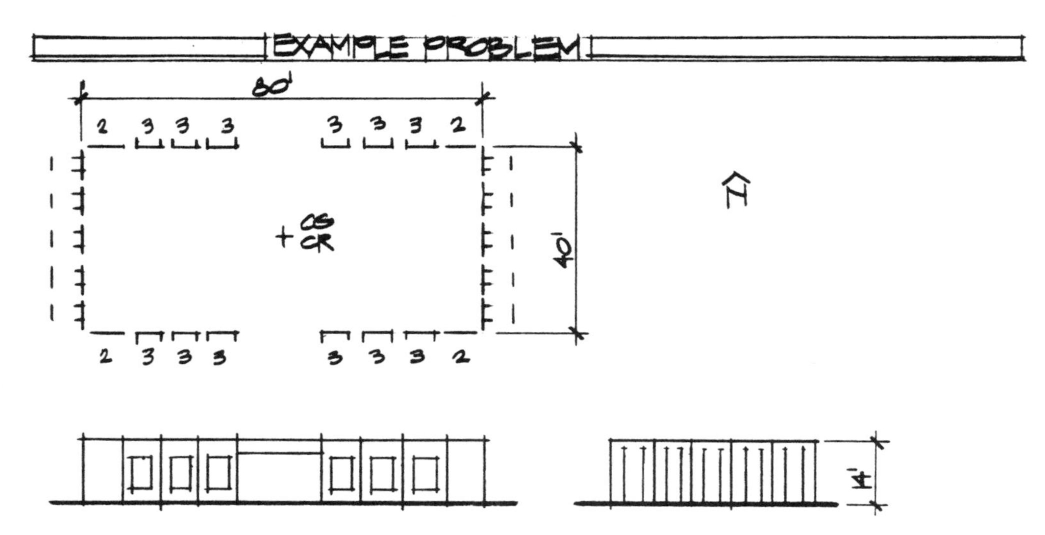

THE WALL PANELS AT THE ROOF LINE ARE NEITHER FULLY FIXED NOR PINNED. IF THE PANELS ARE DOWELED INTO THE THE CAST IN PLACE TOPPING THEY ARE ASSUMED FIXED. IF THERE IS NO CAST IN PLACE TOPPING AND THE CONNECTION CONSISTS OF WELD PLATES AND THE PANEL IS NON LOAD BEARING THE CONDITIONED IS NEARER THAT OF A PINNED CONDITION OR CANTILEVER. IF THE WALL IS LOAD BEARING IT USUALLY CONSIDERED AS FIXED. SOME DESIGNERS WILL ALWAYS ASSUME THE TOP FIXED. IF THE SHEAR WALLS ARE ALWAYS LONG WITHOUT OPENINGS AS IN AN APARTMENT BUILDING. SOME DESIGNERS WILL BASE THE RIGIDITY ONLY ON THE SHEAR DEFLECTION NEGLECTING THE FLEXURAL COMPONENT.

FOR A FIXED TOP THE EQUATIONS ARE

$$R = \frac{1}{\Delta} \qquad \Delta = \Delta_f + \Delta_s = \frac{h^3}{12EI} + \frac{1.2h}{AG}$$

FOR A PINNED TOP THE EQUATIONS ARE

$$R = \frac{1}{\Delta}, \ \Delta = \Delta_f + \Delta_s = \frac{h^3}{3EI} + \frac{1.2h}{A \cdot G},$$

FOR CONCRETE $G = 0.4 E_c$

$V_b = ZIKCSW_b$

$Z = 1$ FOR SEISMIC ZONE IV

$I = 1$ FOR AN OFFICE BUILDING

$K = 1.33$ FOR A SHEAR WALL (BOX SYSTEM) BUILDING

$CS = 0.14$ FOR A ONE STORY BUILDING IS USUALLY TAKEN AS 0.14

W_b ROOF = 6#/FT² ROOFING, 1#/FT² INSULATION, 1#/FT² MISC., 36#/FT² DBL. TEES

$= 42\text{\#/FT}^2 \left(\frac{40 \times 80}{1000}\right) = 134.4$ KIPS

$\text{WALLS} = \frac{14}{2} \times \frac{2(80+40)}{1000} \times 70\text{\#/FT}^2$ (ASSUMED WT.) $= 117.6$ KIPS

$V_b = 1 \times 1 \times 1.33 \times 0.14(134.4 + 117.6) = 47$ KIPS

$M_{TOR} = 0.05 \times 80' \times 47 = 188$ FT.-KIPS

WALL PANEL #1

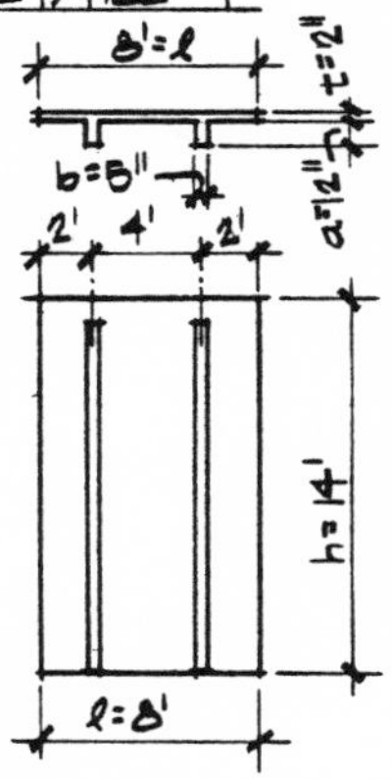

$I = \frac{2 \times (8 \times 12)^3}{12} = \frac{t\ell^3}{12} = 147456$

$2 \times 5 \times 12'' \times 24^2 = 2Ay^2 = 69120$

$\frac{2ab^3}{12} = \frac{2 \times 12 \times 5^3}{12} = 250$

$= 216826 \text{ IN}^4$

$A = (2 \times 8) + (2 \times 5 \times 12) = 136 \text{ IN}^2$

LET $E_c = 1$ $\quad G = 0.4E_c = 0.4$ $\quad h = 14 \times 12 = 168$

$\alpha = \alpha_f + \alpha_s = \frac{h^3}{12EI} + \frac{1.2h}{A\,0.4E} = \frac{168}{12 \times 216826} + \frac{1.2 \times 168}{136 \times 0.4} = 3.717$

$K = \frac{1}{\alpha} = \underline{0.269}$

WALL PANEL #2

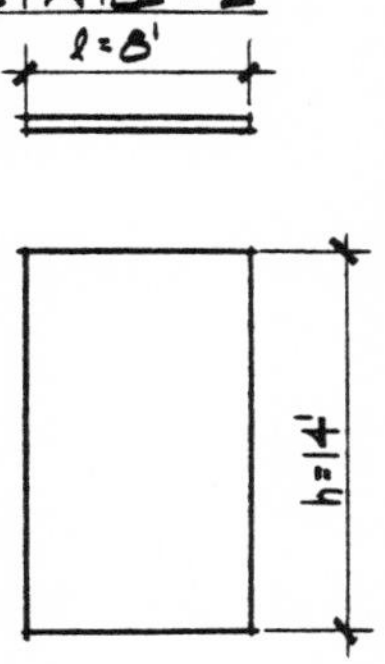

$I = \frac{t\ell^3}{12} = \frac{5 \times (8 \times 12)^3}{12} = 368640$

$A = t\ell = 5 \times 8 \times 12 = 480$

$\alpha = \alpha_f + \alpha_s = \frac{168}{12 \times 368640} + \frac{1.2 \times 168}{480 \times 0.4} = 1.05$

$K = \frac{1}{\alpha} = \underline{0.95}$

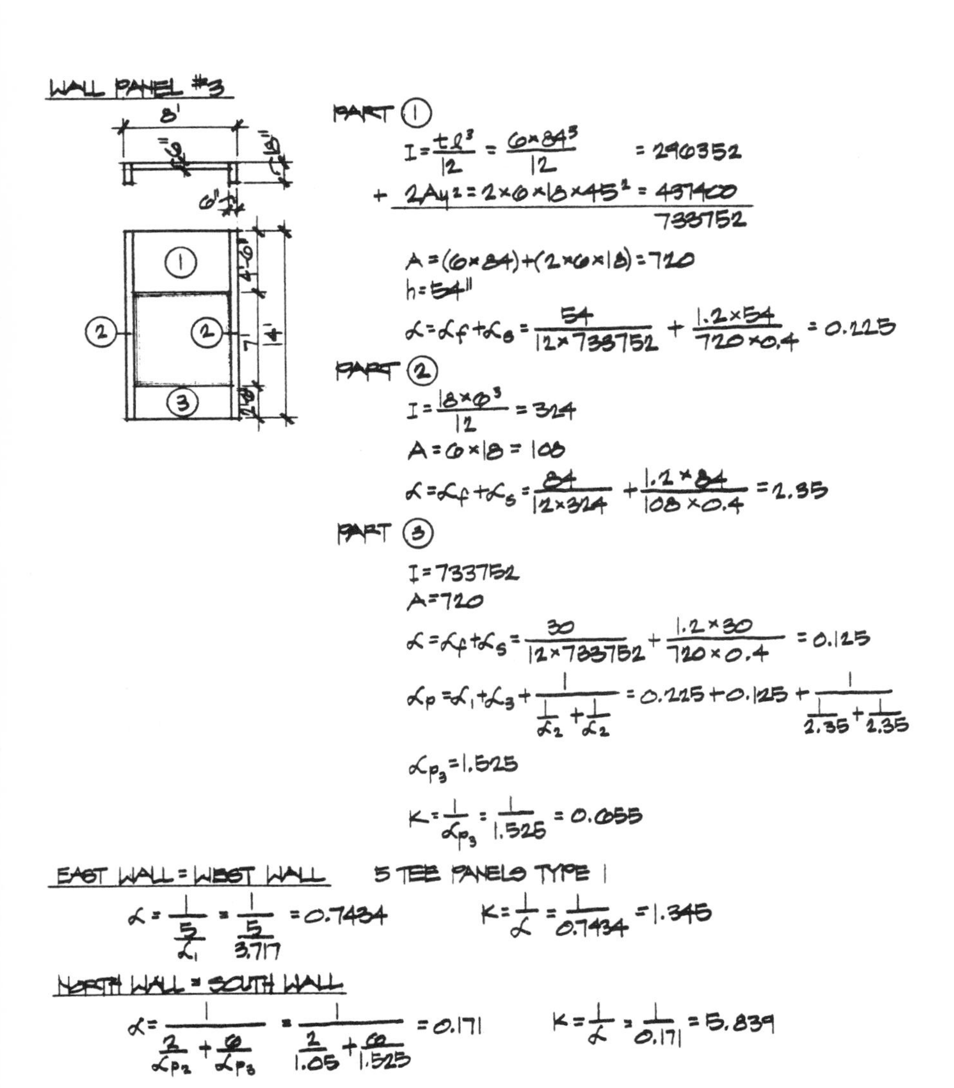
WALL PANEL #3
8'
6"
18"
4'-6"
7'
14'
2'-6"
1
2
2
3
PART 1
I = tl³/12 = 6×84³/12 = 296352
+ 2Ay² = 2×6×18×45² = 437400
733752
A = (6×84)+(2×6×18) = 720
h = 54"
α = αf + αs = 54/(12×733752) + 1.2×54/(720×0.4) = 0.225
PART 2
I = 18×6³/12 = 324
A = 6×18 = 108
α = αf + αs = 84/(12×324) + 1.2×84/(108×0.4) = 2.35
PART 3
I = 733752
A = 720
α = αf + αs = 30/(12×733752) + 1.2×30/(720×0.4) = 0.125
αp = α1 + α3 + 1/(1/α2 + 1/α2) = 0.225 + 0.125 + 1/(1/2.35 + 1/2.35)
αp3 = 1.525
K = 1/αp3 = 1/1.525 = 0.655
EAST WALL = WEST WALL
5 TEE PANELS TYPE I
α = 1/(5/α1) = 1/(5/3.717) = 0.7434
K = 1/α = 1/0.7434 = 1.345
NORTH WALL = SOUTH WALL
α = 1/(2/αp2 + 6/αp3) = 1/(2/1.05 + 6/1.525) = 0.171
K = 1/α = 1/0.171 = 5.839

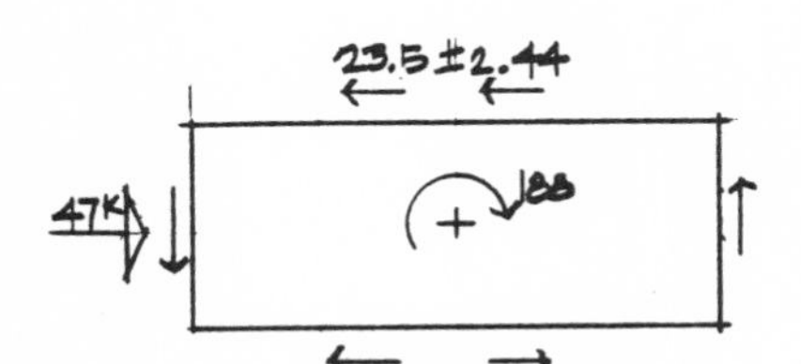

$f_v = V_b \frac{R}{\Sigma R} = \frac{5.839}{2 \times 5.839} \times 47 = 23.5$

$F_T = M_T \frac{Rd}{\Sigma Rd^2}$

	R	d	Rd	Rd^2	$\frac{Rd}{\Sigma Rd^2} \times M_T$
NORTH	5.839	20'	116.78	2335.6	2.44
SOUTH	5.839	20'	116.78	2335.6	2.44
EAST	1.345	40'	53.8	2152	1.12
WEST	1.345	40'	53.8	2152	1.12
				8975	

DESIGN EAST & WEST WALLS FOR 23.5 + 1.12 KIPS = 24.62 KIPS
NORTH & SOUTH WALLS FOR 23.5 + 2.44 = 25.94 KIPS

EAST & WEST WALLS

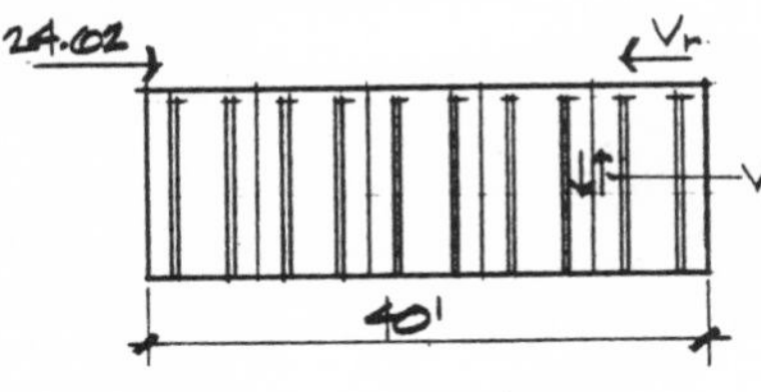

HORIZ. SHEAR STRESS IN A PANEL PER FT.

$= \frac{24.62}{40'} = V_r = 615$ LB/FT

$615 \times U = 615 \times 2.0 = 1230$ LB/FT

ALLOWABLE $= \phi \lambda 2\sqrt{f'_c} = .85 \times .85 \times 2 \times \sqrt{3000}$
$= 102$ PSI OR $102 \times 2 \times 12 = 2448$ LB/FT

$2448 > 1230$ ∴ O.K.

$v_V = v_r = 1230$ LB/FT

VERT. SHEAR FOR THE PANEL $= \frac{1230 \text{ LB/FT} \times 14 \text{ FT}}{1000} = 17.220$ KIPS/PANEL EDGE

USE 3 WELD CONNECTORS

$\frac{17.2}{3} = 5.73$ KIPS/CONNECTOR

USE #4

$V_u = \frac{2 A_s \phi f_y}{1.414} = \frac{2 \times 0.2 \times .9 \times 40}{1.414} = 10.18$ KIPS/CONN.

5.73 REQ. < 10.18 ALLOWABLE ∴ O.K.

WELD LENGTH $= \frac{10.18 K}{42 \times 1/3 \times 1/2} = 1.46$ USE MIN. 2" OF WELD

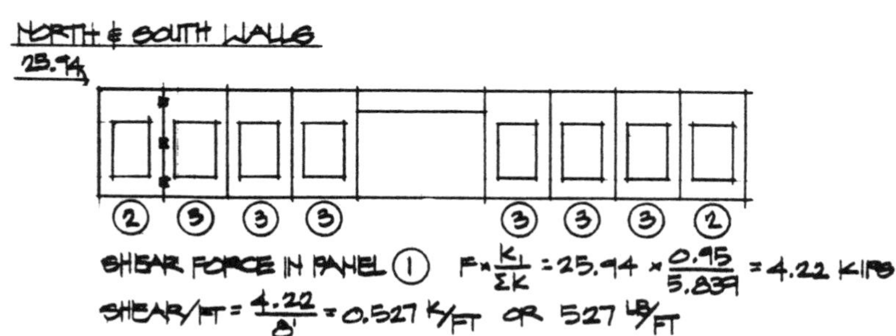

SHEAR FORCE IN PANEL (1) $F = \frac{K_1}{\Sigma K} = 25.94 \times \frac{0.95}{5.839} = 4.22$ KIPS

SHEAR/FT $= \frac{4.22}{8'} = 0.527$ K/FT OR 527 LB/FT

ALLOWABLE $= \frac{\phi \lambda 2\sqrt{f'_c} \times t \times 12''}{U} = \frac{.85 \times .85 \times 2\sqrt{5000} \times 5 \times 12}{2}$

$= 3065$ LB/FT > 527 ∴ O.K.

$V_u = V_r = 527 \times U = 527 \times U = 527 \times 2 = 1054$ LB/FT

FOR THE TOTAL HEIGHT 1054×14 FT $= 14756$ LB.

MIN. SHEAR KEY AREA $= \frac{V_u}{\text{ALLOWABLE SHEAR STRESS}} = \frac{14756}{.85 \times 0.5 \times 2\sqrt{5000}}$

$= \frac{14756}{102} = 144$ IN²

USE 3 SHEAR KEY GROUPS $\frac{144}{3} = 48''$/GROUP

MIN. SHEAR KEY AREA BASED UPON SHEAR FRICTION

ALLOWABLE ACI $\begin{cases} 800 \text{ PSI} \times \lambda = 800 \times .85 = 680 \\ \text{OR} \\ \lambda \times 0.2 f'_c = .85 \times 0.2 \times 5000 = 850 \end{cases}$

PCI $1000 \times \lambda^2 = 1000 \times .85^2 = 722$

$\frac{V_u}{680} = \frac{14756}{680} = 21.7$ IN²

USE 3 GROUPS $\frac{21.7}{3} = 7.2$ IN²

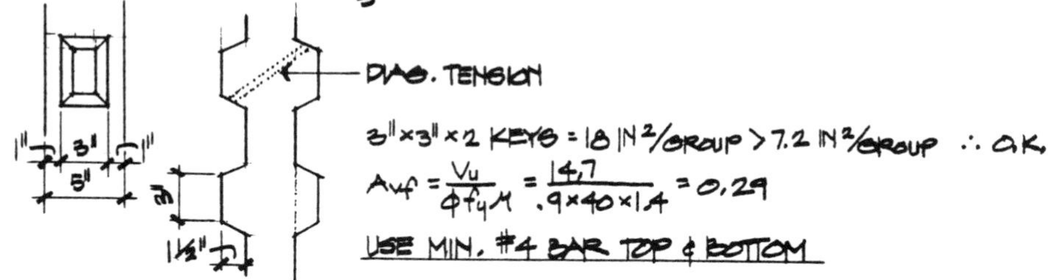

$3'' \times 3'' \times 2$ KEYS $= 18$ IN²/GROUP > 7.2 IN²/GROUP ∴ O.K.

$A_{vf} = \frac{V_u}{\phi f_y \mu} = \frac{14.7}{.9 \times 40 \times 1.4} = 0.29$

USE MIN. #4 BAR TOP & BOTTOM

CHECK OVER TURNING N-S PANELS

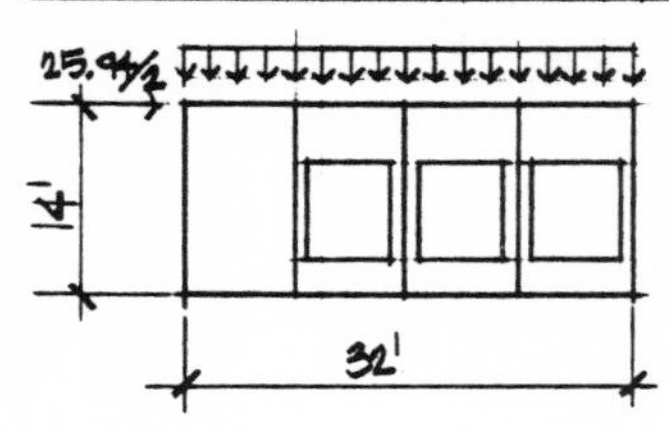

$M_{OT} = \frac{25.94}{2} \times 14' = 181$ FT.-KIPS

WT. PANEL (2) $= \frac{5}{12} \times \frac{115 \times 14}{1000} \times 8 = 5.36$ KIPS

WT. PANEL (3) $= \frac{6 \times 18}{144} \times 14 \times \frac{2 \times 115}{1000} = 2.41$

$+ \frac{6}{12} \times \frac{115 \times 7}{1000} \times (4.5 + 2.5) = 2.81$

5.225 KIPS

$M_R = (5.36 \times 28) + (5.22 \times 20) + (5.22 \times 12) + (5.22 \times 4) = 338$ FT-KIPS

$\frac{M_R}{M_{OT}} = \frac{338}{181} = 1.86 > 1.5$ HENCE NO HOLD DOWN IS REQUIRED THERE IS NO TENSION IN THE PANEL

CHECK PANEL 3 MULLIONS FOR SHEAR

$F \frac{K}{\Sigma K} = 25.94 \times \frac{0.655}{5.839} = 2.90$ KIPS

TWO MULLIONS HENCE $\frac{2.90}{2} = 1.45$ KIPS/MULLION

MULLION AREA = 6" × 18" $\frac{1.45}{6 \times 18} = 0.0134$ OR 13.4 PSI

ALLOWABLE $= \frac{\lambda 2\sqrt{f'_c}}{U} = \frac{0.85 \times 2 \times \sqrt{5000}}{2} = 60$ PSI

$60 > 13.4 \therefore$ OK

CHECK FOR MOMENT

$V = 1.45$ KIPS

$M = \frac{Vh}{2} = \frac{1.45 \times 7'}{2} \times 12'' = 60.9$ IN-KIPS

$60.9 \times U = 60.9 \times 2 = 121.8$ IN-KIPS

ASSUME: $d = 3''$ f_y 60 STEEL

$a = d - \sqrt{\frac{-2M}{\phi .85 f'_c b} + d^2} = 3 - \sqrt{\frac{-2 \times 121.8}{.9 \times .85 \times 5 \times 18} + 3^2} = 0.663$

$A_s = \frac{.85 f'_c b a}{f_y} = \frac{.85 \times 5 \times 18 \times 0.633}{60} = 0.845$

USE 3-#5 BARS $3 \times .31 = 0.93 > 0.845$ OK.

On the next several pages, we shall take a pictorial excursion through several types of buildings comprised of precast and prestressed concrete components. All of the buildings shown are under construction, and allow one to see how the precast components fit together to form the structural frame or enclosure.

FAIRCHILD SEMICONDUCTOR - SAN JOSE

This 3 level structure features 18" deep by 12' wide double tee wall panels, 65 feet high. 41" deep by 12' wide double tees span from the exterior panels to interior precast concrete beams and columns. Lateral support for the building is furnished by the double tee wall panels which are interconnected by flange weld plates to form a long stiff wall diaphragm. Structures such as this can be designed to perform elastically to resist maximum design earthquake forces, thus negating the necessity for providing ductility. This was one of two identical industrial buildings built here in 1976.

KAISER HOSPITAL PARKING STRUCTURE - WALNUT CREEK

Precast concrete columns and prestressed concrete beams form a vertical load carrying frame for this garage structure, shown here under construction in 1974. The deck construction in this type of structure is cast in place concrete, either mild steel reinforced or post-tensioned. The prestressed concrete girders are shored at mid span or at third points concurrent with the floor slab forming operation. Lateral resistance to wind and seismic forces is provided by masonry or cast in place concrete shear walls. Contraction joints and shear walls are both positioned recognizing the effects of long term creep and shrinkage occurring in cast in place post-tensioned structures.

SYNTEX R-6 CHEMICAL RESEARCH BUILDING - PALO ALTO

Architecturally colored and exposed precast concrete spandrels and columns form a vertical load carrying exterior frame supporting 8 foot wide by 18" deep prestressed concrete double tees. Lateral rigidity is provided by interior cast-in-place concrete shear walls, which could have just as easily been provided as precast concrete units, thus saving valuable construction time.

UNION BANK - OCEANGATE BUILDING - LONG BEACH

Details conforming to the Uniform Building Code provisions for cast-in-place concrete ductile moment resisting frames were used in designing this 16 story structure. "U" shaped precast concrete spandrels frame into the exterior precast concrete columns at each level. Reinforcing bars cast in the columns were spliced to field placed reinforcing bars inside the spandrel zone. After placing additional confinement reinforcing, the closures were poured along with the topping over the spread single tee - 8" hollow core deck system. Here we see a 36" single tee being erected by the 200 ton Manitowoc crane used on the project. Use of a fixed tower crane in the core would probably have been more efficient and economical.

The balance of the photos shown on page 135 to 144 depict some of the construction stages of the Site #1A Office Building, Sacramento. Basic lateral resistance is furnished by ductile moment resisting frames which are totally precast in one direction, and combine with cast-in-place concrete girders in the other. Prestressed concrete piling deliver structure loads to the soil. Prestressed concrete double tees form the deck system for the floors and roof.

Precast column column/frames 12 feet wide by 50 feet high shown here being erected. Each 36 ton frame was erected directly over a matching set of prestressed concrete piling.

(Photo - courtesy of Ben Shook, Office of the State Architect, Sacramento.

SITE #1A - SACRAMENTO (cont.)

Precast concrete frame being positioned over 12" square prestressed concrete piling. The frames were plant sandblasted to reveal the light buff aggregate and sand combined with buff cement. Tolerance for pile placement was ±2". A special bearing plate was grouted on the pile prior to erecting the column frame. Note the rebar extending from the bottom of the columns to be embedded in a poured in place pile cap. Note also the notched surface and holes for reinforcing bars to complete the ductile moment resisting frame in the direction transverse to the frame. (Photo - courtesy of Ben Shook, OSA)

SITE #1A, SACRAMENTO (cont.)

Precast concrete frame bents positioned and braced while pile cap forming proceeds. Note cable bracing connected to steel collars around the prestressed concrete piling units. Every frame had to be braced and guyed in 4 directions in this manner. Reinforcing bar design and detailing provides fixity at the pile cap after forming and pouring. (Photo - courtesy of Ben Shook, OSA)

SITE #1A, SACRAMENTO (cont.)

Aerial view showing erected frame bents. Note on the right of the photo where the cast-in-place concrete beams have been formed and poured to provide the ductile moment resisting frame in the direction transverse to the frame. Plant cast prestressed concrete double tees will span the 40 foot gap between frames, and be supported on the cast-in-place concrete beams spanning between frames. (Photo - courtesy of Ben Shook, OSA)

SITE #1A, SACRAMENTO (cont.)

Closer view of precast concrete frames and cable bracing. In the background can be seen the forming operation for the cast-in-place concrete beam spanning between bents to provide frame action in that direction. The field poured girders were made with the same mix design used in the precast concrete frame, and were subsequently field sand blasted, providing a close match with the precast finish.
(Photo - courtesy of Ben Shook, OSA)

SITE #1A, SACRAMENTO (cont.)

Another view of the precast concrete frames. 4 - #11 bars ran the full height of each column, with no reinforcing bar splices. This would have been difficult to achieve in the field if the bents had been cast-in-place concrete. The notched surface at the juncture of the cast-in-place concrete beam and the precast concrete frame could have been replaced by a sandblasted/retarded interface, achieving the same design result. (Photo - courtesy of Ben Shook, OSA)

SITE #1A, SACRAMENTO (cont.)

Placing reinforcing for the cast-in-place concrete frame girders. 2 - #11 bars are threaded through each top hole while one - #11 bar extends through each bottom hole. These holes were then carefully grouted solidly to assure an air-tight closure without entrapped air bubbles, to preclude against the possibility of long term corrosion of the bars in the column sleeve. Then, the beam forming was completed and the cast in place concrete girder was poured. Design was performed in accordance with the Equivalent Lateral Force provisions of the Uniform Building Code, with K=0.67. Note closed loop stirrup reinforcement being placed by workers. (Photo - courtesy of Ben Shook, OSA)

SITE #1A, SACRAMENTO (cont.)

Plant cast prestressed concrete double tees being erected. In the upper photo, notice the extension of the cast-in-place girder beyond the outside face of the exterior frame, to adequately develop the field placed girder reinforcing.

SITE #1A, SACRAMENTO (cont.)

Double tee floor framing installation in progress.

SITE #1A, STATE OFFICE BUILDING, SACRAMENTO

Two views showing partially completed structural frame with various precast and cast-in-place concrete elements in place.

STRAND STRESSING - BASALT ROCK PLANT, NAPA

Bed foreman, Sherman Pitcock (left), watches the pressure gage indicating the force being imparted in the strand by the hydraulic jack. Quality control inspector Mike Shepherd measures the resulting elongation of the strand as a check against gage pressure. This stressing operation is being conducted during the routine set-up of the double tee bed for daily production of 8 foot wide x 24 inch deep double tees for a State Office Building in Sacramento.

PRECAST PRESTRESSED CONCRETE

What is Prestressed Concrete? Prestressed concrete is the introduction of an internal compressive force in a concrete member which counteracts the tensile forces produced by external loads imposed either during handling or in service. This internal force is introduced by pulling high strength cables, or strands as they are more correctly called, to within 70 percent of more of the ultimate strength of the strand and then releasing this force into the concrete. Prestressing may be accomplished in two ways:

1. Pretensioning: The strands are stressed before the concrete is poured and hardened around them. When the concrete has reached a sufficiently high compressive strength to withstand the force imposed by the strand force (release strength) the strands are cut or burnt, and the prestress force is released into the member. This is the method used to produce plant cast precast prestressed concrete members.

2. Post-Tensioning: The concrete element is cast with conduit, or ducts sufficiently large to receive bundles of strand, wires, or high strength rods which are called tendons. The tendons are stressed after the concrete has reached sufficient strength to withstand the force imposed by the tendon. Post-tensioning is usually done with cast in place concrete at the jobsite. Section 21 discusses how this method of prestressing interfaces with plant cast precast-pretensioned concrete.

So now we have factory cast high quality high strength concrete combined with high strength steel strand - both materials stressed to very high percentages of their respective ultimate strengths, yet remaining within their respective elastic ranges. In most design situations, the concrete remains in the uncracked range, thereby more efficiently using the available depth and making possible shallower depths and longer spans for horizontal elements, as well as for prestressed concrete wall panels. Precast concrete structural members are usually prestressed, whereas architectural precast concrete cladding units and load bearing panels are not.

Precast concrete elements which are pretensioned are: double tee floor and wall units, single tees, inverted tee girders, channel units, rectangular beams, hollow core plank floor and wall units, piling, columns, and solid flat slabs.

The concept of prestressing is to introduce precompression in the area of the beam cross-section where tension will be produced by superimposed loads. As the member is subjected to dead and live loads, the resulting tension is counteracted by the precompression induced by the force imparted to the concrete by the prestressing strands. Thus, one offsets the other, with the end product being a net compression stress throughout the beam, or a small value of tension on the bottom of a deck member, usually less than the modulus of rupture (tensile strength of concrete).

It is common for precast concrete to have high compressive strengths of up to 7000 psi for hard rock concrete, with 6000 psi being routinely attainable. Lightweight aggregate concrete can be made to achieve 5000 psi 28 day strengths. In tension, the strength of concrete, or modulus of rupture, is approximately $7.5\sqrt{f'c}$. The tensile strength for sand-lightweight aggregate concrete is 85% of this value. By prestressing, the inefficiency of the concrete in tension is eliminated and replaced with high strength steel strand which has an ultimate tensile strength of 270,000 psi.

The pretensioning force is introduced into the concrete by the process of placing high tensile steel strands in the mold before the concrete is placed, and stressing the strands with hydraulic jacks. The concrete is placed, vibrated and screeded; then it is cured by means of live steam or radiant heat. After 10-12 hours the concrete reaches a minimum compressive strength of approximately 3500 to 4000 psi. As the concrete hardens, it bonds to the tensioned steel strands. Test cylinders are broken to ascertain the strength, and if it is equal to or above the required minimum for stress transfer (release strength), then the strands are cut, releasing the tension into the precast elements. The strands are held firmly by bond developed by the concrete at the ends of the beams. The tension in the strands thus imparts a compressive force in the concrete (precompression).

A diagram of a typical stressing bed is as follows:

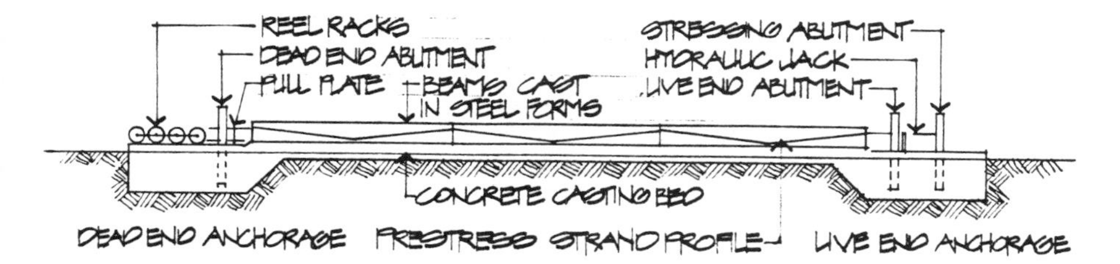

Plant cast prestressed concrete products are cast in long continous beds of up to 800 feet or more for some products, such as piling and hollow core plank. The length of the bed decreases with increasing complexity of the unit and the amount of reinforcing required, with 400 feet being an average. Long lines enable the prestressed concrete manufacturer to cast a multiple of units end to end, with bulkhead separators in between individual units of specified lengths for a particular project. The maximum bed length is the longest length which a standard casting crew can set up and pour on a daily cycle, optimizing plant labor.

HOW PRESTRESSING WORKS:

LOAD
COMPRESSION
TENSION
e_c
e_c
$f_c = 7.5\sqrt{f'_c}$
COMPRESSION
N.A. (NEUTRAL AXIS)
TENSION
$f_t = 7.5\sqrt{f'_c}$ (MODULUS OF RUPTURE)
STRAIN
STRESS

(a) NON-REINFORCED CONCRETE BEAM

LOAD
COMPRESSION
CRACK
STEEL REINFORCING
e_{c_c}
e_s
e_{c_T}
$.85 f'_c$
f'_c
COMPRESSION
N.A.
TENSION
CRACKED SECTIONS; ALL TENSILE STRESSES CARRIED BY REINFORCING
STRAIN
STRESS AT ULTIMATE CONDITIONS

(b) REINFORCED CONCRETE BEAM

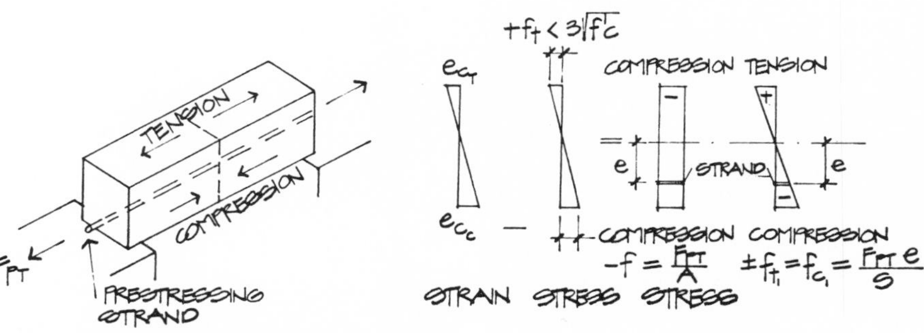

F_{PT} = PRESTRESSING FORCE

(c) PRESTRESSED CONCRETE BEAM AT RELEASE OF PRESTRESS

COMPRESSION (−) TENSION (+)

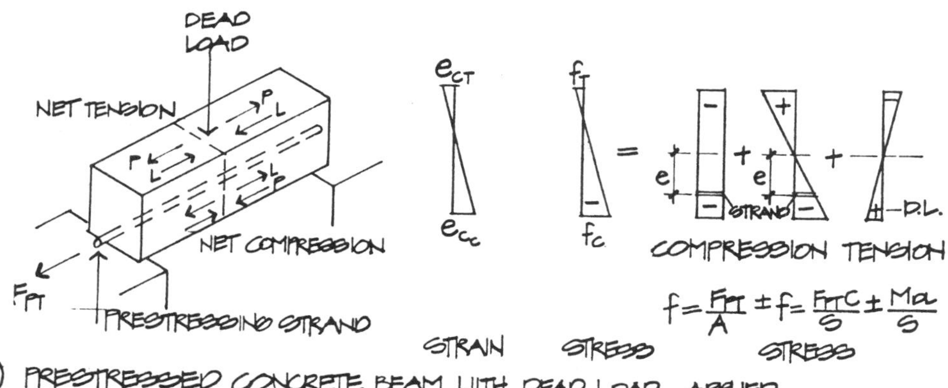

(d) PRESTRESSED CONCRETE BEAM WITH DEAD LOAD APPLIED

NOTE: THIS CONDITION MUST BE EVALUATED FOR EACH OF THE FOLLOWING:

(1) REMOVAL FROM THE FORM AND INPLANT HANDLING INCLUDING IMPACT
(2) SHIPPING AND ERECTION INCLUDING IMPACT
(3) IN FINAL POSITION WITHOUT TOPPING
(4) IN FINAL POSITION WITH TOPPING.

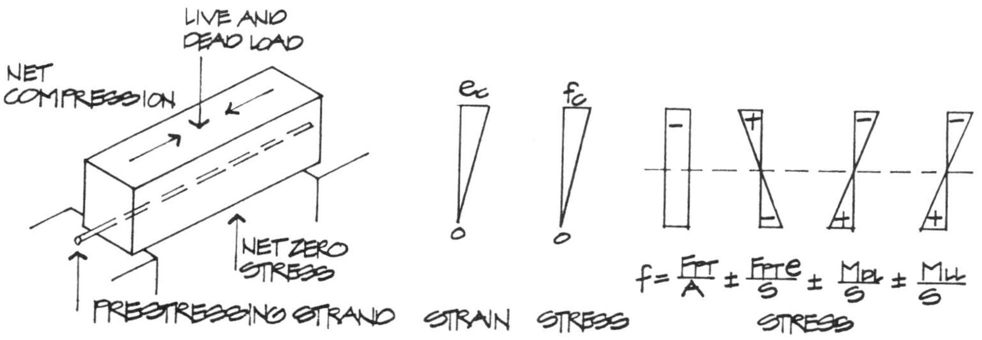

(e) PRESTRESSED CONCRETE BEAM WITH DEAD AND LIVE LOAD APPLIED

COMPRESSION (−) TENSION (+)

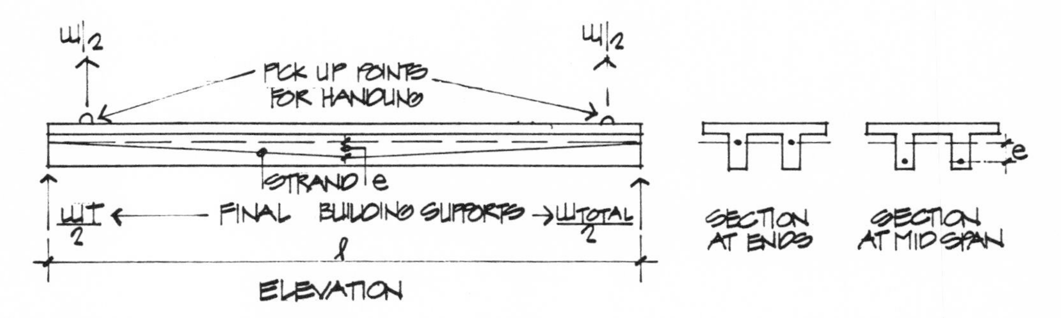

(f) STRAND PLACEMENT
FOR MOST EFFICIENT DESIGN, STRANDS SHOULD BE DEPRESSED

Ideally, handling at the fifth points during handling produces equal negative and positive bending moments. However, most prestressed concrete products are lifted and supported near the ends since the member is designed to support service loads at these points. Also, storage of prestressed concrete deck elements, other than those with especially designed cantilevers, should be at the ends in order that excessive camber and deflection growth due to creep not occur.

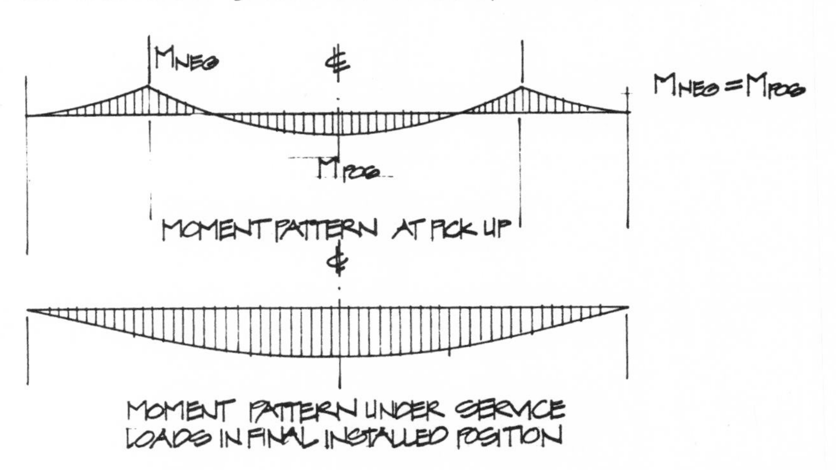

In closing, it is important to remember that other factors also affect the design of prestressed concrete members, such as shear, deflection, and ultimate strength capacity. Since these other design considerations are covered in pre-requisite concrete design courses and in the Second Edition of the PCI Design Handbook, they will not be repeated here. The component design sections show detailed analysis of these important design aspects relating to prestressed concrete members in example form. The intent in this section is to explain to the student and designer the facets of prestressed concrete member analysis that are unique compared to ordinary reinforced concrete.

POST TENSIONING AND ITS INTERFACE WITH PRECAST CONCRETE

As opposed to pretensioning, where the strands are tensioned before the concrete is poured and set around the strands, post-tensioning describes the method for prestressing where the cables are tensioned after the concrete is poured and cured. Post-tensioning is primarily used for cast-in-place construction and precast segmental construction. In the casting operation, ducts are placed in the proper position in the forms and concrete cast around them. The ducts are sufficiently large to receive a bundle of 12 to 15 strands, which are stressed when the concrete has reached sufficient strength to withstand the stresses imposed. After stressing, the remaining space in the duct is usually filled with grout to form what is called a "bonded tendon". The other condition of leaving the void ungrouted is referred to as an "unbonded tendon." The term "tendon" refers to a group of strands that are stressed together in a post-tensioning group in a duct. One advantage of post-tensioning is that the prestress force can be applied in stages during construction as dead loads are applied. Post-tensioning also lends itself to continuous or cantilevered flexural member design since connection limitations at column joints do not exist and the excess of prestress force required at cantilevered design conditions may be terminated at the first interior span. In addition to 1/2" Ø-270^k strand, tendons are also comprised of other size strands, button ended wire, or high strength bars. When unbonded tendons are used, the Uniform Building Code requires that a minimum area of bonded reinforcing be provided in addition to the required post-tensioning.

Occasionally, precast concrete manufacturers get involved with post-tensioning either as a part of their plant operations, or as an interface condition to be recognized in the overall structural design of the building. The following categories outline the areas of potential involvement with post-tensioning:

1. When production facilities or stressing abutments and anchorages cannot handle the magnitude of the prestressing forces involved, the elements are fabricated with mild steel reinforcing, or partially pretensioned, and then post-tensioned after stripping and when the concrete has reached sufficient strength to receive the post-tensioning force.

2. When span or headroom requirements limit the available depth of the section such that an abnormally high value of release strength is required, then the section is post-tensioned.

3. Segmental construction for bridges with plant cast longitudinal segments, such as being provided for the Dumbarton Bridge project.

4. Long span segmental construction - spans of from 180' to 800' with plant cast transverse segments.

5. Large diameter cylinder piling plant post-tensioned to 750 - 800 psi.

6. Architectural precast concrete walls post-tensioned vertically to form shear resisting elements.

7. Plant cast precast segments post-tensioned vertically, horizontally, or both to build liquid storage tanks.

8. Parking structures, where the beams and columns are plant cast precast-prestressed and the deck slab is cast-in-place post-tensioned.

DESIGN CONSIDERATIONS

Several areas of design of post-tensioned concrete differ somewhat from pretensioned concrete; these areas should be recognized by the designer:

1. The anchorage zone at the end of the member is subjected to large confined compressive forces that, if not reinforced properly, could cause longitudinal cracking, or bursting of the concrete at the end of the member. As opposed to pretensioned concrete, where the force buildup is gradual along the development length of the strand, and where sufficient space exists between strands to keep stress transfer uniformly distributed, all of the force of the post-tensioning tendon is delivered to a small area by the anchorage bearing plate at the end of the member, as required by Section 2618(1) of UBC-79.

2. Sufficient area must be provided to accomodate the physical dimensions of the bearing plate required with the multi-strand wedge system that will be used.

3. Ducts should be of sufficient gage to withstand the pressures imposed by vibrating

equipment, and not collapse. Temporary PVC pipe installed during casting and removed afterward will also serve to prevent the occurence of collapsing post-tensioning ducts.

4. Segment interface areas are match cast (mating segment is cast against the element it will subsequently be united with in the field). Interface details are selected that will prevent leakage during the grouting operation.

5. The conduit may be positioned in a parabolic profile to more efficiently use the prestress force when compared with the shape of the externally applied moment diagram.

6. In addition to the stress losses considered in pretensioned designs (elastic shortening, strand relaxation, creep, and shrinkage), slippage at anchorage fixtures and conduit friction losses must also be considered.

7. Creep and shrinkage movements in cast-in-place concrete post-tensioned structures will be considerably greater than those associated with plant cast precast and prestressed component buildings. Post-tensioned deck slabs, for example, will exhibit creep and shrinkage movements more than double those associated with plant cast decks. This should be taken into account in positioning shear walls, and selection of expansion joint locations. Post tensioning jacking points should be located at positions in the floor where temporary gaps can be left, to be poured after some initial percentage of creep and shrinkage have occurred. This will cause some delay in project completion, however.

POST-TENSIONING PROCEDURES

Unbonded tendons are coated with a corrosion inhibitor and wrapped with paper or tape to prevent bond with the concrete. The inhibitor also acts as a lubricant and permits the tendon to be stressed within the concrete.

Bonded tendons are placed in a flexible or rigid galvanized metal duct. The ducts are cast in the concrete but the tendons are usually not placed in the duct until just before the post tensioning operation. If the prestressing steel is placed in the duct prior to casting the concrete then it may require coating with a corrosion inhibitor.

If the concrete is to be steam cured, then the tendons are not placed in the ducts until after the steam curing is completed. Prior to placing the tendons the ducts are flushed with air under pressure or water with 0.1 pound of slaked lime or quick lime per gallon of water. Once the strand is installed, if left to set more than 10 days it must be coated with a corrosion inhibitor. If less than 10 days this is not necessary. Grout is pumped in at 100 psi pressure - water content not exceeding 5 gallons of water/ sack of cement - admixture may also be used.

A complete overview of post-tensioning is presented in the POST-TENSIONING MANUAL published by the Post-Tensioning Institute in Phoenix, Arizona.

DUMBARTON BRIDGE - SOUTH SAN FRANCISCO BAY

150 foot spans are formed by post-tensioning 2 - 75 foot long - 70 ton plant cast segments together. The 7'-6 deep hollow delta girders were plant cast in match cast half segments in Kabo-Karr's Visalia facility, and hauled to the jobsite. This marriage of plant pre-casting and job site post-tensioning provided the most economical solution for this project.

AMBASSADOR COLLEGE SCIENCE AND FINE ARTS BUILDINGS
PASADENA

Classically sculptured precast concrete panels grace these twin buildings. The honeycombed units, supported at the second floor, are vertical load bearing units. They feature smooth white concrete contrasted with recessed areas of exposed black granite and solar bronze glazing.

TEXTURES AND FINISHES - ARCHITECTURAL PRECAST CONCRETE

Precast concrete cast in steel, concrete, wood or fiberglass forms possesses a smooth surface which may be desirable depending upon the architect's desired design effect. The concrete may be white (white cement and white sand), natural gray or colored with a pigment. With smooth concrete, minute imperfections readily show up under bright light and in dark shadows. A smooth as-cast surface is very absorbent and is easily stained by oil, smog, soot, dust and handling. To offset this, the surface is often textured or sculptured or the coarse aggregate is exposed. By exposing the aggregate, a large portion of the surface becomes coarse aggregate. Coarse aggregate has the greatest resistance to absorption, abrasion, and staining from adverse weathering. Depending upon the color and shape of the aggregate, many interesting effects are possible. The various surface textures used with precast concrete are discussed in the following paragraphs.

MARK TAPER FORUM
MUSIC CENTER-LOS ANGELES

Bas Relief panels in white cement cast by using rubber form liners which have nine different patterns that are repeated around the building facade.

MARBLE OR GRANITE FACING

Precast concrete panels can be faced with marble or granite during the casting operation. Though expensive, these materials have the greatest resistance to absorption, abrasion and staining. Marble and granite are beautiful and resist adverse weathering. The stone sheets are drilled and metal anchors attached with epoxy glue or expansive inserts. If the panel is not insulated, a polyethylene bond breaker sheet is applied and the concrete backing is cast. The metal anchors are embedded in the concrete. Lifting and connection inserts are cast in the backup concrete. If the panel is to be insulated, the insulative core is installed between the stone and concrete. Since the insulation acts as a bond breaker, the polyethylene sheet is omitted. The metal anchors must be long enough to pierce the insulation for proper embedment in the concrete. Bond breakers are used to compensate for the different volumetric changes of the stone and concrete caused by temperature, shrinkage and creep. The marble or granite is usually from 1 to 1 1/2 in. thick.

PACIFIC GAS & ELECTRIC
HEADQUARTERS (NEW) SAN FRANCISCO

Granite faced panels clad this landmark office structure.

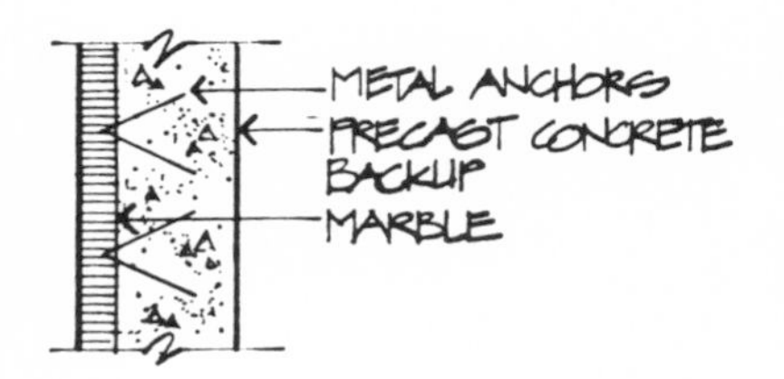

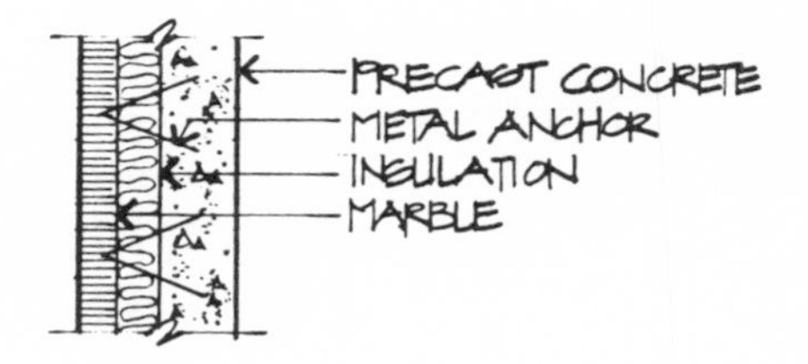

DETAILS - NATURAL STONE FACED PANELS

SANDBLASTED FINISH

Precast concrete units can be lightly sandblasted. The depth of sandblasing is usually only sufficient to expose sand particles and a portion of the coarse aggregate. To achieve greater depths, a light retarder is used along with the sandblasting. Subject to the softness of the coarse aggregate the sandblasting media may also consist of walnut shells. Walnut shells will cut the cement and the sand but only slightly affect the coarse aggregate. The depth of sandblasting must be defined and worked out between the pre-caster and the architect prior to the bidding. Sandblasting produces a lightly etched surface which exposes the aggregate and defuses light reflection.

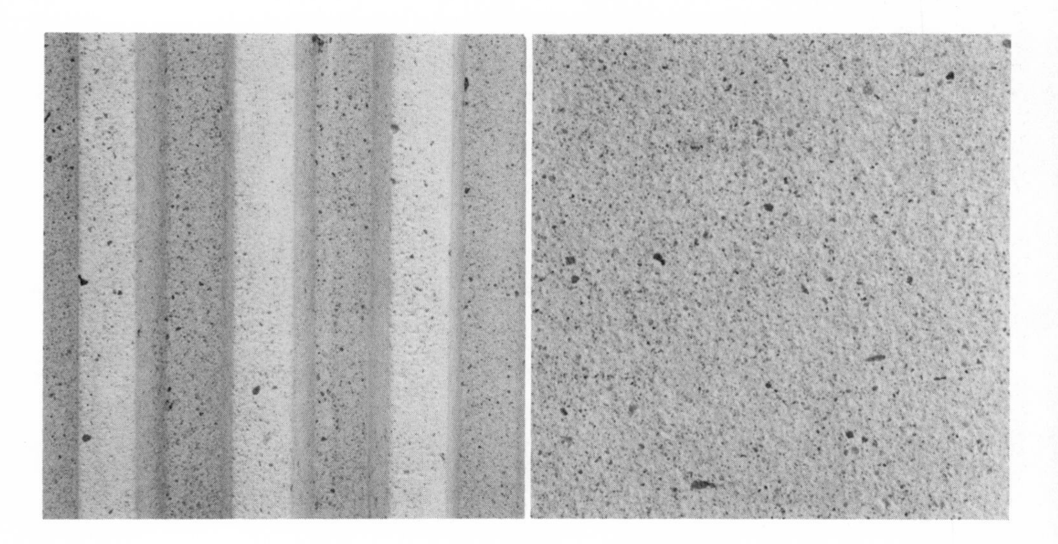

CUTTER LABORATORIES - BERKELEY

Light sandblasting presents a uniform and pleasant finish on the ribbed and smooth portions of the cladding panels on this project.

EXPOSED AGGREGATE (RETARDED) FINISH

Exposed aggregate surfaces are usually obtained with the use of retarders. The unit may be sandblasted afterwards but this is uncommon. If the surface to be exposed is in contact with the forms the retarder is brushed, rolled or sprayed over a release agent previously applied to the forms. After removal from the form the release agent-retarder, cement paste plus sand is washed with water and brushed off the unit. High pressure water jets are also used. If the surface to be exposed is not in contact with the mold, the retarder is sprayed on after trowel finishing.

Retarders are available in light, medium or heavy grades, formulated to give various depths of etch. Retarders prevent the set of the cement, and deter the hardening process at the surface. If left on and not washed off, the cement will eventually set and cure; hence it must be washed off. The usual time is 48 to 72 hours for form applied retarder and 24 hours for direct applied retarder. However, some retarders may be left on a week or more before removal.

MERIDIAN HILL PARK
WASHINGTON, D.C.

This facade, constructed in 1920, demonstrates the durability and excellent weathering characteristics of exposed aggregate concrete. John Earley used strict control over the water content and brushed the surfaces of the still green concrete to expose the aggregate.

FRACTURED SURFACE

A fractured surface can be obtained by placing cable or chains close together on a form bed. After the unit is removed from the form, the cable or chain is pulled out leaving an imprint along with fracturing between the ridges. Another method is to bush hammer or chip the ridges to fracture the surface. When large ribs are cast in the face of the panel, these may be split in the yard by using a mason's chisel positioned at the base of the rib and struck with a heavy hammer or maul. This finish is called "split rib." Care must be taken in forming this finish in that the rib should only be fractured by striking in one direction, so that light striking the panel from an angle results in a uniform shading effect. Another method of achieving a fractured surface appearance is to coat the smooth rib area with retarder prior to casting. Then by sandblasting the panel after stripping, a fractured rib appearance is attained.

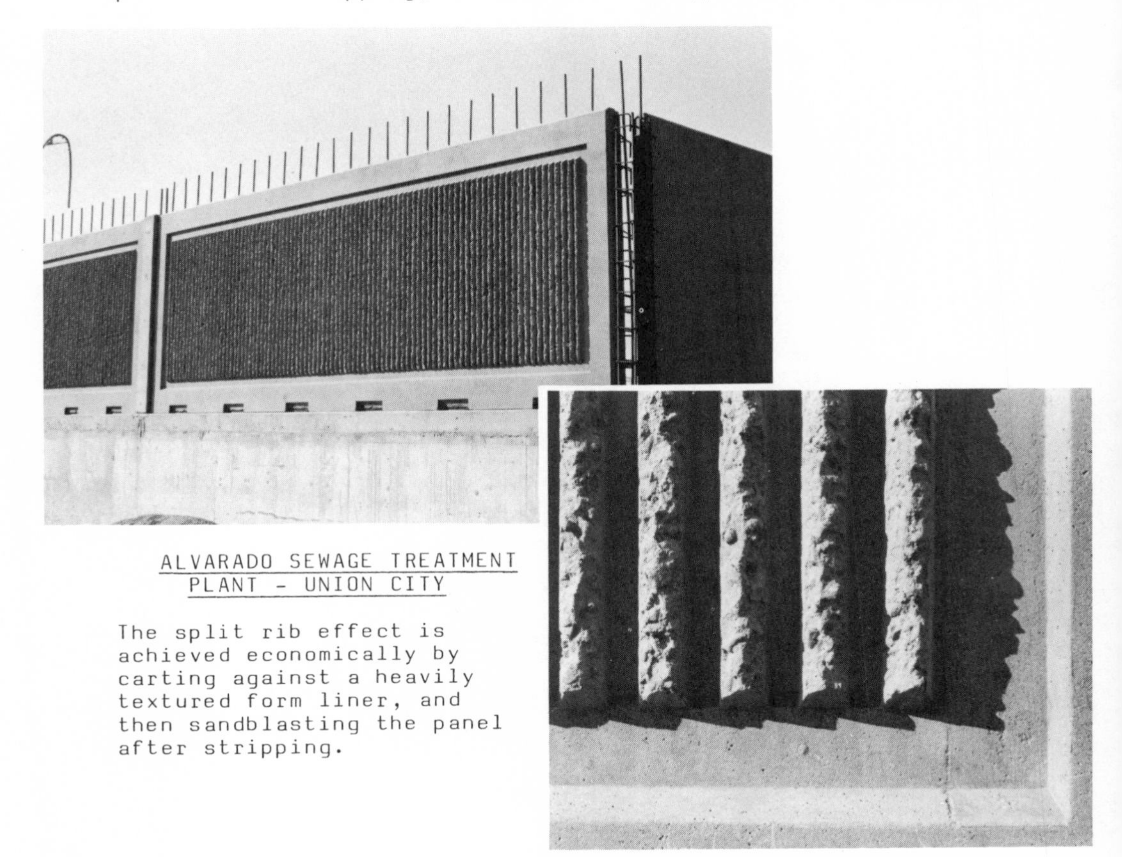

ALVARADO SEWAGE TREATMENT PLANT - UNION CITY

The split rib effect is achieved economically by carting against a heavily textured form liner, and then sandblasting the panel after stripping.

SCULPTURED SURFACE

Since concrete is plastic in the freshly mixed state, unlimited sculptural effects are possible. Forms may be fabricated of wood or fiberglass. To create forms for complex shapes, wood, clay, plaster or polystyrene patterns are made and sprayed with fiberglass resin and chopped fiberglass. The form may be reinforced or stiffened with wood, metal or fiberglass ribs. Once completed, the form is sanded and gel coated. One pattern may be used for more than one form. Wood and fiberglass forms may be used up to 50 times subject to the shape. For greater numbers of casts, concrete or metal molds should be made. Prior to casting, the forms are coated with a release agent to insure ease of removal. For a single non-repeatable unit, polystyrene sheets can be carved into a negative pattern and used as a form liner. The polystyrene pattern is sprayed with a fiberglass resin to seal the surface, permitting ease of removal. In the stripping process, if the polystyrene sticks to the concrete, it is simply pulled off. One time patterns are also made by casting impressions of sand patterns created by an artist.

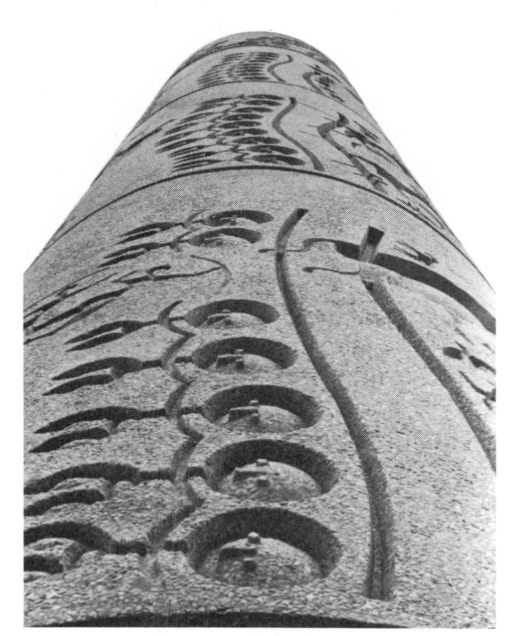

LIBRARY STAIR TOWER - CIVIC CENTER, INGLEWOOD

Sculptured forms combined with an exposed aggregate finish achieve the bold expression desired by the architect. (Photo by Eugene Phillips)

ACID ETCH FINISH

When the architect wants a very light surface texture which just barely breaks the surface skin, an acid etched finish may be selected. Acid etching consists of brushing a solution consisting of one part muriatic acid to 2 or 3 parts of water on the surface of the concrete 2 or 3 days after casting. The surface is then cleaned off with water, resulting in a sand finish. A minor portion of the surface coarse aggregate may be revealed if a 100% acid solution is used. Acid etching results in a very uniform "sanded" appearance.

After the acid is applied with brushes, a jet of water is used to clean the surface, revealing the sand.

BRICK OR TRAVERTINE FINISH

Brick or travertine tiles, from 1/2" to 3/4" in thickness may be embedded in the face of a precast panel to provide the same end result these materials present when installed on the job site. When used as brick facing elements with grouted joints, the tiles are laid down on a rubber form liner pre-indented with joints simulating raked joints. A layer of grout is next laid on the back of the tiles, using adequate vibration to assure that the grout is firmly deposited in the joints. Finally, the concrete backup is poured, after placement of reinforcing and positioning of connection hardware and handling inserts. When the tiles are laid up without joints, then they are mechanically anchored to the precast concrete back-up. Brick tiles are manufactured with a positive dovetailed key to provide anchorage. Travertine tiles are drilled by the stonecutter with angled holes to receive special stainless steel spring clips to anchor the files to the back up concrete. Subsequent cleaning is usually required after casting unjointed tiles. Brick faced panels are usually cleaned by light sandblasting. (Obviously, acid washing would etch the travertine surface.) High pressure water cleaning is also employed.

HIBERNIA BANK MOCK-UP
SAN FRANCISCO

Brick tiles cast in a precast concrete back-up are shown in the mock-up in Western Art Stone Company's yard in Brisbane. These panels were subsequently installed on this building at Front and California Streets.

FORM LINERS

Form liners can be used to simulate wood boards, wood grain, small corrugations, rope, cable or other small patterns. The form liners may be rubber, plastic or metal. Rubber form liners are preferred due to ease of stripping, cutting and reuse. Interesting effects can also be produced by placing various items such as beans or wood sticks on the molds. Beans will expand due to the moisture and give an interesting texture to the concrete. Rock salt sprinkled on the bottom of a smooth mold creates an interesting effect also.

FACING MIXES

Often times, the cost of the coarse aggregates selected for exposed aggregate finishes is high, when compared to the cost of locally available granitic, limestone, or river gravel coarse aggregates. When this situation arises in finish selection, the concrete may be cast in two different mixes, with the portion to be exposed with the expensive aggregate forming a thin outer layer. The depth of etch is directly related to the aggregate size. Where a light etch is desired, the coarse aggregate size should not exceed 3/8". Where a medium etch is desired the coarse aggregate size should be 5/8" or 3/4". For deep exposure, a 1" stone should be used. Facing mixes are usually 1½" thick, except for very large coarse aggregate sizes. The architect should also be aware of variations caused by form configuration.

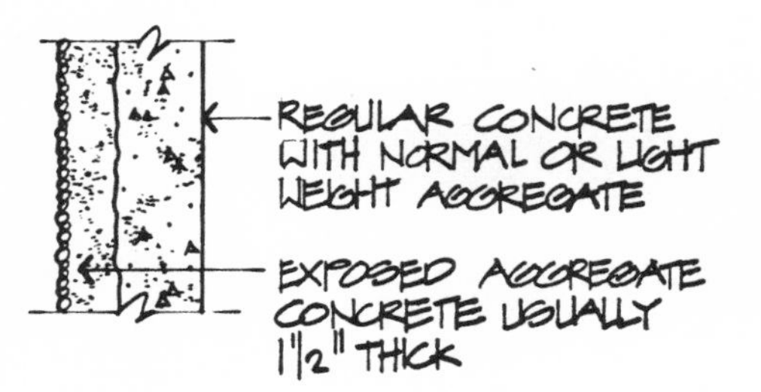

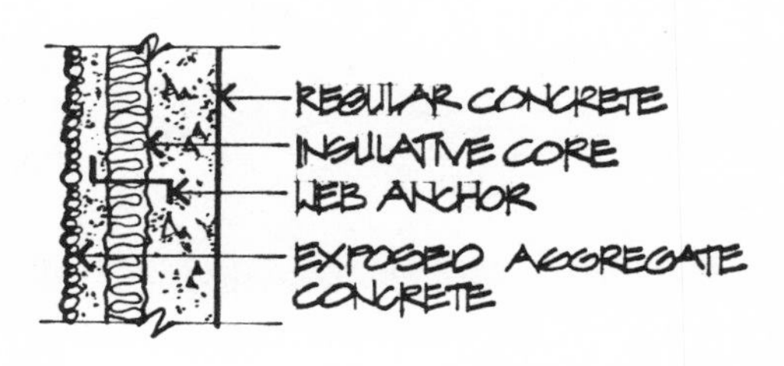

ANHEUSER BUSCH PLANT EXPANSION - FAIRFIELD

14" square pile - 110 feet long, being lifted from the horizontal to the vertical by picking with both crane lines. By picking at 3 points as shown, approximately equal positive and negative bending moments are induced in the pile, which are well below the value which would produce cracking.

PRODUCT HANDLING

The precast concrete manufacturer is responsible for handling precast and prestressed concrete units from casting to the time when they are erected and permanently built into the structure. In the situation where the precast manufacturer sells his product F.O.B. jobsite, he is responsible to that point where the buyer unloads the truck at the jobsite. In some instances for F.O.B. jobsite contracts, the precaster may want to provide handling information to be used by the purchaser's erector, even though his responsibility may not extend this far. In any event, proper handling of precast and prestressed units consists of the procedures and picking/support points to be used so as not to produce cracking. Product lifting devices are also designed with sufficient factors of safety to guarantee shop and field personnel safety, as well as product integrity. Design loads used to calculate handling loads and stresses include allowances for impact which reflect the magnitude of the dynamic loading to which the element is being subjected. This chapter will discuss handling devices and inserts, some of the variables in handling decisions made for stripping, storage, shipping, and erection, and finally, a step-by-step procedure to use in analyzing handling.

LIFT LOOPS AND MANUFACTURED INSERTS

Inserts are devices cast into the precast concrete unit which are used for handling the unit, connecting the unit to the building system or both. When both handling and connection inserts are used, then dual usage is desirable to reduce cost. This is however often difficult to accomplish.

Strand lift loops are used only as lifting devices. Strand in short sections is readily available as waste from pretensioning operations.

As a lifting device, a short piece of looped strand is set in the concrete; the looped strand is usually shoved in to the required depth after the concrete has been screeded, thus not interfering with the screeding operation. When the precast section is thin, and embedment is critical, the looped strand is usually anchored to the reinforcing cage with tie wire.

Where the precast concrete is to receive a poured in place concrete topping, the protruding strand can be either burned off, or depending on the topping thickness, left extended from the member. This is the usual case for beams, double tees, inverted tee beams, piling, and flat roof or floor panels. If the surface of the unit is to be exposed, then a lift loop is set in a pocket using a polystyrene or wood blockout. After casting the concrete the blockout is removed. After the precast unit has been erected the lift loop is burned off and the pocket patched and finished to match the panel surface. This method is often used for wall panels; however many precasters prefer to use manufactured inserts.

LIFT LOOP CONFIGURATIONS

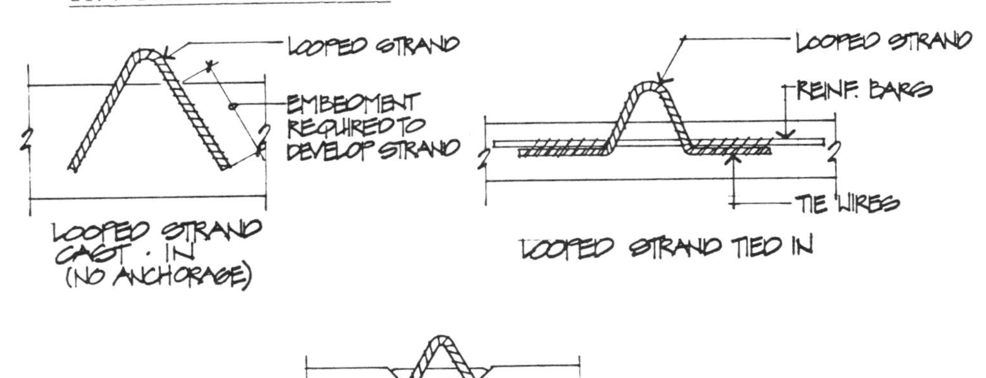

The size of strand lift loops is determined by calculating the load to the lifting point, including impact, and then applying a factor of safety of 4 to the ultimate capacity of the strand. For example, if waste 1/2" diameter 270k strand were used as a lift loop, the ultimate capacity of the strand is $0.153 \times 270 = 41.3^k$. Applying a factor of safety of 4, the maximum safe load to the lift loop (including impact is 10^k. The development length of the strand in the concrete may conservatively be taken at 50 diameters. In special cases, the shear cone method may be used to check the concrete strength of the lift loop assembly. This method is explained in the following paragraph on manufactured inserts.

* Manufactured inserts are available for use with precast concrete. The manufacturers' brochures list both the ultimate capacities, and the safe working loads with a factor of safety of 4 applied. Two basic types of inserts are used for precast concrete. They are the coil thread type and the standard thread type. Examples of each are shown below.

* Factors of safety may vary with actual application circumstances.

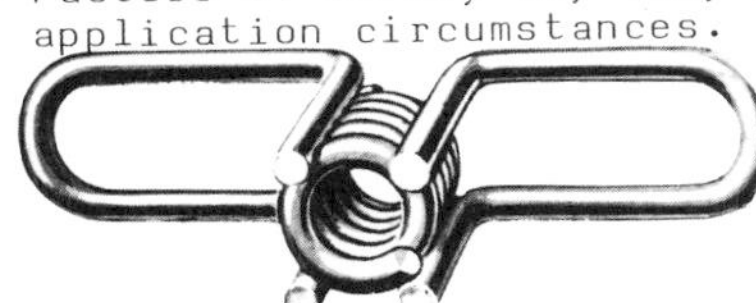

COIL INSERT
(Accepts coil thread bolts)

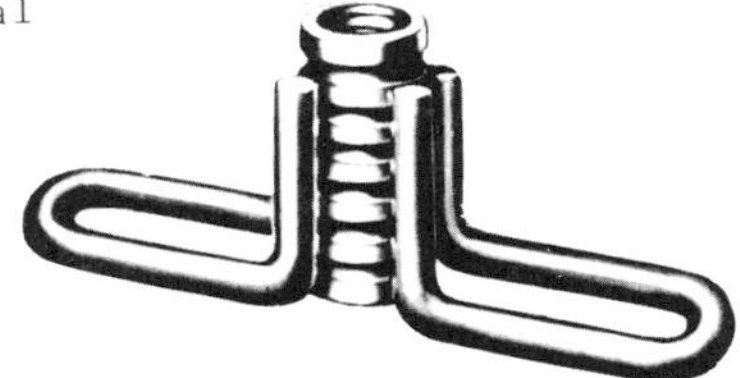

FERRULE INSERT
(Accepts standard thread machine bolts)

A limited amount of technical information on inserts appears on page 5-48 of the PCI Design Handbook. More complete information is contained in excellent literature available from the following manufacturers in California:

Acme Wire Products Co., Garden Grove
Burke Company, San Mateo
Superior Concrete Accessories, Santa Fe Springs

The manufacturers have conducted tests on various types of inserts and using different concrete strengths; the results of these tests are contained in the literature.

These inserts are made in various diameters from 3/8" to 1" or larger in some types. They are also made in various embedment configurations to maximize concrete pull-out strength for thick, as well as thin sections. When the insert is in full tension, and there is sufficient concrete surrounding the insert, the following shear pattern develops in the concrete.

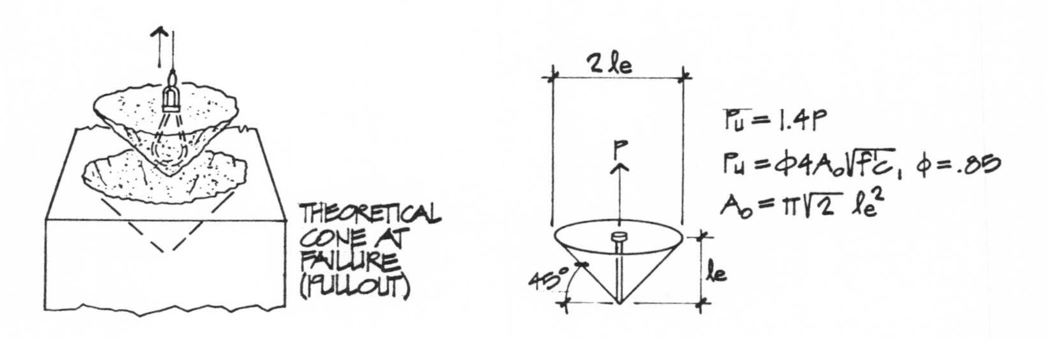

However if there is insufficient concrete, only a partial shear cone is developed. The shaded area in the cone shows the extent to which the area is reduced.

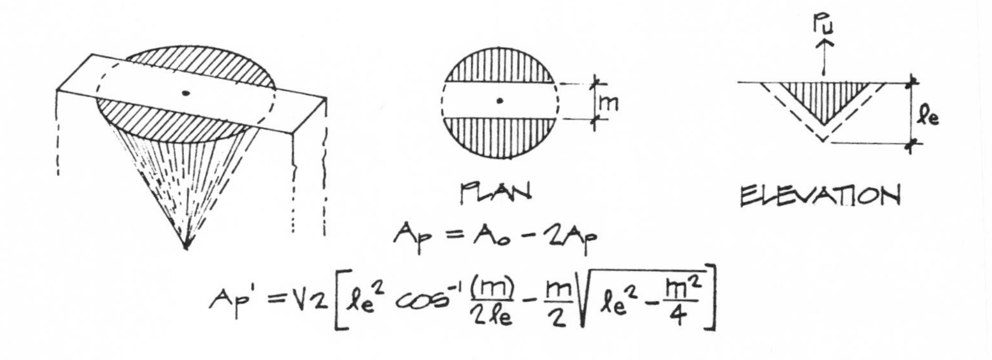

$$A_p = A_o - 2A_p$$

$$A_p' = \sqrt{2}\left[l_e^2 \cos^{-1}\frac{(m)}{2l_e} - \frac{m}{2}\sqrt{l_e^2 - \frac{m^2}{4}}\right]$$

Charts are available in the P.C.I. Design Handbook which give reduction factors for partial shear cones for both studs and inserts. (see pp 5-42 to 5-46)

In the normal handling of precast concrete units, lifting slings are used which produce combined shear and tension on the insert. Also, when rotating a precast element into its final erected position, a maximum critical condition of combined shear and tension will be produced. This combined loading condtion should be checked for both the steel (insert) capacity and the concrete capacity in accordance with the interaction relationship given on p. 5-47 of the P.C.I. Design Handbook.

When the panel is lifted by an edge insert the following stress pattern occurs:

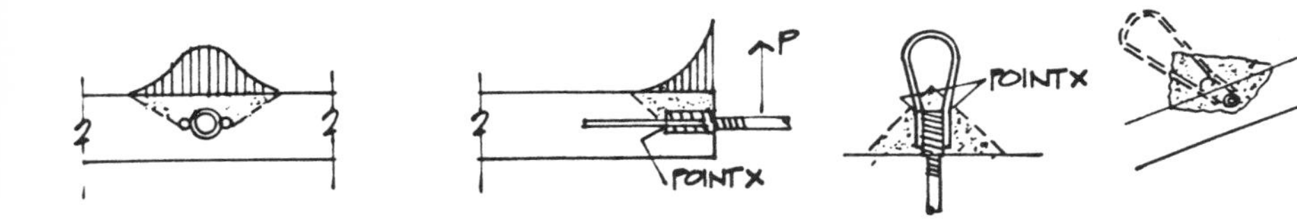

The insert will tend to bend about point "X" and the dotted line indicates the outline of probable concrete failure.

One final note of caution regarding lifting accessories: only strand or manufactured inserts are to be used; bent reinforcing bars are not to be used under any circumstances for lifting and handling precast concrete elements.

STRIPPING

The relative ease (or difficulty) of the stripping operation varies with the shape of the member. During the stripping operation, care must be taken to keep moment, shear and torsional stresses to a minimum. The design loads to consider during stripping are (1) dead load of the member, (2) suction (atmospheric pressure) between the form and the contact surface of the members, and (3) mechanical bonding between the form and some projection, indentation or other surface variation in the member.

Wall panels usually require more consideration during the stripping operation than do other member shapes because (1) the stripping and handling loads perpendicular to the plane of the walls are much greater than the in-place wind and seismic loads, (2) wall panels are cast face-down as shallow members with relatively little depth-of-section to develop resisting moments, (3) suctional loads tend to be relatively large, and (4) decorative surface variations tend to increase the degree of mechanical bonding between the form and the decorative surface cast against the form. In many instances, the reinforcement required for stripping and handling loads is more than adequate for the loads which must be resisted after the panel is erected. The handling decision to be made for stripping is to pick the precast element at a sufficient number of points so as to keep tensile stresses within allowable limits, and not require reinforcing in excess of that required to satisfy service loads in the final installed condtion. The other structural shapes (beams, columns, etc.) are relatively easy to strip because (1) the stripping loads are usually less than the in-place loads for which the primary reinforcing is designed, (2) the sections are relatively deep and capable of developing large resisting moments, (3) the suction is released when the prestress load is transferred to the concrete, and (4) decorative and other surface variations tend to be minimal.

STORAGE

The support system used in storage is very important. The effect of differential settlement or movement of two-point and three-point support systems in storage and shipping of prestressed precast concrete members is shown below

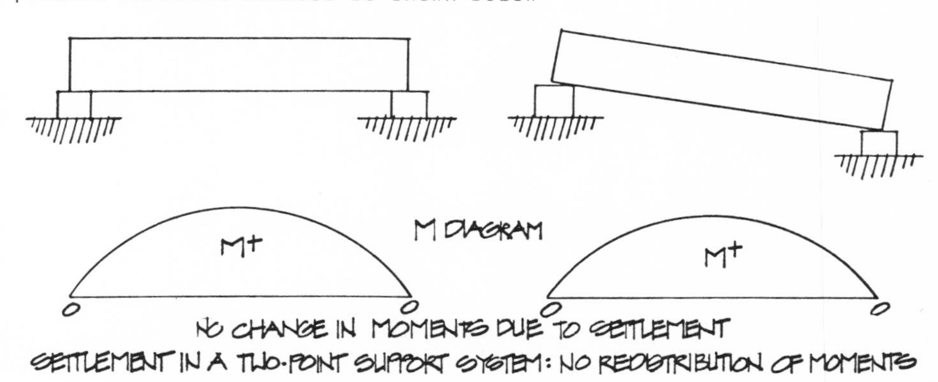

SETTLEMENT IN A TWO-POINT SUPPORT SYSTEM: NO REDISTRIBUTION OF MOMENTS

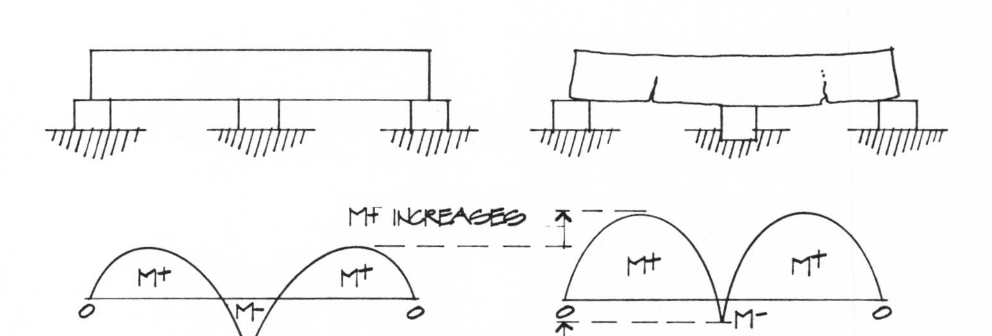

CHANGES IN MOMENT DUE TO SETTLEMENT

A two-point, simple support system is normally used to support all shapes of precast concrete members. If a three (or more) point support system were used, one of the supports might settle, and the resultant redistribution of stresses could severely damage the member. If a support in a two-point system settles, there is no redistribution of stresses, only a slight, and negligible, tilting of the supported member as shown in the diagram. However permanent in-storage support systems are often built into solid poured foundations, such as in the plants of many pile manufacturers, permitting multiple point supports. An alternative method of support is to use continuous longitudinal supports, such as for thin solid prestressed slabs. In general, the method of support during storage will mirror the support system used for transportation. For example, if two points of support horizontally satisfy shipping handling stresses, then this is the logical method for storage. If long thin panels must be shipped on edge, they will be stored this way also. Long prestressed concrete piling are stored and shipped on 4 points of support, with the shipping rig being comprised of special pivot supports.

The fabricator's storage yard is valuable real estate. If the members are designed in such a manner that they can be stacked on top of one another, then some flexibility and economy of total production time might be realized in some situations. For example, stacking might allow the producer to begin a continuous operational cycle at an early date, stack the members, and develop a readily available inventory from which to quickly supply the erector as necessary. When this is done, the total construction time is governed more by the rate of erection than by the rate of fabrication. If the members cannot be stacked, or the fabricator does not have a sufficiently large storage yard, both the fabricator and the erector may have to run a slow and costly start-and-stop operation; i.e., fabricate some units, stop fabrication until the units are moved out of the way, start up again, etc.

Slabs, double tees, columns and flat wall panels of the same shape (width and length) are usually stacked; single tees and deep girders are usually stored side by side (unstacked); and decorative-surfaced wall panels are usually stored in a vertical (tilted) position as separate units which are protected from contact with other potentially damaging items. Prestressed flexural members should be blocked close to the ends to minimize the amount of camber growth in storage.

SHIPPING

Members must be designed to resit the impact loads encountered while being transported to the jobsite. The manner of placement (stacked, side-by-side, or specially braced) and the shipping position (vertical, flat or tilted) influences the magnitude and direction of these important design loads. Members are usually (1) loaded on the transporting vehicle (usually a tractor-drawn trailer) in the same position in which they are stored, (2) carried from the fabricating plant to the jobsite on the trailer, (3) lifted from the trailer and re-positioned (re-oriented) as necessary while still suspended from the crane, and (4) moved from the trailer to the final in-place position in the building where the member is anchored, often temporarily, as quickly as possible. The lifting, reorienting and temporary anchoring processes are repeated until the trailer is unloaded. Whenever possible, on-site storage and subsequent "double" handling operations are avoided. (If the transporting operation is defined as encompassing movement from the fabricating plant to the in-place position (with temporary connections) then any transporting system which requires on-site storage and re-lifting may be described as an inefficient transportation system.) The exception to this statement is a pile driving operation with several drivers spread out on the site.

The shape of the member often determines how many members may be placed on one trailer. If the member shapes can be stacked, then each trailer may be able to carry a full trailer capacity load, and the number of trips may be minimized. If the members cannot be stacked, or if they require special bracing, then each trailer may not be able to carry a full load, and more trips will have to be made. Normally, full trailer capacity can be achieved with all shapes except decorative surfaced wall panels.

In summary for shipping, other than special situations such as the pivot rigs used for shipping piling, only 2 points of support are available on the truck. This makes shipping critical for handling in most cases.

ERECTION

Handling in erection consists of rotating the precast element from the horizontal, or sideways position, into the vertical erected position for wall elements, or merely transferring from truck to final position for deck members. Prior thought should be given to the erection process in product design to assure that necessary erection loops or inserts are provided to facilitate translation of precast wall elements into the vertical position. Care must also be taken to assure that erection slings do not create an angle of greater than 45° so as not to induce excessive shear in erection inserts.

HANDLING PROCEDURES

The following procedure is useful as a checklist in designing precast and prestressed concrete elements for handling:

1. Calculate section properties of the cross-section of the precast concrete element through the portion resisting bending.

2. Calculate member weight and location of center of gravity.

3. Check concrete stresses for handling during stripping, shipping, and erection. Concrete stresses during handling should be kept below the modulus of rupture (fr), divided by a factor of safety of 1.5.

$$f_t = \frac{fr}{F.S.} = \frac{7.5\sqrt{f'c}}{1.5} = 5\sqrt{f'c}$$

f'c is the concrete strength in the member at the time it is being analyzed.

For sand lightweight concrete, multiply the above value by 0.85

(0.85 x 5 f'c = 4.25 f'c)

Use the following impact factors in calculating service load moments for the various handling conditions:*

Stripping	1.5	* Other producers may use different values; these are not code requirements.
Shipping	2.0	
Erection	1.25	

The following general equations are useful in determining design moments for various handling conditions:

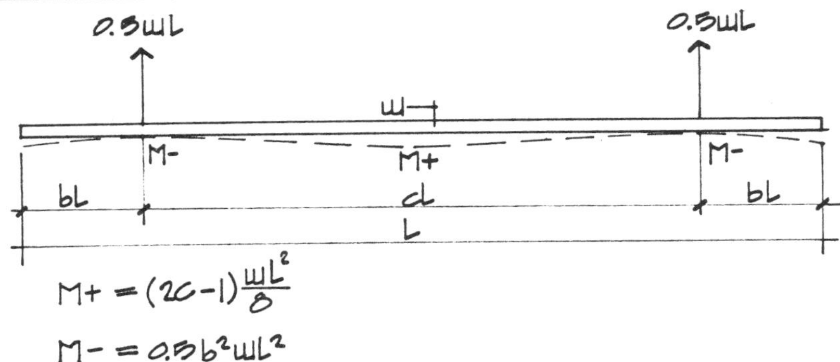

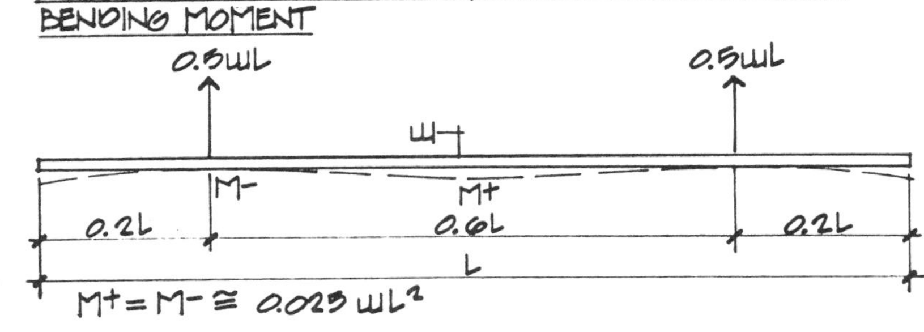

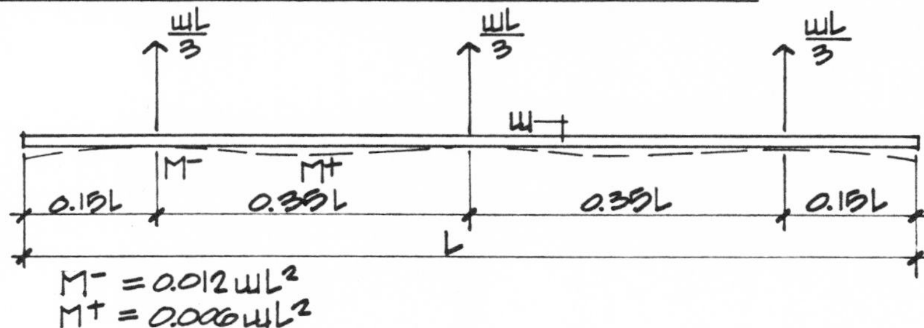

(4) 4 POINT PICK FOR STRIPPING · EQUAL REACTIONS.
EQUAL NEGATIVE AND POSITIVE BENDING MOMENTS

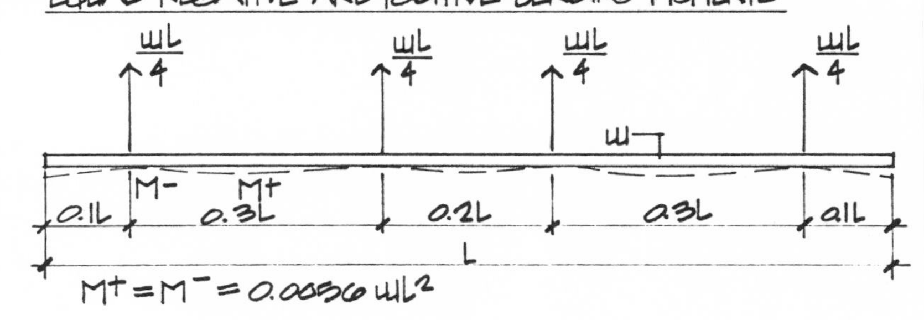

(5) GENERAL EQUATION · 2 POINT PICK FOR ERECTION

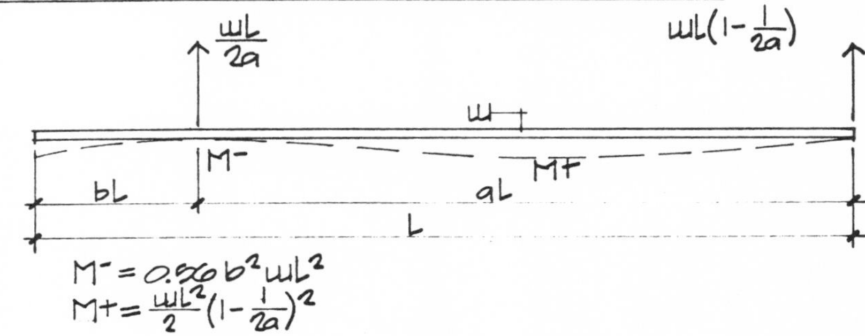

(6) 2 POINT PICK FOR ERECTION · EQUAL NEGATIVE AND POSITIVE BENDING MOMENTS

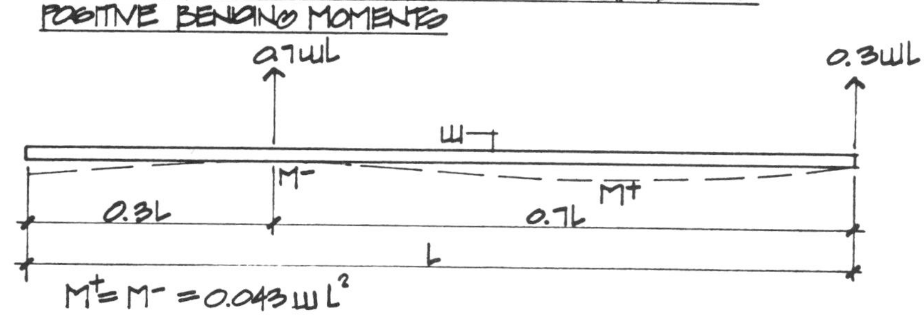

(7) 2 POINT PICK FOR ERECTION · LOWER PICK POINT AT BOTTOM 2 PT. STRIPPING INSERT LOCATION

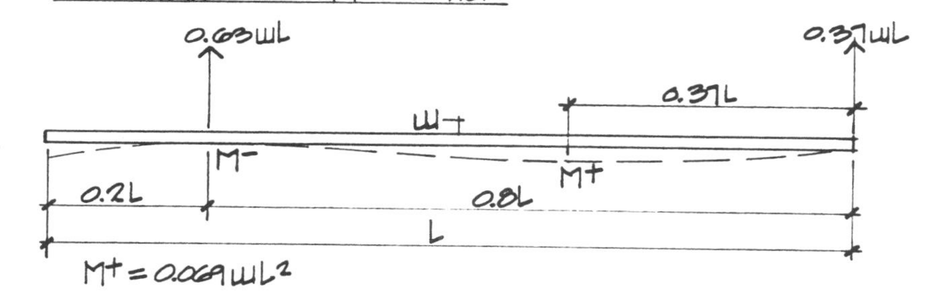

(8) 3 POINT PICK FOR ERECTION AT 4 POINT STRIPPING INSERT LOCATIONS.

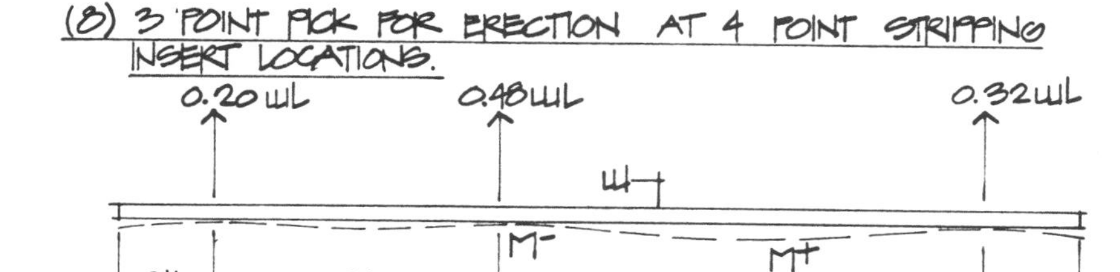

$M^+ = 0.019\ wL^2$
$M^- = 0.021\ wL^2$

(9) 3 POINT PICK FOR ERECTION · AT TOP AND BOTTOM 4 PT. STRIPPING INSERT LOCATIONS

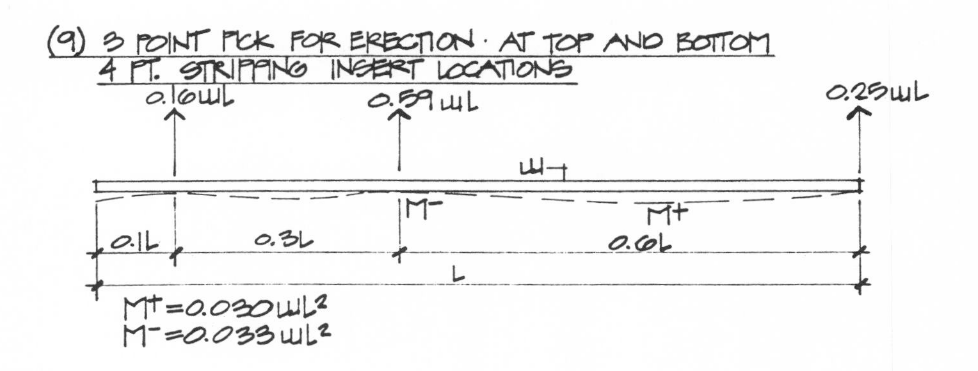

(10) 3 POINT PICK FOR ERECTION · EQUAL LOWER END REACTIONS USING 2 PT. STRIPPING INSERT LOCATIONS

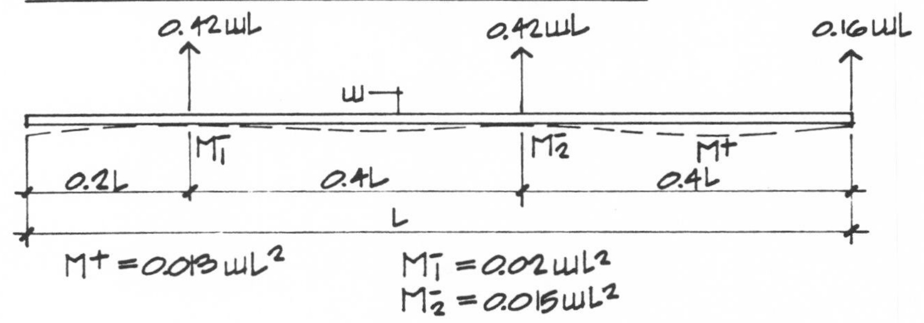

(11) 3 POINT PICK FOR ERECTION - TOP PLUS THIRD POINT STRIPPING LOCATIONS

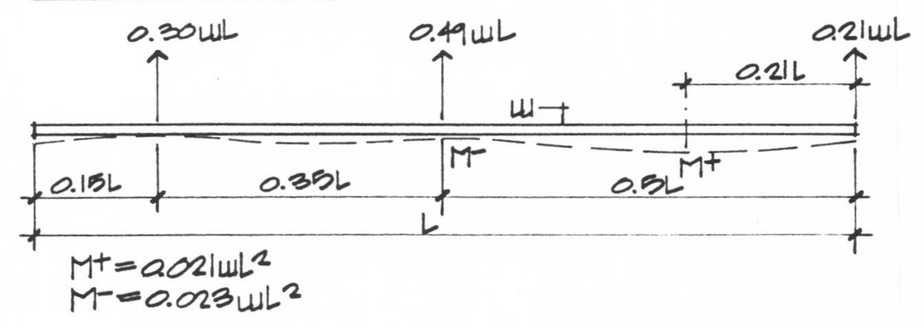

DOROTHY CHANDLER PAVILION - LOS ANGELES MUSIC CENTER

Built in 1965, the architectural precast concrete cladding elements shown here demonstrate the beauty and permanence synonymous with this wonderful material.
(Photo by Eugene Phillips)

GENERAL DESIGN PROCEDURE

ARCHITECTURAL PRECAST CONCRETE CLADDING UNITS

This chapter presents a logical procedure to use in sizing and detail designing non-load bearing architectural precast concrete cladding panels. It is written to provide a basis for decisions required in designing these elements, many of which must be made in the preliminary stages of the project.

1. Preliminary Size Determination: Try to keep panel weights below 11 tons per unit. This is compatible with the handling capacity of available erection equipment, and would also facilitate vertical shipment on easel trailers, a common method for handling long flat units that would otherwise be cracked if shipped flat, or would require excessive reinforcing needed only for handling.

   If possible, analyze the site access available for cranes and trucks. Tight sites or complicated spread-out buildings may require even further reduced weights of units for feasible erection.

2. Applicable Design Criteria:

   a. Uniform Building Code - 1979 Edition
      (1) Seismic - ($F_p = ZIC_pW_p$) Section 2312(j)3C

      Reinforcing design:

      wall panel element: $C_p = 0.3$ (applied normal to flat surface)

      Connection design:

      connector body: $C_p = 0.4$ (applied in any lateral direction)
      connector fastener: $C_p = 1.2$

   The 1976 edition of the Uniform Building Code required an arbitrary value of $C_p = 2.0$ to be applied in a lateral direction on the panel for connector design. Although this value is higher than the lateral force coefficients given above, the code did not make provisions to assure that the connector behaves in a ductile manner. The revised provisions in UBC-79 for panel connections are intended to assure that distortion will be accommodated by the body of the connector, and not by brittle failure of welds, and inserts or studs embedded in the concrete. In this way, the connector body by yielding can form an energy absorbing mechanism to accommodate extreme seismic conditions and large values of building drift. However, some jurisdictions still require that the 2.0 g criteria be applied. Even in these instances provision should be made in the connector for ductile behavior to be possible under extreme loading. The 0.4 g loading to connector bodies is an elastic load, with the steel element being designed by the allowable stress method. The 1.2 g load to fasteners is also considered an elastic load, with the fasteners being taken to ultimate strength or yield by using appropriate load factors, and comparing this with the nominal strength of the fastener, modified by appropriate capacity reduction factors.

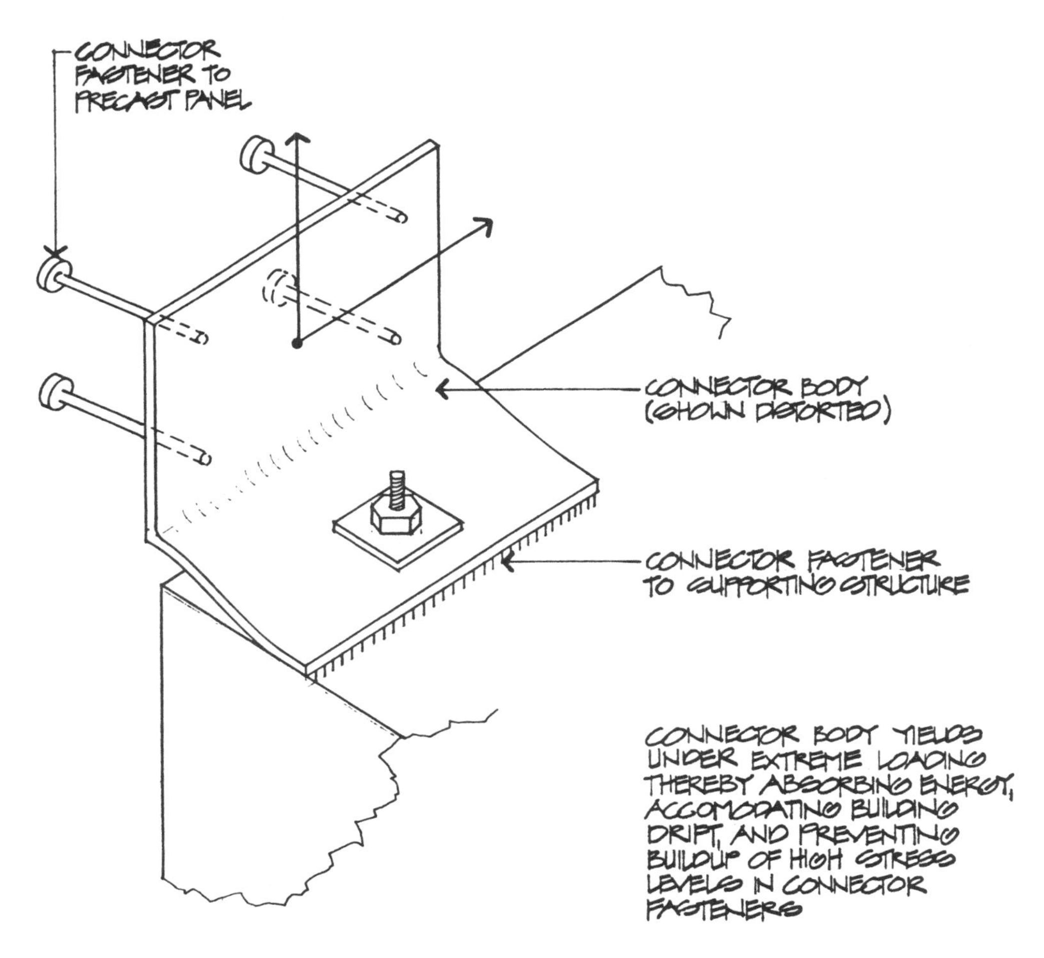

DETAIL SHOWING DEFORMATION OF CONNECTOR BODY DURING EXTREME SEISMIC DISTURBANCE

SCHEMATIC PRESENTATION OF PANEL LOADINGS (UBC-79)

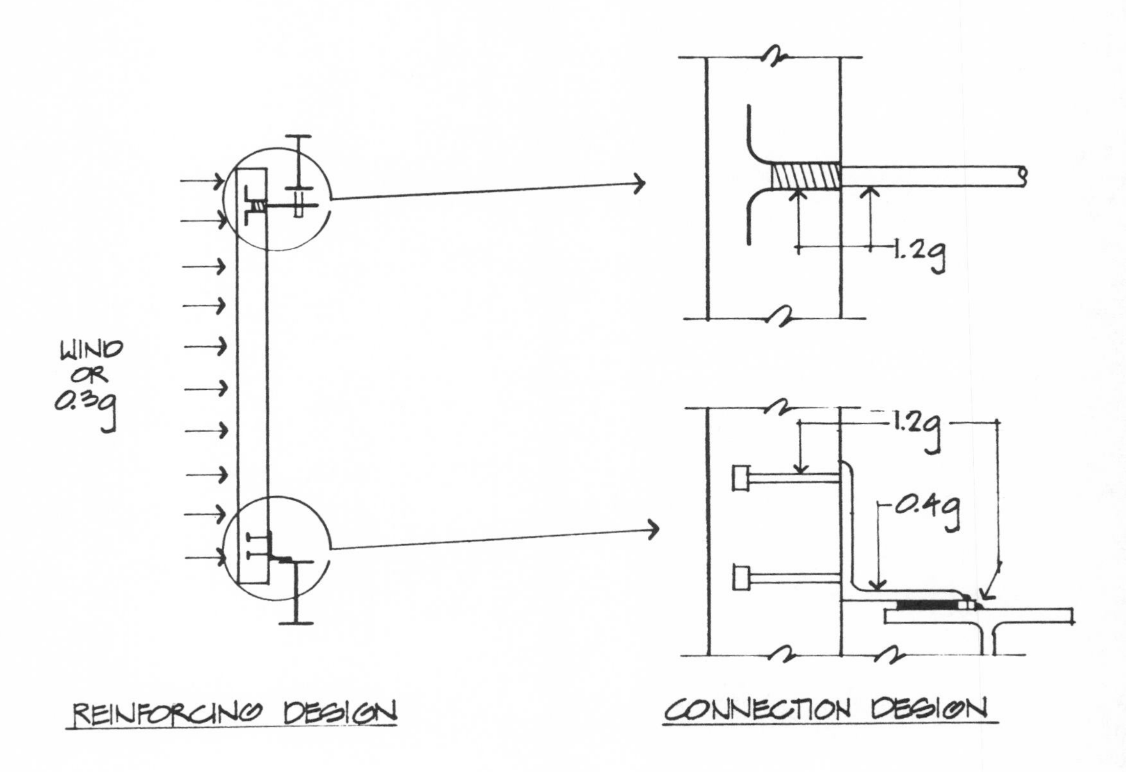

(2) Wind - see Table 23-F and Figure No. 4

b. Concrete strength 5,000 psi minimum for durability

c. Cover 3/4" minimum, or as required to attain fire rating greater than 2 hours. See UBC-79, Table 43-A, and Sec. 2607(o)

d. Reinforcing

#3 & #4 bars: ASTM A 615-40

#5 & larger: ASTM A 615-60

Welded Wire Fabric: ASTM A 185

Note: Tack welding of cages permitted only with all 40 ksi yield bars.

3. Determine thickness of panel elements

   a. As required for final design condition

      (1) Seismic

      (2) Wind

   b. As required for required minimum fire resistance ratings set forth in Table 43-B, UBC-79

   c. As determined by handling (shipping will usually be most critical)

      Note: for panel thicknesses less than 5", use one row of reinforcing; for panel thicknesses 5" or greater, use 2 layers of reinforcing.

4. Design panel reinforcing

   a. Calculate section properties of panel element, usually a cross-section through the portion resisting bending.

   b. Calculate panel weight and location of center of gravity.

   c. Analyze handling - determine reinforcing requirements for stripping, shipping, and erection. The formulae given on pages 173 to 176 are useful in determining design moments for various handling conditions:

      Keep concrete stresses during handling below the modulus of rupture (fr) divided by a factor of safety of 1.5 ($5\sqrt{f'c}$)
      For sand lightweight concrete, multiply the above value by 0.85
      Use the following impact factors in calculating service load moments for the various handling conditions: *

Stripping	1.5	* Other producers may use different values; these are not code requirements
Shipping	2.0	
Erection	1.25	

Summarizing the handling criteria for mild steel reinforced panels:

Stripping: f'ci = 2000 psi
Mdesign = 1.5 Mactual
Ft allowable = 0.225 ksi (normal wt.)
= 0.190 ksi (light wt.)

$$As\ req'd = \frac{Mdesign}{1.0\ d}$$

where: M is in FT-K
d is in inches
As is in sq. in.

Shipping: f'ci = f'c = 5000 psi
Mdesign = 2.0 Mactual
Ft allowable = 0.350 ksi (normal wt.)
= 0.300 ksi (light wt.)

As req'd = $\frac{\text{Mdesign}}{1.44\ d}$ for fy = 40 ksi rebar

As req'd = $\frac{\text{Mdesign}}{1.76\ d}$ for fy = 60 ksi rebar or WWF

Erection: Mdesign = 1.25 Mactual
(other criteria same as shipping)

(for prestressed concrete panels, keep maximum tensile stress below $3\sqrt{fc}$)

d. Locate Panel Connection Points to resist lateral and vertical florces. Check panel reinforcing for the most critical of the following conditions:

(1) Wind

(2) Seismic - 0.3 g

(3) Temperature steel requirements - UBC-79, Sec. 2607(n). Adjust/increase the number of lateral panel connection points to optimize the amount of reinforcing used in the panel.

In detailed connection design, only 2 bearing connections are used. As many lateral connections as required may be provided.

e. Check any bowing induced by thermal gradients, and provide additional lateral supports as required (usually critical for thin flat panels only). Design charts are presented on pages 184 and 185.

5. Design Panel Connections

a. Using fixed and lateral connection points previously selected, analyze the distribution of a unit load of 1^k applied in any direction at the center of gravity of the panel element.

i.e.
1^k down
1^k longitudinal (left or right)
1^k lateral (in or out)

Load analysis sketches are presented on pages 186, 187 and 188 to help in visualizing these loading conditions on typical cladding panel configurations.

b. By the principle of superposition, determine the service load to the connector bodies and the service load to the connector fasteners, or:

*P conn. body = 1.0g down + 0.4 g in any lateral direction

P conn. fastener = 1.0g down + 1.2 g in any lateral direction.

*Note that the Uniform Building Code does not require designing for any component of vertical acceleration. However, high narrow elements, such as column covers, on tall buildings will be subjected to vertical acceleration during an earthquake. For this reason, it is recommended to design the bearing connectors of these elements for some value of vertical acceleration, say 0.3 g to 0.5 g, to be combined with gravity loading.

Multiplying by the actual panel weight "W" will give the actual service load on the connection to be used in the design. The above service loads must be multiplied by appropriate load factors in order to design the connections for the appropriate material capacities. From UBC-79, Section 2609(d), the required ultimate strength shall be:

$$UP = 0.75\ (1.4D + 1.87E)$$

PCI recommends the use of an overload factor of 1.3 to be applied to the ultimate loads obtained from the above equation (p. 5-3, PCI Design Handbook). Therefore, the above critical ultimate load then becomes:

$$UP = 1.3 \times .75\ (1.4D + 1.87D) = 1.4D + 1.8E$$

PCI also recommends the following capacity reduction factors be used when the above ultimate load criteria are followed:

ϕc = 0.85 (concrete)
ϕst = 1.0 (steel in tension)
ϕsv = 0.75 (steel in shear)

The Design Charts in Chapter 5 of the PCI Design Handbook incorporate these values in the design charts for connectors as governed by concrete and steel strengths. (See pp 5-22 to 5-27; 5-40 to 5-49)

6. Other Design Criteria to be checked

a. Assure that the movement of the building under wind or seismic loading can be accommodated in the connections and the joint widths and configurations selected. (2 times story drift from wind or 3/k times the seismic story drift or 1/2", whichever is greater - UBC-79, Sec. 2312).

b. Design any required temporary bracing as required by CAL-OSHA, Article 29.

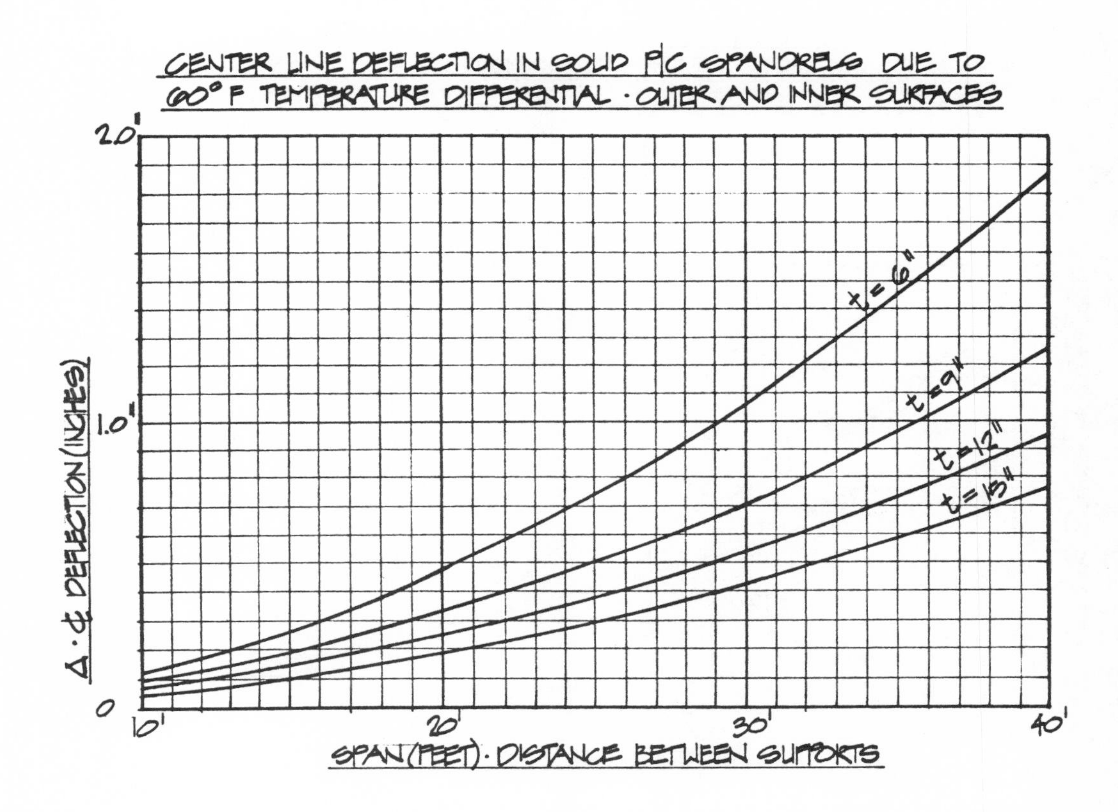

Long flat cladding elements without intermediate supports will be subject to bowing due to temperature gradients such as those produced by hot sunlight on panels covering air conditioned buildings. Conversely, cold exterior surfaces of flat cladding elements enclosing heated buildings would cause an inward movement. The magnitude of this movement which is acceptable is subject to designer judgment and the ability of glazing or window wall infill to accommodate these movements. The above chart gives an estimate of the amount of movement that would be encountered for a 60°F thermal gradient.

UNIFORM RESTRAING FORCE REQUIRED (W) IN KIPS PER FOOT PER FOOT OF SPANDREL DEPTH · FOR SOLID FLAT P/C SPANDRELS· TO RESIST FORCE GENERATED BY 60° THERMAL GRADIENT.

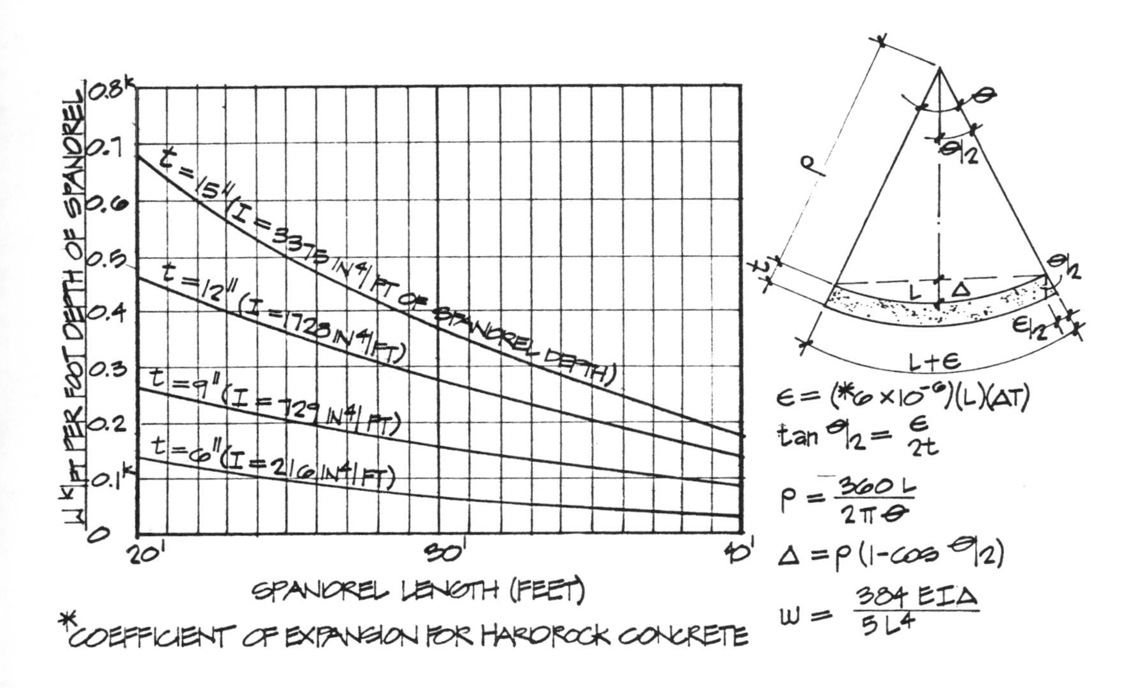

The diagram to the right shows the geometrical relationship which is the basis for the design charts shown here and on the previous page. While being empirical, and ignoring thermal lag due to the mass of the concrete, especially in thicker panels, it gives the designer an idea of the relative order of magnitude of movements and intermediate restraining forces. The above chart furnishes restraining force values which can be correlated to the specific design situation. Remember, that if you restrain these precast elements at intermediate points, be sure that the connections can develop the forces imposed by thermal gradients.

LOAD ANALYSIS SKETCHES FOR TYPICAL SPANDRELS

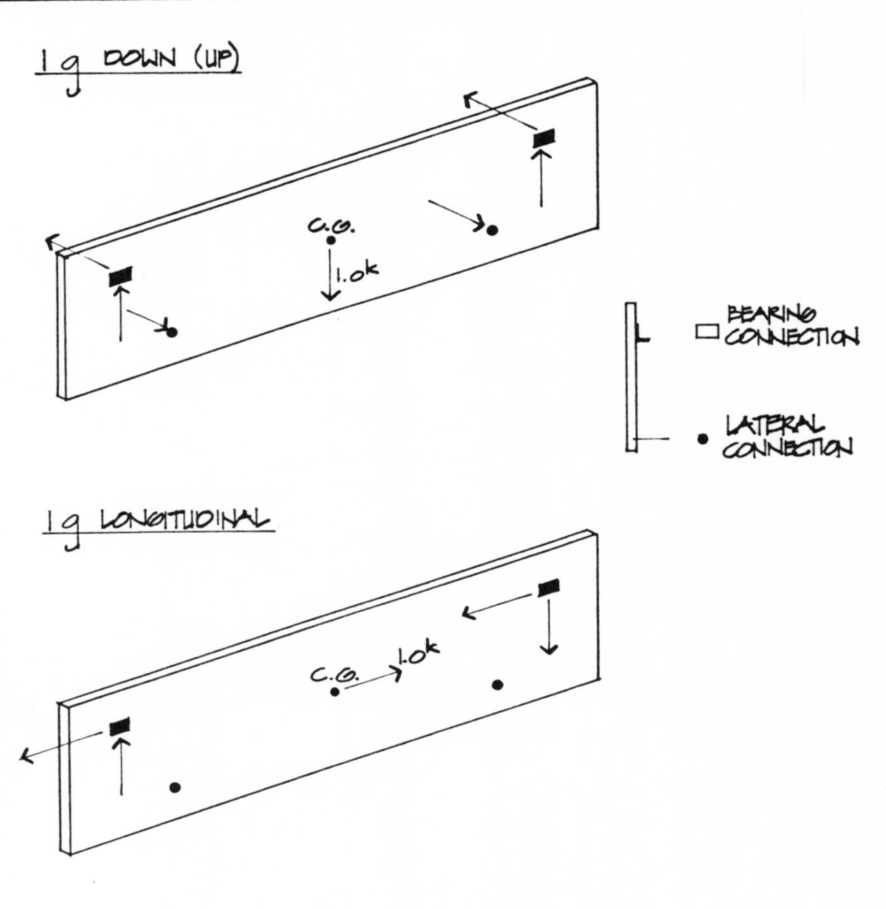

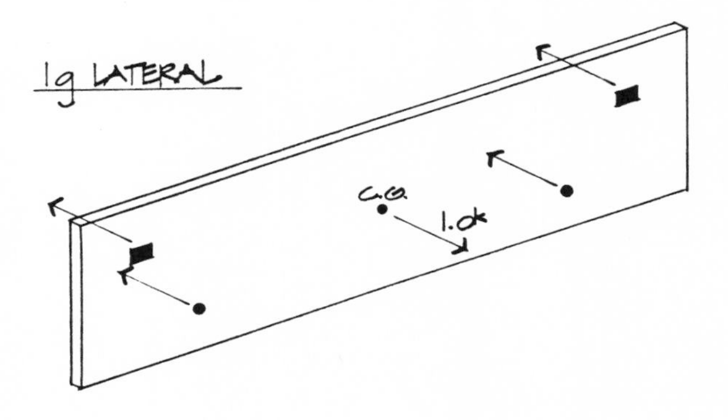

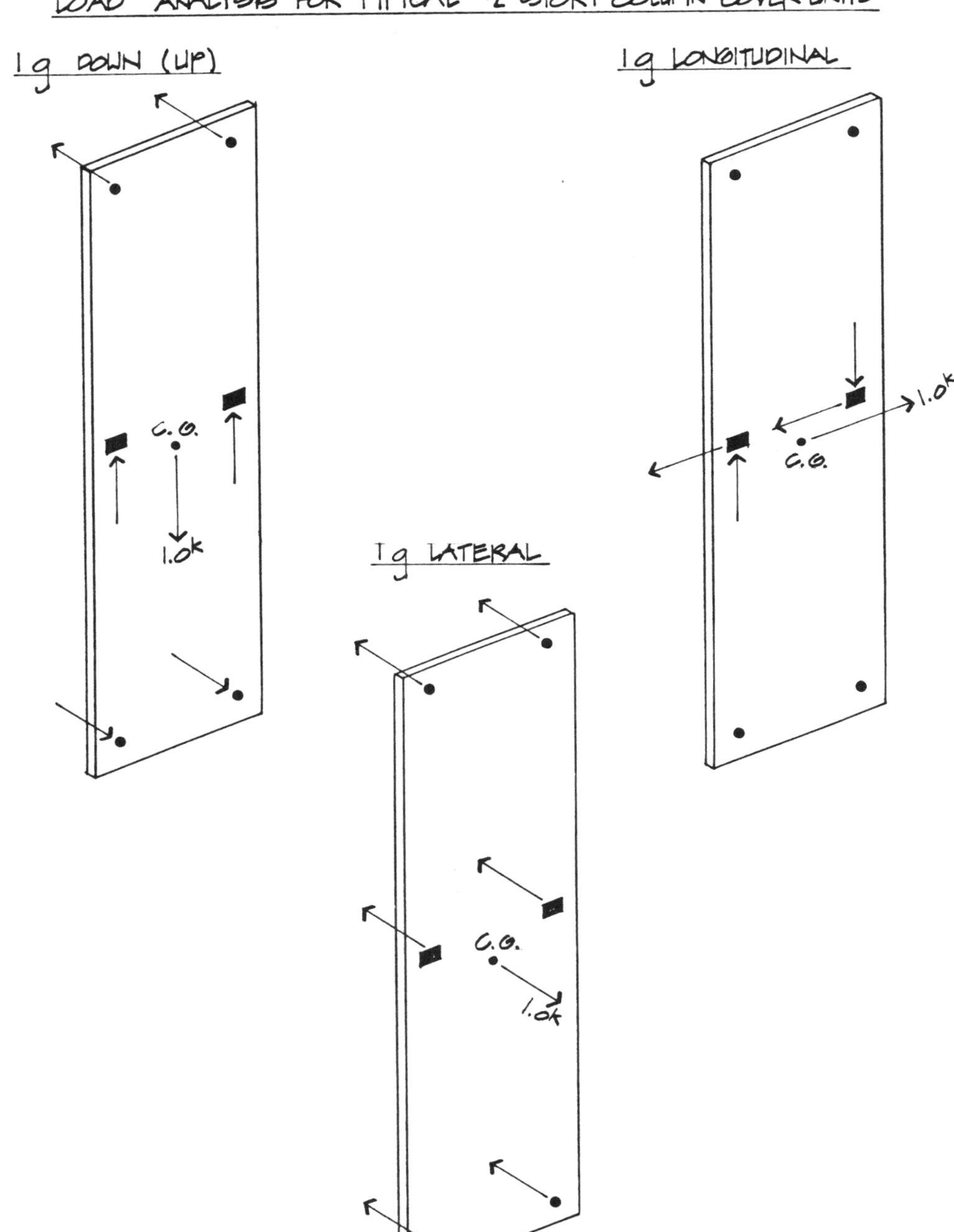
LOAD ANALYSIS FOR TYPICAL 2 STORY COLUMN COVER UNITS
1g DOWN (UP)
C.G.
1.0k
1g LONGITUDINAL
1.0k
C.G.
1g LATERAL
C.G.
1.0k

LOAD ANALYSIS SKETCHES FOR TYPICAL WINDOW PANELS

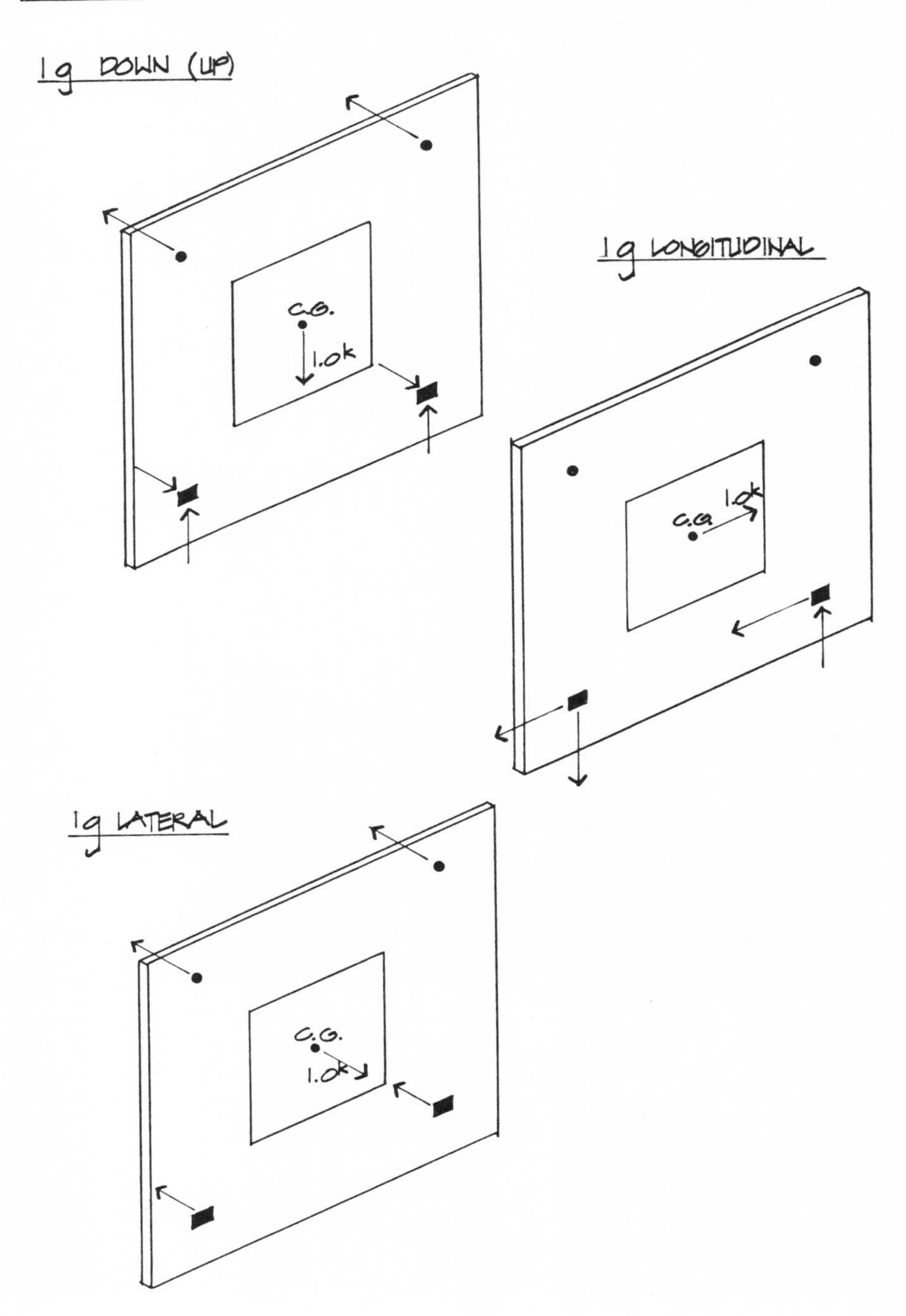

ALVARADO SEWAGE TREATMENT PLANT - UNION CITY

Standard 8 foot wide by 24" deep double tee being erected on the roof of this project. Standard production information exists for this shape, both in the form of manufacturer's literature and load tables in the PCI Design Handbook. In the absence of such information, the design procedures given in this chapter can save time and unnecessary trial designs.

GENERAL DESIGN PROCEDURE: STRUCTURAL PRESTRESSED CONCRETE FLEXURAL MEMBERS

The intent of this chapter is to give the designer a step-by-step procedure to use in designing a prestressed concrete flexural member. It is presented in such a manner that the design may be developed from the beginning point where all that is known is the design span and the design loading. Normally, some of the preliminary rough design "rules of thumb" and preliminary design relationships may be skipped when manufacturer's load tables, or the load-span tables given in Chapter 2 of the PCI Design Handbook are available. In all cases, however, when beginning the design of prestressed concrete flexural members, the known factors are:

a. Span
b. Superimposed dead and live loads and the loading arrangement
c. The strength of the concrete
d. The grade of the prestressing steel
e. Allowable concrete and strand stresses at release of prestress and in service
f. Approximate prestress losses
g. Camber/deflection requirements

The unknown factors are:

a. The size of the beam
b. The number of strands
c. The size of the strands
d. The strand profile/positioning in the member

The procedure gives a logical procedure to methodically zero in on the final design in as few steps as possible, with a minimum of trial designs. In the absence of manufacturer's product literature, the preliminary depth calculation quide-lines are used to select the size of the beam. Then, the strand information is developed with the balance of the preliminary design procedure. The rest of the process is self explanatory, and is given below:

DESIGN PROCEDURE - PRESTRESSED CONCRETE FLEXURAL MEMBERS

NOTATION

f_b = bending stress or bending stress - bottom fiber

f_t = tensile stress or bending stress - top fiber

f'_{ci} = concrete strength at release of prestress

f'_c = 28 day conc. cylinder strength

M_B = bending moment resisted by basic P/C section

MC = bending moment resisted by composite section

<u>Basic Section Moduli</u>

S_t = section modulus at top fiber

S_b = section modulus at bottom fiber

<u>Composite Section Moduli</u>

S_{tc} = section modulus at top fiber of P/C section

S_{bc} = section modulus at bottom fiber

F_o = initial strand tension, kips

F_i = prestress force in member at release, kips

F_f = final prestress force in member after losses, kips

f_{pu} = ultimate strength of prestressing steel

f_{ps} = stress in prestress strand at nominal (ultimate) strength of member

f_{se} = effective stress in prestressing steel after all losses have occurred.

e_c = strand eccentricity at ℄ span

e_e = strand eccentricity at the end of the member

A_{ps} = area of prestressing strand

A = area of prestressed concrete section

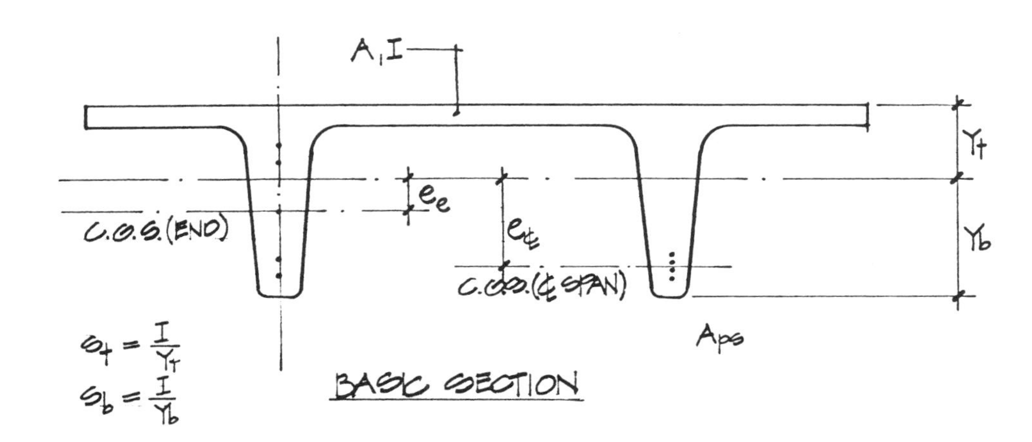

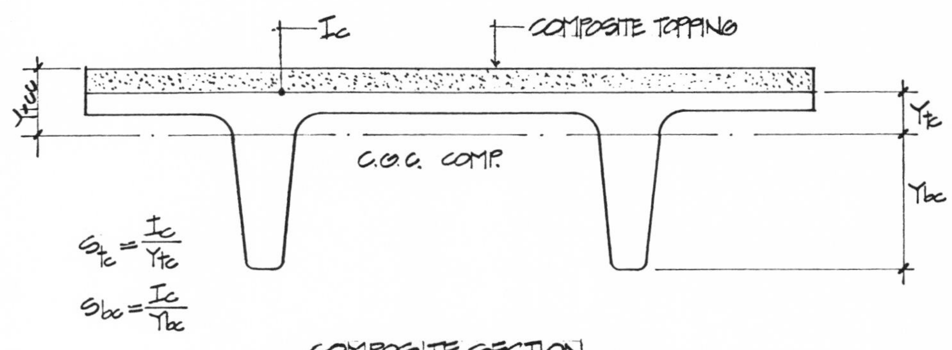

COMPOSITE SECTION

1. Preliminary Depth Calculation:

   Span/Depth Ratios should be as follows:

   a. Single Tee

      (1) Floor — L/D = 23
      (2) Roof — L/D = 29

   b. Double Tee

      (1) Floor — L/D = 26
      (2) Roof — L/D = 34

   c. Single or Double Tee - H20-44 Highway Loading

      L/D = 18

   d. Single or Double Tee - Plaza Load (S_{D+L} = 350 PSF)

      L/D = 12

   e. Hollow Core Plank

      (1) Floor — L/D = 40
      (2) Roof — L/D = 46

   f. Girders

      L/D = 18

2. Approximate S_{bc} in^3 required = $6 \times M_{SD+L}$ (k-ft.)

3. Procedure

   a. Calculate section properties for the trial section, if not already known.

   b. Determine $f_b = \frac{M_B}{S_b} + \frac{M_C}{S_{bc}}$

c.* If $f_b > f_t + \frac{f'ci}{2}$, strands must be depressed.

$(f_t = 6\sqrt{f'c})$

d.* If $f_t + \frac{f'ci}{2} < \frac{MBM}{6S_b} + \frac{MTPG}{S_b} + \frac{MC}{S_{bc}}$, then a deeper section must be used.

for f'c = 5000 PSI (f'ci = 3500 PSI)

$$f_t + \frac{f'ci}{2} = 2.2 \text{ KSI}$$

for f'c = 6000 PSI (f'ci = 4200 PSI)

$$f_t + \frac{f'ci}{2} = 2.6 \text{ KSI}$$

e. Of course, manufacturer's load/span tables, where available, allow us to select the proper section without requiring the above preliminary trials.

f. Find effective prestress force required. (At this point in the design, we know everything but the required effective prestress force. A very close approximation for prestress eccentricity "e" may be assumed based upon strand profile and required concrete cover at the bottom of the member.)

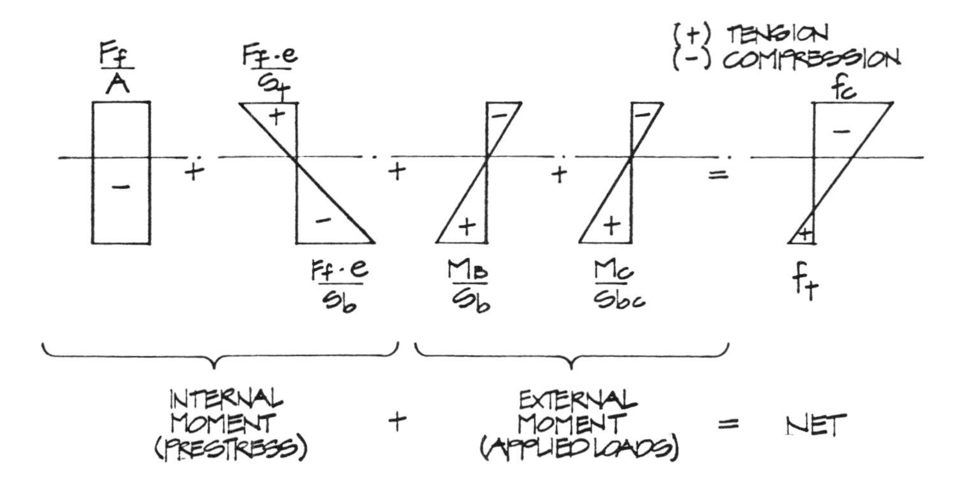

* See derivations at the end of this chapter.

BASIC RELATIONSHIPS

$$f^{top} = -\frac{F_f}{A} + \frac{F_f \cdot e}{S_t} - \frac{M_B}{S_t} - \frac{M_c}{S_{tc}} = f_c \qquad (1)$$

$$f_{bottom} = -\frac{F_f}{A} - \frac{F_f \cdot e}{S_b} + \frac{M_B}{S_b} + \frac{M_c}{S_{bc}} = f_t \qquad (2)$$

WORKING WITH EQUATION (2):

$$-F_f\left(\frac{1}{A} + \frac{e}{S_b}\right) = f_t - \frac{M_B}{S_b} - \frac{M_c}{S_{bc}}$$

$$F_f = \frac{\frac{M_B}{S_b} + \frac{M_c}{S_{bc}} - f_t}{\frac{1}{A} + \frac{e}{S_b}}$$

g. Select number of strands and check service load stresses.

h. When end stresses cannot be satisfied and it is not feasible to break bond on the strands at the ends, or use top end mild steel, recalculate required prestress, substituting: $M_{0.4L} = 0.96M_{℄}$

$e_{0.4L} = 0.90e_c$ for shallow drape

$e_{0.4L} = 0.80e_c$ for steep drape

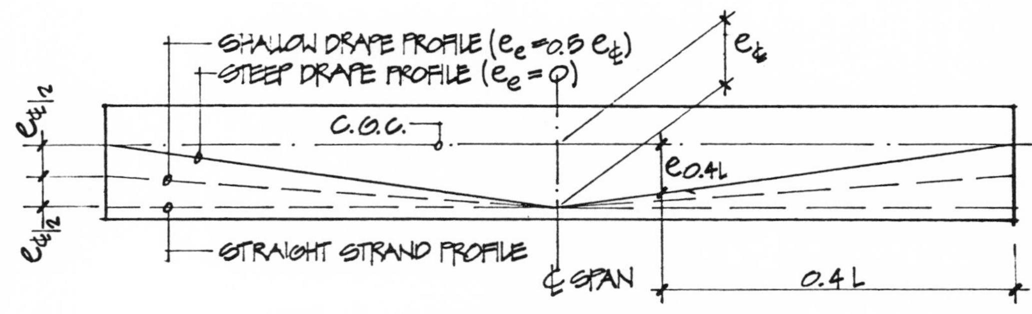

$e_{0.4L} = 0.9\, e_{℄}$ FOR SHALLOW DRAPE
$e_{0.4L} = 0.8\, e_{℄}$ FOR STEEP DRAPE

STRAND PROFILES

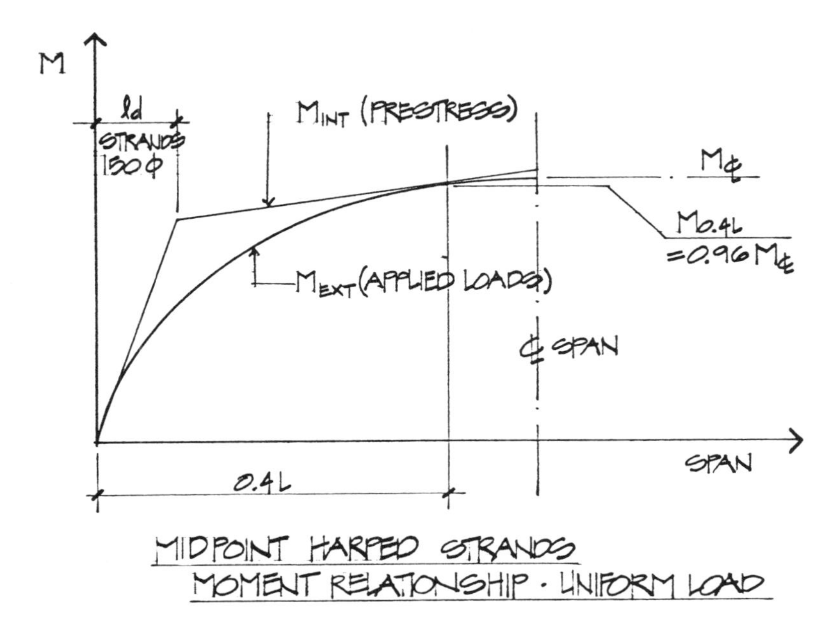

MIDPOINT HARPED STRANDS
MOMENT RELATIONSHIP · UNIFORM LOAD

4. Detailed Design (See UBC-79 Section 2618)

Check service load stresses at critical points:

a. Concrete Stresses:

At Release

Compression	0.60 f'ci
Tension (except at ends)	$3\sqrt{f'ci}$
Tension at ends (Simply supported members)	$6\sqrt{f'ci}$

Final Design Condition

Compression	0.45 f'c
Tension	
Hollow core slabs) Flat slabs)	$6\sqrt{f'c}$
Beams & stemmed) members)	$*12\sqrt{f'c}$

* When deflections are checked in accordance with bilinear strain relationships. (see page 3-25-PCI Design Handbook)

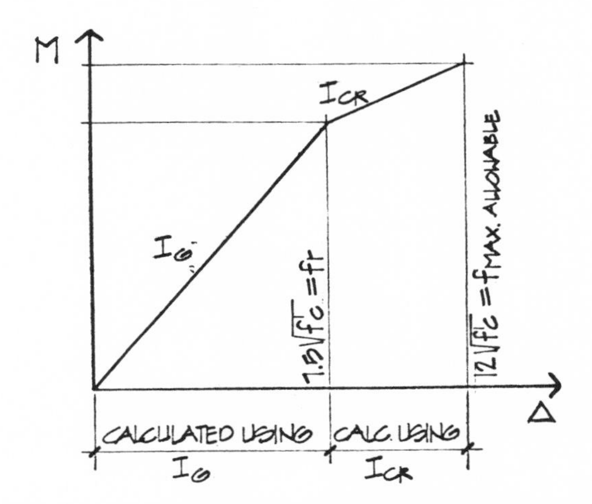

BILINEAR MOMENT-DEFLECTION RELATIONSHIP

Stress Losses

Stress losses in pretensioned members result from four principle causes:

1. Elastic Shortening - resulting from the deformation and resultant strain decrease when the prestress force "P" is released into the cross sectional area "A"

2. Relaxation of Steel - decrease in stress in the prestressing strand due to creep.

3. Creep of Concrete - time dependent phenomena which is the redistribution of the concrete molecules causing shortening under sustained load. Just as for relaxation of the steel, the magnitude of creep depends upon the stress level at the various positions in the cross section.

4. Shrinkage of Concrete - shortening due to moisture loss from concrete.

For most typical design applications, stress losses at release can be conservatively taken as 10% of the initial prestress; final losses at 78% for hard rock concrete and 75% for sand-lightweight concrete. For members with high span-depth ratios and prestress levels or heavy sustained loads, a more detailed investigation should be carried out. One such method is given in the 2nd Edition of the PCI Design Handbook on p. 3-29.

b. Strand Stresses

Initial Tendon Stress: $0.70\ f_{pu}$

@ release $F_i = 0.9 \times 0.70\ f_{pu} \times A_{ps}$

Final Condition

Hardrock concrete (22% losses)

$(F_f = f_{se} \times A_{ps}$ $\quad F_f = 0.78 \times .70\ f_{pu} \times A_{ps}$

Lightweight concrete (25% losses)

$F_f = 0.75 \times .70\ f_{pu} \times A_{ps}$

For 1/2" Ø - 270 ksi strand: (A_{ps} = 0.1531 in ^{2}/ strand)

$F_o = 0.70 \times 0.1531 \times 270 = 28.9\ ^k$/ strand

$F_i = 0.9 \times 28.9 = 26.0\ ^k$

$F_f = 0.78 \times 28.9 = 22.5\ ^k$ (H.R. Conc.)

(or) $F_f = 0.75 \times 28.9 = 21.7\ ^k$ (sand-Lt. Wt.) (Conc.)

Check Stresses at Critical Points

@ span for straight strands
@ 0.4 L for depressed strand

$$f^{top} = -\frac{F_f}{A} + \frac{F_f \cdot e}{S_t} - \frac{MB}{S_t} - \frac{Mc}{S_{tc}}$$

$$f_{bot} = -\frac{F_f}{A} - \frac{F_f \cdot e}{S_b} + \frac{MB}{S_b} + \frac{Mc}{S_{bc}}$$

Sign convention:

(-) compression
(+) tension

b. Check ultimate strength (See PCI Handbook, pp 3-4; 3-53)
c. Check shear-proportion web reinforcement (See PCI Handbook, pp 3-20 to 3-22; 3-58 to 3-65)
d. Check camber and deflection (See PCI Handbook, pp 3-23 to 3-38; 3-66 to 3-68)

How to take advantage of inherent capacity of prestressed concrete designs often overlooked - without requiring high un-economical concrete release strengths or final design strengths:

1. Use transformed section properties. This can increase "I" and "S" by 10% or more for deep flexural members.

2. Break bond at ends of some strands so that top end tension ($>6\sqrt{f'c}$) does not govern design, or dictate a higher release strength.

   Caution: Take strand development length into account when calculating flexural resistance and shear strength near end of member. Bottom mild steel may be required plus additional stirrups at end.

3. Use mild steel or non-prestressed strand in bottom to satisfy ultimate strength requirements, when additional prestressing steel would result in excessively high release strengths.

4. Use strain compatibility in calculating fps for use in ultimate strength equation.
   ϕ Aps fps (d-a/2) = Mu

5. Let bottom tension under all loads approach $12\sqrt{f'c}$, consistent with allowable deflections. (Use bilinear moment-deflection relationship).

6. Take allowable live load reduction where permitted. (See UBC-79, Section 2306)

7. Use midpoint or third point temporary shoring under long span or heavily loaded composite flexural members to take advantage of the composite section to carry the deck dead loads as well as live loads; use top and bottom mild steel as required to increase "I" and satisfy deflection/camber requirements.

PRESTRESSED CONCRETE FLEXURAL DESIGN

DERIVATION OF PRELIMINARY DESIGN RELATIONSHIPS SERVICE LOAD STRESSES

1. FOR STRAIGHT STRANDS

@ BOTTOM END RELEASE:

$$0.9\left(-\frac{F_o}{A}-\frac{F_o \times e}{S_b}\right)=-0.6 f'_{ci} \qquad (1)$$

@ BOTTOM ℄ SPAN - MAXIMUM SERVICE LOAD CONDITION

$$0.78\left(-\frac{F_o}{A}-\frac{F_o \times e}{S_b}\right)+\frac{M_B}{S_b}+\frac{M_C}{S_{bc}}=f_t \qquad (2)$$

$$-\frac{F_o}{A}-\frac{F_o \times e}{S_b}=-\frac{0.6}{0.9}f'_{ci}=\frac{f_t-\frac{M_B}{S_b}-\frac{M_C}{S_{bc}}}{0.78}$$

$$f_t+\frac{(0.6)(0.78)}{(0.9)}f'_{ci}=\frac{M_B}{S_b}+\frac{M_C}{S_{bc}}$$

$$f_t+f'_{ci}/1.923=\frac{M_B}{S_b}+\frac{M_C}{S_{bc}}$$

$$\therefore f_t+\frac{f'_{ci}}{2}\cong\frac{M_B}{S_b}+\frac{M_C}{S_{bc}}$$

2. FOR DEPRESSED STRANDS - MAXIMUM HARP

$e_{℄}=e$ $\qquad M_{0.4L}=0.96M_{℄}$

$e_{END}=0$

$e_{0.4L}=0.8e$

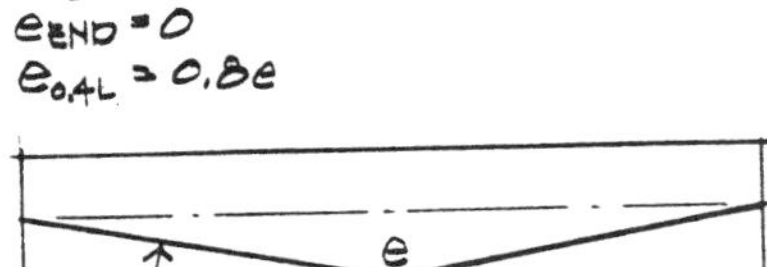

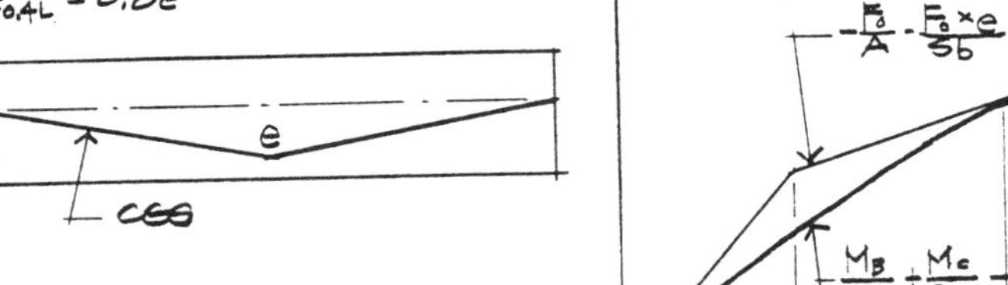

@ ℄ RELEASE (BOTTOM)

$$0.9\left(-\frac{F_o}{A}-\frac{F_o \times e}{S_b}\right)+\frac{M_{BM}}{S_b}=-0.6 f'_{ci} \qquad (1)$$

@ 0.4L (BOTTOM) - MAXIMUM SERVICE LOAD CONDITION

$$\frac{0.78\left(-\frac{F_o}{A}-\frac{F_o \times e}{S_b}\right)}{1.1}+\frac{0.96M_B}{S_b}+\frac{0.96M_C}{S_{bc}}=f_t \qquad (2)$$

$$-\frac{F_o}{A}-\frac{F_o \times e}{S_b} = -\frac{0.6}{0.9} f'_{ci} - 1.111\frac{M_{BM}}{S_b} = 1.410\, f_t - 1.354\frac{M_B}{S_b} - 1.354\frac{M_c}{S_{bc}}$$

$$-0.667 f'_{ci} - 1.111\frac{M_{BM}}{S_b} = 1.410 f_t - 1.354\frac{M_{BM}}{S_b} - 1.354\frac{M_{TPS}}{S_b} - 1.354\frac{M_c}{S_{bc}}$$

$$\frac{0.243}{(1.354-1.111)}\frac{M_{BM}}{S_b} + 1.354\frac{M_{TPS}}{S_b} + 1.354\frac{M_c}{S_{bc}} = 1.410 f_t + 0.667 f'_{ci}$$

DIVIDE BOTH SIDES BY 1.382

$$0.176\frac{M_{BM}}{S_b} + 0.980\frac{M_{TPS}}{S_b} + 0.980\frac{M_c}{S_{bc}} = 1.020 f_t + 0.483 f'_{ci}$$

$$\boxed{\therefore \frac{M_{BM}}{6S_b} + \frac{M_{TPS}}{S_b} + \frac{M_c}{S_{bc}} \cong f_t + \frac{f'_{ci}}{2}}$$

SOLID PRECAST CONCRETE WALL PANELS

TRANSAMERICA PYRAMID - SAN FRANCISCO

Precast concrete cladding panels form the exterior of this San Francisco landmark. White cement and light colored coarse and fine aggregates achieve harmony with the San Francisco tradition as being the alabaster city. The panels were lightly sandblasted in the plant, and then erected on this 48 story - 853 foot high tower. In the background can be seen the tower crane used to erect precast panels on 601 Montgomery, under construction in the summer of 1979.

Solid precast concrete wall panels are modular elements used to both form the envelope of the building and to express the character of the exterior design. Solid precast concrete wall panels fall into 3 general categories:

(1) Non-load bearing cladding panels, designed to support only their own weight and wind or seismic forces perpendicular to the panel. Cladding panels may be designed as spandrel panels, solid wall panels, or pierced window panels.

(2) Non-load bearing shear walls, designed to transfer lateral wind or seismic forces from horizontal diaphragms to the foundation or another panel element.

(3) Load-bearing wall panels, designed to support vertical loads from the building framing system. Load-bearing panels may also be designed to transmit lateral forces to the building foundations.

With precast concrete the architect has the opportunity to bring out the full sculptural potential of the material and to freely express his design concept for the building. The three dimensional form of precast concrete is limited only by its function and the creativity of the designer. Due to its industrialized manufacturing process and the repetition resulting from correct modular design, forms become possible with precast concrete which are not economically feasible with cast-in-place concrete. Precast concrete is in reality the dream material for the architect.

ADVANTAGES

Plant cast precast concrete exterior panels have several advantages which justify their use.

(1) Fast construction. Units are factory cast on a daily cycle, and are stockpiled while other necessary site work and building framework is completed. Quick erection of precast panel units assures a savings in interim financing costs with early completion realized.

(2) Quality control with plant cast units made in precision made molds achieving realization of close tolerances and uniform finish quality.

(3) Durability. High strength dense concrete quarantees "rock of ages" life expectancy.

(4) Finishes. A variety of surface textures and finishes are attainable.

(5) Economy. Resulting from modular design and multi-use of forms.

(6) Fire Resistance. Is inherent with concrete construction.

(7) <u>Low Maintenance</u>. Proper selection of exposed aggregate finishes coupled with proper panel design and drip details eliminate long term staining and streaking from accumulated surface dirt in polluted air associated with city environments.

(8) <u>Energy Efficient</u>. Construction with the inherent thermal lag in concrete elements.

Exterior finishes are available in a wide range of textures from smooth to deeply exposed aggregate. Ranges of texture in between are achieved by acid etching to lightly reveal the sand and aggregate at the surface, and light or medium exposed aggregate by sandblasting or with the use of retarders sprayed on the mold before casting. After stripping, the aggregate is revealed by washing the panel surface with high pressure water hoses. The retarder material prevents the cement hydration reactions from occurring at the surface, thereby allowing this powdery material to be washed away. Retarded finishes leave the coarse aggregate smooth, whereas sandblasting dulls the stone. The cement matrix may be white, or grey, or colored by the use of coloring admixtures added to the mix. Smooth grey finish on flat surfaces should be avoided due to the difficulty in preventing streaking and mottling of the surface, especially with steam curing. Textured surfaces may be obtained with the use of form liners.

<u>AMBASSADOR COLLEGE AUDITORIUM - PASADENA</u>

Column covers, soffit panels, and parapet panels are precast concrete with white cement and milky quartz aggregate with a medium sandblasted finish.

ARCHITECTURAL DESIGN CONSIDERATIONS

Prior to the design of wall panels the architect should visit one or more precasters who produce wall panels or architectural precast concrete. He should visit their manufacturing plants, as well as projects being erected, if possible. In doing this, the designer can become familiar with the manufacturing process including the fabrication of the molds, the problems in casting and finishing, the methods of plant and jobsite handling and the methods used in connecting the panels to the structure. This is very important in order to fully understand the material and its proper utilization.

In the development of the working drawings the architect should work closely with the precaster. The engineering of the panels is done by the architect in conjunction with his structural engineer. The precaster will subsequently prepare shop drawings and necessary calculations to assure that the panel will be properly handled from manufacture through shipping, to erection. He will also design the handling and erection inserts. These shop drawings are submitted to the architect for shape approval prior to the manufacturing of molds. For large projects, a mock-up of a typical panel is first fabricated and approved by the architect as to finish and design. This mock-up serves as the standard throughout manufacture.

The precaster will either handle the erection of the units with his own crews or will subcontract and supervise the erection. Hauling may also be subcontracted, but nonetheless, the precaster is responsible for the panels until they are installed on the building and accepted by the architect through his general contractor.

In designing wall panels the following must be considered; (a) form design and fabrication, (b) concrete placement, (c) lifting the units from the forms (stripping(, (d) plant handling during finishing and storing, (e) transporting the units to the jobsite, (f) erection and (g) connection to the structure including temporary bracing. In the manufacturer's analysis of the units the handling inserts and panel reinforcing are designed for: (1) form removal, (2) in plant handling, (3) transporting, (4) erection, (5) connection to the structure and (6) in place loading. Often. the stresses on the panel during form removal, handling and erection exceed those caused by in place loads. The inserts and reinforcing must be designed for each of the above situations. Dual use of inserts is considered wherever possible to reduce costs.

(a) Form Design

For simple shapes or where there are many casts and the form cost is justified, steel forms are used. Where shapes are complex, wood, fiberglass or concrete forms are used. With fiberglass forms, up to 70 reuses are possible with minor rework depending upon the complexity of the shape. The majority of wall panels and architectural precast concrete units are cast in either fiberglass or concrete forms. Where the shape requires it the form may be in parts, assembled and disassembled for each cast. The primary consideration is the architect's design concept and how to accomplish this design concept in the most efficient manner. The final design should consider maximum form reuse, keeping the number of different shapes to a minimum, ease of removal from the form, and the quality of the edges or changes in direction of the various planes desired.

The concrete unit is normally cast in a horizontal flat position with the exposed, textured or sculptured face down and the flat inside face up. As such, the vertical faces of the form must be sloped (possess draft) for ease of removal from the mold. Where plug sections of forms, or back formers are used, these must also be sloped for both ease of removal and to avoid pockets or voids when casting the concrete. An exception to this rule is when the side forms are in removable sections. If the panel is pretensioned, consideration must be made for shortening of the concrete during detensioning. Outside forms present few problems, but inside forms unless removed with cause the unit to crack during the detensioning due to binding on the form. The magnitude of the draft is subject to the width of the section (surface friction vs strength of the section.

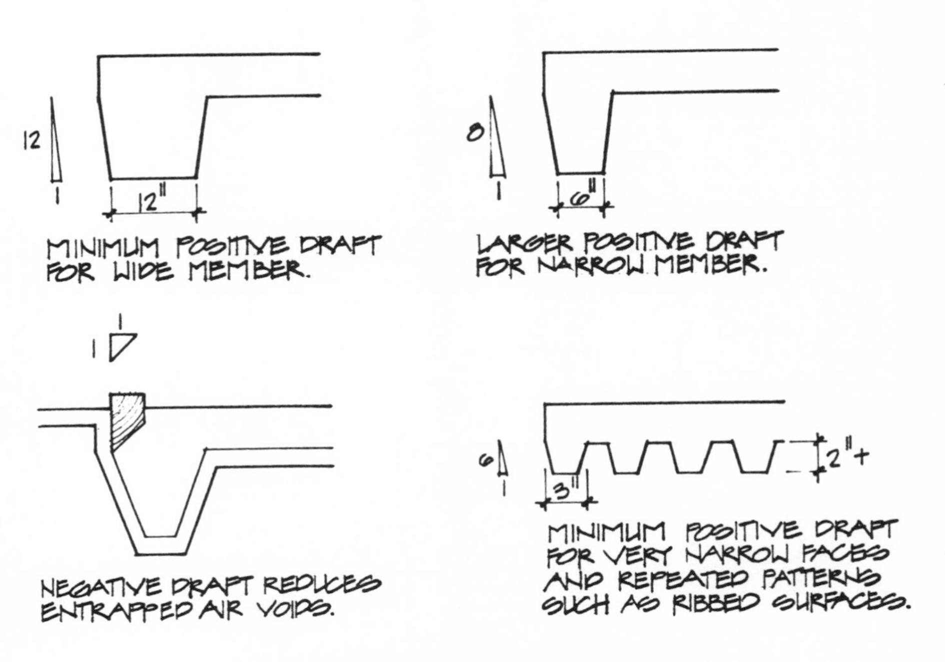

RECOMMENDED MINIMUM DRAFTS

Where edge grooves are required, removable side forms or expendable form materials are used which are pulled out after the concrete unit is removed from the basic form.

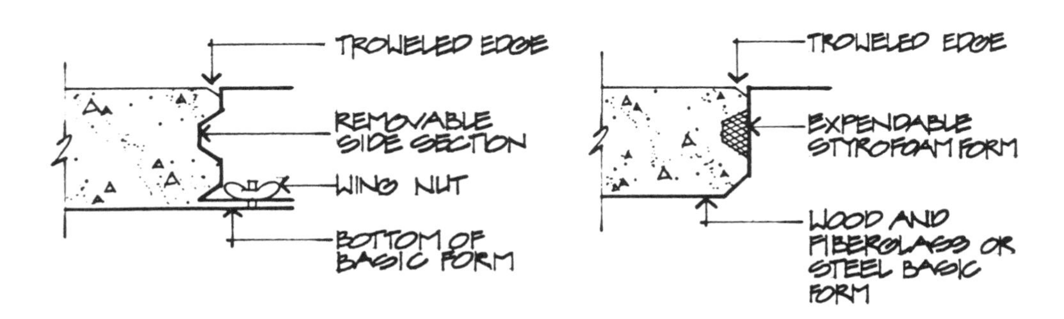

In the final design of the shape, sharp edges and corners should be avoided and replaced with curved or chamfered edges because of possible edge damage in form removal and in handling. When the edge is sharp, only fine aggregate collects there and weakens the edge since larger aggregate is not present. Also, voids occur due to the plugging action of larger aggregate. As concrete shrinks during curing or otherwise changes volume, stresses build up at sharp corners which can cause cracking as well as binding in the form. As such all inside corners should have a radius or chamfer. The recommended minimum is 3/8".

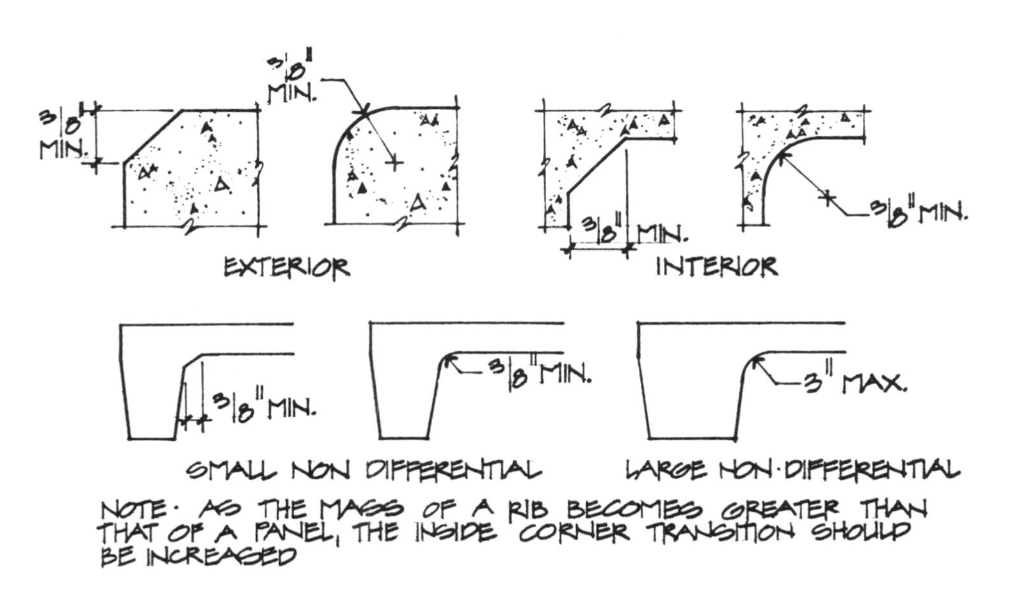

The casting surface of the forms must be smooth and adequately reinforced to prevent form deflection. The forms are usually made up and coated with fiberglass to the shape of the desired section. However, complex forms are made by a sprayed on fiber-glass-fiber reinforced coating applied to a pattern of the desired shape. The form is released from the pattern and hand finished with a gel coat and sanded. It is also reinforced during and after the spray-up process. Another excellent method for making molds of complex shapes is to cast a concrete mold around a full-sized model of the desired panel.

b. Concrete Placement

The concrete strength used for non-prestressed wall panels is usually 5 ksi. This high strength results from the minimum cement content desirable from the standpoint of durability, usually 6½ to 7 sacks per C.Y. If the wall panel is pretensioned, sand-light weight concrete of 5 ksi or normal weight concrete of either 5 or 6 ksi is used. The maximum coarse aggregate size is 3/4". For exposed aggregate concrete, the coarse aggregate where the surface is to be exposed may range from 3/8" to 1/2" subject to the desired design effect. Usually the concrete is of the same mix throughout the panel. Where an expensive, special exposed aggregate is to be used, two mixes are often used within the same panel. The special aggregate mix (face mix) is placed as a thin layer in only those areas to be exposed. While still fresh, a more economical back-up mix is poured, with care being taken to blend the two to prevent separation due to laitance. When two different mixes are used the designer should note the demarcation line on the drawings.

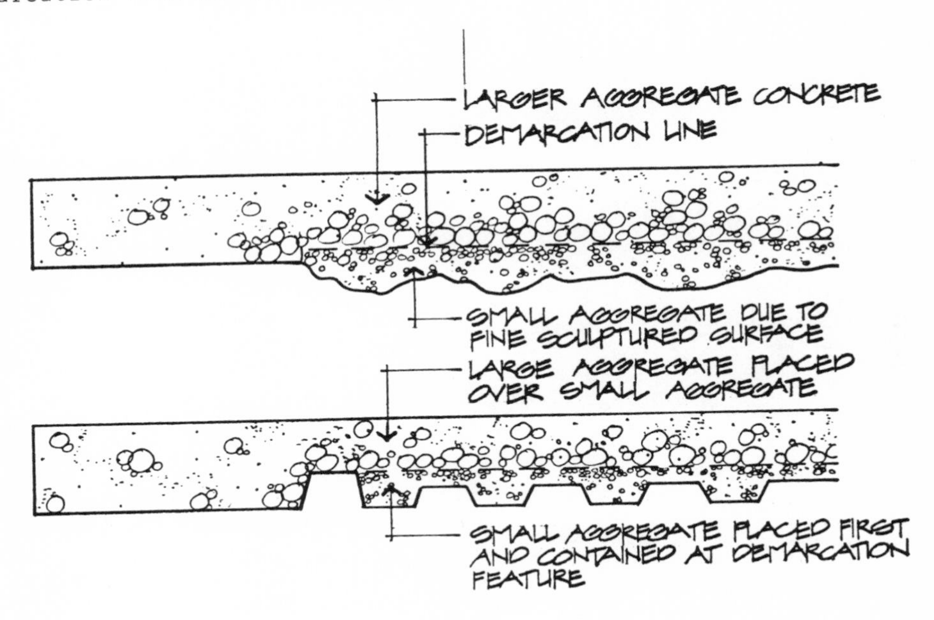

Where part of the surface is to receive a fine sculptural detail, two mixes can be used with the mix in the sculptural area containing a smaller maximum coarse aggregate size. While concrete is being placed in the forms, either internal vibration with the use of spud vibrators or external pneumatic mold vibrators are used to assure dense compacted concrete.

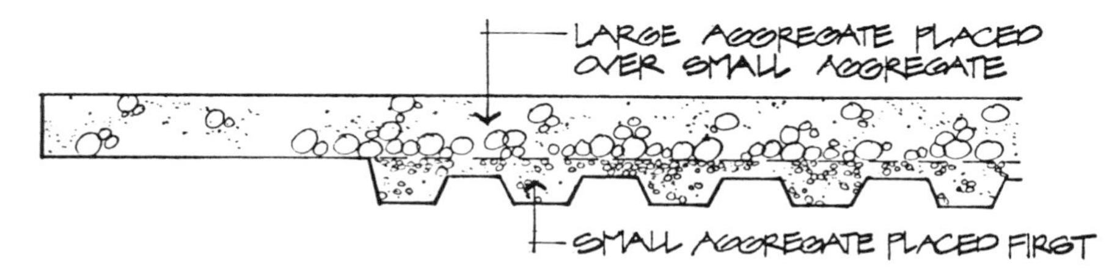

AMBASSADOR COLLEGE AUDITORIUM - PASADENA

Note the fine detail and excellent alignment of the precast column covers and soffit panels, achieved with quality precision molds and high vibration of low slump concrete with small coarse aggregate.(Photo by Eugene Phillips)

(c) Removal From the Form and Handling

The primary consideration in the design of wall panels is the architect's design concept and how to accomplish this design concept in the most efficient manner. This is the underlying principle behind the design of any wall panel. In finalizing the panel design, consideration must be given to the handling of the panel from stripping through to erection at the jobsite. The stages consist of the position the panel is cast in, how it is to be removed from the form and how it is handled from finishing to storage, to shipment and to erection. This sequence is studied and the panel is structurally designed for each step and the inserts positioned accordingly. In each of these steps, the panel is kept equilibrium and balanced about its center of gravity. At no time during handling may the stresses in any portion exceed that which would produce cracking. Also a factor of safety for impact and handling must be considered. The panel and each section is designed as an uncracked section neglecting the contribution of the reinforcing steels. (See Section 24 for handling criteria) Whenever possible, four pick up points are used in the horizontal flat position and two in the vertical position. Sometimes the size, weight or configuration is such that more than four pick up points are necessary. In stripping the panel from the mold, the pick up points must be in balance with equal load to each sling; otherwise the panel will bend or rotate and bind in the form with possible cracking and form distortion.

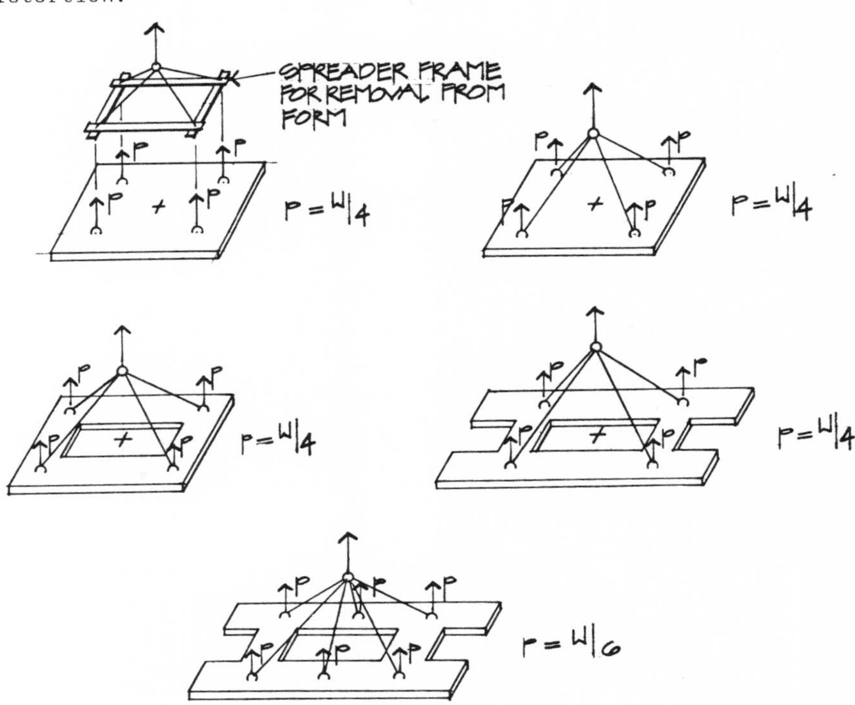

Some precasters will use spreader beams, traveling blocks, slings with fixed cable lengths or vacuum lift systems for lifting the units out of the molds. Any openings or major projections must be such that beam sections within the panel can be used to transfer the panel weight without cracking to the pick up points. The beam section must be large enough to keep the stresses induced by handling, bending tension and direct tension within allowable limits.

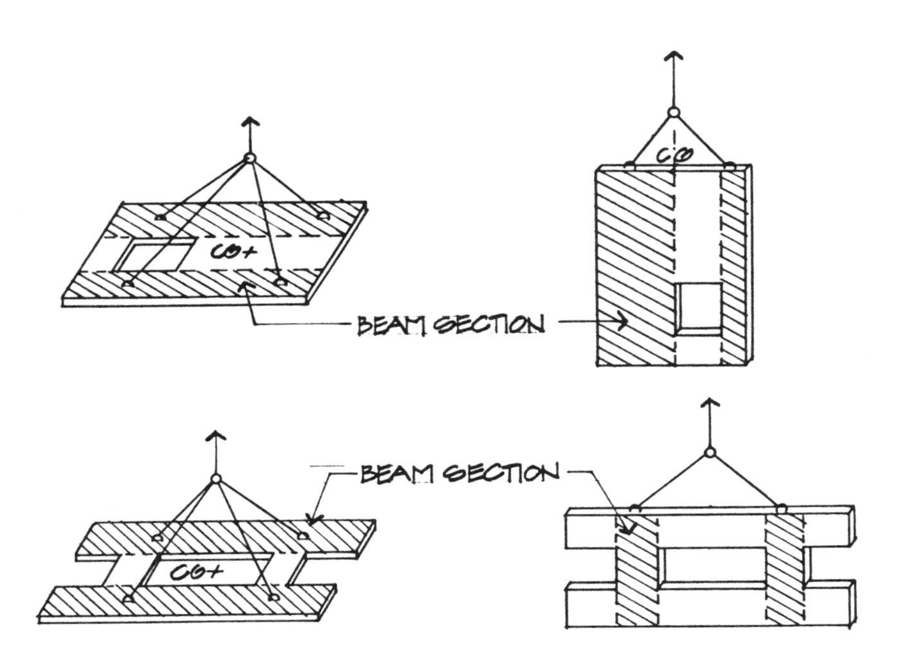

The use of strong backs in handling the panels is undesirable and should be avoided. When designing the panel shape the architect must consider handling.

Some panel positions occurring during the total handling process are shown on the following pages.

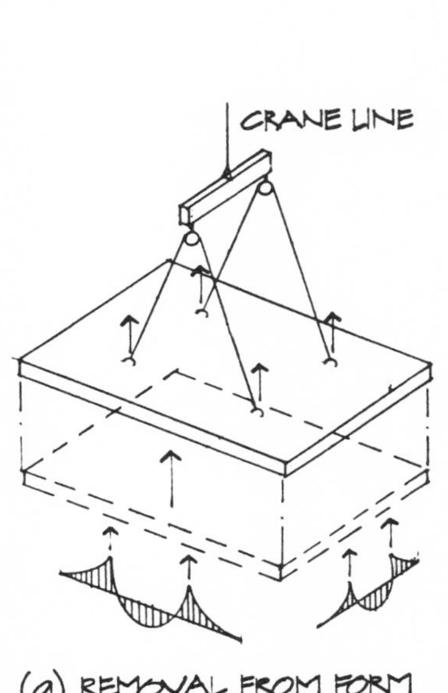

(a) REMOVAL FROM FORM
SCULPTURED PANEL

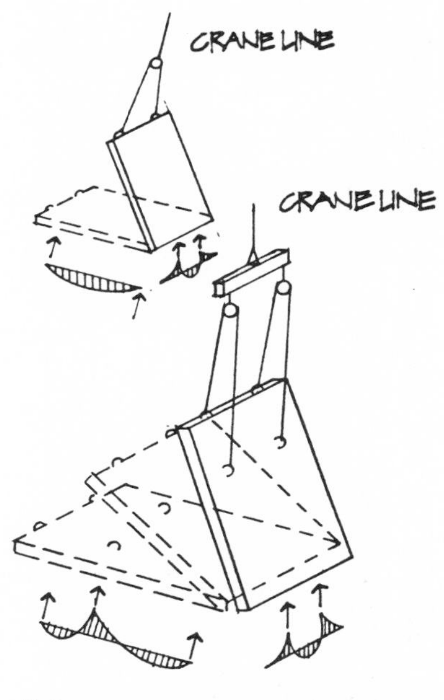

(b) REMOVAL FROM FORM
PLAIN PANEL · TILT UP

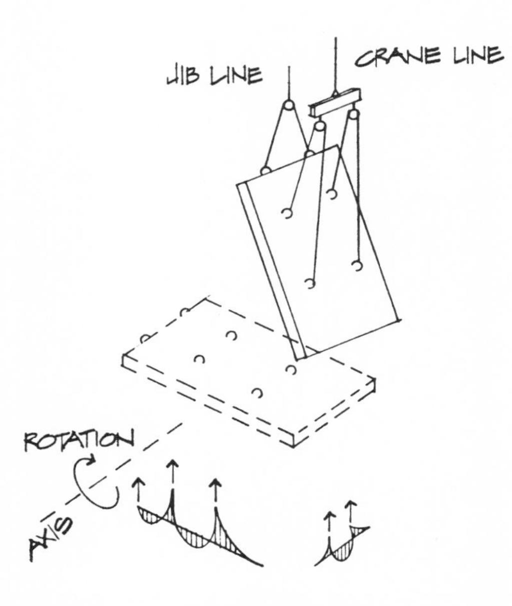

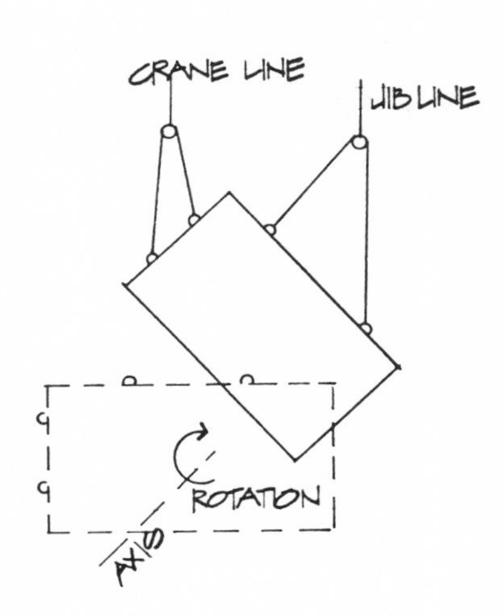

(c&d.) PANEL ROTATION

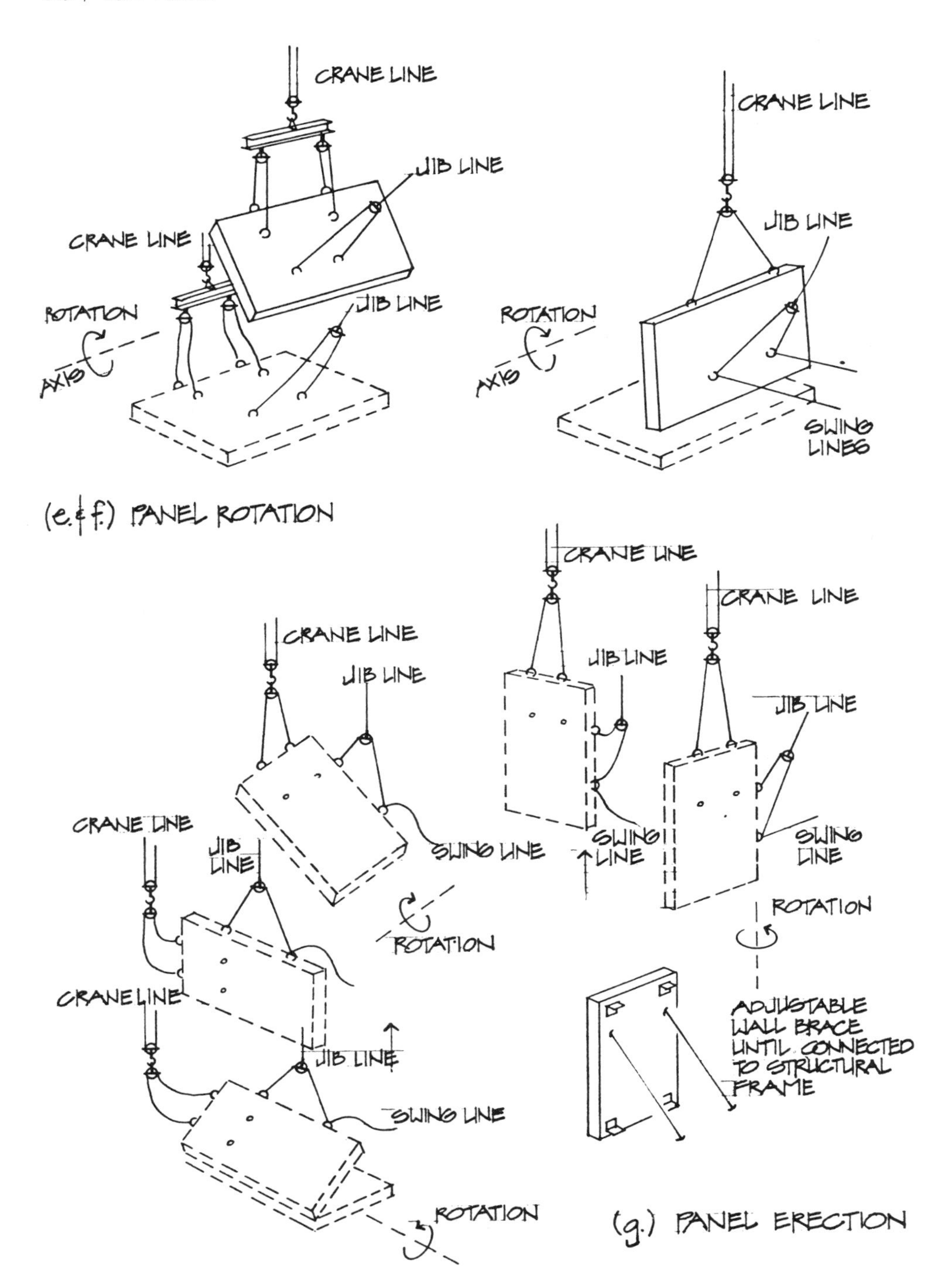
CRANE LINE
JIB LINE
CRANE LINE
ROTATION
AXIS
JIB LINE
CRANE LINE
JIB LINE
ROTATION
AXIS
SWING LINES
(e.&f.) PANEL ROTATION
CRANE LINE
CRANE LINE
JIB LINE
JIB LINE
CRANE LINE
JIB LINE
SWING LINE
SWING LINE
JIB LINE
SWING LINE
CRANE LINE
JIB LINE
ROTATION
CRANE LINE
ROTATION
JIB LINE
ADJUSTABLE WALL BRACE UNTIL CONNECTED TO STRUCTURAL FRAME
SWING LINE
ROTATION
(g.) PANEL ERECTION

STRUCTURAL DESIGN CONSIDERATIONS

On the following sixty-odd pages, the design process described on pp. 177-188 is applied to various typical architectural and structural wall elements. Here will be demonstrated in practical example form the current philosophy and procedure recommended to be used in designing reinforcing and connections for solid precast concrete wall panels.

EMBARCADERO CENTER - SAN FRANCISCO

Four high rise towers comprise this landmark complex, all cladded with 2 story precast concrete panels. Steel molds were used to produce these panels.

595 MARKET STREET - SAN FRANCISCO

34 foot by 5'-3" deep spandrels band this 30 story building, completed in 1979. The old gives way to the new as the modern architecture of the Financial District invades the Mission Street Redevelopment Area.

SPANDREL PANELS

Spandrel panels are long narrow horizontal elements that span from column to column at the elevation of the floor and roof planes. They enclose the floor framing system and fill the space between the finished floor to the bottom of the fenestration or to the parapet height, as in open parking structures. The spandrel may be a non-load bearing cladding element, or may be designed to carry the tributary load from the floor system. The space between units is usually filled with glass or window units, as in office buildings, educational buildings, residential buildings, condominiums, apartments, and hotels. Spandrels are supported near or at their ends, on column corbels, or by edge beams or the floor system itself. One aspect of the structural design of spandrels often overlooked is the allowance for bowing due to thermal differentials between the outer and inner surfaces of the panel. The resulting movement should be able to be accommodated by the windows or other intermediate infill materials; conversely, if intermediate lateral connectors are used, they should be able to develop the forces induced by restraint of the thermal gradient. The following design example demonstrates the development of intermediate connectors to restrain against the thermal forces induced.

595 MARKET STREET - SAN FRANCISCO

Typical spandrel arrangement. Precast concrete was made with limestone aggregate and white cement, with a light sandblasted finish.

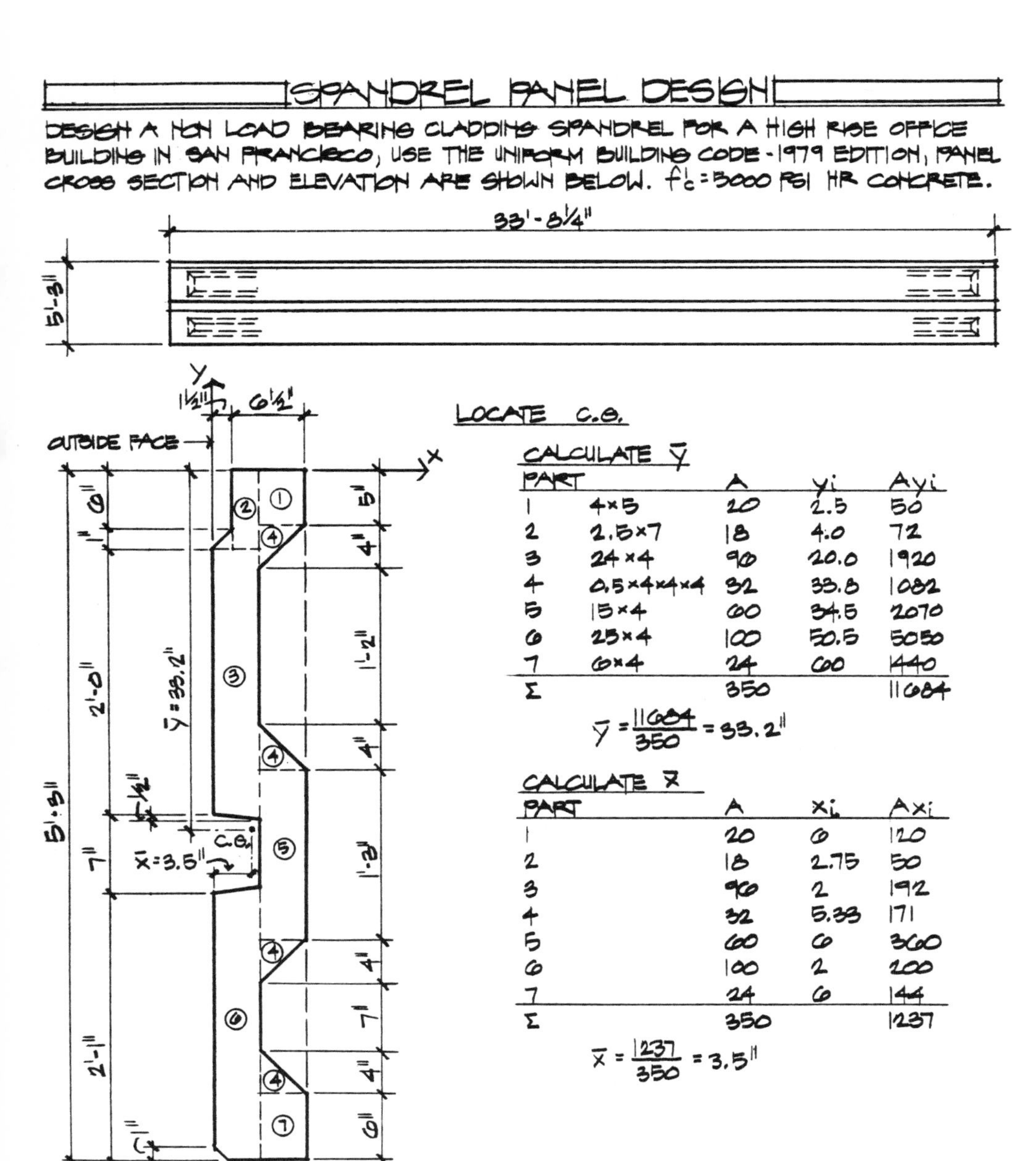

CALCULATE ȳ

PART		A	yi	Ayi
1	4×5	20	2.5	50
2	2.5×7	18	4.0	72
3	24×4	96	20.0	1920
4	0.5×4×4×4	32	33.8	1082
5	15×4	60	34.5	2070
6	25×4	100	50.5	5050
7	6×4	24	60	1440
Σ		350		11684

$$\bar{y} = \frac{11684}{350} = 33.2''$$

CALCULATE x̄

PART	A	xi	Axi
1	20	6	120
2	18	2.75	50
3	96	2	192
4	32	5.33	171
5	60	6	360
6	100	2	200
7	24	6	144
Σ	350		1237

$$\bar{x} = \frac{1237}{350} = 3.5''$$

DETERMINE SECTION PROPERTIES

PART	$\bar{x} - x_i$	$(x - x_i)^2$	A	$A(\bar{x} - x_i)^2$	I_o
1	2.5	6.25	20	125	27
2	0.75	0.56	18	10	10
3	1.5	2.25	96	216	128
4	1.83	3.35	32	107	28
5	2.5	6.25	60	375	80
6	1.5	2.25	100	225	133
7	2.5	6.25	24	150	37

$I_{yy} = \Sigma A(\bar{x} - x_i)^2 + I_o = 1651 \text{ IN}^4$

$S_{OUTSIDE} = \frac{1651}{3.5} = 472 \text{ IN}^3$

$S_{INSIDE} = \frac{1651}{4.5} = 367 \text{ IN}^3$

PANEL WEIGHT - NORMAL WEIGHT CONCRETE

$\frac{352}{144} \times 0.150 = 0.370 \text{ K/FT}$

USE 0.40 K/FT TO ACCOUNT FOR END BLOCKS

CHECK HANDLING

STRIPPING — TRY 2 PT. PICK AT 1/5 PTS.

$M^+ = M^- = 1.5 \times 0.025 \times 0.40 \times \overline{33.7}^2 = 17.0^{K \cdot FT}$

$f_t = \frac{17.0 \times 12}{367} = 0.550 \text{ KSI} > 5\sqrt{f'_{ci}}$ N.G.

$\therefore$ USE 4 PT. PICK — .1L ↑ .3L ↑ .2L ↑ .3L ↑ .1L (L)

$M^+ = M^- = 1.5 \times 0.0050 \times 0.40 \times \overline{33.7}^2 = 3.82^{K \cdot FT}$

$f_t = \frac{3.82 \times 12}{367} = 0.124 \text{ KSI} < 5\sqrt{f'_{ci}}$ O.K.

$A_s \text{ REQ'D} = \frac{3.82}{1 \times 5.8} = 0.66 \text{ IN}^2$ — 6-#3 REQ'D E.F.

STORE AND SHIP IN VERTICAL POSITION SINCE STRESSES ARE EXCESSIVE WITH 2 POINT SUPPORT

REINFORCING IN BOTTOM OF PANEL TO SATISFY SHIPPING STRESSES

BLOCK PANEL 3'-0" IN FROM EACH END

$M = 2.0 \times (2 \times 0.8 - 1) \times (0.125 \times 0.40 \times \overline{33.7}^2 = 68^{K \cdot FT}$

$A_s = \frac{68}{1.44 \times 60} = 0.79 \text{ IN}^2$ — USE 4-#4 IN BOTTOM

DESIGN PANEL ELEMENT TO RESIST LATERAL FORCES
(ASSUME PANEL CONNECTORS 6" FROM EACH END)

WIND @ 30 PSF

$w = 0.030 \times 5.25 = 0.158\ ^{K}/_{FT}$

$M = 0.125 \times 0.158 \times \overline{32.7}^2 = 21.1\ ^{K \cdot FT}$

$f_{T\ INSIDE} = \dfrac{21.1 \times 12}{367} = 0.691$ KSI N.G.

TRY LATERAL CONNECTORS AT MIDSPAN TOP & BOTTOM

$M^- = 0.125 \times 0.158 \times \overline{16.5}^2 = 5.4\ ^{K \cdot FT}$

$f_{T\ OUTSIDE} = \dfrac{5.4 \times 12}{472} = 0.137$ KSI $< 5\sqrt{f'_c}$ O.K.

$A_s = \dfrac{5.4}{1.44 \times 5.8} = 0.65$ IN^2 6 - #3 REQ'D. E.F.

SEISMIC - 0.3g ⟵⟶

$w = 0.3 \times 0.40 = 0.120\ ^{K}/_{FT}$ < WIND O.K.

TEMPERATURE REQUIREMENTS

$A_s = 0.002 \times \frac{1}{2} \times 352 = 0.35$ IN^2 E.F. < WIND O.K.

CHECK SPANDREL FOR BOWING DUE TO 60° THERMAL GRADIENT (IGNORE BENEFICIAL EFFECT OF LONG TERM DIFFERENTIAL SHRINKAGE)

$I = \dfrac{1651}{5.25} = 314\ ^{IN^4}/_{FT}$

EQUIVALENT THICKNESS ≅ 7"

FROM CHARTS FOR LATERAL SUPPORT SPACING OF 16.5', $\Delta = 0.3"$ WHICH IS EXCESSIVE

∴ USE LATERAL SUPPORTS AT 8'-0" ± O.C.
$\Delta \cong 0.08"$ O.K.

RE-ANALYZE WIND LOADING REINFORCEMENT REQUIREMENTS

$M = 0.11 \times 0.158 \times \overline{8.0}^2 = 1.11\ ^{K \cdot FT}$

$A_s = \dfrac{1.11}{1.44 \times 5.8} = 0.13$ IN^2 ∴ HANDLING GOVERNS

DETERMINE TEMPERATURE STEEL REQUIREMENTS FOR TRANSVERSE SECTION

$A_s = 0.002 \times \dfrac{352}{5.25} = 0.13\ ^{IN^2}/_{FT}$

USE #3 @ 10" O/C ($A_s = 0.11 \times {}^{12}/_{10} = 0.13\ ^{IN^2}/_{FT.}$)

REINFORCING SUMMARY

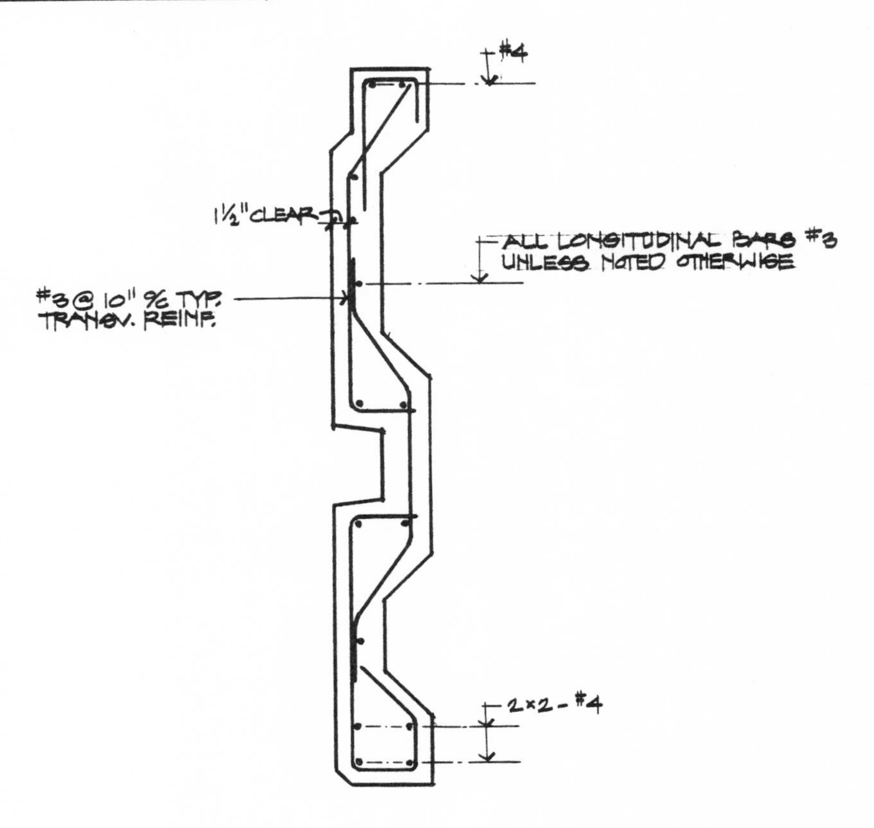

CONNECTION DESIGN

DISTRIBUTION FACTORS FOR LOADS TO LATERAL CONNECTORS

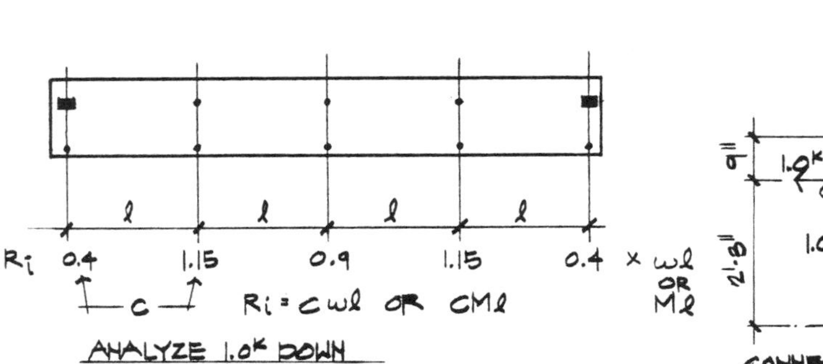

ANALYZE 1.0K DOWN

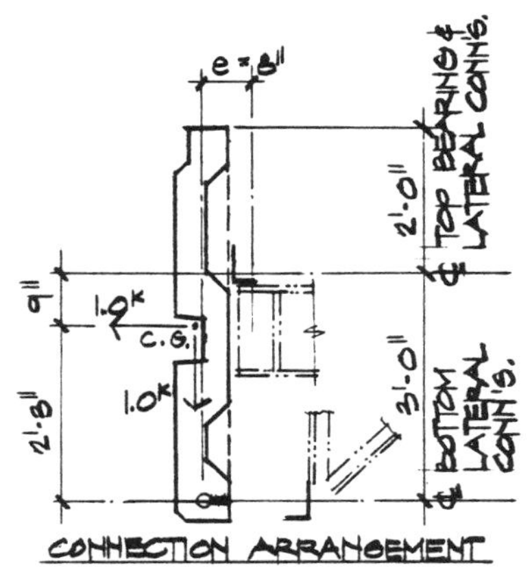

CONNECTION ARRANGEMENT

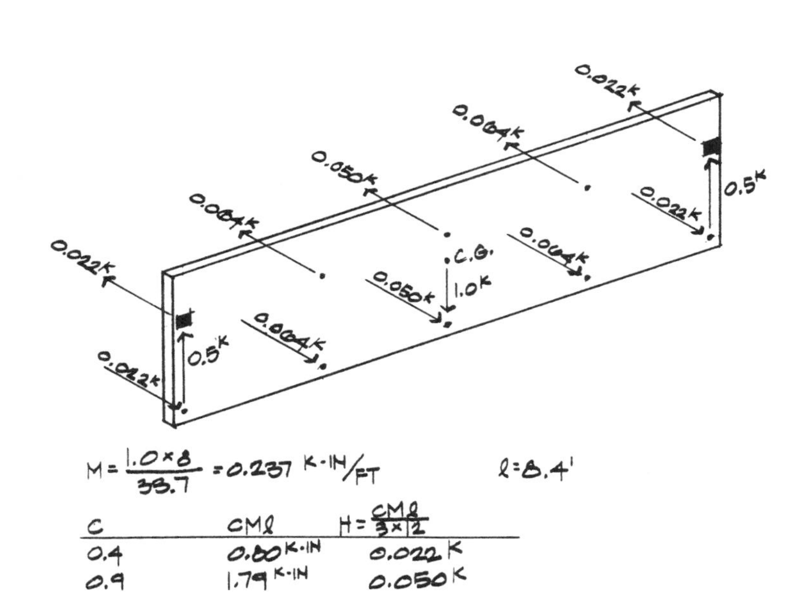

$M = \frac{1.0 \times 8}{33.7} = 0.237$ K-IN/FT $\qquad \ell = 8.4'$

C	CMℓ	$H = \frac{CM\ell}{3 \times 12}$
0.4	0.80 K-IN	0.022 K
0.9	1.79 K-IN	0.050 K
1.15	2.29 K-IN	0.064 K

ANALYZE 1.0K LONGITUDINAL

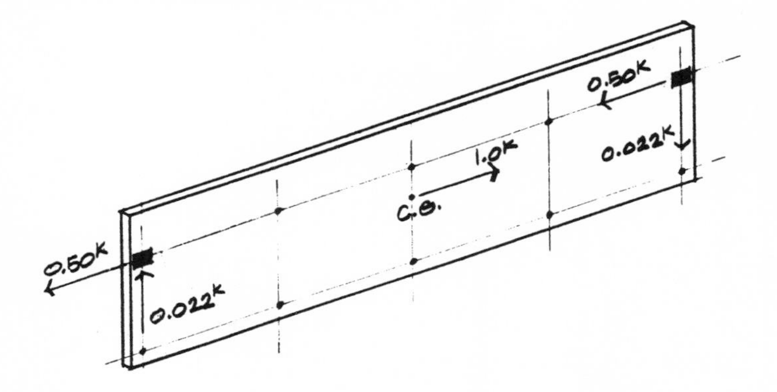

$\Sigma M_{C.G.} = 0$

$2R_v \times 16.8 = 1 \times 0.75$

$R_v = 0.022^K$

ANALYZE 1.0K LATERAL

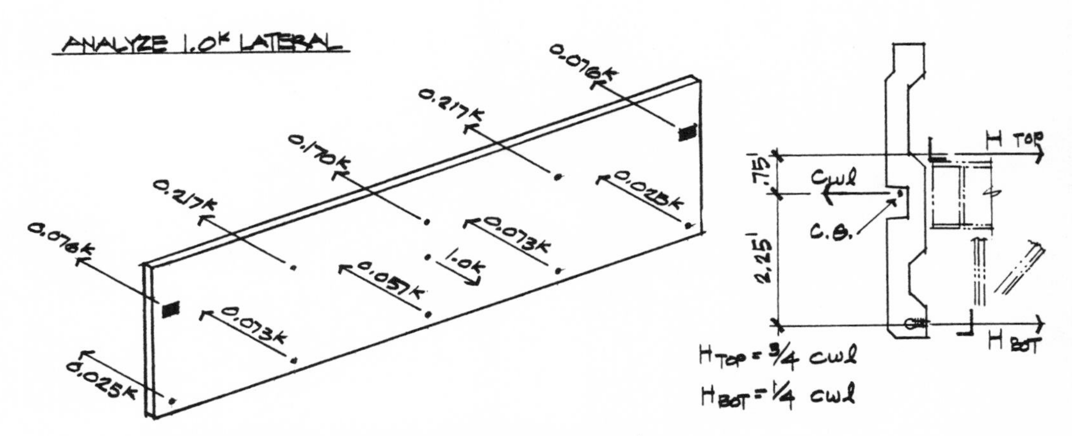

(SEE SHEET 5 FOR DISTRIBUTION FACTORS)

$w = \frac{1.0}{33.7} = 0.030\ ^K/FT. \qquad l = 8.4'$

c	cwl	H_{TOP}	H_{BOT}
0.4	0.101^K	0.076^K	0.025^K
0.9	0.227^K	0.170^K	0.057^K
1.15	0.290^K	0.217^K	0.073^K

PANEL WT. $W = 0.4^{K}/FT \times 33.7' = 13.5^{K}$

DESIGN CONNECTOR FASTENER - BOTTOM LATERAL CONN.

$W(1.0^{K}\ DN + 1.2^{K}\ IN) = P_{CRITICAL}$

OR: $(13.5^{K})(0.064 + 1.2 \times 0.073) = 2.04^{K}$ COMPRESSION
$(13.5^{K})(0.064 - 1.2 \times 0.073) = 0.32^{K}$ TENSION

* 5/8" φ FERRULE LOOP INSERT O.K. $P_{ALLOW.} = 3.0^{K}$
ACME WIRE PRODUCTS - FL-625

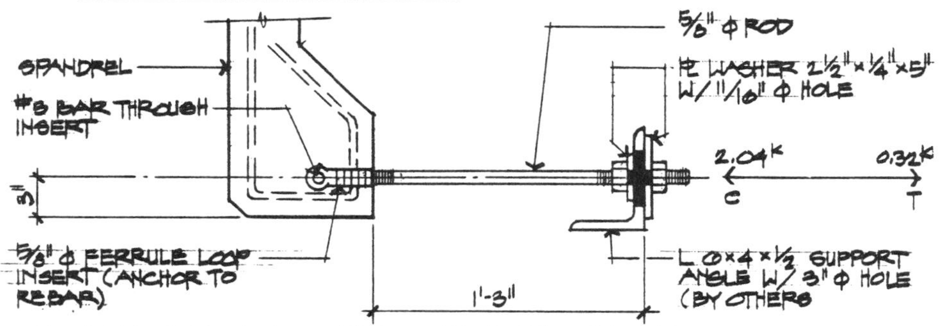

CALCULATE SIZE OF COIL THREAD STOCK REQUIRED AS GOVERNED BY COMPRESSION (BODY OF CONNECTOR)

$(13.5^{K})(0.064 + 0.4 \times 0.073) = 1.26^{K}$ COMPRESSION
USE 5/8" φ ROD (SEE CHART BELOW)

					$L/r = 200$, $F_a = 8.27$		$L/r = 100$, $F_a = 10.47$		$L/r = 50$, $F_a = 22.60$		TENSION $F_t = 26.70$
ϕ_{NOM} "	ϕ_{NET} "	A_{NET} □"	† I IN4	$r = \sqrt{I/A}$	L_u	P_{ALLOW}	L_u	P_{ALLOW}	L_u	P_{ALLOW}	P_T **
½	.42	.1385	.0015	0.10"	20"	1.2K	10"	2.3	5"	3.2	3.7
¾	.63	.3079	.0077	0.16"	32	2.5	16	5.1	8	6.9	8.2
1	.83	.5410	.0233	0.21"	42	4.3	21	8.9	10	12.2	14.4
1¼	1.08	.9161	.0668	0.27"	54	7.0	27	15.0	14	20.7	24.4

$F_y = 33$ KSI (ASTM A-307) $K = 1.0$
ALLOWABLE STRESSES INCREASED BY 33%

** ALLOWABLE INSERT VALUE WILL GOVERN

† $I = 0.049087\ d^4$

* ALL EMBEDDED INSERTS DESIGNED WITH A FACTOR OF SAFETY OF 4 WITH RESPECT TO ULTIMATE CAPACITIES BASED UPON ACTUAL TEST DATA.

DESIGN CONNECTOR FASTENER · TOP LATERAL CONNECTION

W(1.0^K DN + 1.2^K IN/OUT) P CRITICAL

OR: $(13.5^K)(0.64 + 1.2 \times 0.217) = 4.37^K$ TENSION

$(13.5^K)(0.64 - 1.2 \times 0.217) = 2.65^K$ COMPRESSION

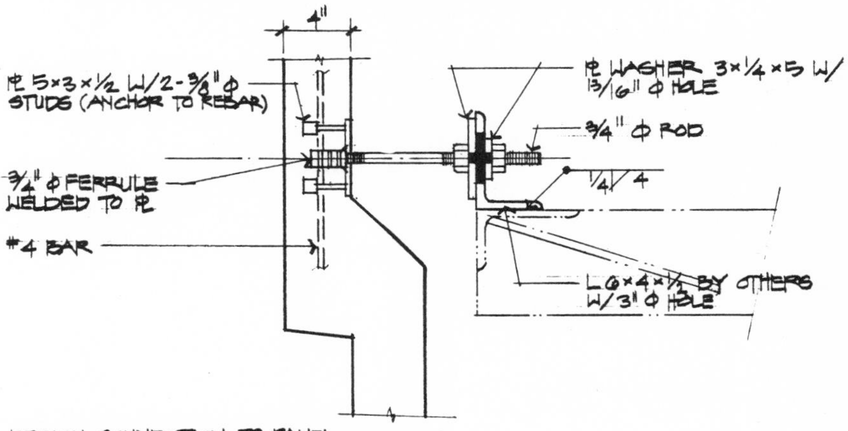

DESIGN CONNECTION TO PANEL

$UP = 0.75(1.4D + 1.7L + 1.87E)$
$= 0.75(1.4 \times 0.064 + 1.87 \times 1.2 \times 0.217)(13.5^K) = \underline{5.82^K}$

OR: $UP = 0.9D + 1.43E$
$= (0.9 \times 0.064 + 1.43 \times 1.2 \times 0.217)(13.5^K) = \underline{5.80^K}$

TRY PL 5×3×3/8 W/2-3/8"φ × 2 1/2" LONG STUDS

AS GOVERNED BY CONC. STRENGTH
(SEE PCI DESIGN HANDBOOK, P. 5-42)

$\ell_e = 2.5''$ $y = 0$ $x = 4''$ $\therefore \phi P_c = 15^K$

AS GOVERNED BY STUD CAPACITY
(SEE PCI DESIGN HANDBOOK, P. 5-41)

$\phi P_s = 2 \times 6.0 = \underline{12.0^K}$ GOVERNS

PL THICKNESS REQ'D. USE $f_b = 0.9F_y = (0.9)(36) = 32.4$ KSI

$f_b = \frac{M}{S} = \frac{PL/8}{1/6\, b t^2}$

$32.4 = \frac{(1/8)(4.37)(4)}{(1/6)(3)t^2}$ $t^2 = 0.135$ $t = 0.37''$

USE 3/8" PL

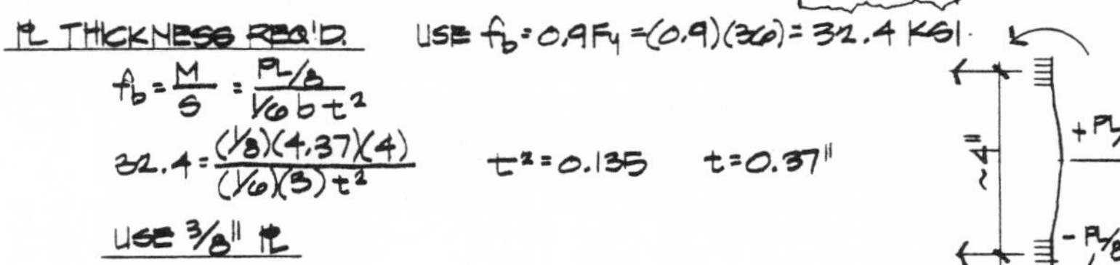

ROD CONNECTOR TO PL

USE 3/4" FERRULE WELDED TO PL (P_t = 8.2^K)

CHECK PUNCHING SHEAR - C = 2.65^K

$U_P = 0.75(1.4 \times .064 - 1.87 \times 1.2 \times 0.217)(13.5^K)$

$v_{bo} = \frac{C}{\phi b_o d} = \frac{4.02}{(0.85)(2\times9+2\times7)(4)} = 0.037$ KSI

$v_c = 4\sqrt{f'_c} = 0.283$ KSI O.K.

4.02^K

2" = d/2

4"

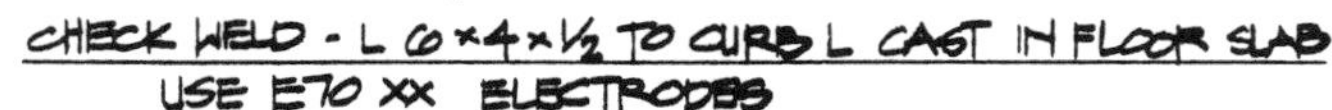

CHECK WELD - L 6×4×1/2 TO CURB L CAST IN FLOOR SLAB

USE E70 XX ELECTRODES

(SEE P. 4-68, AISC MANUAL FOR STEEL CONSTRUCTION, 7TH EDITION)

$P = 4.37^K$ $\ell = 4"$ $C_1 = 1.0$ $K = 0$

$a\ell = 4.5"$ $\therefore a = \frac{4.5}{4} = 1.125$

$C = 0.275$

$D = \frac{P}{CC_1\ell} = \frac{4.37}{(0.275)(1)(4)} = 3.97$

4.37^K

4.5"

4"

∴ USE 1/4" WELDS

(33% INCREASE IN WELD STRESS ALLOWED BUT NOT TAKEN)

DESIGN CONNECTOR BODY - L 6×4×t

$P_{CRITICAL} = W(1.0^K\ DN + 0.4^K\ IN/OUT)$

$= (3.5^K)(0.064 + 0.4 \times 0.217) = 2.03^K$

TRY 6" LONG ANGLE (ASSUME K = 0.8")

$F_b = \frac{M}{S} = \frac{(2.03)(3.2)}{1/6 \times 6 \times t^2} = 1.33 \times 0.6 \times 36$

$t^2 = 0.23$

$t = 0.48"$

USE L 6×4×1/2×0'-6" LONG

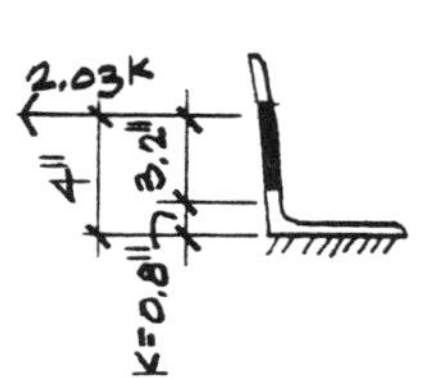

CHECK ANGLE STRESS FOR P = 1.2g

$f_b = \frac{4.37 \times 3.2}{1/6 \times 6 \times 0.5^2} = 55.9$ KSI > F_y O.K.

∴ BODY OF CONNECTOR EXHIBITS DUCTILE BEHAVIOR AT MAXIMUM DESIGN FORCE ON CONNECTOR FASTENER.

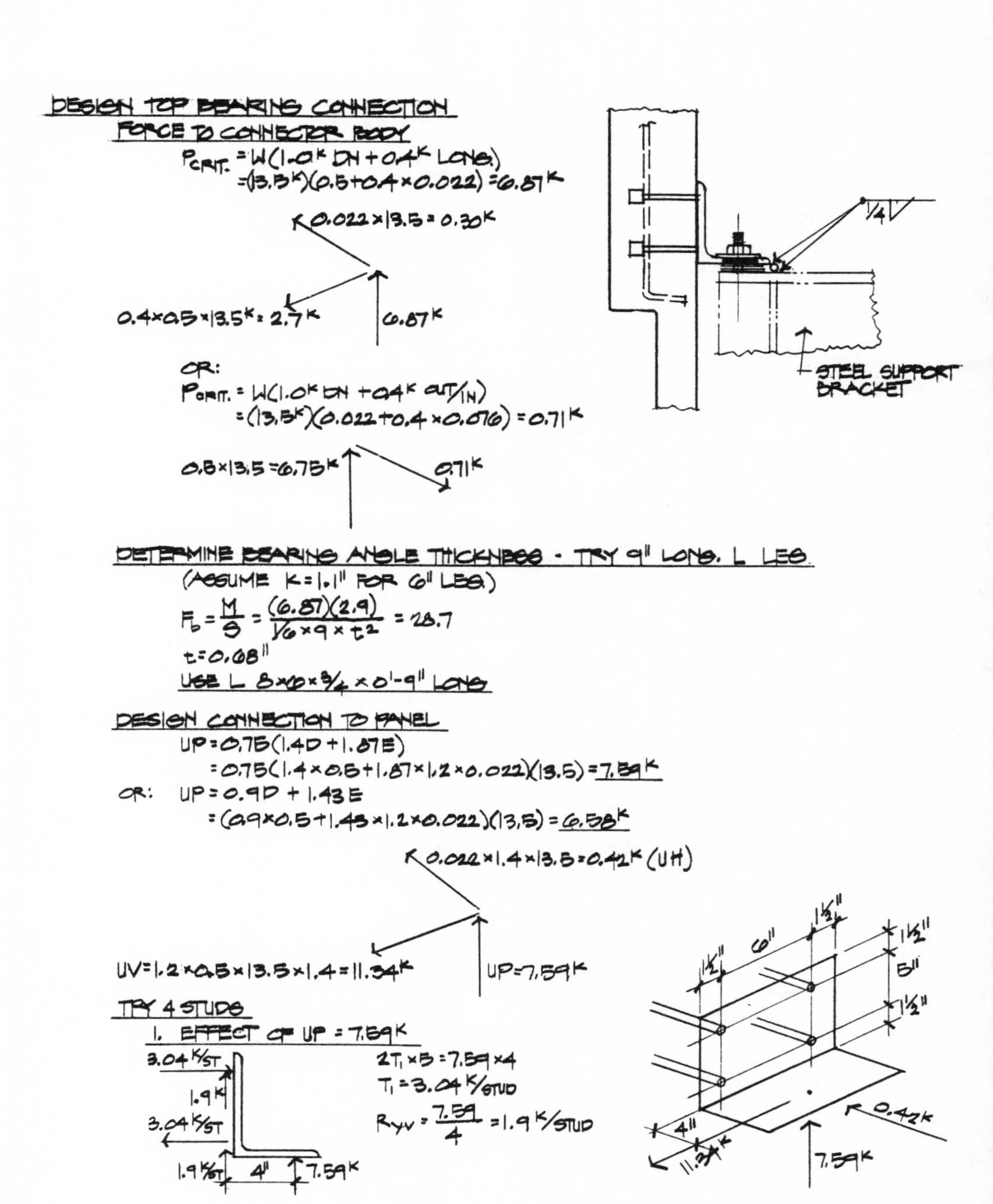
DESIGN TOP BEARING CONNECTION
FORCE TO CONNECTOR BODY
P CRIT. = W(1.0K DN + 0.4K LONG)
=(13.5K)(0.5+0.4×0.022) = 6.87K
0.022×13.5 = 0.30K
0.4×0.5×13.5K = 2.7K
6.87K
OR:
P CRIT. = W(1.0K DN + 0.4K OUT/IN)
=(13.5K)(0.022+0.4×0.076) = 0.71K
0.5×13.5 = 6.75K
0.71K
1/4
STEEL SUPPORT BRACKET
DETERMINE BEARING ANGLE THICKNESS - TRY 9" LONG, L LEG
(ASSUME K=1.1" FOR 6" LEG)
Fb = M/S = (6.87)(2.9) / 1/6×9×t² = 28.7
t = 0.68"
USE L 8×6×3/4 × 0'-9" LONG
DESIGN CONNECTION TO PANEL
UP = 0.75(1.4D + 1.87E)
= 0.75(1.4×0.5+1.87×1.2×0.022)(13.5) = 7.59K
OR: UP = 0.9D + 1.43E
= (0.9×0.5+1.43×1.2×0.022)(13.5) = 6.58K
0.022×1.4×13.5 = 0.42K (UH)
UV = 1.2×0.5×13.5×1.4 = 11.34K
UP = 7.59K
TRY 4 STUDS
1. EFFECT OF UP = 7.59K
3.04K/ST
1.9K
3.04K/ST
1.9K/ST
4"
7.59K
2T1×5 = 7.59×4
T1 = 3.04K/STUD
Ryv = 7.59/4 = 1.9K/STUD
1 1/2"
6"
1 1/2"
1 1/2"
5"
1 1/2"
0.42K
4"
11.34K
7.59K

2. EFFECT OF ECCENTRIC SHEAR $UV = 11.34^K$

DUE TO SHEAR

$$R_{xv} = \frac{11.34}{4} = 2.84\ ^K/_{STUD}$$

DUE TO TORSION

$$R_{xT} = \frac{Pey}{\Sigma x^2 + y^2} = \frac{(11.34)(4)(2.5)}{61} = 1.86\ ^K/_{ST}$$

$$R_{yT} = \frac{Pex}{\Sigma x^2 + y^2} = \frac{(11.34)(4)(3)}{61} = 2.23\ ^K/_{ST}$$

$$\Sigma x^2 + y^2 = (4)(3)^2 + (4)(2.5)^2 = 61$$

3" R_y R R_x 2.5"   e=4" 11.34^K

EFFECT OF ECCENTRIC SHEAR CAUSING BENDING (e = 4")

$2T_2 \times 6 = (11.34)(4)$ $\therefore T_2 = 3.78\ ^K/_{STUD}$

$$R = \sqrt{(R_{xv}+R_{xT})^2 + (R_{yv}+R_{yT})^2} = \sqrt{(2.84+1.86)^2 + (1.90+2.23)^2} = 6.26\ ^K/_{STUD}$$

CRITICAL DESIGN CONDITION IS $\begin{cases} V_u = 6.26\ ^K/_{STUD} \\ P_u = 3.04 + 3.78 = 6.82\ ^K/_{STUD} \end{cases}$

TRY 4-3/4"φ × 6" LONG STUDS

AS GOVERNED BY STUD CAPACITY (SEE PCI DESIGN HANDBOOK, 2ND EDITION PP 5-24, 5-40, & 5-41)

$\phi P_s = 23.9\ ^K/_{STUD}$ $\phi V_s = 19.9\ ^K/_{STUD}$

$$\left(\frac{P_u}{\phi P_s}\right)^2 + \left(\frac{V_u}{\phi V_s}\right)^2 = \left(\frac{6.82}{23.9}\right)^2 + \left(\frac{6.26}{19.9}\right)^2 = 0.08 + 0.10 = 0.18 < 1.0 \quad O.K.$$

AS GOVERNED BY CONCRETE CAPACITY

EDGE DISTANCE $d_e = 4"$ $f'_c = 5000$ PSI

$\phi P_c = 31.0\ ^K/_{STUD}$ $\phi V_c = 8.3\ ^K/_{STUD}$

$$\left(\frac{P_u}{\phi P_c}\right)^{4/3} + \left(\frac{V_u}{\phi V_c}\right)^{4/3} = \left(\frac{6.82}{31}\right)^{4/3} + \left(\frac{6.26}{8.3}\right)^{4/3} = 0.13 + 0.69 = 0.82 < 1.0 \quad O.K.$$

∴ USE 4-3/4"φ × 6" LONG STUDS

BEARING CONNECTION DETAIL

$d_e = 4"$

#4 REBAR

L 8×6×3/4 ×0'-9" LONG W/ 4-3/4"φ × 6" LONG WELDING STUDS (3"φ HOLE IN O.S.L.)

3/4"φ × 3" THREADED STUD, 4×4 PL WASHER, & HEX NUT.

1/4

SHIMS

P/C SPANDREL

STEEL SUPPORT BRACKET

AMBASSADOR COLLEGE AUDITORIUM - PASADENA

The column cover units shown here were erected progressively from the ground level, with bottom "blind" connections and top welded connections being made to the structural steel columns as erection progressed. See the connections pages at the end of this chapter for some typical "blind" connections.(Photo by Eugene Phillips)

333 MARKET STREET - SAN FRANCISCO

2 story "V" shaped column cover units clad this 33 story building, emphasizing the vertical mode and forming a pleasing contrast with the continuous window wall and glazing infill.

COLUMN COVER UNITS

While spandrel units give the exterior of the building a banded appearance, the use of column cover units emphasize the verticality of the structure. These units are made in multi-story lengths, and are sometimes prestressed concentrically to keep handling stresses within allowable, and to achieve economy in the reinforcing of the member. A key element in the location of the bearing connections of column cover units is to position these connections as closely as possible to the center of gravity of the column cover unit. In this manner, the counter balancing moment of forces required to counteract longitudinal seismic effects on the column cover is minimized. Another aspect of structural design of column cover connector bearing assemblies is that there is often a considerable distance between the attachment point on the column cover and the support location on the building frame, requiring a torsion resistant steel assembly to form the connection bracket. There are a variety of ways to "hide" the structural column with the column cover unit, and attachments may be made to the flanges of the column section, or to the adjacent floor framing. One arrangement is demonstrated in the following design example.

TRANSAMERICA PYRAMID - SAN FRANCISCO

Architectural precast concrete column covers clad the complex structural steel framing at the building's base, providing beauty as well as protection from corrosion and fire.

COLUMN COVER DESIGN

DESIGN A 2 STORY COLUMN COVER UNIT FOR A HIGH RISE OFFICE BUILDING IN SAN FRANCISCO. USE THE UNIFORM BUILDING CODE - 1979 EDITION, COLUMN COVER CROSS SECTION & ELEVATION ARE SHOWN BELOW. f'_c = 5000 PSI HR CONCRETE.

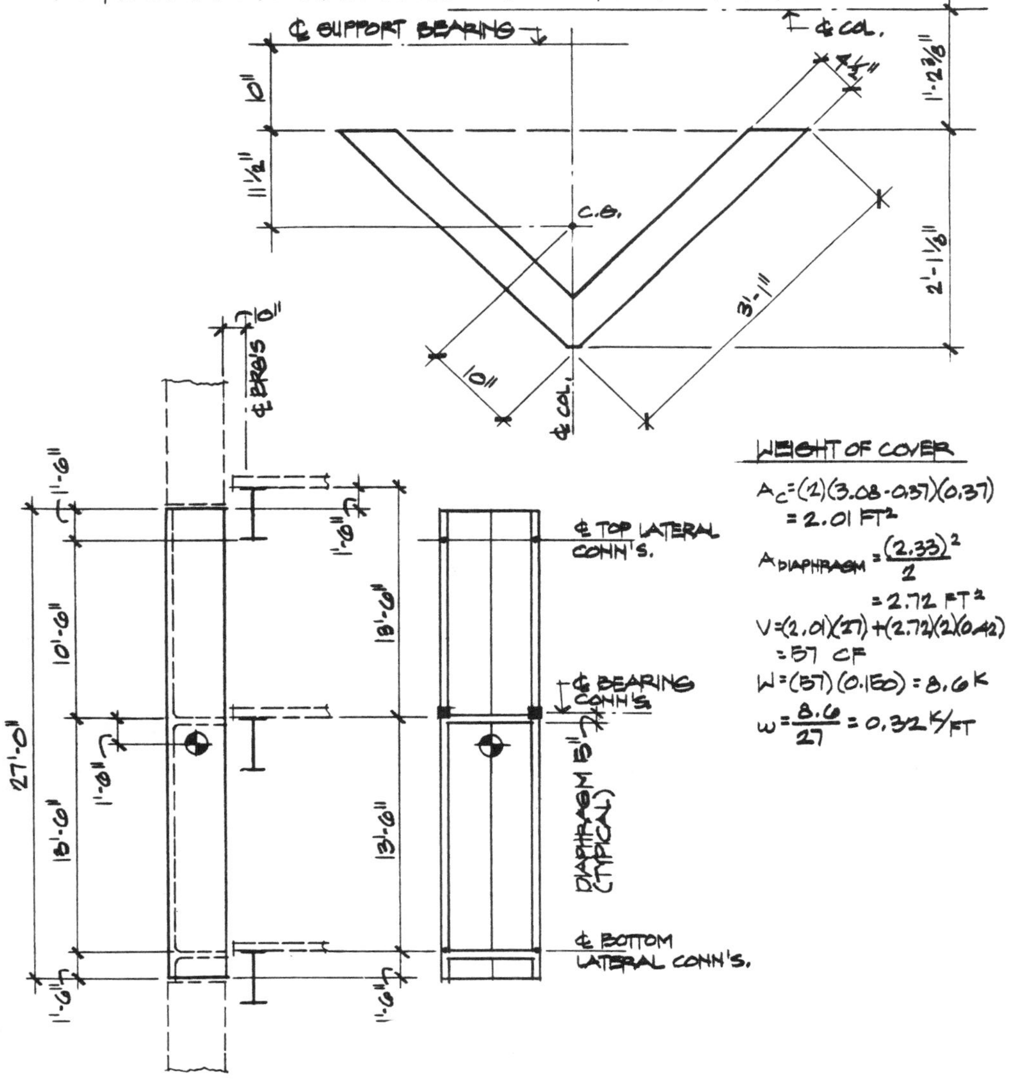

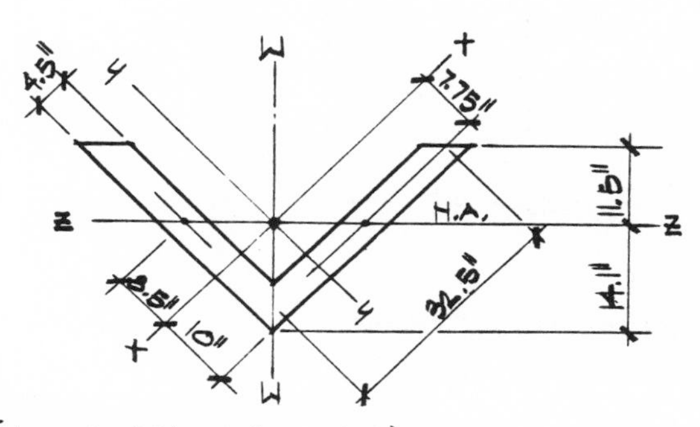

SECTION PROPERTIES

(SEE P. 6-30 - AISC MANUAL - 7TH EDITION)

$$I_x = I_y = \left(\tfrac{1}{3}\right)\left[(4.5)(34.75-10)^3 + (34.75)(10)^3 - (30.25)(10-4.5)^3\right]$$

$$= \left(\tfrac{1}{3}\right)(68{,}224 + 34{,}750 - 5033) = \underline{32{,}647 \text{ IN}^4}$$

$$K = \pm \frac{(30.25)(34.75)(30.25)(34.75)(4.5)}{(4)(34.75+30.25)} = \pm 19{,}125 \text{ IN}^4$$

$$I_z = I_x \sin^2\theta + I_y \cos^2\theta + K \sin 2\theta$$
$$= (2)(32{,}647)(\sin 45°)^2 + (-19{,}125)(\sin 90°)$$
$$= 32{,}647 - 19{,}125 = \underline{13{,}520 \text{ IN}^4}$$

ABOUT Z-Z AXIS (STRIPPING)

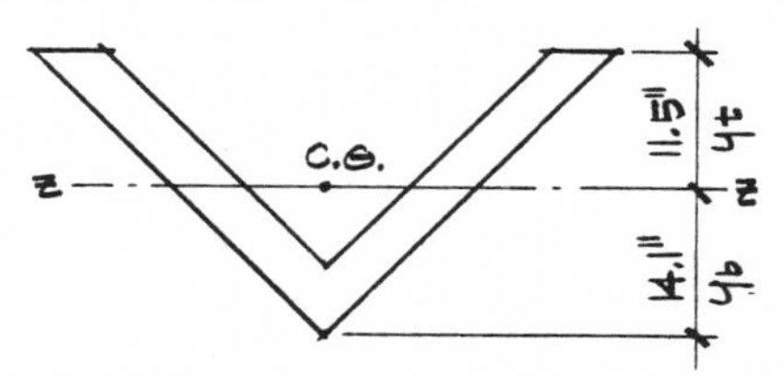

$I_z = 13{,}520 \text{ IN}^4$

$$S_t = \frac{13{,}520}{11.5} = 1176 \text{ IN}^3$$

$$S_b = \frac{13{,}520}{14.1} = 959 \text{ IN}^3$$

ABOUT X-X AXIS (STORE & SHIP)

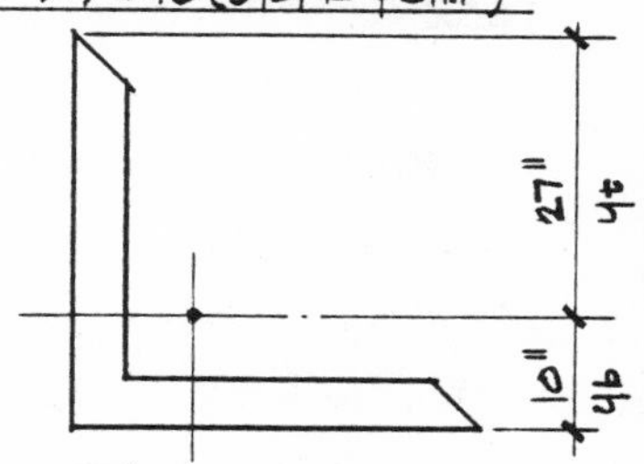

$I_x = 32{,}647 \text{ IN}^4$

$$S_t = \frac{32647}{27} = 1209 \text{ IN}^3$$

$$S_b = \frac{32647}{10} = 3265 \text{ IN}^3$$

STRIPPING USE 2 PT. PICK AT 1/5 PTS.

$M = 1.5 \times 0.025 \times 0.32 \times \overline{27}^2 = 8.75^{K\text{-}FT}$

$f_t = \frac{8.75 \times 12}{959} = 0.109 \text{ KSI} < 5\sqrt{f'_{ci}}$ O.K.

$A_{s \text{ REQ'D}} = \frac{8.75}{1.0 \times 20} = 0.44 \text{ IN}^2$ ∴ 2-#4 T&B

SHIPPING

$M = 2.0 \times 0.025 \times 0.32 \times \overline{27}^2 = 11.66^{K\text{-}FT}$

$f_t = \frac{11.66 \times 12}{1209} = 0.116 \text{ KSI} < 5\sqrt{f'_c}$ O.K.

$A_{s \text{ REQ'D}} = \frac{11.66}{1.44 \times 33} = 0.25 \text{ IN}^2$ ∴ 1-#4 O.K. + MESH

ERECTION

TRY LIFTING AT ONE END ONLY

$M = 1.25 \times 0.125 \times 0.32 \times \overline{27}^2 = 36.5^{K\text{-}FT}$

$f_t = \frac{36.5 \times 12}{3268} = 0.134 \text{ KSI} < 5\sqrt{f'_c}$ OK.

$A_{s \text{ REQ'D}} = \frac{36.5}{1.44 \times 33} = 0.77 \text{ IN}^2$ ∴ 3-#4 + MESH

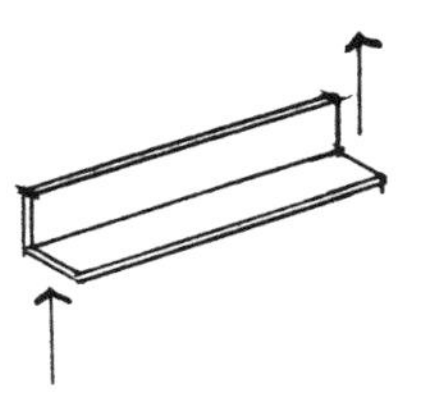

SERVICE LOAD CONDITION

SEISMIC: $0.3g = 0.3 \times 0.32 = 0.096$ K/FT

WIND: $0.030 \times 4.3 = 0.129$ K/FT

$M = 0.125 \times 0.129 \times \overline{13.5}^2 = 2.94^{K\text{-}FT}$

∴ LESS CRITICAL THAN HANDLING

TEMP. $A_s = (0.002)(4.5)(12) = 0.16 \text{ IN}^2/\text{FT}$

USE 6×6 W5.5 × W5.5 WWF

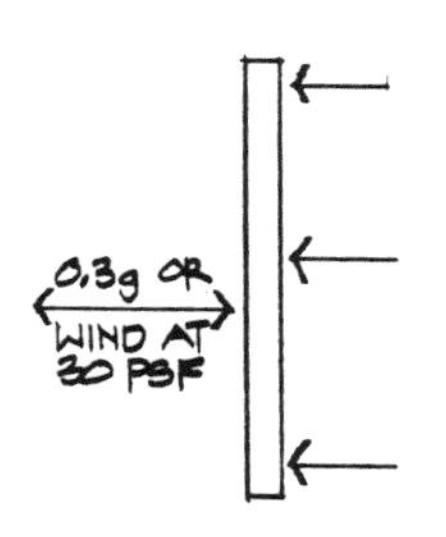

DIAPHRAGM REINFORCING

$A_s = (0.0025)(5)(12) = 0.15 \text{ IN}^2$

USE 2-#4 E.W.

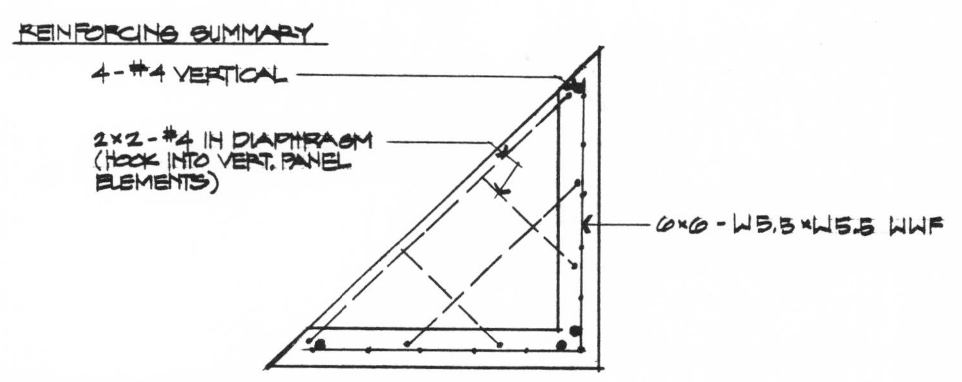
REINFORCING SUMMARY
4-#4 VERTICAL
2×2-#4 IN DIAPHRAGM
(HOOK INTO VERT. PANEL ELEMENTS)
6×6-W5.5×W5.5 WWF

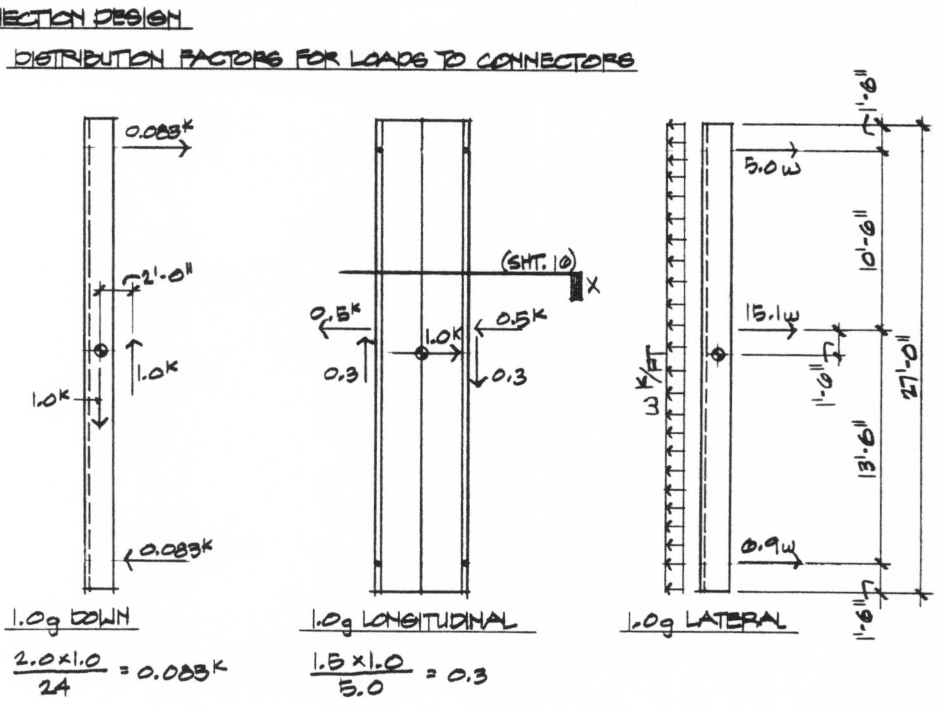
CONNECTION DESIGN
DISTRIBUTION FACTORS FOR LOADS TO CONNECTORS
0.083K
2'-0"
1.0K
1.0K
0.083K
1.0g DOWN
2.0×1.0 / 24 = 0.083K
(SHT. 10)
X
0.5K
0.5K
1.0K
0.3
0.3
1.0g LONGITUDINAL
1.5×1.0 / 5.0 = 0.3
wK/FT
5.0w
15.1w
0.9w
1'-6"
10'-6"
1'-0"
13'-6"
27'-0"
1'-6"
1.0g LATERAL

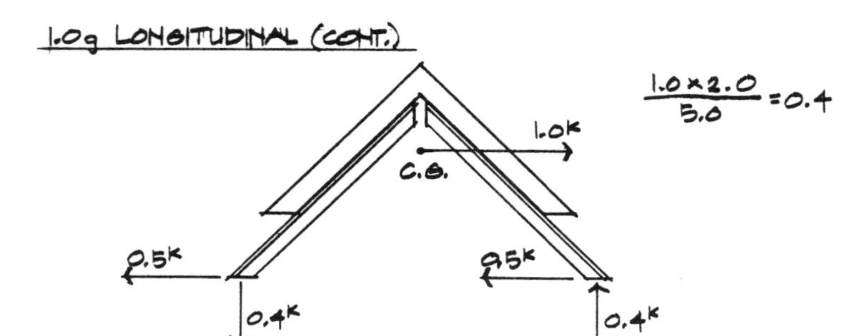

DESIGN CONNECTOR FASTENER - BOTTOM & TOP LATERAL CONN.

$W(1.0^K \text{ DN} + 1.2^K \text{ IN/OUT}) = P_{CRITICAL}$

$8.0^K \times .088 + 1.2 \times 6.9 \times 0.32\ ^K/_{FT} = 0.71 + 2.65 = 3.36^K$

OR: $\frac{3.36}{2} = 1.68\ ^K/_{CONN.}$

ROD SIZE

$A_{REQ'D} = \frac{1.68}{20} = 0.084\ IN^2$

USE ½" ϕ ROD

$(A_S = 0.139\ IN^2)$

ANGLE ON PANEL

$M = \frac{PL}{4} = \frac{1.68 \times 6}{4} = 2.52^{K\text{-}IN}$

$S = \frac{M}{1.33 F_b} = \frac{2.52}{22 \times 1.33} = 0.086\ IN^3$

USING 6" L LEG 4" LONG

$\frac{1}{6} \times 4 \times t^2 = 0.086$

$t = 0.36''$

USE L 6×6×⅜×0'-4"

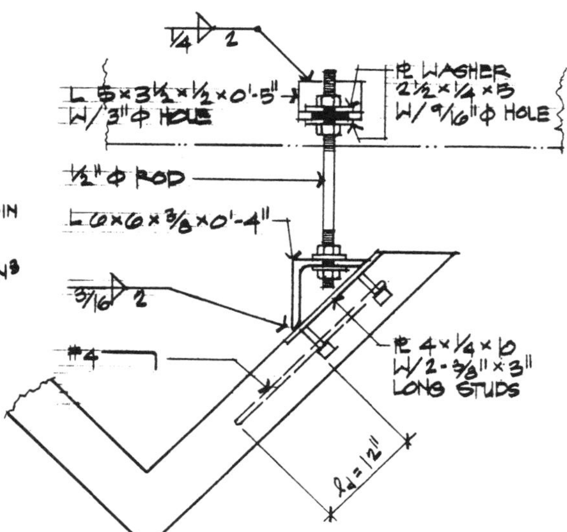

INSERT PL IN PANEL

$UP = 0.75(1.4D + 1.7L + 1.87E) = (0.75)(1.4 \times 0.71 + 1.87 \times 2.05) = 4.46^K$

OR: $\frac{4.46}{2} = 2.23\ ^K/_{CONN.}$

$UP = 0.9D + 1.43E = (0.9)(0.71) + (1.43)(2.05) = 4.43^K$

OR: $\frac{4.43}{2} = 2.22\ ^K/_{CONN.}$

$V_u = P_u = 0.707 \times 2.23^K = 1.58^K$

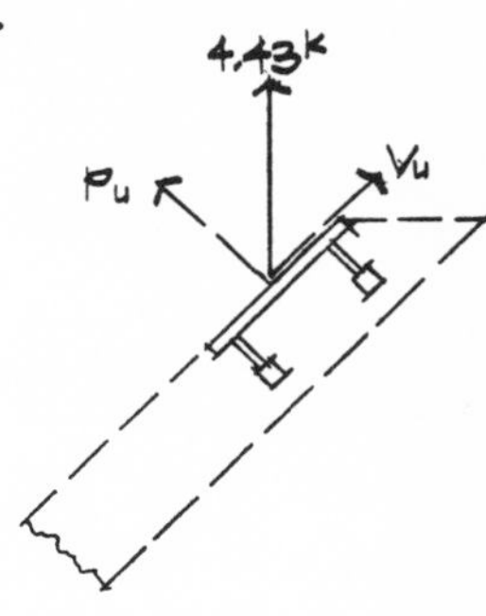

TRY 2 - 1/4"φ × 3" LONG STUDS

($d_e = 2''$)

(SEE PCI DESIGN HANDBOOK PP. 5-24, 5-40, & 5-41)

AS GOVERNED BY STUD CAPACITY

$\phi V_s = 2.2\ ^K/_{STUD}$

$\phi P_s = 2.7\ ^K/_{STUD}$

$$\left(\frac{P_u}{\phi P_s}\right)^2 + \left(\frac{V_u}{\phi V_s}\right)^2 = \left(\frac{0.5 \times 1.58}{2.7}\right)^2 + \left(\frac{0.5 \times 1.58}{2.2}\right)^2 = 0.22 < 1.0 \quad O.K.$$

AS GOVERNED BY CONCRETE CAPACITY

$\phi V_c = 2.3\ ^K/_{STUD}$

$\phi P_c = 6.4^K$

$$\left(\frac{P_u}{\phi P_c}\right)^{4/3} + \left(\frac{V_u}{\phi V_c}\right)^{4/3} = \left(\frac{0.5 \times 1.58}{6.4}\right)^{4/3} + \left(\frac{0.5 \times 1.58}{2.3}\right)^{4/3} = 0.30 < 1.0 \quad O.K.$$

INSERT PL THICKNESS - USE 1/4" PL × 4" × 10"

WELD: $\frac{1.68}{3 \times .93} = 0.6''$

USE 2" - 3/16" WELD EA. SIDE - EA. END

ANGLE BRACKET TO BEAM

$M = 1.68 \times 3.25 = 5.46^{K\text{-}IN}$

$S = \frac{5.46}{22 \times 1.33} = 0.19\ IN^3$

$\frac{1}{6} \times 5 \times t^2 = 0.19 \quad \therefore t = 0.48$

USE L 5×3½×½

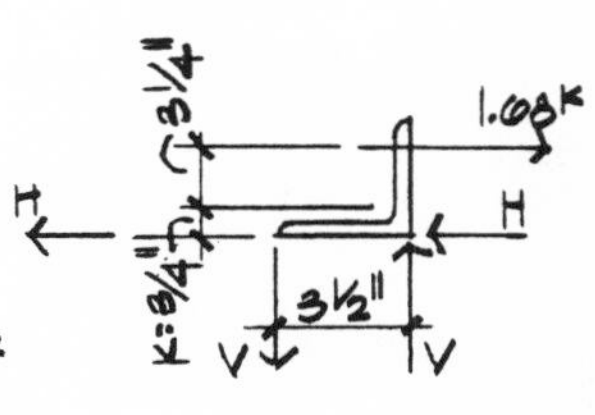

WELD: $H = 0.84^K \quad V = 1.92^K \quad R = \sqrt{H^2 + V^2} = 2.1^K$

USING 1/4" WELDS $L = \frac{2.1}{4 \times .93 \times 1.33} = 0.4''$

USE 2" (HEEL & TOE)

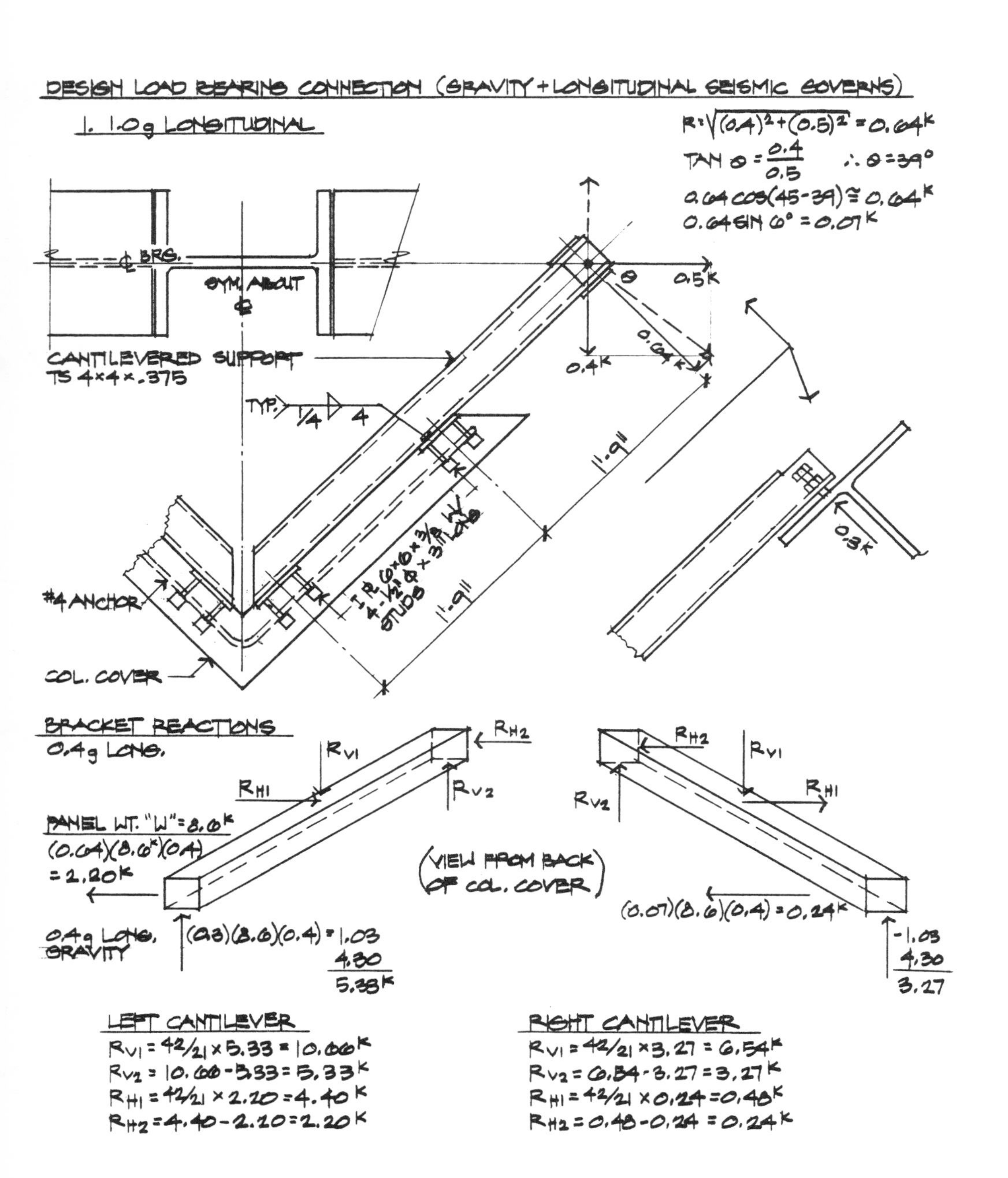
DESIGN LOAD BEARING CONNECTION (GRAVITY + LONGITUDINAL SEISMIC GOVERNS)
1. 1.0g LONGITUDINAL
R=√((0.4)²+(0.5)²) = 0.64K
TAN θ = 0.4/0.5 ∴ θ=39°
0.64 COS(45-39) ≅ 0.64K
0.64 SIN 6° = 0.07K
℄ BRG.
SYM. ABOUT ℄
θ
0.5K
0.64K
0.4K
CANTILEVERED SUPPORT TS 4×4×.375
TYP. 1/4 4
1'-9"
1'-9"
1 PL 6×6×3/8 W/ 4-1/2 Φ×3" LG STUDS
0.3K
#4 ANCHOR
COL. COVER
BRACKET REACTIONS
0.4g LONG.
RV1
RH1
RH2
RV2
PANEL WT. "W"=8.6K
(0.64)(8.6K)(0.4) = 2.20K
(VIEW FROM BACK OF COL. COVER)
RH2
RV1
RH1
RV2
(0.07)(8.6)(0.4)=0.24K
0.4g LONG. GRAVITY
(0.3)(8.6)(0.4) = 1.03
4.30
5.33K
-1.03
4.30
3.27
LEFT CANTILEVER
RV1 = 42/21 × 5.33 = 10.66K
RV2 = 10.66-5.33 = 5.33K
RH1 = 42/21 × 2.20 = 4.40K
RH2 = 4.40-2.20 = 2.20K
RIGHT CANTILEVER
RV1 = 42/21 × 3.27 = 6.54K
RV2 = 6.54-3.27 = 3.27K
RH1 = 42/21 × 0.24 = 0.48K
RH2 = 0.48-0.24 = 0.24K

DESIGN BRACKET SECTION

DESIGN MOMENTS

$M_V = 5.33 \times 21 = 112^{K\text{-}IN}$

$M_H = 2.20 \times 21 = 46^{K\text{-}IN}$

$S_{REQ.} = \frac{112 + 46}{1.33 \times 24} = 4.95\ IN^3$

CHECK CONDITION - D.L. ONLY

$M_V = 4.30 \times 21 = 90^{K\text{-}IN}$

$S_{REQ.} = \frac{90}{24} = 3.75\ IN^3$

USE TS 4×4×.375 ($S = 5.10\ IN^3$)

DESIGN CONNECTOR FASTENERS

BRACKET REACTIONS - 1.2g LONGITUDINAL

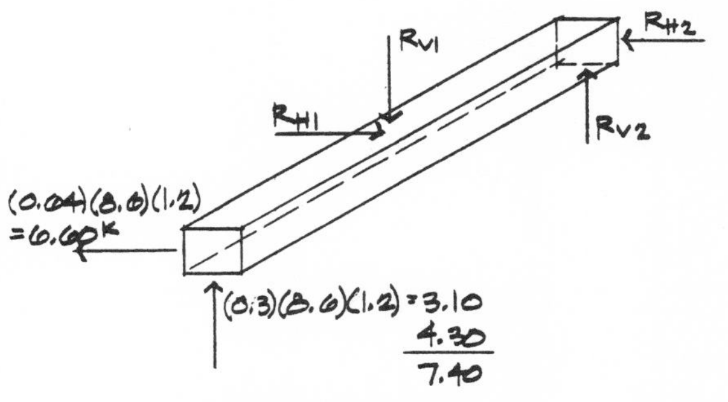

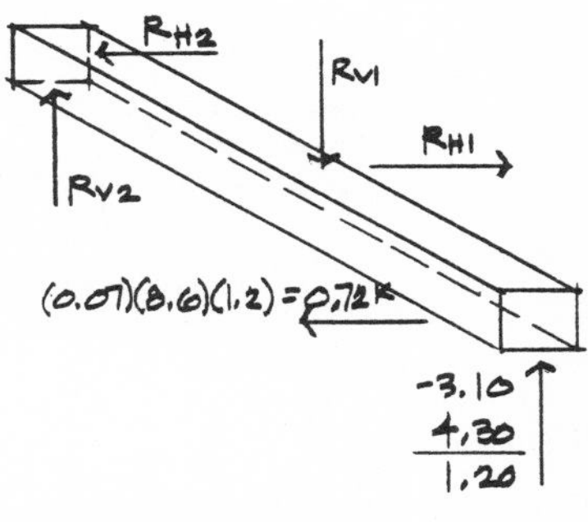

LEFT CANTILEVER

$R_{V1} = 42/21 \times 7.40 = 14.80^K$

$R_{V2} = 14.80 - 7.40 = 7.40^K$

$R_{H1} = 42/21 \times 0.6 = 13.20^K$

$R_{H2} = 13.20 - 6.60 = 6.60^K$

RIGHT CANTILEVER

NOT CRITICAL

$UV_{CRIT} = 0.75(1.4D + 1.7L + 1.87E) = (0.75)(1.4 \times 42/21 \times 4.3 + 1.87 \times 42/21 \times 3.10)$

$= \underline{17.7^K}$

$UH_{CRIT} = 0.75 \times 1.87 \times 13.20 = \underline{18.5^K}$

BRACKET REACTIONS - 1.2g LATERAL

PANEL WT. $W = 0.32\ ^{K}/_{FT}$

$P = (0.5)(15.1)(0.32\ ^{K}/_{FT})(1.2) = 2.90^{K}$

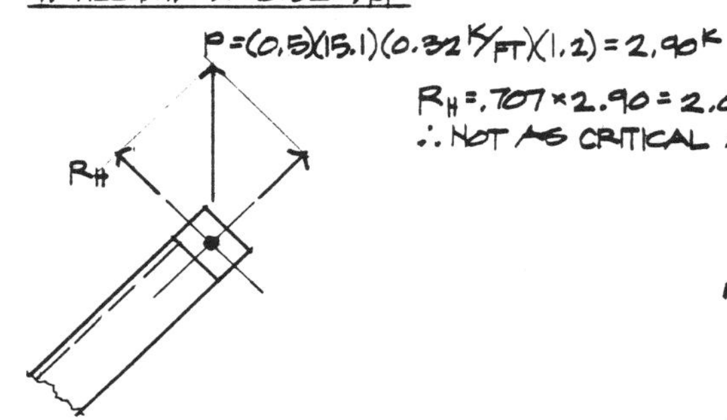

$R_H = .707 \times 2.90 = 2.05^{K}$

∴ NOT AS CRITICAL AS 1.2g LONGITUDINAL

UP

UV

#4 ANCHOR

DESIGN BRACKET INSERT ℙ'S

$UV = 17.7^{K}$ $UP = 18.5^{K}$

TRY ℙ 6×6×3/8 W/4-1/2"φ ×3" LONG STUDS

(SEE PCI DESIGN HANDBOOK PP. 5-24, 5-40, & 5-41)

AS GOVERNED BY STUD CAPACITY

$\phi V_S = 8.8\ ^{K}/_{STUD}$

$\phi P_S = 10.6\ ^{K}/_{STUD}$

$$\left(\frac{P_u}{\phi P_S}\right)^2 + \left(\frac{V_u}{\phi V_S}\right)^2 = \left(\frac{18.5}{4 \times 10.6}\right)^2 + \left(\frac{17.7}{4 \times 8.8}\right)^2 = 0.19 + 0.25 = 0.44 < 1 \quad \text{O.K.}$$

AS GOVERNED BY CONCRETE CAPACITY

ϕV_c: 2 @ $d_e = 2"$ $2 \times 2.8^{K} = 5.6^{K}$

2 @ $d_e = 5"$ $2 \times 9.0^{K} = 18.0^{K}$ } $23.6^{K} = \phi V_c$

ϕP_c: 2 @ $d_e = 2"$ $2 \times 8.6^{K} = 17.2^{K}$

2 @ $d_e = 5"$ $2 \times 13.3^{K} = 26.6^{K}$ } $43.8^{K} = \phi P_c$

$$\left(\frac{P_u}{\phi P_c}\right)^{4/3} + \left(\frac{V_u}{\phi V_c}\right)^{4/3} = \left(\frac{18.5}{43.8}\right)^{4/3} + \left(\frac{17.7}{23.6}\right)^{4/3} = 0.32 + 0.68 = 1.00 \quad \text{O.K.}$$

USE ℙ 6×6×3/8 W/4-1/2"φ STUDS 3" LONG

WELD - TS TO INSERT PL

$$\frac{14.8 \times 2}{4} = 7.4^K$$

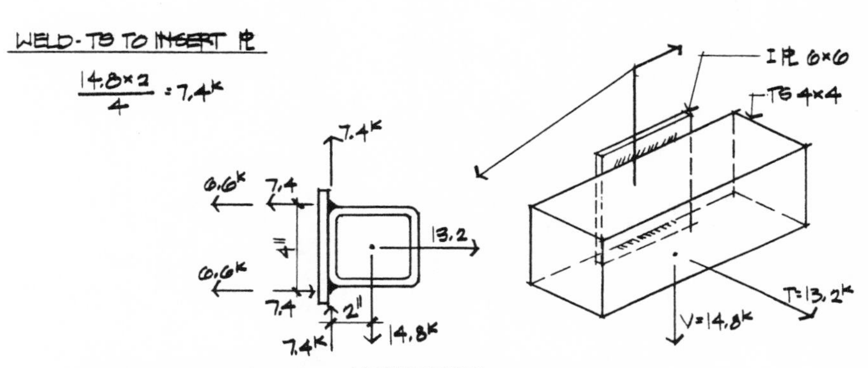

RESULTANT TO TOP WELD: $\sqrt{(7.4)^2 + (14)^2} = 15.83^K$

USING 1/4" WELDS:

$$L = \frac{15.83}{1.33 \times 4 \times .93} = 3.2"$$ USE 4" - 1/4" WELD

LEVELING BOLT AT BRACKET SUPPORT

$P = 4.3^K$ USE 3/4" ϕ BOLT × 4" LONG. ($P_{ALLOW} = 7.36^K$)

SEE P. 4-3 AISC MANUAL

WELD AT BRACKET SUPPORT

$T = 6.6^K$ $L = \frac{6.6}{1.33 \times .93 \times 4} = 1.33$ USE 3" - 1/4" WELD EA. SIDE W/ BACKING BAR

REBAR ANCHOR INTO SLAB

$$A_s = \frac{6.6^K}{20 \times 1.33} = 0.24 \text{ IN}^2$$

USE #4 BAR THROUGH SLOTTED HOLE IN TUBE - EMBED INTO C.I.P. DECK SLAB

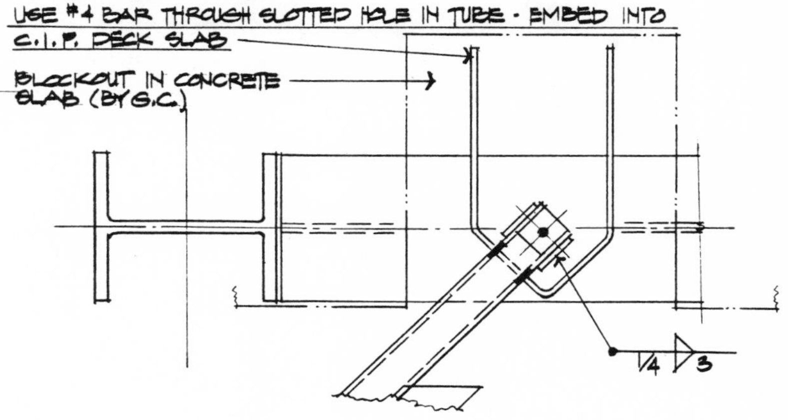

601 MONTGOMERY - SAN FRANCISCO

Pierced window panels are the traditional solution to high rise exteriors in precast concrete. Uniformity of finish and a high degree of quality control are exemplified in these single unit plant cast cladding panels.

WINDOW PANEL UNITS (PIERCED WALL PANELS)

These units are used as either load bearing or non-load bearing cladding units for buildings, with the load bearing applications normally limited to a 4 story, or 50 foot height. However, the large panel building systems for housing have employed these elements as shear panels in seismic zones for many years in Europe. In California, window panel units have been used principally as cladding on multistory structures, with lateral and vertical load resistance being supplied by the building frame. The openings in the precast units are normally cast with a PVC neglet, which serves to receive a zipper type neoprene gasket and subsequent glazing in the field. Problems with trades on site have traditionally prevented glazing being done in the precast concrete fabricator's yard. Often, the shape of the panel is designed such that the projecting elements are used as sun shades, thus reducing solar heat gain and glare. Panel sizes are limited by the usual shipping and erection constraints. Multistory cladding elements with widths of 8' to 12' are subject to seismic drift limitations on the design of the top and bottom connectors. For this reason, in seismic zones, the tendency is toward single story single opening "punched units. Multi window units are subject to shipping limitations and require sufficient project size to justify special tilt truck frames to ship units up to 14' in width. Pierced panels are seldom prestressed due to the difficulty in achieving concentric stressing and elimination of consequent bowing due to eccentric prestressing. The following example demonstrates the proper structural design of the reinforcing and connections of a typical single opening window panel.

601 MONTGOMERY - SAN FRANCISCO

Only precision molds and factory casting techniques can provide the degree of quality demonstrated by these precast concrete cladding units.

WINDOW PANEL DESIGN

DESIGN AN ARCHITECTURAL PRECAST CONCRETE WINDOW PANEL FOR A HIGH RISE OFFICE BUILDING IN SAN FRANCISCO. USE THE UNIFORM BUILDING CODE - 1979 EDITION. PANEL CROSS SECTION AND ELEVATION ARE SHOWN BELOW. 5000 PSI HARD ROCK CONCRETE WILL BE USED FOR THESE PIERCED, NON-LOAD BEARING CLADDING ELEMENTS.

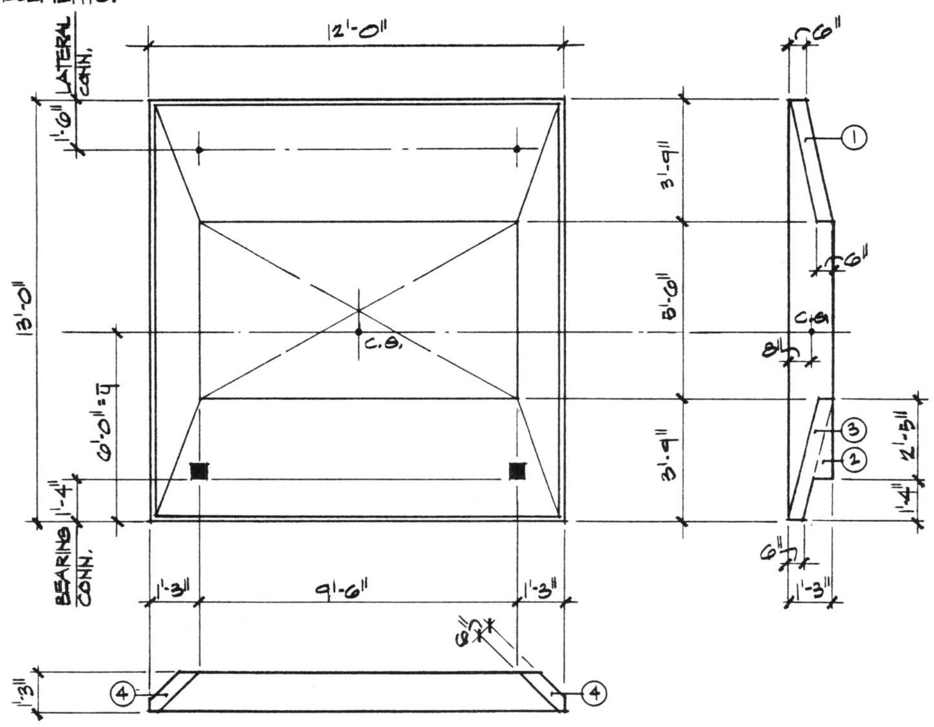

LOCATE C. G.

PART		A ×	L	= VOL ×	y_{bot}	= Vy
1	6×45	270	144	38880	11.13	432734
2	½×6×29	87	144	12528	2.14	26810
3	6×45	270	144	38880	1.88	73094
4	2×6×19	228	111	25308	6.50	164502
				115596		697,140

$$WT. = \frac{115,596}{1728} \times .150 = 10.0^K \qquad \bar{y}_{bot} = \frac{697,140}{115,596} = 6.03'$$

CHECK HANDLING

STRIP @ FINAL SUPPORT POINTS

SECTION PROPERTIES OF VERTICAL MULLIONS
(SEE P. 6-25 AISC MANUAL 7TH EDITION)

$$S = \frac{bd(b^2 \sin^2 a + d^2 \cos^2 a)}{6(b \sin a + d \cos a)}$$

$$S = \frac{(6)(15)[(6)^2(.707)^2 + (15)^2(.707)]}{(6)[(6)(.707) + (15)(.707)]} = 131 \text{ IN}^3$$

$I = 1.5$ $\qquad A = (6)(15) = 90 \text{ IN}^2$

SPAN LENGTH ≅ 9'-3" $\qquad w = \frac{90}{144} \times .150 = 0.094 \text{ K/FT}$

$M = 1.5 \times .125 \times 0.094 \times \overline{9.25}^2 = 1.51^{K\text{-}FT} (18^{K\text{-}IN})$

$f_t = \frac{18}{131} = 0.137 \text{ K/IN}^2 < 5\sqrt{f'_{ci}} = 5\sqrt{2000} = 0.225$ O.K.

∴ HANDLING NOT CRITICAL. IN ORDER TO SIMPLIFY OFF-LOADING AT TIGHT ACCESS POINTS AT THE JOBSITE IN MID-CITY STREETS, IT MAY BE DESIREABLE TO SHIP THESE PANELS VERTICALLY ON AESEL TRAILERS. THIS DECISION WOULD ALSO DICTATE THE MODE IN WHICH THE PANELS WOULD BE STORED.

REINFORCING REQUIRED TO SATISFY STRIPPING

$A_s = \frac{1.51}{1.0 \times 11} = 0.14 \text{ IN}^2$ $\qquad$ ∴ 1-#4 T&B

SERVICE LOAD CONDITION (WIND ON VERT. MULLION GOVERNS)

$w \cong 0.030 \times 6.0 = 0.180 \text{ K/FT}$

$M = 0.125 \times 0.180 \times \overline{9.25}^2 = 1.93^{K\text{-}FT}$

$f_t = \frac{1.93 \times 12}{131} = 0.176 \text{ K/IN}^2 = 5\sqrt{f'_c}$ O.K.

$A_s = \frac{1.93}{1.44 \times 11} = 0.12 \text{ IN}^2$ $\qquad$ ∴ 1-#4 T&B

TEMPERATURE REINFORCING

$A_s = 0.0025 \times 90 = 0.23 \text{ IN}^2$ $\qquad$ 1-#4 T&B IN MULLION

SILL: $\quad A_s = 0.0025 \times 400 = 1.00 \text{ IN}^2$ $\quad$ ∴ USE 6-#4 ($A_s = 1.20$)

HEAD: $\quad A_s = 0.0025 \times 270 = 0.68 \text{ IN}^2$ $\quad$ ∴ USE 4-#4 ($A_s = 0.80$)

TRANSVERSE: $\quad A_s = 0.0025 \times 6 \times 12 = 0.18 \text{ IN}^2$ $\quad$ ∴ USE #4 @ 12" % ($A_s = 0.20$)

REINFORCING SUMMARY

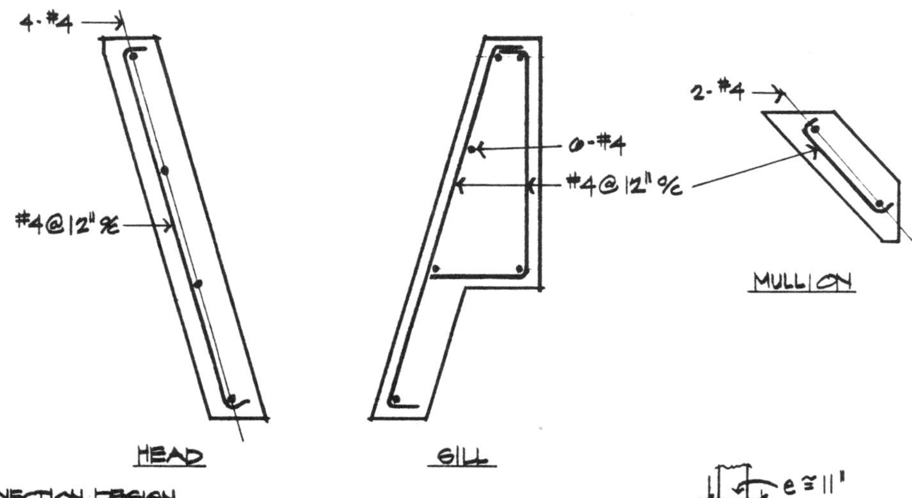

CONNECTION DESIGN

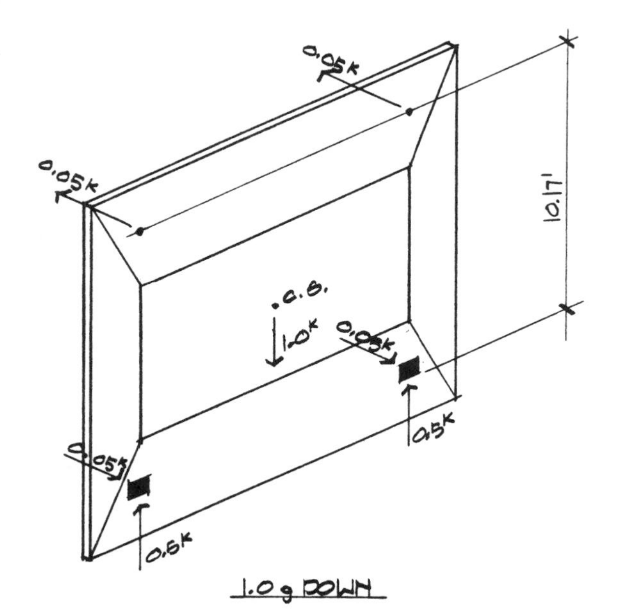

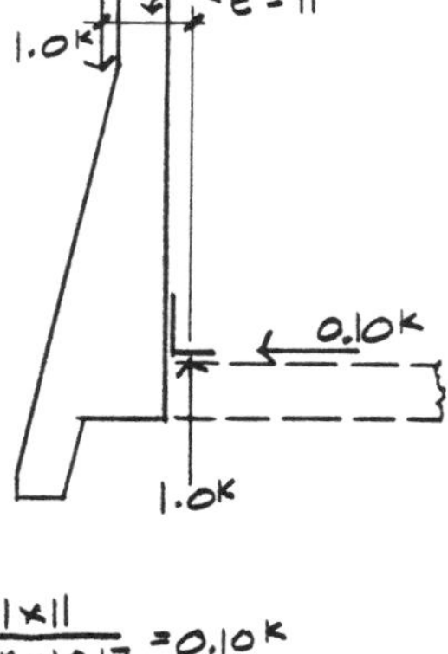

$$\frac{1 \times 11}{12 \times 10.17} = 0.10^k$$

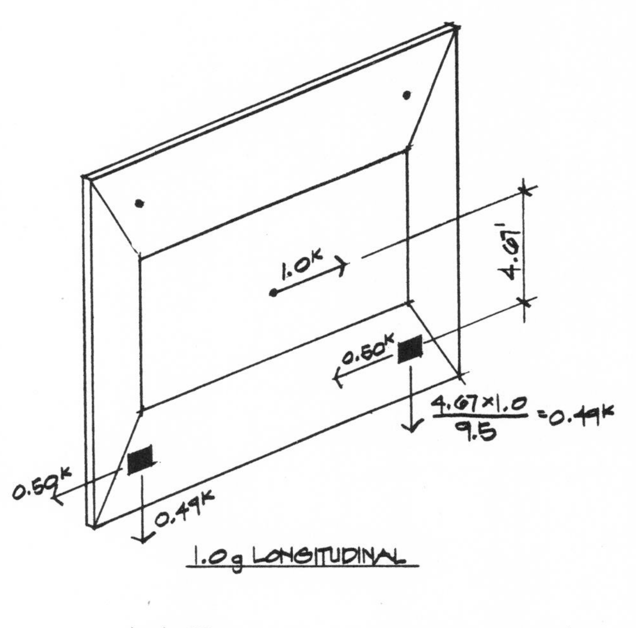

1.0g LONGITUDINAL

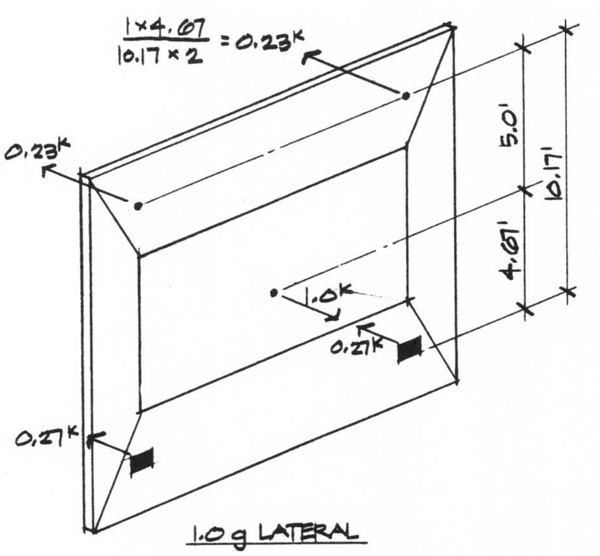

1.0g LATERAL

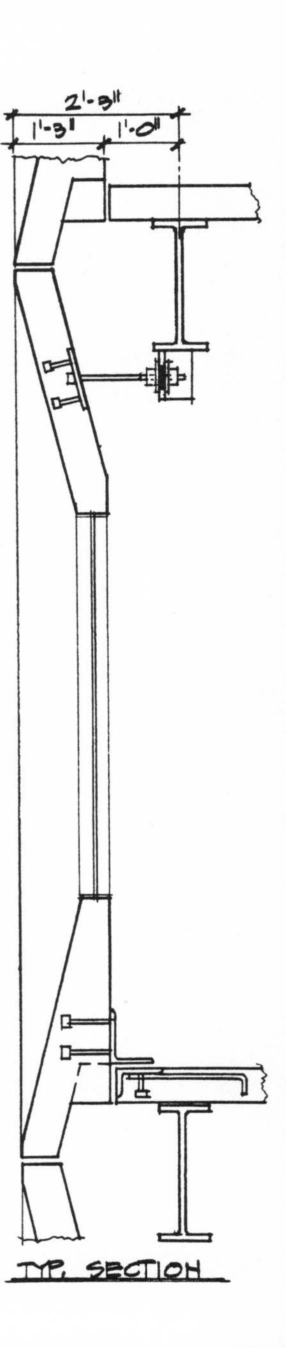

TYP. SECTION

DESIGN CONNECTOR FASTENER - TOP LATERAL CONNECTION

W(1.0^K DOWN + 1.2^K IN/OUT) = $P_{CRITICAL}$

10.0(0.05 + 1.2 × 0.23) = 3.26^K $L_u \cong$ 16"

ROD SIZE

$A_{REQ'D} = \frac{3.26}{20} = 0.163$ IN^2 USE 3/4" φ ROD

P_{ALLOW} = 5.1^K COMP.   8.2^K TENSION

INSERT PL IN PANEL HEAD

UP = 0.75(1.4D + 1.7L + 1.87E)
= 0.75(1.4×0.5 + 1.87×1.2×2.3) = 4.4^K

OR: UP = 0.9D + 1.43E
= 0.9×0.5 + 1.43×1.2×2.3 = 4.4^K

TRY PL 6×3×3/8 W/2 - 3/8" φ × 4" LG. STUDS

(SEE PCI DESIGN HANDBOOK, 2ND EDITION PP. 5-41 & 5-42)

ϕP_c = 15^K (P. 5-42)

ϕP_s = 2×6 = 12^K (P. 5-41) O.K.

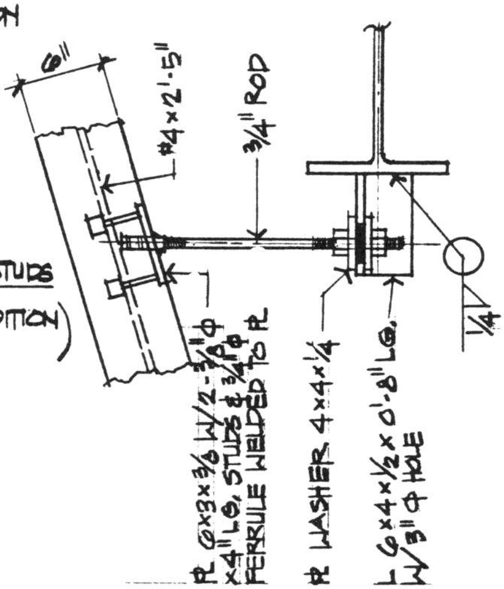

PL THICKNESS REQ'D

$F_b = \frac{M}{S} = 0.9\,F_y = \frac{PL/8}{1/6\,bt^2}$

$(0.9)(36) = \frac{(3.26)(5)(0.125)}{(1/6)(3)(t^2)}$ $t^2 = 0.126$ $t = 0.35"$

USE 3/8" PL

PUNCHING SHEAR

$v_{bo} = \frac{C}{\phi b_o d} = \frac{4.4}{(0.85)(42)(6)} = 0.021$ KSI $< 4\sqrt{f'_c}$ O.K.

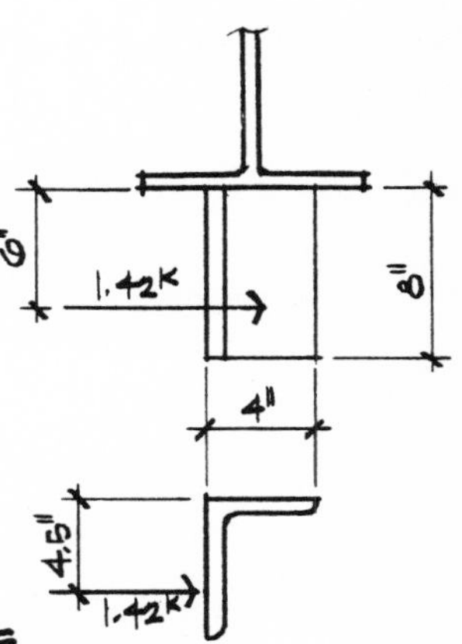

DESIGN CONNECTOR BODY - AT WF SPANDREL

W(1.0g DN + 0.4g IN/OUT)

$= 10.0(0.05 + 0.4 \times 0.23) = 1.42^K$

TRY L6x4 W/ 3" ϕ HOLE

L THICKNESS REQ'D

$$F_b = \frac{M}{S}$$

$$1.33 \times 22 = \frac{1.42 \times 4.5}{1/6 \times 5 \times t^2}$$

$$t^2 = 0.26$$

$$t = 0.51''$$

USE L6x4x½x0'-8" LG. W/ 3" ϕ HOLE

WELD REQ'D TO BOTTOM FLANGE OF WF SPANDREL

$W(1.0g\ DN + 1.2g\ OUT/IN) = 3.26^K$

$M = (3.26)(6) = 19.56^{K\text{-}IN}$

SECTION MODULUS OF WELD GROUP

4"
6"
x = 0.8"

PART	L	x	Lx
4"	8	2	16
6"	12	0	0
	20		16

$$\bar{x} = \frac{16}{20} = 0.8''$$

PART	I_o	Ld^2
4"	10.7	11.5
6"	0	7.7

$$\Sigma I_x = 30\ IN^3$$

$$S_{MIN} = \frac{30}{3.2} = 9.4\ IN^2$$

$$f_{MAX} = \frac{V}{A} + \frac{M}{S} = \frac{3.26}{20} + \frac{19.56}{9.4} = 2.24\ K/IN$$

USE ¼" WELD (0.93 × 4 = 37.2 K/IN)

DESIGN BEARING CONNECTION

FORCE TO CONNECTOR BODY = W(1.0g DN + 0.4g LONG.)

= 10.0(0.50 + 0.4 × 0.49) = 6.96^K

USE L 8×6×3/4

DETERMINE ANGLE LENGTH REQ'D

6.96^K 3.55" K=1.1" 4.9"

0.05×0.4×10=0.2^K

0.4×0.5×10 = 2.0^K

6.96^K

$M = (6.96)\left(\frac{4.9}{2}\right) = 17.05^{K\text{-}IN}$

$F_b = \frac{M}{S} = \frac{17.05}{1/6 \times L \times (0.75)^2} = 28.7 \qquad \therefore L = 6.34''$

USE L 8×6×3/4×0'-8" LG.

DESIGN CONNECTION TO PANEL

UP = 0.75(1.4D + 1.7L + 1.87E) = 0.75(1.4×0.5 + 1.87×1.2×.49)(10) = 13.5^K

OR: UP = 0.9D + 1.43E = (0.9×0.5 + 1.43×1.2×0.49)(10) = 12.9^K

UH = 1.4×0.05×10 = 0.7^K

UV = 1.4×0.50×1.2×10 = 8.4^K

TRY 4 STUDS

1. EFFECT OF UP = 13.5^K

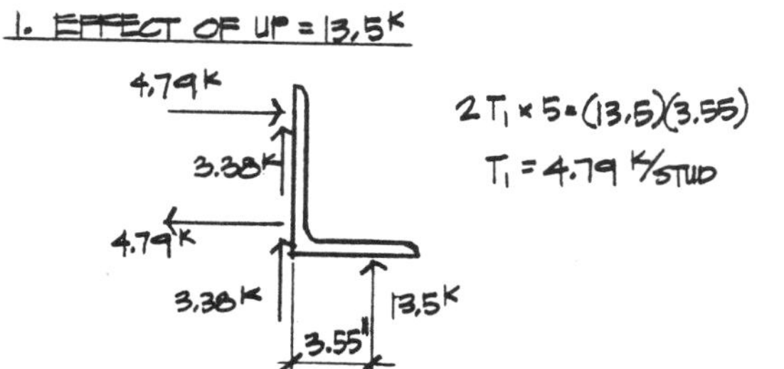

$2T_1 \times 5 = (13.5)(3.55)$

$T_1 = 4.79$ K/STUD

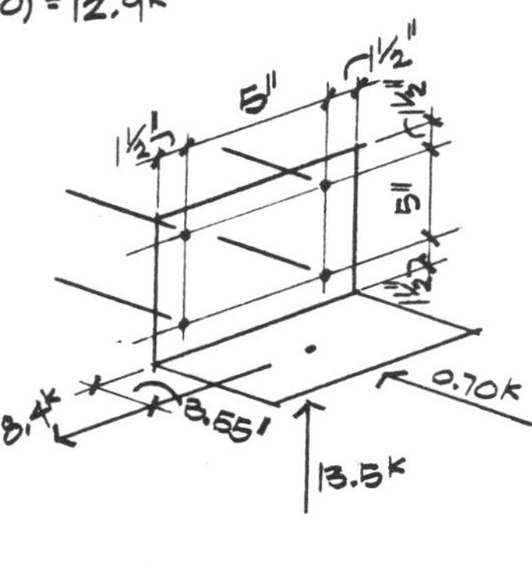

$R_{yv} = \frac{13.5}{4} = 3.38$ K/STUD

2. EFFECT OF ECCENTRIC SHEAR $V_u = 8.4^k$

DUE TO BENDING ($e = 3.55''$)

$$2T_2 \times 5 = (8.4)(3.55)$$

$$T_2 = 2.98 \text{ k/STUD}$$

DUE TO SHEAR

$$R_{xv} = \frac{8.4}{4} = 2.1 \text{ k/STUD}$$

DUE TO TORSION

$$R_{xT} = \frac{Pe\,y}{\Sigma x^2 + y^2} = \frac{(8.4)(4)(2.5)}{50} = 1.68 \text{ k/STUD}$$

$$R_{yT} = \frac{Pe\,x}{\Sigma x^2 + y^2} = \frac{(8.4)(4)(2.5)}{50} = 1.68 \text{ k/STUD}$$

R_y, R, R_x, $e = 4''$, $2.5''$, 8.4^k, $2.5''$

$$\Sigma x^2 + y^2 = (4)(2.5)^2(2) = 50$$

$$R = \sqrt{(R_{xv} + R_{xT})^2 + (R_{yv} + R_{yT})^2} = \sqrt{(2.1 + 1.68)^2 + (3.38 + 1.68)^2}$$

$$= 6.31 \text{ k/STUD}$$

CRITICAL DESIGN CONDITION IS: $\begin{cases} V_u = 6.31 \text{ k/STUD} \\ P_u = 4.79 + 2.98 = 7.77 \text{ k/STUD} \end{cases}$

TRY 4-5/8" ϕ STUDS × 6" LONG.

AS GOVERNED BY STUD CAPACITY (SEE PCI DESIGN HANDBOOK 2ND EDITION PP. 5-40, & 5-41, & 5-24)

$\phi P_s = 16.6$ k/STUD $\qquad \phi V_s = 13.8$ k/STUD

$$\left(\frac{P_u}{\phi P_s}\right)^2 + \left(\frac{V_u}{\phi V_s}\right)^2 = \left(\frac{7.77}{16.6}\right)^2 + \left(\frac{6.31}{13.8}\right)^2 = 0.22 + 0.21 = 0.43 < 1.0 \quad \text{O.K.}$$

AS GOVERNED BY CONCRETE CAPACITY

EDGE DISTANCE $d_e = 8''$ $\qquad f'_c = 5000$ PSI

$\phi P_c = 46.4$ k/STUD $\qquad \phi V_c = 14.1$ k/STUD

$$\left(\frac{P_u}{\phi P_c}\right)^{4/3} + \left(\frac{V_u}{\phi V_c}\right)^{4/3} = \left(\frac{7.77}{46.4}\right)^{4/3} + \left(\frac{6.31}{14.1}\right)^{4/3} = 0.09 + 0.34 = 0.43 < 1.0 \quad \text{O.K.}$$

∴ USE 4-5/8" ϕ × 6" LONG STUDS

DESIGN WELD - BEARING ANGLE TO INSERT ℘ IN FLOOR SLAB

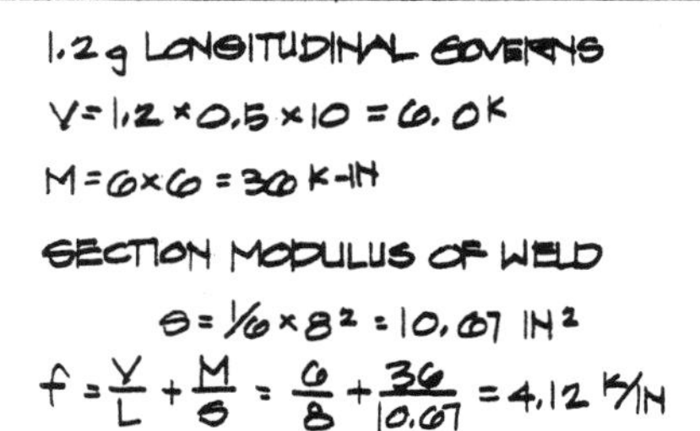

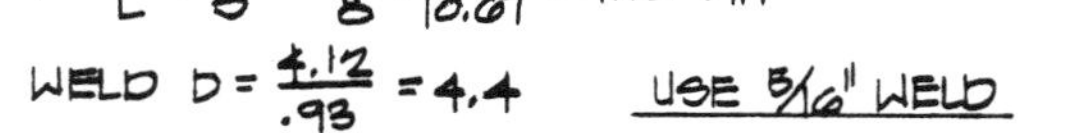

BEARING CONNECTION DETAIL

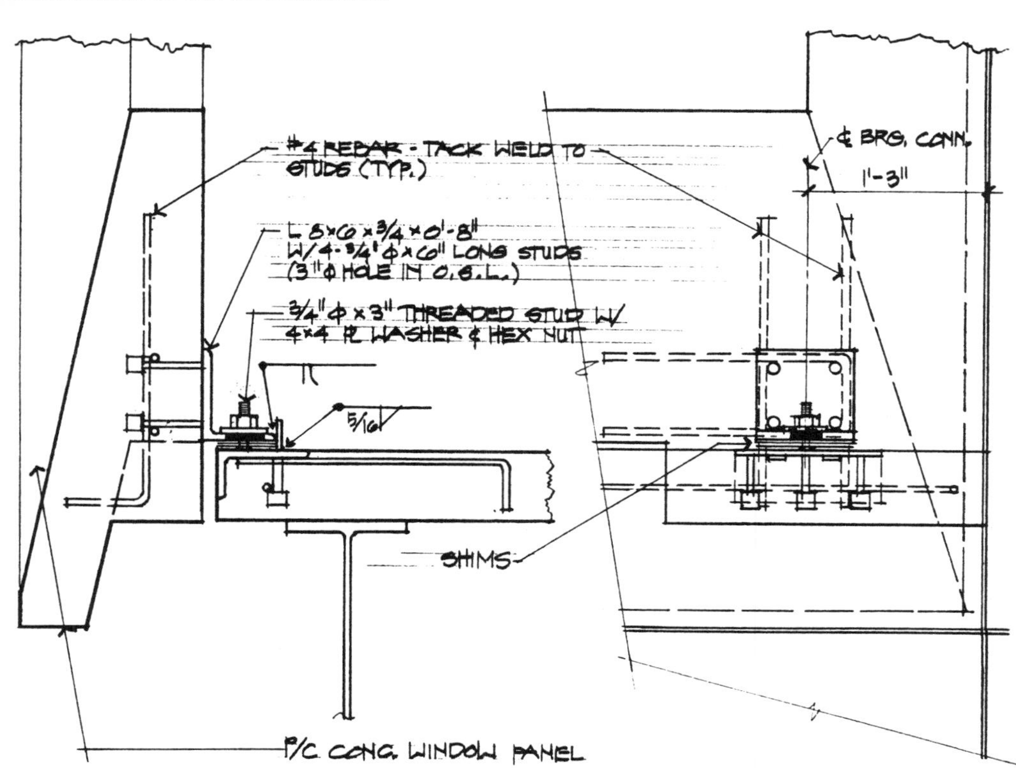

PANEL ERECTION - 601 MONTGOMERY, SAN FRANCISCO

Panel being lowered into position by roof mounted tower crane

Iron worker adjusts top "in-and-out" connection as panel is aligned

Bottom bearing connection with oversized hole to allow for field placing tolerance required for positioning treaded stud

CUTTER LABORATORIES - BERKELEY

Parapet panels, spandrels, column cover panels and 24 foot high unbraced wall panels completely clad this steel frame building. Permanence, speed of construction and attractive appearance, as well as low initial cost influence owners to select precast concrete cladding units.

SOLID WALL PANELS - NARROW HIGH PANELS

Solid wall panels may be designed as cladding elements, or as load bearing-shear resisting elements in the building. They also lend themselves to design as prestressed concrete elements, and may be cast in the form of standard structural shapes, such as single or double tees. Sandwich panel design is very advantageous, especially with the increasingly high cost of energy required to heat and cool building interiors. This design aspect of solid panels is covered on pp. 287-302. Panels may be interconnected horizontally and vertically when they are used to transmit vertical and lateral loads to another panel element or the foundation. When used as cladding elements, the joints between elements are left open, with subsequent caulking installed in the field. They may be used to enclose an entire wall, or spaced with areas of fenestration in between. Normally, these panels are placed vertically one or two stories high, with 8' being used as a common panel width and building module. 8' wide units may be shipped without special permits. Two aspects of design of non-load bearing high wall panels merit some additional explanation. First, the thickness of these solid, relatively long units is such that it is usually always more economical to store and ship the units edgewise, so as to negate the requirement for having additional reinforcement or prestressing required only for controlling stresses associated with flat storage and shipment. The other unusual aspect of design of high panels is that the top lateral connections (the bottom connections are usually always the bearing connections) must be designed to restrain the panel in the longitudinal direction, as well as in the lateral (in-and-out) direction. However, this top connection must also be designed to accommodate building drift under extreme seismic conditions. This is achieved by designing the connector body to behave elastically at 0.4 g, but to yield at forces approaching 1.2 g. The next example demonstrates these items.

ALVARADO SEWAGE TREATMENT PLANT - UNION CITY

These wall panel units transmit seismic lateral forces to the foundations, as well as providing support for the double tee roof.

NARROW, HIGH WALL PANEL DESIGN

DESIGN A HIGH CLADDING PANEL FOR THE EXTERIOR OF A STEEL FRAMED LABORATORY BUILDING IN BERKELEY. USE THE UNIFORM BUILDING CODE - 1979 EDITION. NO INTERMEDIATE SUPPORT POINTS ARE AVAILABLE SUCH THAT THE TOP CONNECTIONS MUST BE DESIGNED TO RESIST THE EFFECT OF SEISMIC FORCES PARALLEL TO, AS WELL AS PERPENDICULAR TO, THE PLANE OF THE PANEL. f'_c = 5000 PSI NR CONCRETE. PANEL CROSS SECTION AND ELEVATION ARE SHOWN BELOW.

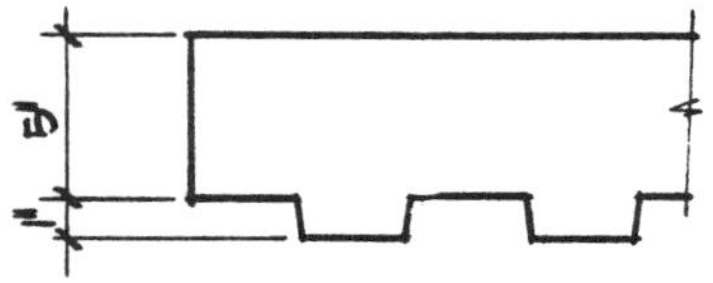

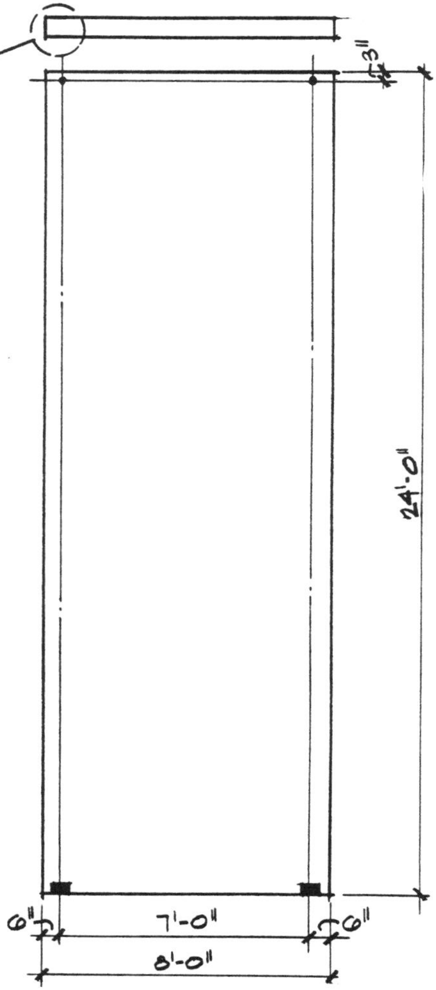

5" 1" 3" 24'-0" 6" 7'-0" 6" 8'-0"

FOR HANDLING CALCULATIONS USE:

$A = 5 \times 12 = 60 \ \text{IN}^2/\text{FT}$

$S = \frac{1}{6} \times 12 \times 5^2 = 50 \ \text{IN}^3/\text{FT}$

$WT. = \frac{5.5}{12} \times .150 = 0.070 \ \text{K}/\text{FT}^2$

PANEL WT.

$(0.07)(8)(24) = 13.5^K (W)$

OR:

$\frac{13.5}{24} = 0.56 \ \text{K}/\text{FT} \ (w)$

STRIPPING TRY 4 POINT PICK

L/10 3L/10 2L/10 3L/10 L/10

$M^+ = M^- = 0.0050 \times 1.5 \times 0.07 \times \overline{24}^2 = 0.339$ K-FT/FT

$f_t = \frac{M}{S} = \frac{0.339 \times 12}{50} = 0.081$ K/IN² O.K.

$A_{s\ REQ'D} = \frac{0.339}{(1)(2.5)} = 0.136$ IN²/FT

OR: $8 \times 0.136 = 1.088$ IN² PER 8' PANEL

REQ'D: 4×4 - W4.0 × W4.0 + 2 - #3 LONG. ($A_s = 1.18$ IN²)

SHIPPING - TRY 2 POINTS FLAT

2L/10 6L/10 2L/10

$M^+ = M^- = 0.025 \times 2.0 \times 0.07 \times \overline{24}^2 = 2.016$ K-FT/FT

$f_t = \frac{2.016 \times 12}{50} = 0.484$ K/IN² $> 5\sqrt{f'_c} = 0.350$ KSI N.G.

∴ STORE AND SHIP VERTICALLY

FINAL DESIGN CONDITION

WIND = 15 PSF

SEISMIC = $0.3g = (0.3)(70) = 21$ PSF ∴ GOVERNS

$M = 0.125 \times 0.021 \times \overline{24}^2 = 1.512$ K-FT/FT

$f_t = \frac{1.512 \times 12}{50} = 0.363 \simeq 5\sqrt{f'_c}$ O.K.

$A_{s\ REQ'D} = \frac{1.512}{1.70 \times 2.5} = 0.344$ IN²/FT

0.021 K/FT²

REINFORCING SUMMARY

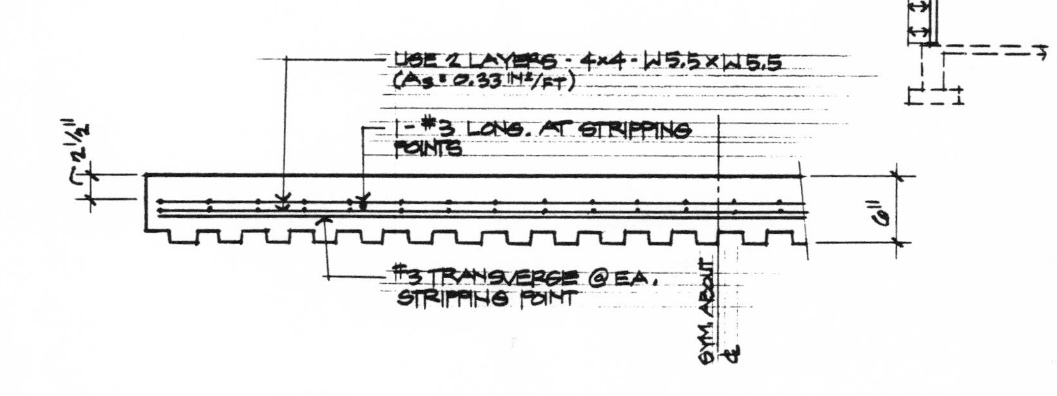

SCHEMATIC HANDLING SUMMARY

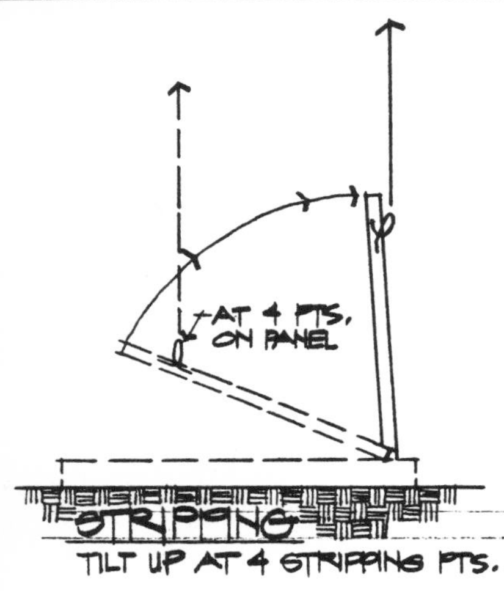

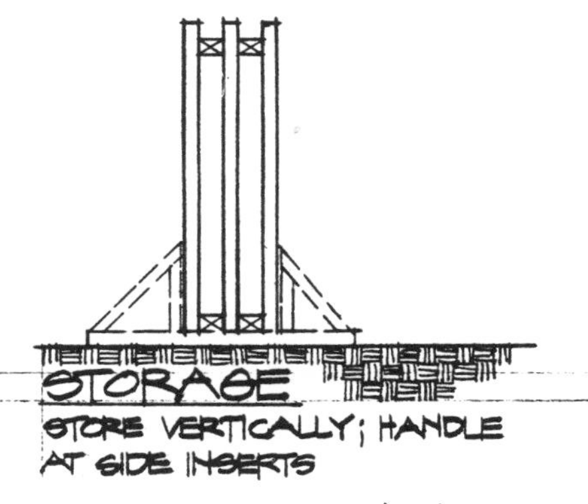

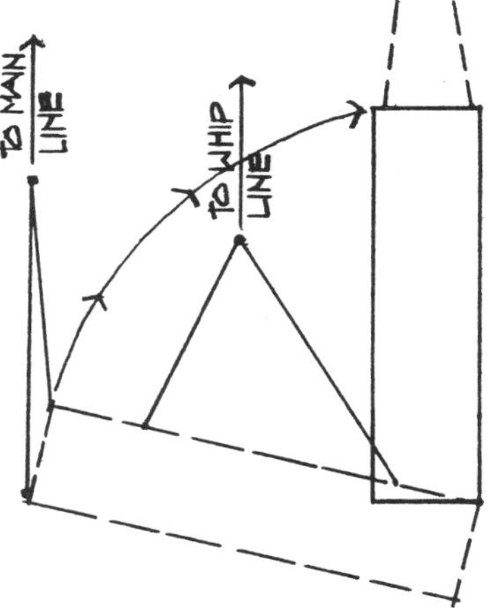

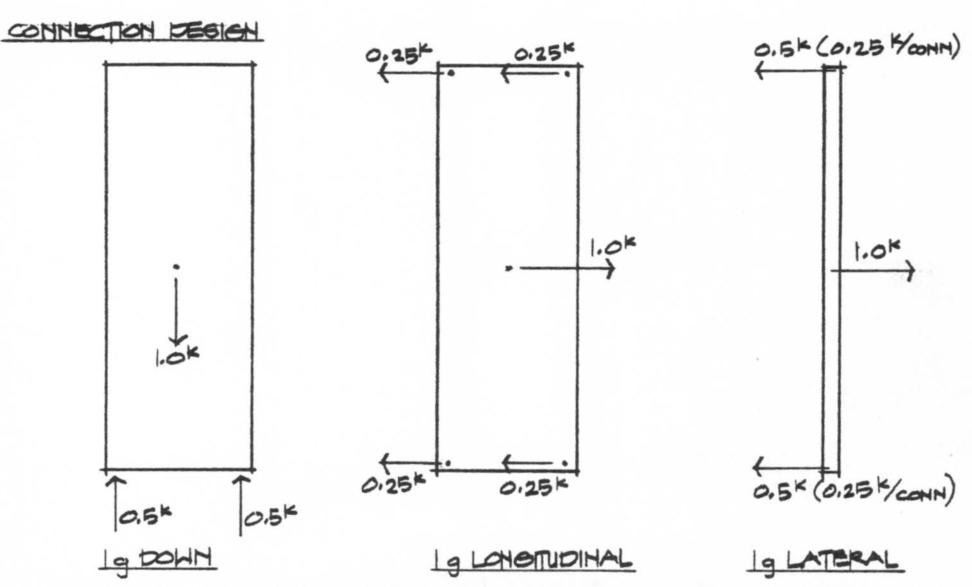

DESIGN TOP BOTTOM CONNECTIONS TO BEHAVE ELASTICALLY UNDER 0.4g LONGITUDINAL LOADING, BUT TO YIELD AND ABSORB ENERGY AT 1.2g LOADING.

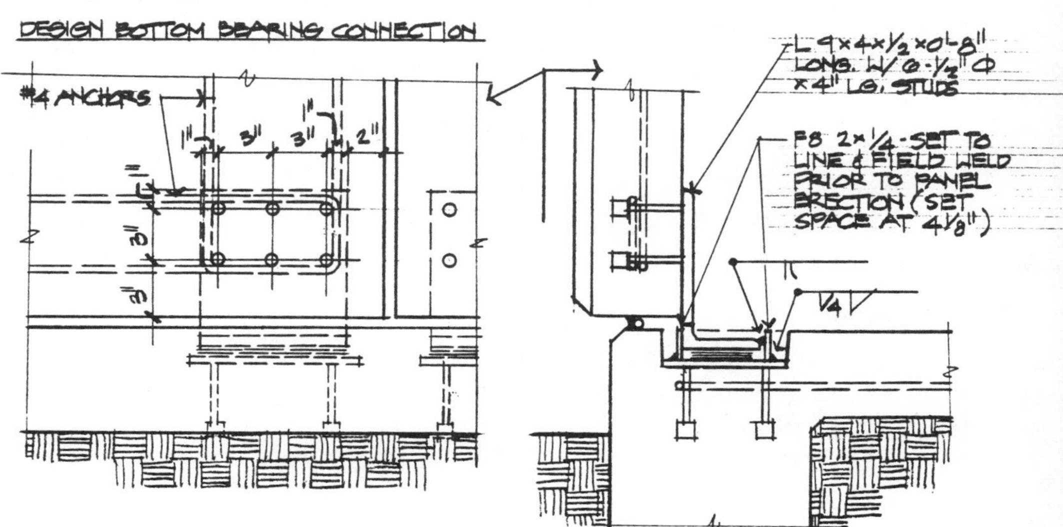

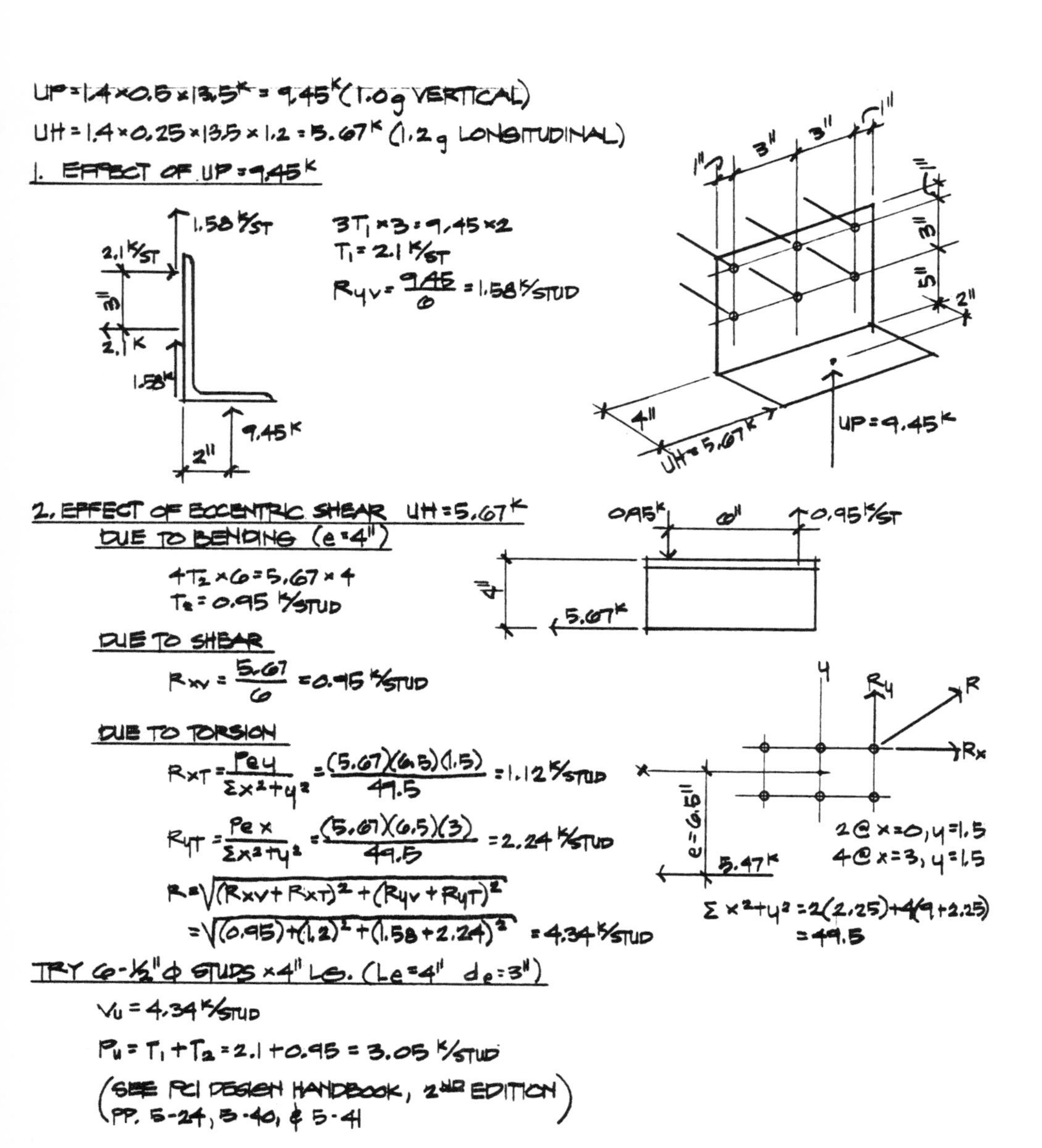

$UP = 1.4 \times 0.5 \times 13.5^K = 9.45^K$ (1.0g VERTICAL)

$UH = 1.4 \times 0.25 \times 13.5 \times 1.2 = 5.67^K$ (1.2g LONGITUDINAL)

1. EFFECT OF UP = 9.45^K

$3T_1 \times 3 = 9.45 \times 2$

$T_1 = 2.1\ ^K/_{ST}$

$R_{yv} = \frac{9.45}{6} = 1.58\ ^K/_{STUD}$

2. EFFECT OF ECCENTRIC SHEAR UH = 5.67^K

DUE TO BENDING (e = 4")

$4T_2 \times 6 = 5.67 \times 4$

$T_2 = 0.95\ ^K/_{STUD}$

DUE TO SHEAR

$R_{xv} = \frac{5.67}{6} = 0.95\ ^K/_{STUD}$

DUE TO TORSION

$R_{xT} = \frac{Pey}{\Sigma x^2 + y^2} = \frac{(5.67)(6.5)(1.5)}{49.5} = 1.12\ ^K/_{STUD}$

$R_{yT} = \frac{Pex}{\Sigma x^2 + y^2} = \frac{(5.67)(6.5)(3)}{49.5} = 2.24\ ^K/_{STUD}$

$R = \sqrt{(R_{xv} + R_{xT})^2 + (R_{yv} + R_{yT})^2}$

$= \sqrt{(0.95) + (1.2)^2 + (1.58 + 2.24)^2} = 4.34\ ^K/_{STUD}$

$\Sigma x^2 + y^2 = 2(2.25) + 4(9 + 2.25)$
$= 49.5$

TRY 6-½"φ STUDS × 4" LG. ($L_e = 4"$ $d_e = 3"$)

$V_u = 4.34\ ^K/_{STUD}$

$P_u = T_1 + T_2 = 2.1 + 0.95 = 3.05\ ^K/_{STUD}$

(SEE PCI DESIGN HANDBOOK, 2ND EDITION
PP. 5-24, 5-40, & 5-41)

AS GOVERNED BY STUD CAPACITY

$\phi P_s = 10.6$ K/STUD $\quad \phi V_s = 8.8$ K/STUD

$$\left(\frac{P_u}{\phi P_s}\right)^2 + \left(\frac{V_u}{\phi V_s}\right)^2 = \left(\frac{3.05}{10.6}\right)^2 + \left(\frac{4.36}{8.8}\right)^2 = 0.08 + 0.24 = 0.32 < 1.0 \quad \text{O.K}$$

AS GOVERNED BY CONCRETE CAPACITY ($d_e = 3''$, $L_e = 4''$, $f'_c = 5000$)

$\phi P_c = 16$ K/STUD $\quad \phi V_c = 5.5$ K/STUD

$$\left(\frac{P_u}{\phi P_c}\right)^{4/3} + \left(\frac{V_u}{\phi V_c}\right)^{4/3} = \left(\frac{3.05}{16}\right)^{4/3} + \left(\frac{4.34}{5.5}\right)^{4/3} = 0.11 + 0.73 = 0.84 < 1.0 \quad \text{O.K.}$$

∴ USE 6-½"φ×4" LONG STUDS AS SHOWN

CHECK 1.2g OUT ON CONNECTION

$UH = 1.4 \times 1.2 \times 0.25 \times 13.5 = 5.67^K$

$5.67 \times 8 = 3T \quad \therefore T = 15.12^K$ OR 5.04 K/STUD

$\phi P_s = 10.6^K$ O.K.

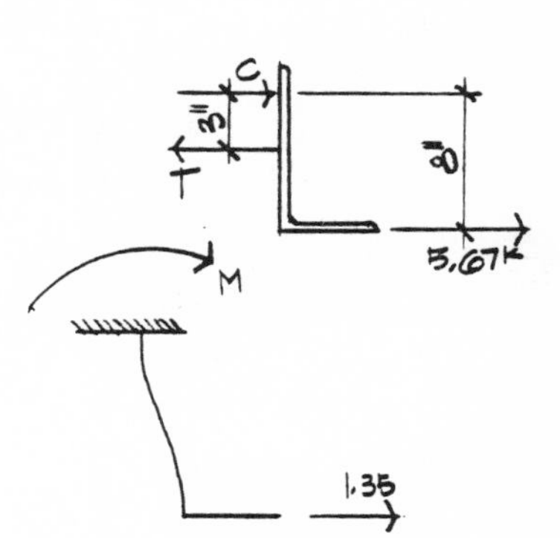

CHECK ANGLE LEG AT 0.4g OUT

$H = 0.4 \times 0.25 \times 13.5 = 1.35^K$

$M = (1.35)(5) = 6.75^{K\text{-}IN}$

$$f = \frac{M}{S} = \frac{6.75}{1/6 \times 8 \times (.5)^2} = 20.3 \text{ KSI} \quad \text{O.K.}$$

CHECK FORCE IN O.S.L. OF L9×4×½ WHEN PANEL ROTATES UNDER BUILDING DRIFT AT 1.2g, ASCERTAIN THAT CONNECTION IS DEVELOPED

$M_p = 3F$

@ YIELD: $\quad z = \frac{bd^2}{4} = \frac{(8)(.5)^2}{4} = 0.5 \text{ IN}^3$

$F_y z = M_p$

$(36)(0.5) = 3F \quad \therefore F = 6^K$ O.K.

DESIGN WELD - L9×4×½ TO INSERT PL IN C.I.P. FLOOR

$M = (5.67)(4) = 22.68^K \quad (UH @ 1.2g = 5.67^K)$

$S = 1/6 \times 8^2 = 10.67$

$$f = \frac{22.68}{10.67} = 2.13 \text{ K/IN}$$

$$D = \frac{2.13}{.93} = 2.28$$

USE ¼" WELD

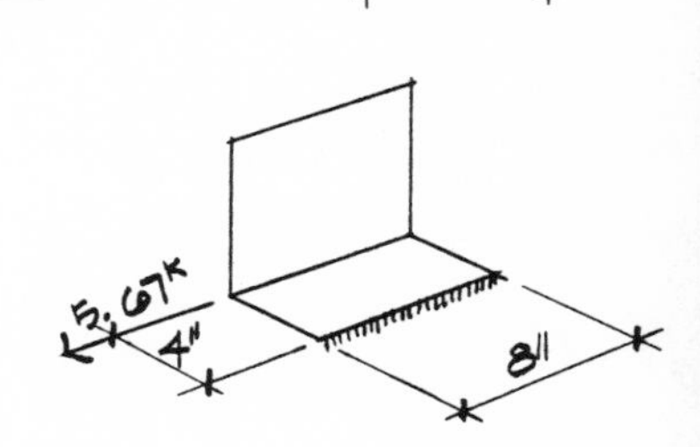

DESIGN TOP CONNECTION

$1.2g$ IN/OUT $= (1.2)(0.25)(13.5) = 4.05^K$

UP $= 1.4 \times 4.05 = 5.67^K$

USE 2-½" φ × 4" LONG STUDS + L 4×4×3/8

$P_u = V_u = .707 \times 5.67 \times .5 = 2.00^K$/STUD

$d_e = 2 \qquad l_e \cong 3"$

$\phi P_c = 8.6^K \qquad \phi P_s = 10.6^K$

$\phi V_c = 2.8^K \qquad \phi V_s = 8.8^K$

$$\left(\frac{P_u}{\phi P_c}\right)^{4/3} + \left(\frac{V_u}{\phi V_c}\right)^{4/3} = \left(\frac{2.0}{8.6}\right)^{4/3} + \left(\frac{2.0}{2.8}\right)^{4/3}$$

$= 0.14 + 0.64 = 0.78 < 1.0$ OK.

THREADED STUD SIZE

$$A_{REQ'D} = \frac{4.05}{1.33 \times 20} = 0.15 \text{ IN}^2$$

USE 5/8" φ STUD

TUBE SIZE

$M = (4.05)(9) = 36.45^{K\cdot FT}$

$$S = \frac{36.45}{1.33 \times 24} = 1.14 \text{ IN}^3$$

USE TS 4×4×.25 ($S = 4.00 \text{ IN}^3$)

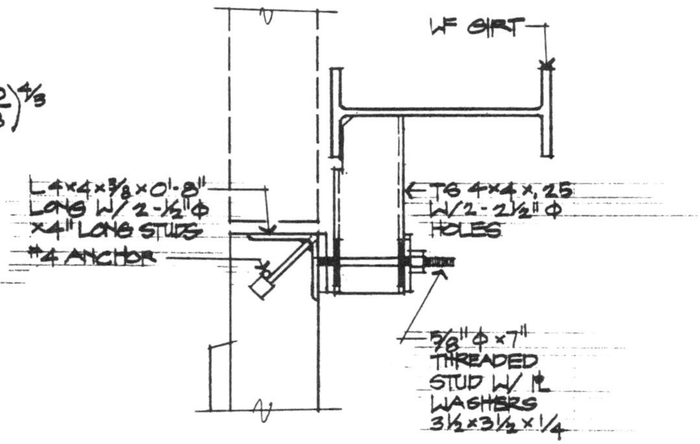

601 MONTGOMERY
SAN FRANCISCO

Panel erection with a roof mounted tower crane allows uninterrupted erection of precast concrete cladding panels despite limited access at the site and congested mid-city streets.

LEVITZ'S FURNITURE WAREHOUSE - CONCORD

24 and 40 foot high double tee wall panels support vertical loads and resist lateral loads from earthquake and wind.

PRESTRESSED SOLID WALL PANELS

Solid precast concrete wall panels are often prestressed, whereas pierced panels are not. This is due to the interference of the openings with the prestressing strands, the tendency of the opening surrounds to bind to the form after release of prestress but before stripping, and the possibility of undesired bowing induced by eccentric prestressing. Prestressed panels are usually flat, ribbed, or made from standard structural shapes, such as the double tee. The panels may be single thickness or sandwich construction. Sandwich panels usually have the structural wythe prestressed, with the wythe covering the insulation "floating." Flat sandwich panels with both outer and inner wythes prestressed are designed taking into account the temperature gradient difference between the two wythes, and the effect of differential drying shrinkage and the resultant induced stresses and bowing. These members are often designed as composite elements, with horizontal shear transfer being designed in the ties between the outer and inner wythes. The usual selections of finishes are available, with the conditional requirement that transverse ribs should be avoided to prevent binding in the forms resulting from prestress elastic shortening strains. Prestressed panels are made in a long line similar to the process used for flexural deck members, with associated efficiencies in plant labor.

An advantage of prestressing is that it is a more economical way to reinforce the wall panel as compared with mild steel reinforcing. Prestressing also helps to keep handling stresses within allowable limits. Also, for units where the average effective prestress exceeds 225 psi, the minimum amounts of reinforcing specified for mild steel reinforced walls are waived, and ties around the prestress strands may be omitted, unless they are required for shear. Prestressing also increases the ultimate flexural capacity of the wall, and in most applications, bending governs the design and not the axial load capacity of the element.

In designing these walls, often the fire resistive requirements dictate the flange thickness of stemmed or double tee units. Thicknesses required for various fire ratings are shown below:

Fire Rating/hours	1	2	3	4
Sand-lightweight concrete	3"	3 3/4"	4 3/4"	5 1/2"
Hardrock concrete	3 1/2"	5"	6 1/4"	7"

Of course, the proper cover distances for reinforcing and strand, as listed in Table 43A of UBC-79 would be observed also, in addition to the overall required wall or flange thickness. UBC-79 also waives arbitrary h/t ratios given in Section 2614(c) provided a detailed structural analysis is made to show structural stability. The moment magnification method given in the code is sufficient analysis for slenderness for $kl/r \leq 100$. For a slenderness ratio greater than 100, the P-Δ effect shall be checked in accordance with Section 2610(k) of UBC-79.

Load bearing prestressed concrete wall panels are normally designed to exhibit no tension under full service load conditions. For tall slender panels, the seismic lateral loading condition of 0.3 g, combined with dead plus live load, will normally govern. The wall element will also have to fall within the envelope created for combined ultimate axial and bending indicated by the interaction diagram for the particular section and reinforcing under consideration. Often overlooked in the design of prestressed panels are the minimum eccentricities specified in Section 2610(d) of UBC-79.

In outline form, the design procedure for analysis of a prestressed concrete wall panel is as follows:

(1) Select flange or wall thickness based upon required fire rating.

(2) For preliminary design, select a section for which kl/r =100. Preliminary sections can be rectangular, or standard structural shapes.

(3) Summarize the loads to the panel, both vertical and lateral.

(4) Determine the most critical ultimate loading condition -

$$1.4D + 1.7L$$
$$0.75(1.4D + 1.7L + 1.7W)$$
$$0.75(1.4D + 1.7L + 1.87E)$$

(5) Refer to interaction diagrams, such as those in the 2nd edition of the PCI Design Handbook on pp 2-58 through 2-61, to select a trial section.

(6) Select prestressing to provide a minimum of 225 psi; check for zero tension under full service loading. Cover over strand should conform to that required for the required fire rating, or weather exposure conditions as outlined in Section 2607(o) of UBC-79.

(7) Evaluate slenderness effects

(8) Check handling

(9) Construct interaction diagram - ascertain that the trial section satisfies ultimate strength design criteria.

The following design example demonstrates the use of this procedure in designing a load bearing double tee wall panel in an industrial building.

DOUBLE TEE WALL PANEL DESIGN

DESIGN AN 8'-0" WIDE LOAD BEARING WALL PANEL FOR AN INDUSTRIAL BUILDING IN FIRE ZONE I (1 HR RATING REQUIRED FOR THE WALL). PHYSICAL GEOMETRY IS SHOWN IN THE SKETCH BELOW:

PRELIMINARY DESIGN

FROM UBC-79 TABLE 43-B, 3½" OF GRADE A HR CONCRETE ARE REQUIRED FOR A 1 HOUR RATING ∴ USE 3½" TH. DT FLANGE

LOADS AND MOMENTS (PER 8' WIDE DT PANEL)

L.S. JOISTS	$\frac{20\ PLF}{4} \times \frac{60}{2} \times 8$	=	1200 LB
METAL DECK	3 PSF × 30 × 8	=	720
B.U. ROOF + INS.	6 PSF × 30 × 8	=	1440
SPRINKLERS	5 PSF × 30 × 8	=	1200
MECH + MISC.	5 PSF × 30 × 8	=	1200
LL	16 PSF × 30 × 8	=	3840
	40 PSF		9600 LB

TEE WEIGHT (EST.) 60 PSF × 8 × 29 = 13,920 LB

$UP^{TOP} = (1.4)(5.760) + (1.7)(3.840) = 14.6^{K}$

$UP_{BOT} = 14.6 + (1.4)(13.920) = 34.1^{K}$

WIND

$1/8 \times 0.015 \times 8 \times \overline{26}^2 = 10.1^{K\text{-}FT}$

SEISMIC (0.3g UBC-79 TABLE 23-J)

$1/8 \times 0.3 \times 0.060 \times 8 \times \overline{26}^2 = 12.2^{K\text{-}FT}$

SUMMARY OF LOADING CONDITIONS

1. D+L

AT TOP $UP = 14.6^{K}$

$UM = (14.6)(4/12) = 4.9^{K\text{-}FT}$

(ASSUME e = 4")

AT BASE (ASSUME 50% FIXED)

$UP = 34.1$

$UM \cong 4.9 \times 0.5 = 2.5^{K\text{-}FT}$

2. D+L+W (MAX @ MID HEIGHT)

$UP = 0.75\left(\frac{14.6 + 34.1}{2}\right) = 18.3^{K}$

$UM = 0.75 \times 1.7 \times 10.1 - 3/4 \times 0.75 \times 4.9 = 10.3^{K\text{-}FT}$

3. D+L+E (MAX. AT MID. HT.)

$U_P = 18.3^K$

$U_M = 0.75 \times 1.7 \times 1.1 \times 12.2 + 3/4 \times 0.75 \times 4.9 = 19.9^{K\text{-}FT}$

(2-58)* CONDITION 3 GOVERNS. FROM INTERACTION CHART FIG. 2.6.3, A 12" DEEP SECTION WITH A 2" FLANGE IS AMPLE.

TRY A 10" DEEP PANEL

$f'_c = 5000$ PSI

$E_c = 4300$

8'-0" ; 6 3/4" ; 2" ; 3 1/2" ; 10" ; 6 1/2" ; 6" ; 5" ; 2" ; CGS 4.00" ; CGC 2.74" ; e = 1.26" ; 3-7/16" φ STRANDS (e = 1.26")

SECTION PROPERTIES

PART	A	y_t	Ay_t	I_o	Ad^2
3.5×96	336	1.75	588	343	329
6/6.5 ×6.5	83	6.75	560	292	1335
	419		1148	ΣI = 2300 IN⁴	

$\bar{y}_T = \frac{1148}{419} = 2.74"$ $\quad S_t = 839$ IN³ $\quad$ (WT. = 56 PSF)

$y_b = 7.26"$ $\quad S_b = 317$ IN³

CHECK PRESTRESSING [MIN. REQ'D. = 225 PSI PER UBC-79 SEC. 7618 (o)]

(3-15)(3-29) $F_{f\,AVG.} = (0.225)(419) = 94.3^K$

FOR 7/16" φ - 270^K STRANDS

$F_\phi = (0.78)(21.7) = 16.9$ K/STR.

$N = \frac{94.3}{16.9} = 5.58$

USE 6-7/16" φ -270^K STRANDS (e = 1.26")

(3-38) SERVICE LOAD CONDITION - D+L+E

$$f_{OUTSIDE} = -\frac{(16.9)(6)}{419} - \frac{(101.4)(1.26)}{317} + \frac{[(3/4)(4.9)+12.2](12)}{317} - \frac{9.6+0.5\times13.9}{419}$$

$$= -0.242 - 0.403 + 0.601 - 0.039 = -0.083 \text{ KSI O.K.}$$

CHECK AVERAGE PRESTRESS INDUCED

$$f_{INSIDE} = -\frac{101.4}{419} + \frac{(101.4)(1.26)}{839} = -0.090$$

$$f_{OUTSIDE} = -\frac{101.4}{419} - \frac{(101.4)(1.26)}{317} = -0.645$$

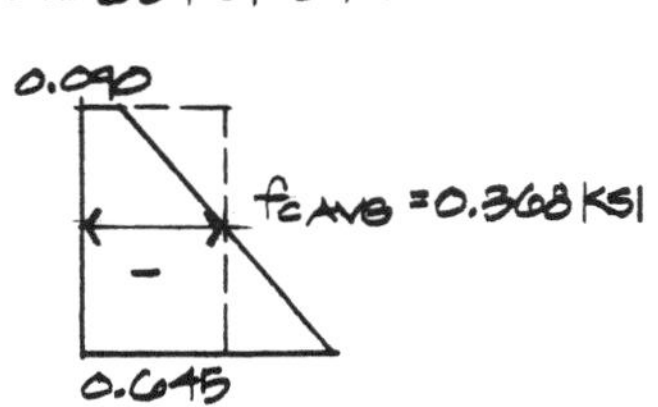

* ( ) PAGE REFERENCE IN 2ND EDITION OF PCI DESIGN HANDBOOK

CHECK HANDLING:

STRIPPING - USE IMPACT FACTOR $I = 1.5$
TRY 2 PT. PICK @ 1/5 PTS

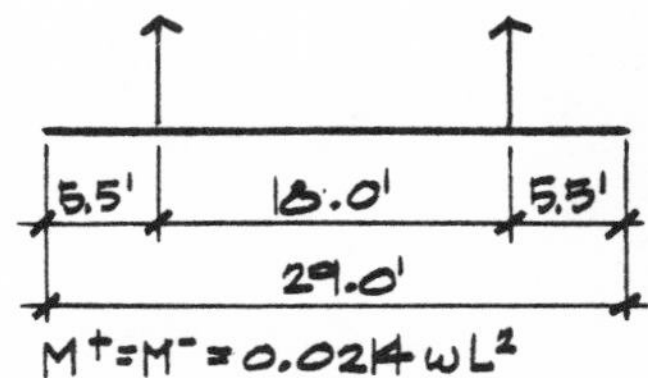

$M^+ = M^- = (0.0214)(0.44)(1.5)(29)^2 = 11.9^{K\text{-}FT}$

$f^{TOP} = -0.090 + \frac{(11.9)(12)}{889} = +0.080$ KSI

MAX. ALLOWABLE $= \frac{f_r}{F.S.} = \frac{(7.5)(\sqrt{3500})}{1.65} = +0.270$ KSI O.K.

$\therefore$ STRIP, STORE, AND SHIP BY BLOCKING AT 1/5 POINTS

(4-33) EVALUATE SLENDERNESS EFFECTS OF PANEL/COLUMN UBC-79, 2610 (ℓ)

RADIUS OF GYRATION OF PANEL

$r = \sqrt{\frac{I}{A}} = \sqrt{\frac{2300}{419}} = 2.34''$

FOR SEMI-FIXED BASE, $K = 0.9$ (SEE AISC MANUAL P. 5-138)

$\frac{K\ell}{r} = \frac{(0.9)(23)(12)}{2.34} = 115 > 100$

(4-35) $\therefore$ ANALYSIS OF PΔ EFFECT MUST BE MADE AS PER UBC-79 SEC. 2610 K. FOR THE ANALYSIS INCLUDE EFFECT OF PANEL BEING ERECTED 2" OUT OF PLUMB.

$\Delta = \frac{5wL^4}{384\,EI} = \frac{(5)(0.202)(23)^4(1728)}{(384)(4300)(2300)} = 0.18''$

$UP\Delta = (18.3)(0.18/12) = \underline{0.3}^{K\text{-}FT}$

ADDITIONAL MOMENT FROM 2" OUT OF PLUMBNESS

$UM^{2''e} = (14.6)(2/12) = 2.4^{K\text{-}FT}$

FINAL MAGNIFIED ULTIMATE DESIGN MOMENT:

$\Sigma UM = 19.9 + 0.3 + 2.4 = 22.6^{K\text{-}FT}$

$UP = 18.3^K$

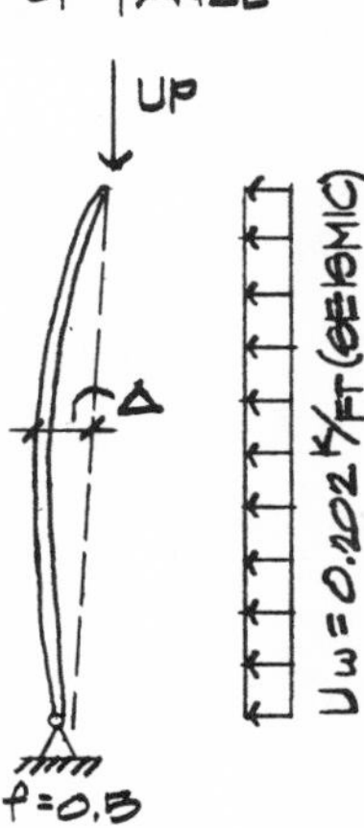

CONSTRUCT INTERACTION DIAGRAM

(3-35) (3-55) (3-5)

1. DETERMINE M_o ($\phi P_n = 0$)

$f_{se} = 147$ KSI $\quad c\bar{\omega}_p = (1.06)\left(\frac{4 \times 0.115}{96 \times 5.5}\right)\left(\frac{270}{5}\right) = 0.05 < 0.08$

$\therefore f_{ps}/f_{pu} = 0.98 \quad \therefore f_{ps} = 264.6$ KSI

$a/2 = \frac{A_{ps} f_{ps}}{1.7 f'_c b} = \frac{(0.46)(264.6)}{(1.7)(5)(96)} = 0.15''$

$\phi M_o = 0.85\,\phi A_{ps} f_{ps} (d - a/2)$

$= (0.85)(0.9)(0.46)(264.6)(5.5 - 0.15) = 498^{K\text{-}IN}$ $\underline{(41.5^{K\text{-}FT})}$

2. DETERMINE PT. WHERE $\phi = 0.7$

$0.7 P_n = 0.1 f'_c A_g = (0.1)(5)(419) = 210^K \quad (P_n = 300^K)$

(3-36)

3. DETERMINE ANOTHER POINT NEAR END OF CURVE

SET $a = 0.5''$ $\quad \therefore c = \frac{a}{\beta_1} = \frac{0.5}{0.8} = 0.625''$; $y' = \frac{a}{2} = 0.24''$

$A_{comp} = (0.625)(96) = 60$ IN² ; $d = 5.5''$; $0.85 f'_c = 4.25$ KSI

$\epsilon_s = \left[\frac{f_{se}}{E_s} - \frac{0.003}{c}(c - d')\right] = \left[\frac{147}{27{,}500} - \frac{0.003}{0.625}(0.625 - 1.5)\right] = 0.00955$ IN/IN

$\epsilon_{ps} = \left[\frac{f_{se}}{E_s} + \frac{0.003}{c}(d - c)\right] = \left[\frac{147}{27{,}500} + \frac{0.003}{0.625}(5.5 - 0.625)\right] = 0.02875$ IN/IN

$f_s = \epsilon_s E_s = (0.00955)(27{,}500) = 262.6$ KSI

(8-17)

$f_{ps} = 0.98 f_{pu} = 264.6$ KSI

$0.85\,\phi P_n = 0.85\,\phi\left[(60)(4.25) - (2)(0.115)(262.6) - (4)(0.115)(264.6)\right] = 62.0\phi^K$

BY INTERPOLATION BETWEEN $\phi = 0.9$ @ $P_n = 0$ & $\phi = 0.7$ @ $P_n = 300^K$

ϕ @ $P_n = 62^K = 0.74 \quad \therefore \phi P_n = (0.74)(62) = \underline{45.9^K}$

$0.85\,\phi M_n = (0.85)(0.74)\left[(255)(2.74 - 0.25) - (60.4)(2.74 - 1.50) + (121.7)(5.50 - 2.74)\right]$

$= 564^{K\text{-}IN}$ $\underline{(47.0^{K\text{-}FT})}$

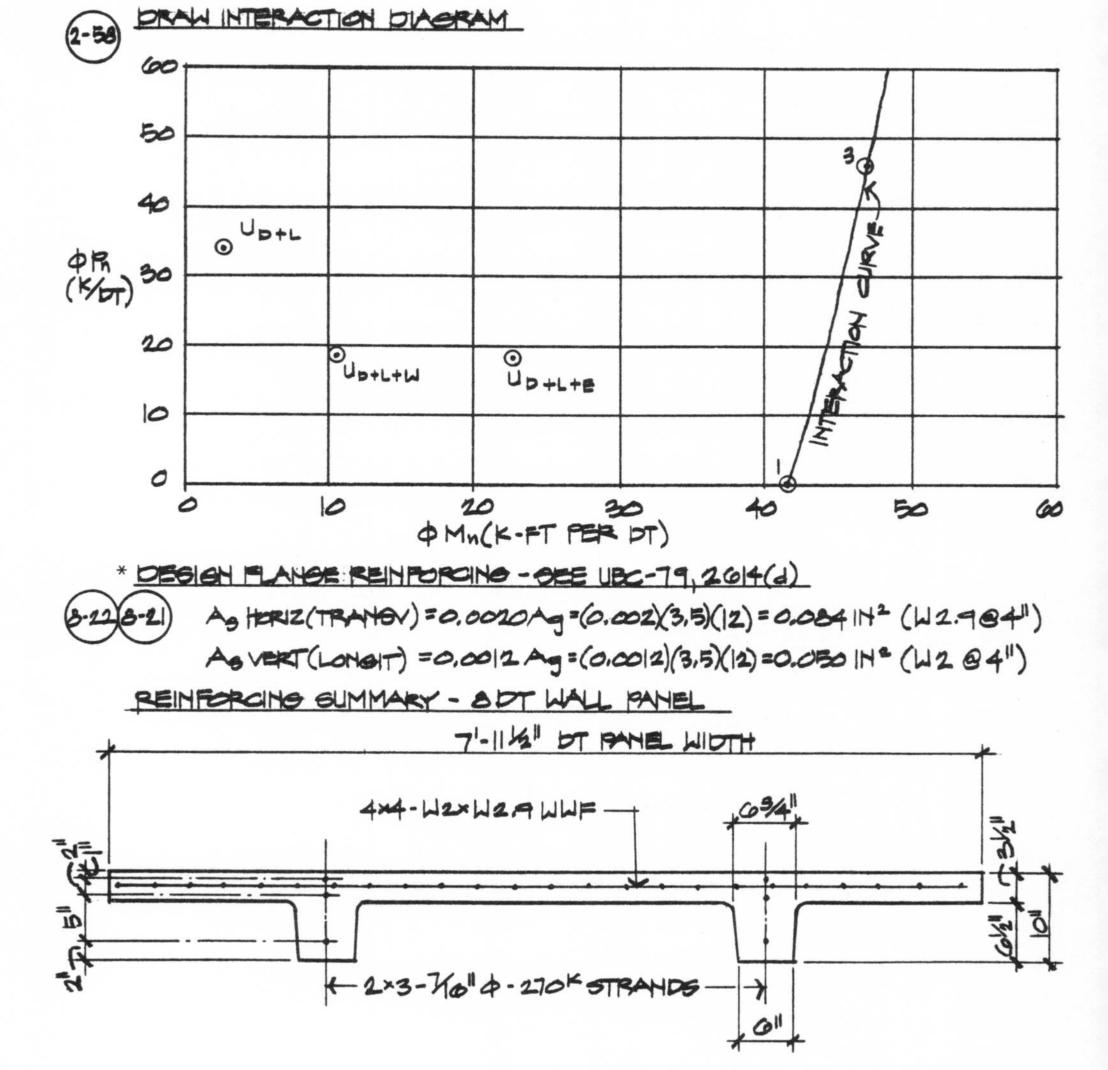

* SEAOC Recommended Lateral Force Requirements - 1975 (Blue Book), in Section 3(C)3 indicate a minimum reinforcing ratio of 0.0025 each way.

CONNECTION DESIGN - LATERAL LOAD RESISTING SOLID WALL PANELS

In 1969, static testing was done in a program co-sponsored by the Prestressed Concrete Manufacturers Association of California, and performed by San Jose State University. These tests verified the capacity of standard shear connectors used to develop the horizontal shear transfer between flanges of double tee members used as floors or walls. The following details demonstrate the use of these connectors for the example double tee wall panel problem presented on the previous pages. The other connection details show the connection details recommended at the roof, and at the foundation.

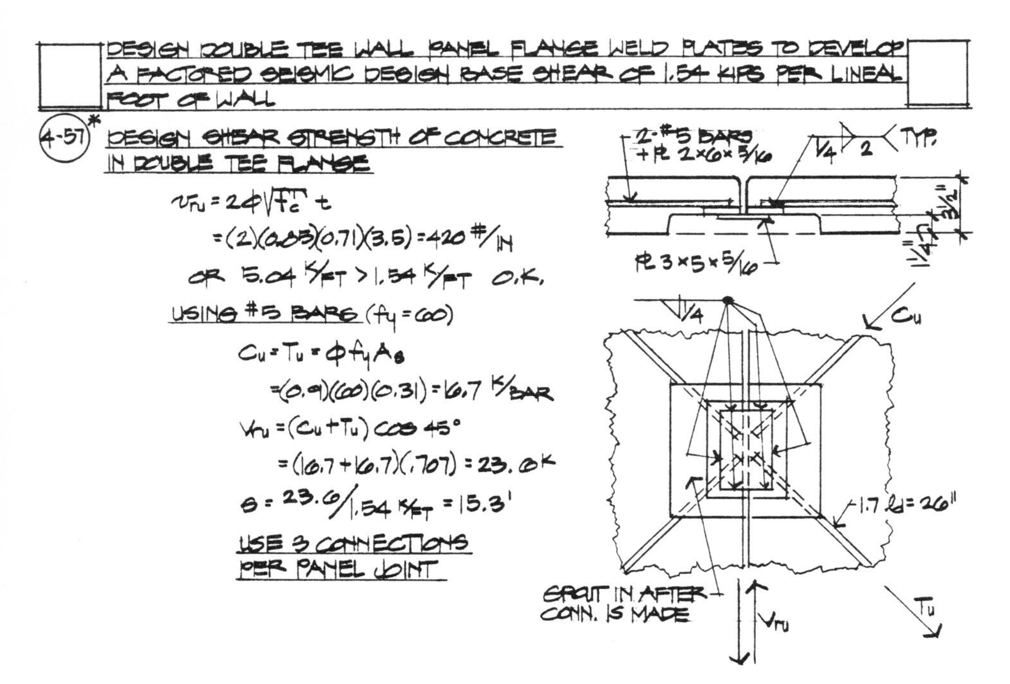

* ◯ PAGE REFERENCE IN 2ND EDITION OF PCI DESIGN HANDBOOK

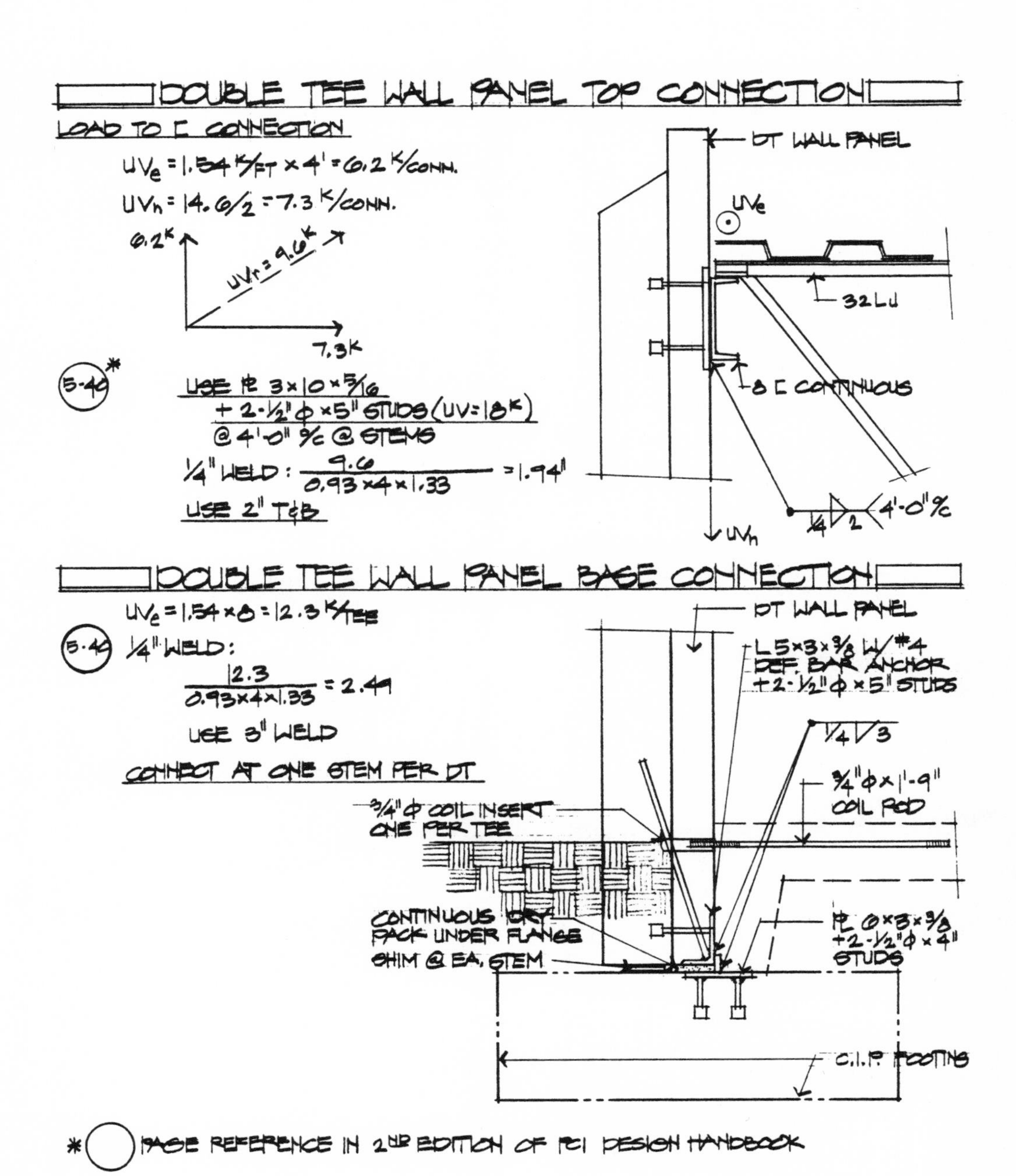
DOUBLE TEE WALL PANEL TOP CONNECTION
LOAD TO [CONNECTION
UVe = 1.54 K/FT × 4' = 6.2 K/CONN.
UVn = 14.6/2 = 7.3 K/CONN.
6.2K
UVr = 9.6K
7.3K
5-46*
USE PL 3×10×5/16
+ 2-1/2"φ ×5" STUDS (UV=18K)
@ 4'-0" o/c @ STEMS
1/4" WELD: 9.6 / 0.93×4×1.33 = 1.94"
USE 2" T&B
DT WALL PANEL
UVe
32LJ
8 [CONTINUOUS
1/4 2 4'-0" o/c
UVn
DOUBLE TEE WALL PANEL BASE CONNECTION
UVe = 1.54×8 = 12.3 K/TEE
5-46
1/4" WELD:
12.3 / 0.93×4×1.33 = 2.49
USE 3" WELD
CONNECT AT ONE STEM PER DT
DT WALL PANEL
L5×3×3/8 W/ #4 DEF. BAR ANCHOR +2-1/2"φ×5" STUDS
1/4 3
3/4"φ×1'-9" COIL ROD
3/4"φ COIL INSERT ONE PER TEE
CONTINUOUS DRY PACK UNDER FLANGE
SHIM @ EA. STEM
PL 6×3×3/8 +2-1/2"φ×4" STUDS
C.I.P. FOOTING
* ◯ PAGE REFERENCE IN 2ND EDITION OF PCI DESIGN HANDBOOK

SOLID WALL PANEL - CONNECTION DESIGN CONFORMING TO UBC-76

The 1979 Uniform Building Code has revised the lateral force coefficients to be used in the design of cladding panels to emphasize the ductile behavior of the connector body under extreme seismic loading conditions, in lieu of the previously used arbitrarily high force coefficients given in the 1976 Uniform Building Code. The previous design examples all conform to this new philosophy, as covered on pp. 177-188 - General Design Procedure for Architectural Precast Concrete Elements. However, many cities and jurisdictions have not yet adopted the 1979 Edition of the Uniform Building Code, and still require the use of a lateral force coefficient Cp = 2.0. The following design example demonstrates the design of connectors for these higher lateral force values.

AMBASSADOR COLLEGE ADMINISTRATION BUILDING - PASADENA

The connections for these intricate honeycomb shaped panels were designed to resist two times the panel's own weight, acting in any lateral direction.

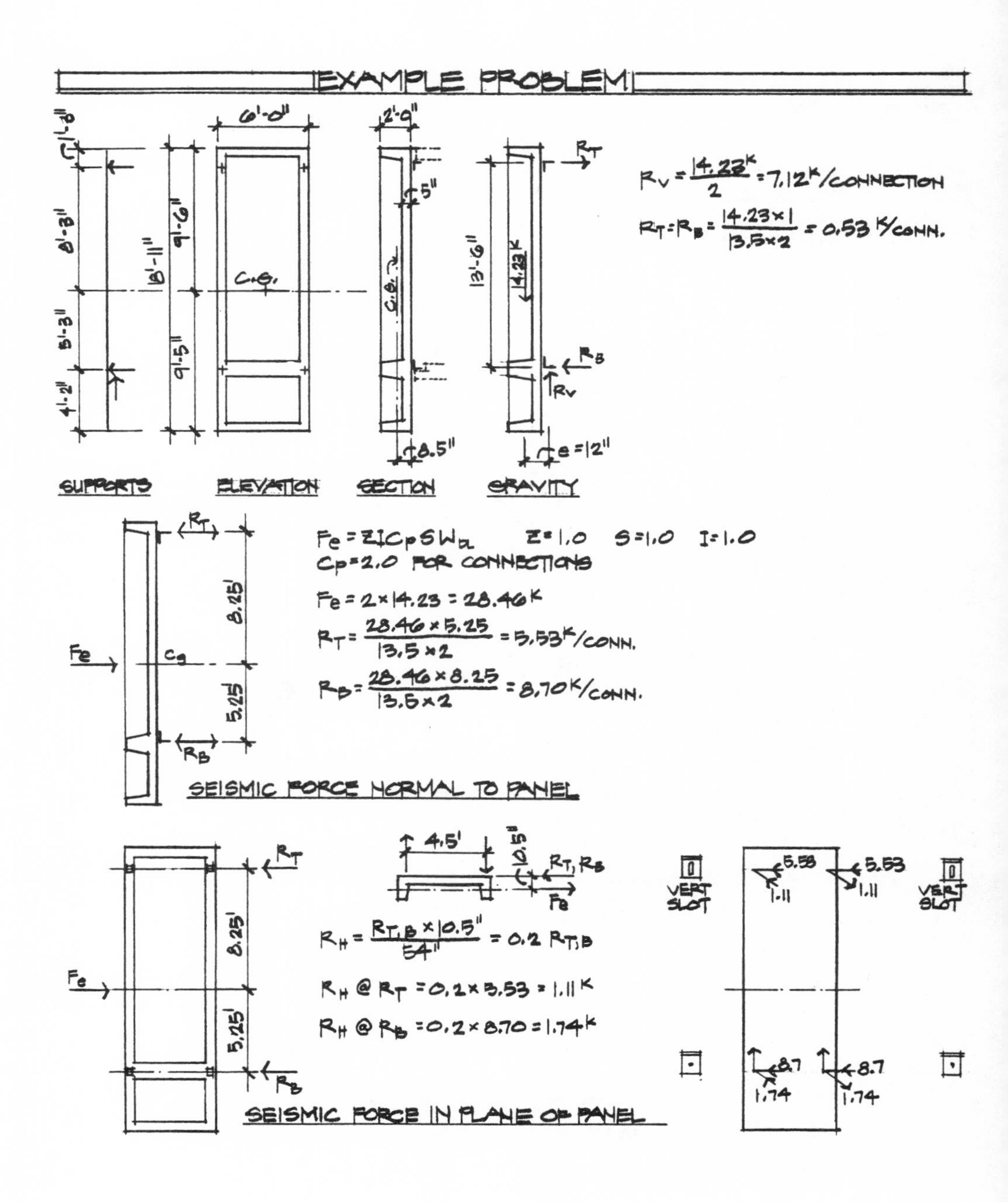
EXAMPLE PROBLEM
SUPPORTS
ELEVATION
SECTION
GRAVITY
6'-0"
2'-9"
1'-8"
8'-3"
5'-3"
4'-2"
18'-11"
9'-6"
9'-5"
C.G.
5"
8.5"
13'-6"
14.23K
e = 12"
RT
RB
RV
RV = 14.23K / 2 = 7.12K/CONNECTION
RT = RB = 14.23×1 / 13.5×2 = 0.53 K/CONN.
Fe = ZICpSWDL Z = 1.0 S = 1.0 I = 1.0
Cp = 2.0 FOR CONNECTIONS
Fe = 2×14.23 = 28.46K
RT = 28.46×5.25 / 13.5×2 = 5.53K/CONN.
RB = 28.46×8.25 / 13.5×2 = 8.70K/CONN.
Fe
Cg
8.25'
5.25'
SEISMIC FORCE NORMAL TO PANEL
4.5'
10.5"
RT, RB
RH = RT,B × 10.5" / 54" = 0.2 RT,B
RH @ RT = 0.2×5.53 = 1.11K
RH @ RB = 0.2×8.70 = 1.74K
VERT SLOT
5.53
1.11
8.7
1.74
SEISMIC FORCE IN PLANE OF PANEL

TOP CONNECTION

CASE 1 GRAVITY + SEISMIC NORMAL

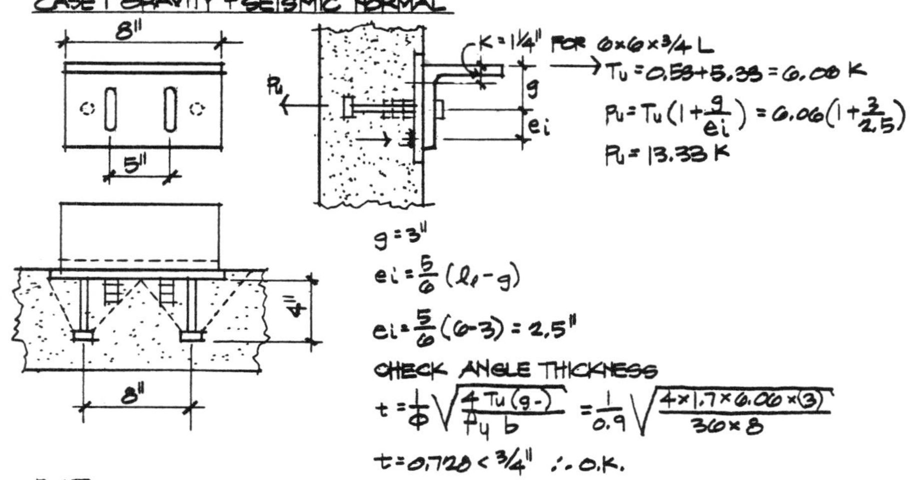

$T_u = 0.53 + 5.33 = 6.06^K$

$P_u = T_u\left(1+\frac{g}{e_i}\right) = 6.06\left(1+\frac{3}{2.5}\right)$

$P_u = 13.33^K$

$g = 3"$

$e_i = \frac{5}{6}(l_1 - g)$

$e_i = \frac{5}{6}(6-3) = 2.5"$

CHECK ANGLE THICKNESS

$t = \frac{1}{\phi}\sqrt{\frac{4\,T_u\,(g-)}{F_y\,b}} = \frac{1}{0.9}\sqrt{\frac{4 \times 1.7 \times 6.06 \times (3)}{36 \times 8}}$

$t = 0.728 < 3/4"$ ∴ O.K.

BOLTS

$\frac{13.33^K}{2} = 6.67\ ^K/_{BOLT}$

TRY ¾" φ BOLTS TENSION AREA = 0.334 IN^2

$0.334 \times 20\ KSI = 6.68^K > 6.67$ OK.

WELDED HEADED STUDS

TRY ½" × 4" STUDS FULL SHEAR CONE

FROM TABLE:

CONC. CAPACITY $18.5^K \times 0.85$ FOR SAND LT. WT. CONC. $= 15.7^K$

STUD CAPACITY 16.6^K

CONCRETE $P_u = 6.67 \times 1.4 = 9.33 < 15.7$ ∴ OK

CASE 2 GRAVITY + SEISMIC IN THE PANEL PLANE

R_T IN THE PLANE $= 5.53^K$ $R_H = 1.11^K + 0.53^K = 1.64^K$

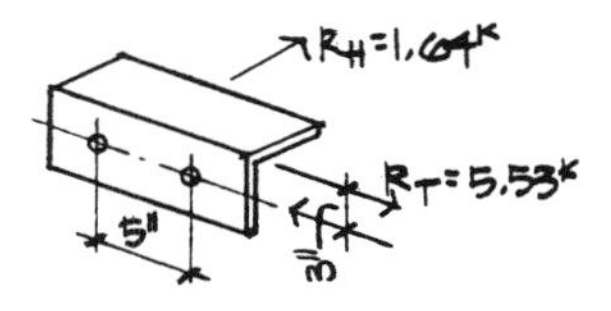

$$SHEAR = \frac{\sqrt{5.53^2 + \left(\frac{5.53 \times 3"}{5"}\right)^2}}{2\ BOLTS} = 3.22\ ^K/_{BOLT}$$

$$P_u = T_u\left(1+\frac{g}{e_i}\right) = \frac{1.64\left(1+\frac{3}{2.5}\right)}{2\ BOLTS} = 1.8^K$$

$f_t = 28\ KSI - 1.6 f_v \le 20\ KSI$ FOR A-307 BOLTS

TENSION AREA $= 0.334\ IN^2$, $\frac{3.22^K}{0.334\ IN^2} = 9.64\ KSI$

$(28\ KSI - 9.64) \times 0.334\ IN^2 = 6.13^K > 1.8^K$ O.K.

CHECK HEADED STUDS $\frac{1}{2}'' \times 4''$

$V_{uc} = 0.85 \times 8.8 = 7.48^K$

$P_{uc} = 0.85 \times 18.5 = 15.72^K$

$P_{us} = 10.6^K$ MAX

$\left(\frac{P_u}{P_{us}}\right)^{4/3} + \left(\frac{V_u}{V_{uc}}\right)^{4/3} \leq 1.0 \qquad \left(\frac{1.8}{10.6}\right)^{4/3} + \left(\frac{3.22}{7.48}\right)^{4/3} = 0.42 < 1 \quad \therefore$ O.K.

BOLTS TO FLOOR BEAM

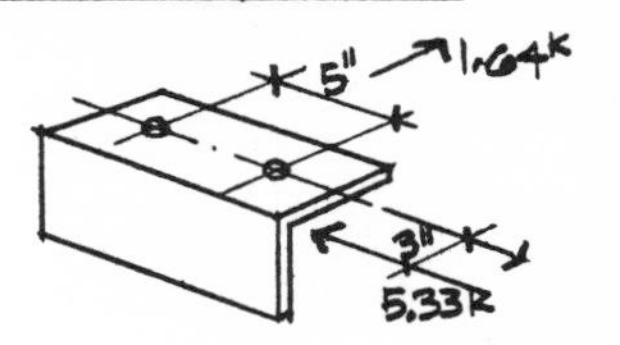

TORQUE $= 5.53 \times 3'' = 16.59^{K\text{-}IN}$

SHEAR $= \frac{16.59^{K\text{-}IN}}{5''} = 3.31^K$

$\frac{1.64}{2} = 0.82^K$

SHEAR $= \frac{5.53}{2} = 2.77^K$

SHEAR PER BOLT $= \sqrt{2.77^2 + (3.31 + 0.82)^2} = 4.97^K$

$\frac{3}{4}''\phi$ BOLTS

ALLOWABLE SHEAR $= 4.42 \times 1.33$ FOR SEISMIC $= 5.87 > 4.97 \quad \therefore$ O.K.

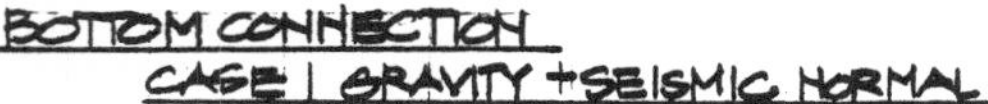

BOTTOM CONNECTION

CASE 1 GRAVITY + SEISMIC NORMAL

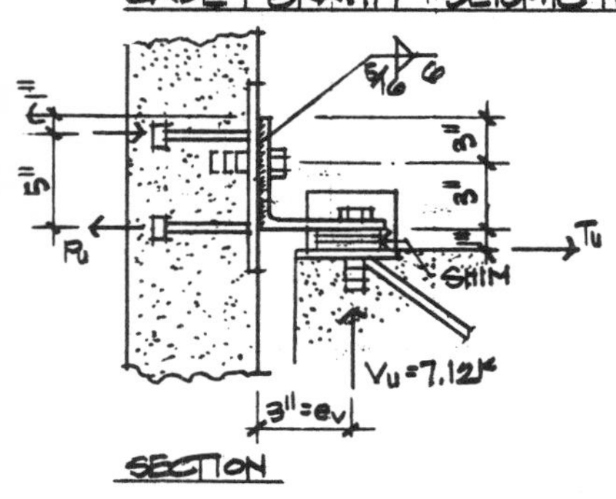

SECTION

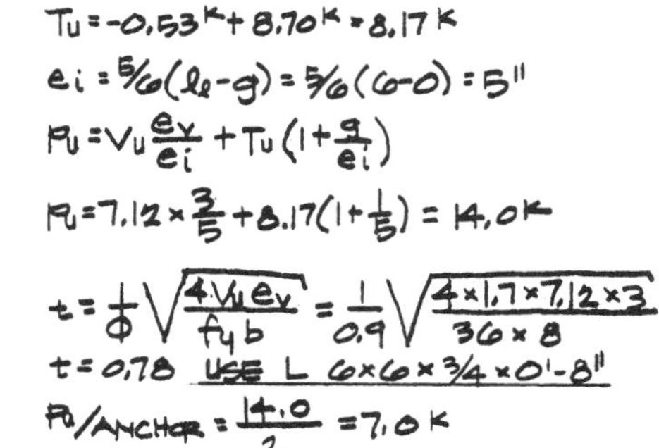

$T_u = -0.53^K + 8.70^K = 8.17^K$

$e_i = 5/6(l_e - g) = 5/6(6-0) = 5''$

$P_u = V_u \frac{e_v}{e_i} + T_u\left(1 + \frac{g}{e_i}\right)$

$P_u = 7.12 \times \frac{3}{5} + 8.17\left(1 + \frac{1}{5}\right) = 14.0^K$

$t = \frac{1}{\phi}\sqrt{\frac{4 V_u e_v}{f_y b}} = \frac{1}{0.9}\sqrt{\frac{4 \times 1.7 \times 7.12 \times 3}{36 \times 8}}$

$t = 0.78$ USE L 6×6×3/4×0'-8"

$P_u/\text{ANCHOR} = \frac{14.0}{2} = 7.0^K$

TEMPORARY BOLT TO PANEL

$e_i = 5/6(l_e - g) = 5/6(6-3) = 2.5''$

$P_u = V_u\left(\frac{e_v}{e_i}\right) = 7.12 \times \frac{3}{2.5} = 8.54^K$

USE 1-7/8"φ A-307 BOLT ALLOWABLE = 9.23 K/bolt

OR 2-3/4"φ A-307 BOLTS ALLOWABLE = 6.7 K/bolt

TRY 5/8"×4" STUDS - FULL CONE

SAND LIGHT WEIGHT CONCRETE

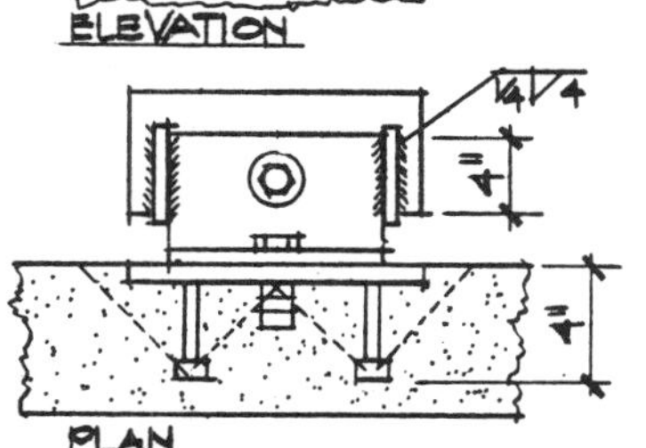

ELEVATION

$V_{u_c} = 13.8 \times 0.85 = 11.73^K$

$P_{u_c} = 18.9 \times 0.85 = 16.0^K$ FROM CHARTS

$P_{u_s} = 16.56^K$

$\left(\frac{P_u}{P_{u_c}}\right)^{4/3} + \left(\frac{V_u}{V_{u_c}}\right)^{4/3} \leq 1.0$

$\left(\frac{7.0}{16.0}\right)^{4/3} + \left(\frac{3.56}{11.73}\right)^{4/3} = 0.54 < 1.0 \quad \therefore \text{O.K.}$

1/4 4

4"

4"

PLAN

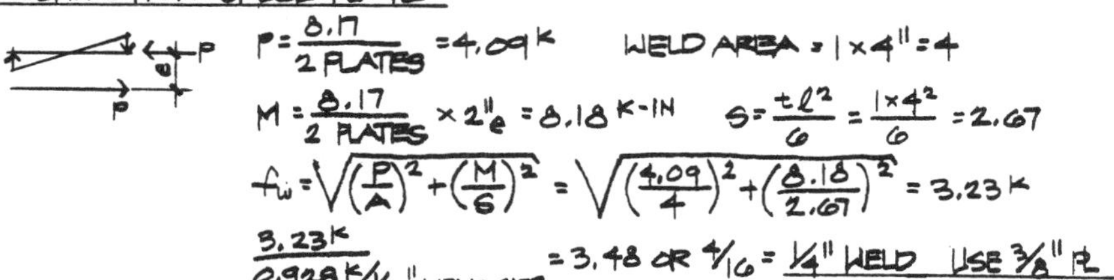

WELD CONNECTION SPLICE PLATE

$P = \frac{8.17}{2 \text{ PLATES}} = 4.09^K$ WELD AREA = 1×4" = 4

$M = \frac{8.17}{2 \text{ PLATES}} \times 2''_e = 8.18^{K\text{-IN}}$ $S = \frac{t l^2}{6} = \frac{1 \times 4^2}{6} = 2.67$

$f_w = \sqrt{\left(\frac{P}{A}\right)^2 + \left(\frac{M}{S}\right)^2} = \sqrt{\left(\frac{4.09}{4}\right)^2 + \left(\frac{8.18}{2.67}\right)^2} = 3.23^K$

$\frac{3.23^K}{0.928^K/_{1/16''} \text{ WELD SIZE}} = 3.48$ OR $4/16$ = 1/4" WELD USE 3/8" PL

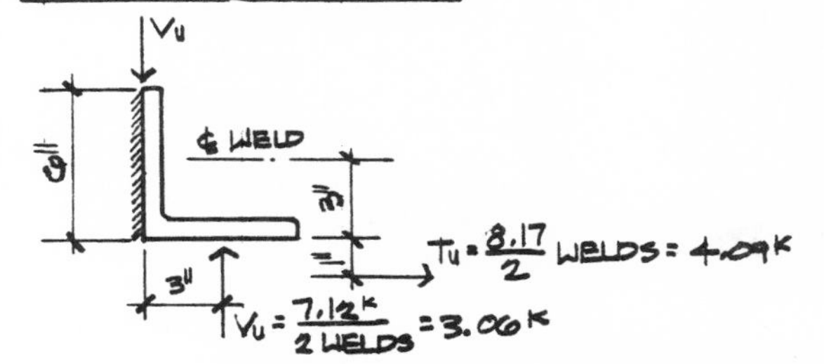

AREA OF WELD = 6"

SECTION MODULUS = $\frac{1 \times 6^2}{6} = 6$

$M = 4.09^K \times 4'' + 3.06^K \times 3''$
$= 25.54^{K\text{-}IN}$

$$f_w = \sqrt{\left(\frac{V_u}{A}\right)^2 + \left(\frac{M_u}{S}\right)^2} = \sqrt{\left(\frac{3.06}{6}\right)^2 + \left(\frac{25.54}{6}\right)^2} = 4.29^K$$

$$\frac{4.29}{0.928\ ^K/_{1/16''} \text{ OF WELD SIZE}} = 4.62 \text{ OR } 5/16'' \text{ WELD SIZE}$$

CASE 2 - GRAVITY + SEISMIC IN THE PANEL'S PLANE

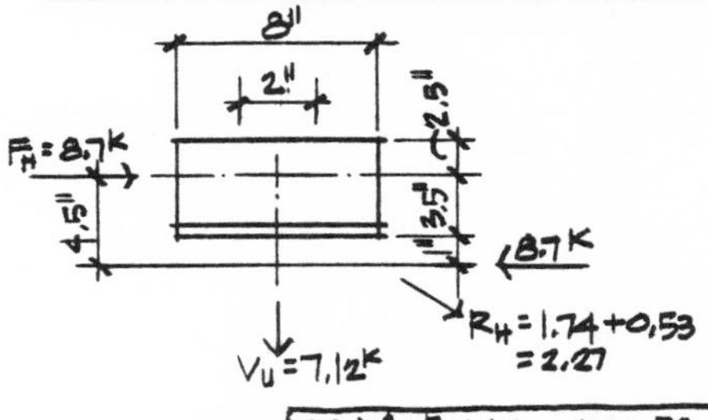

$$\frac{(6 \times 2'') + (2 \times 6'' \times 3'')}{2 + 12} = 2.57'' \text{ SAY } 2.5''$$

$$I_2 = 2Ay^2 + \frac{1 \times L_w^3}{12}$$

$$I_2 = 2 \times 1 \times 6 \times 4^2 + \frac{1 \times 2^3}{12} = 192.6$$

$c = \frac{8''}{2} = 4''$ $\quad M = 8.7^K \times 4.5'' = 39.15^{K\text{-}IN}$

$$f_w = \sqrt{\left(\frac{F_H}{A}\right)^2 + \left[\left(\frac{Mc}{I}\right) + \left(\frac{V_u}{A}\right)\right]^2 + \left(\frac{R_H}{A}\right)^2}$$

$$f_w = \sqrt{\left(\frac{8.7}{14}\right)^2 + \left[\left(\frac{39.15 \times 4}{192.6}\right) + \left(\frac{7.12}{14}\right)\right]^2 + \left(\frac{2.27}{14}\right)^2}$$

$f_w = 1.47^K < 4.64^K \therefore$ O.K.

ALLOWABLE FOR A 5/16" WELD E70XX = 0.928×5 = 4.64K/IN. OF WELD

NOTE: THE CONNECTION WILL BE DRY PACKED IN A POCKET IN THE FLOOR SYSTEM HENCE TRANSFER OF THE 8.7 KIPS TO THE FLOOR PRESENTS NO PROBLEM.

CONNECTION DETAILS

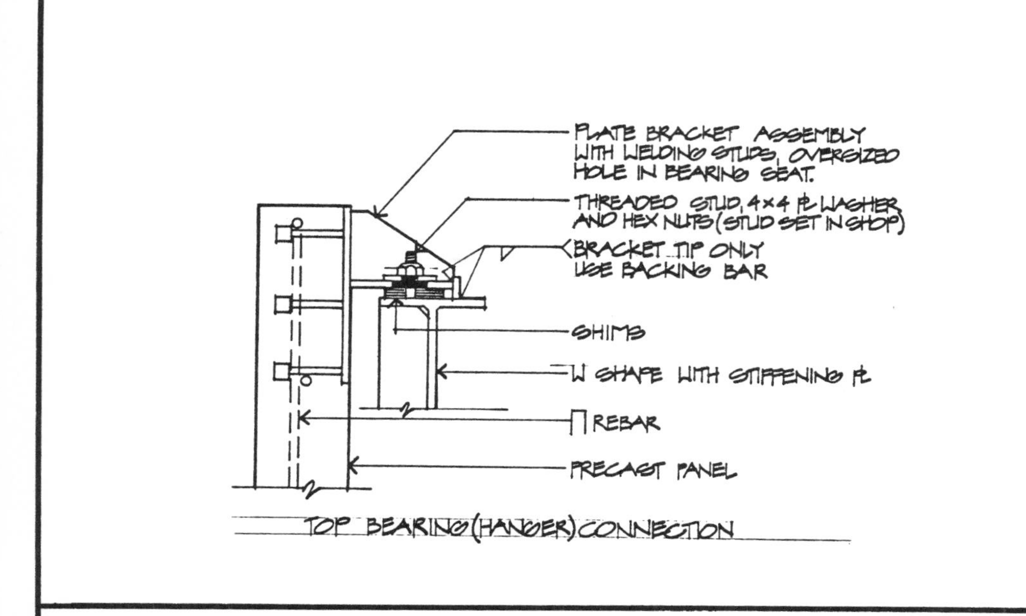

TOP BEARING (HANGER) CONNECTION

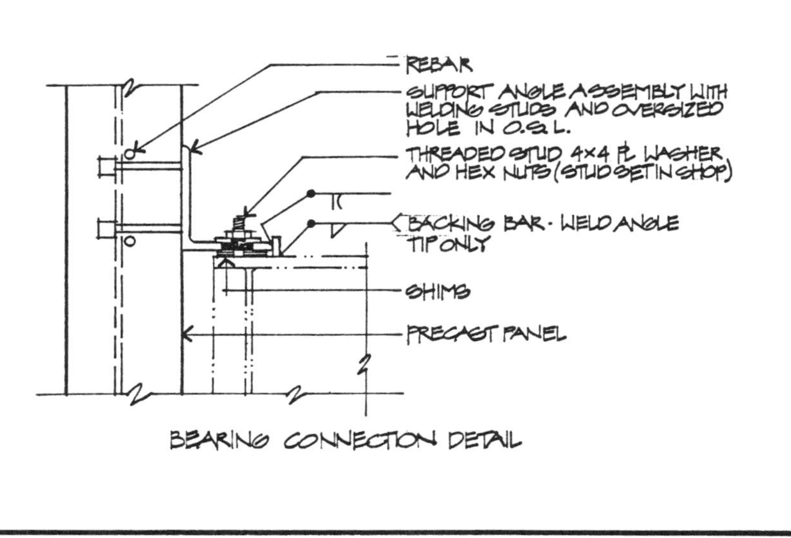

BEARING CONNECTION DETAIL

CONNECTION DETAILS

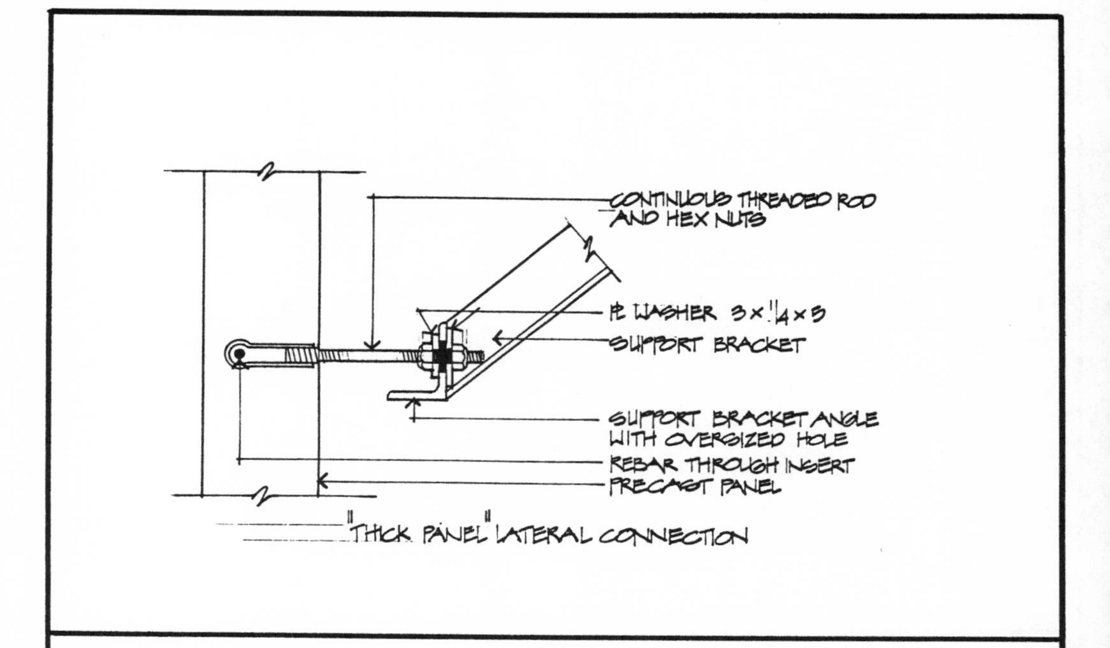

"THICK PANEL" LATERAL CONNECTION

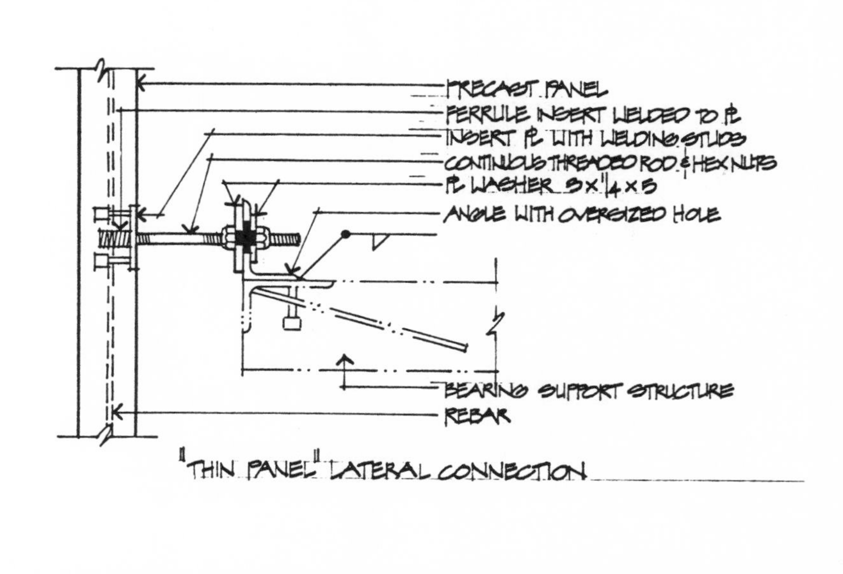

"THIN PANEL" LATERAL CONNECTION

CONNECTION DETAILS

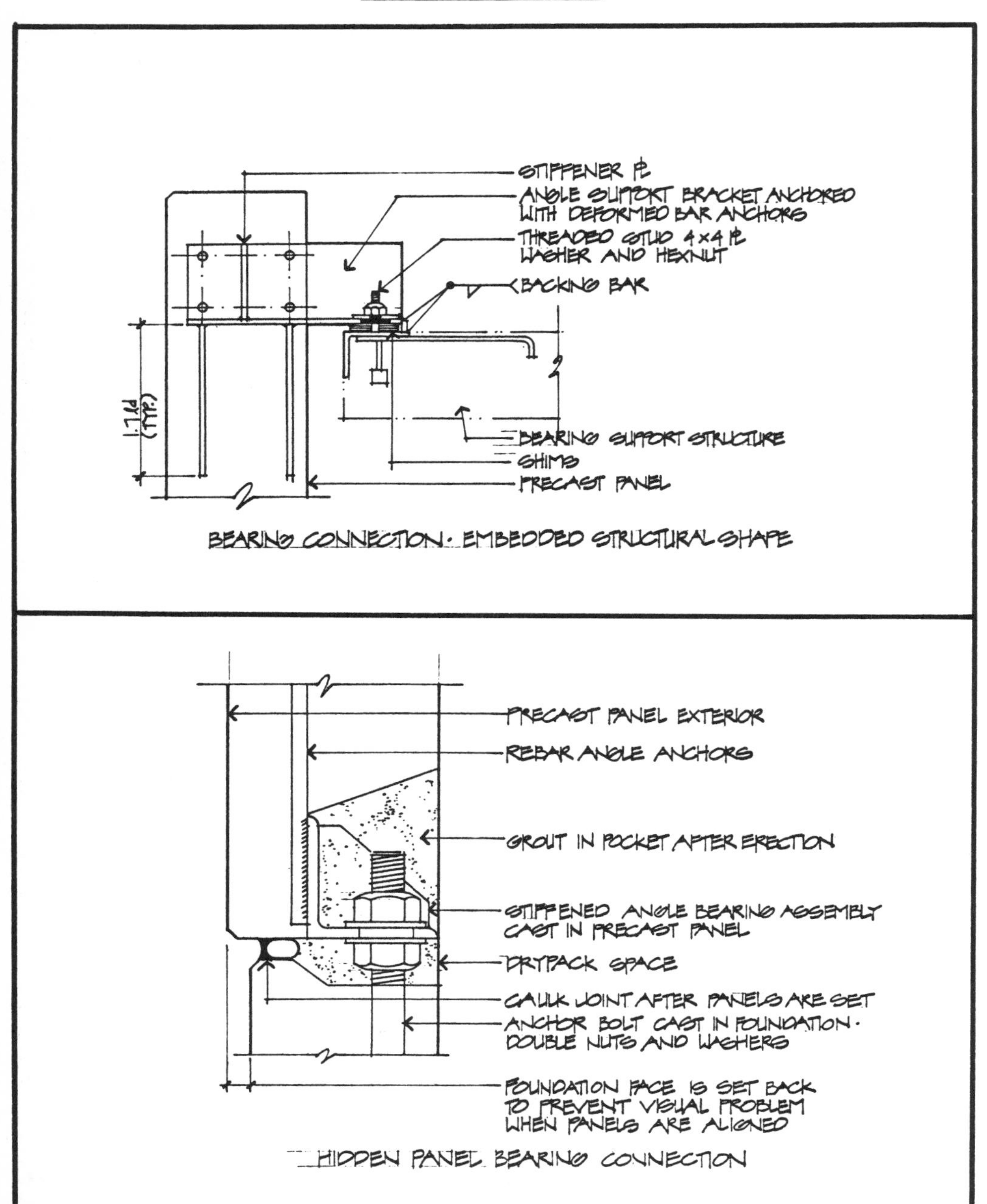

CONNECTION DETAILS

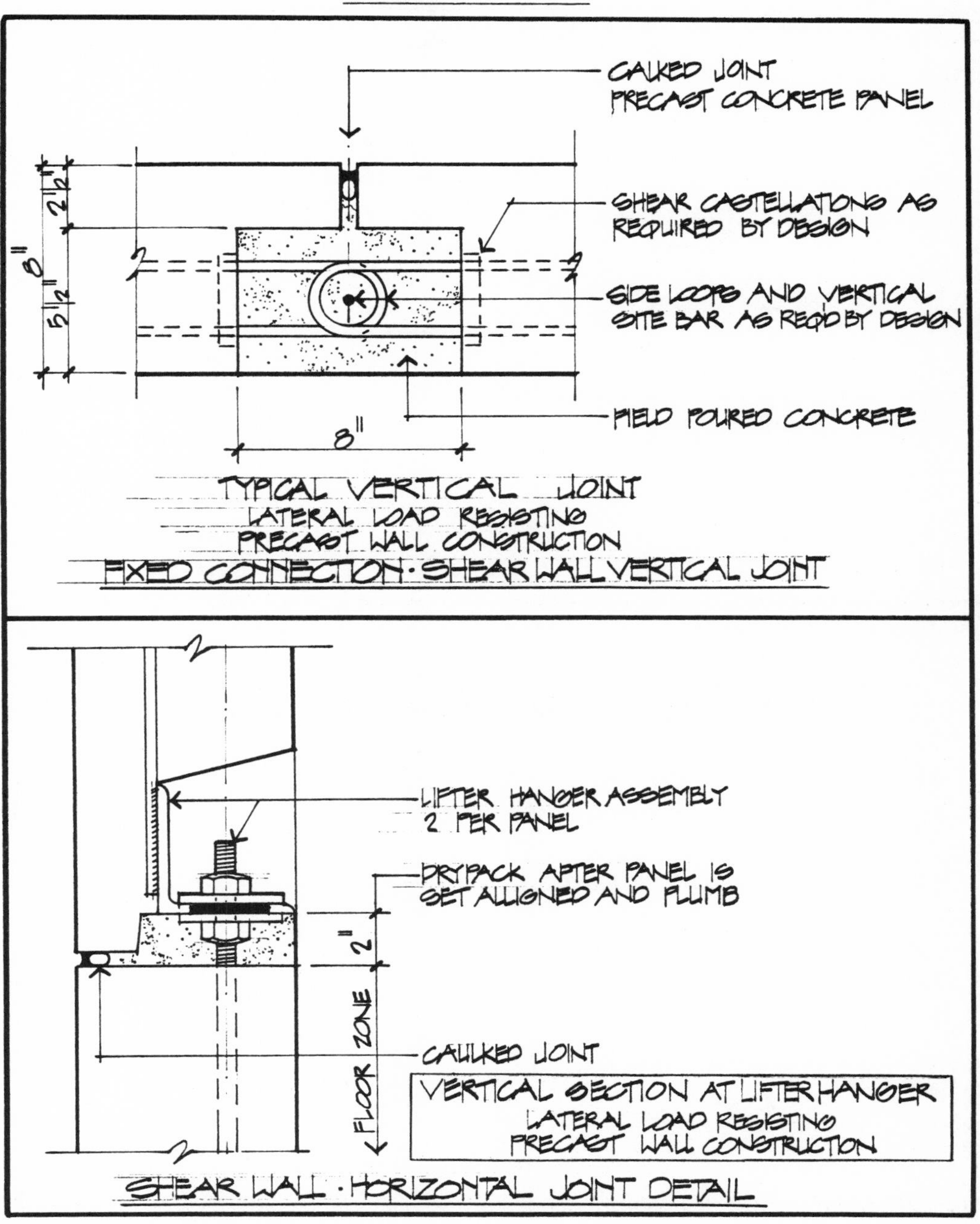

CONNECTION DETAILS

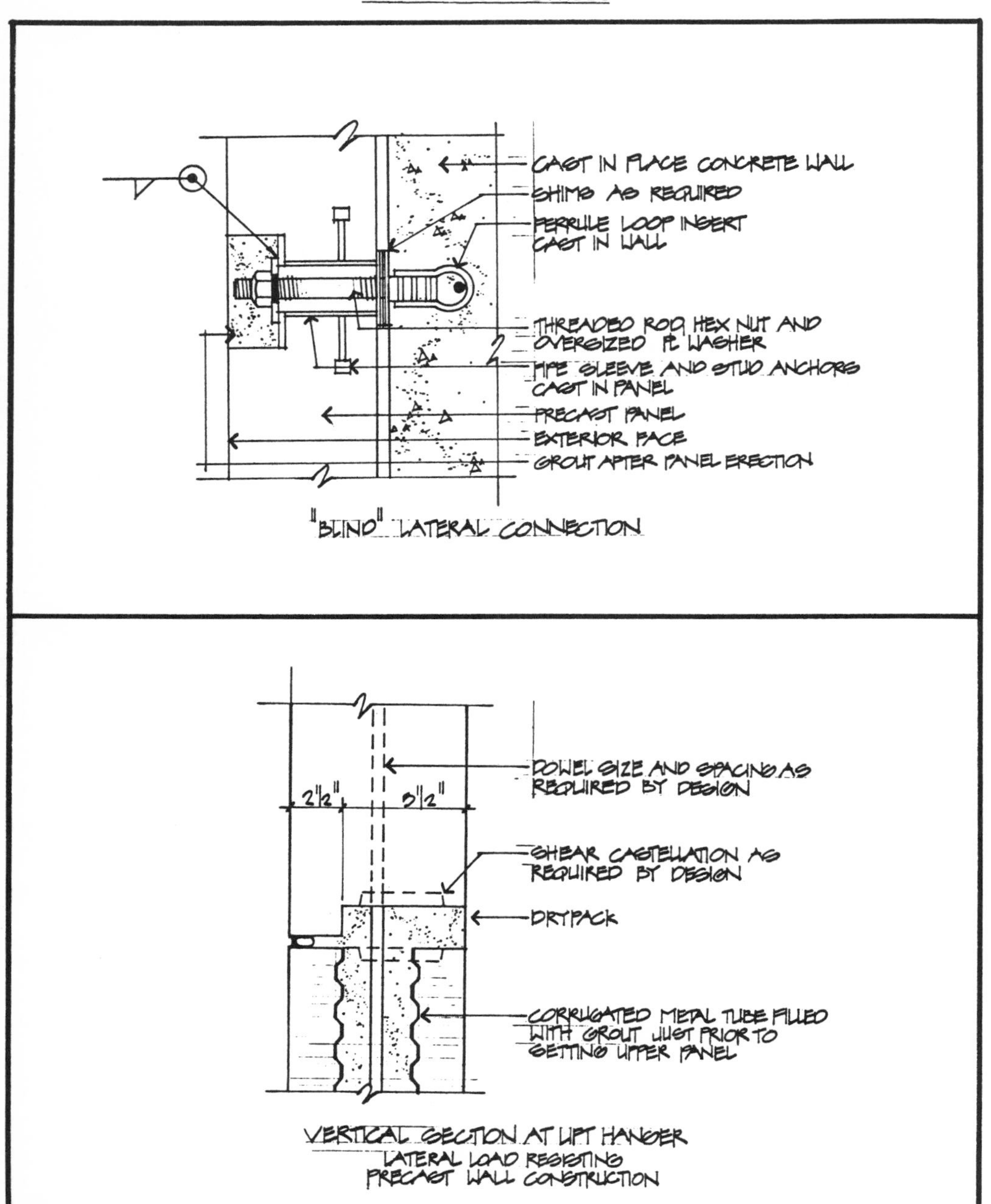

CONNECTION DETAILS

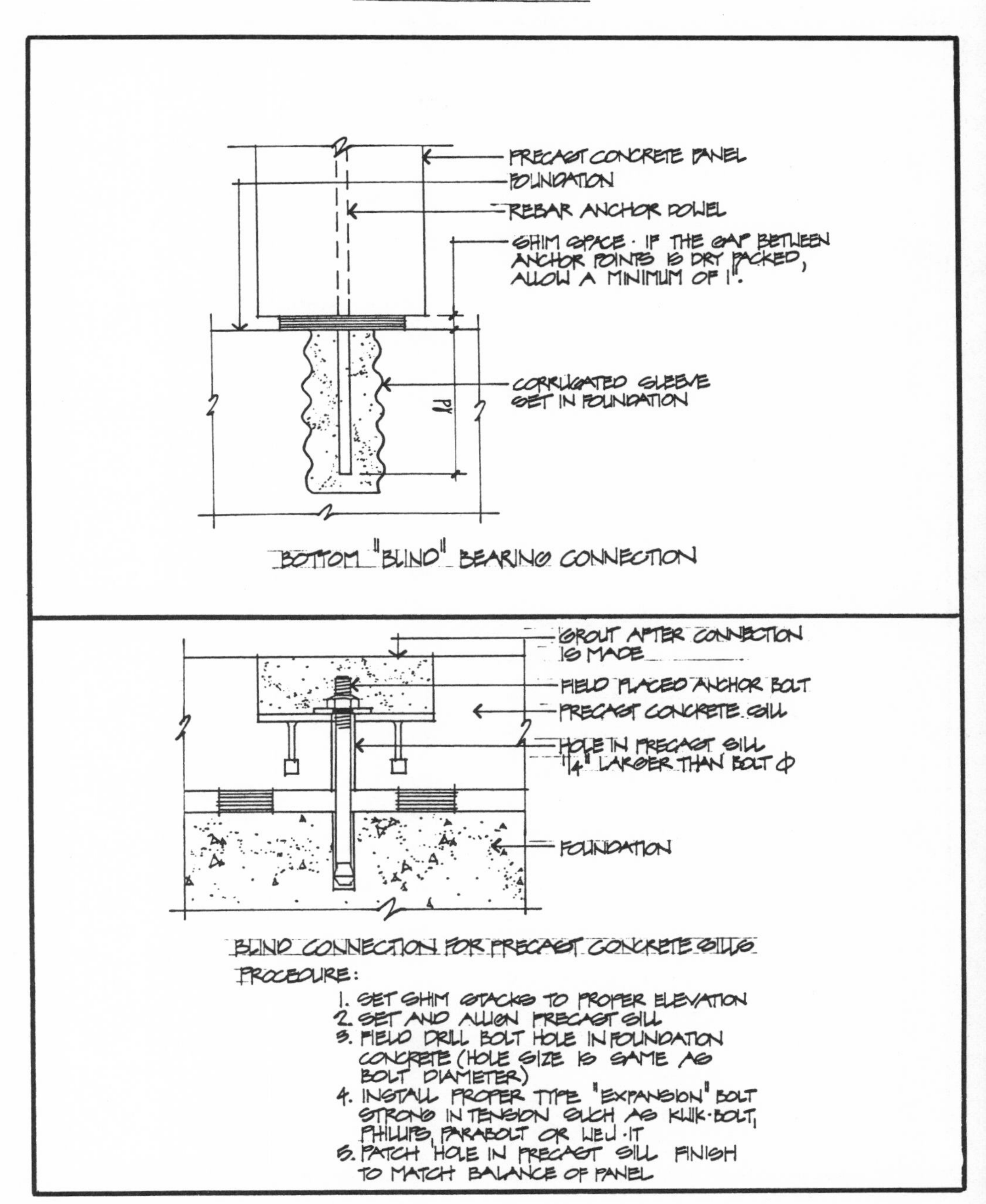

CONNECTION DETAILS

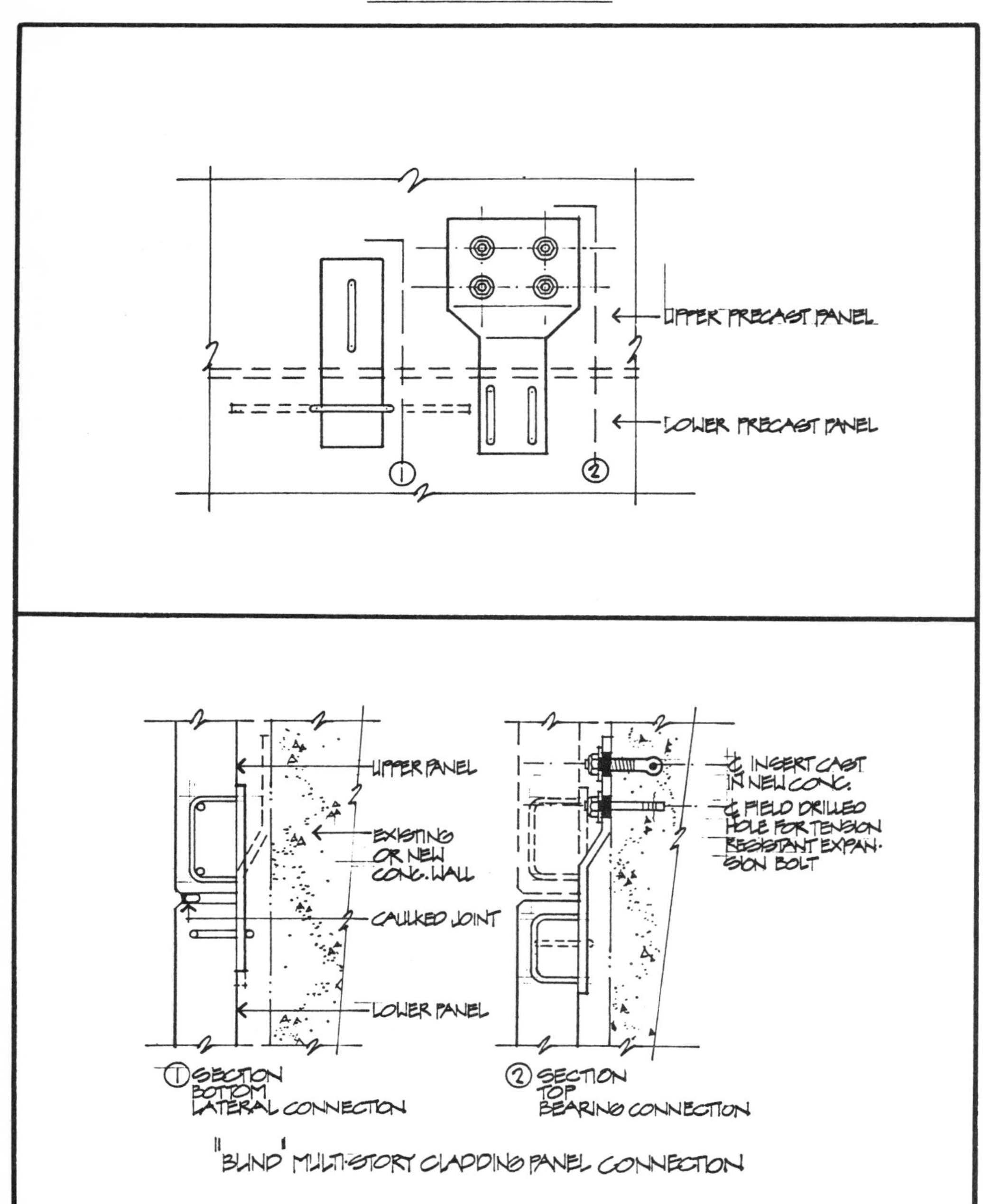

CONNECTION DETAILS

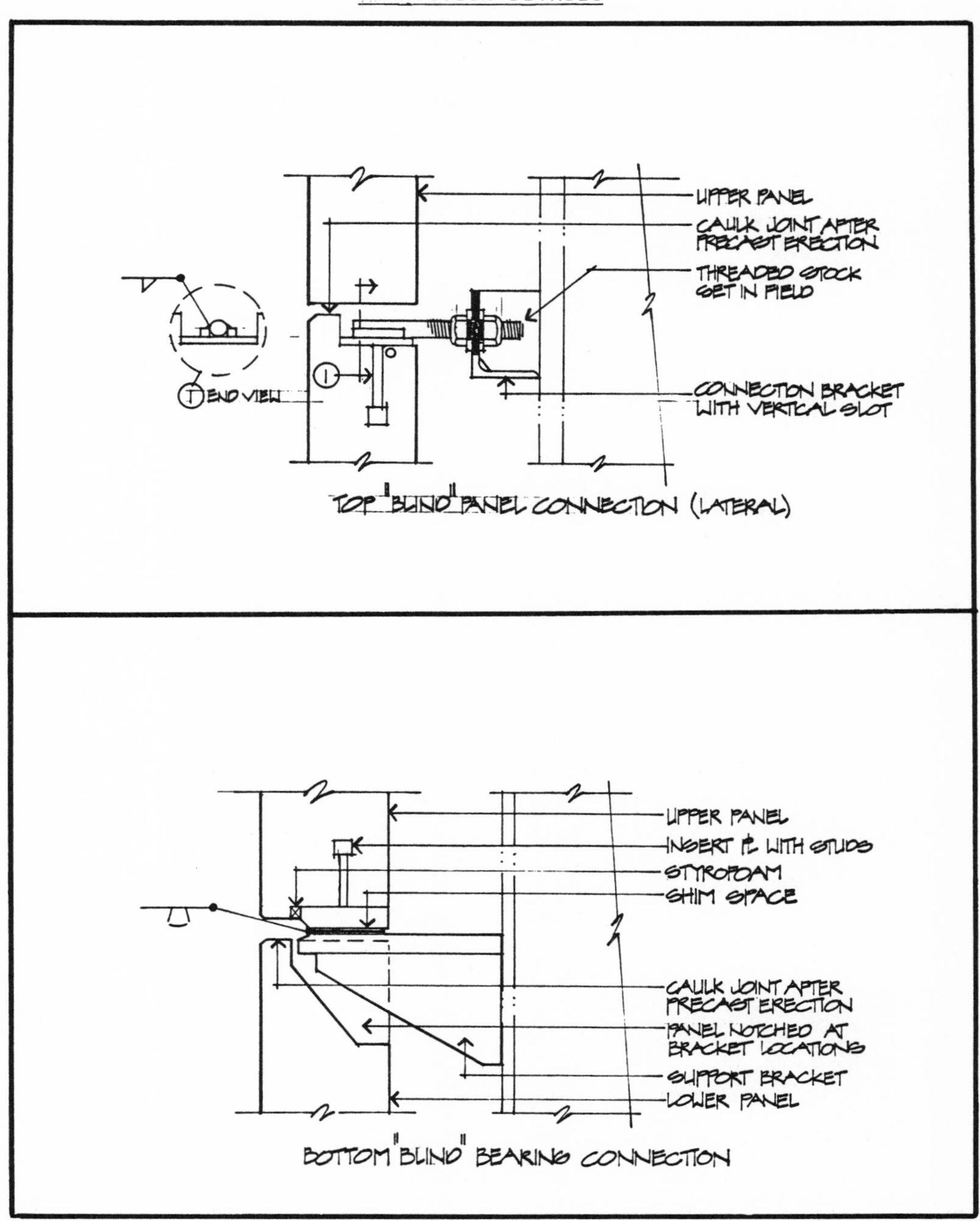

PRECAST CONCRETE SANDWICH PANELS

CONTRA COSTA COUNTY DETENTION FACILITY - MARTINEZ

10 inch thick non-load bearing sandwich panels clad the top two floors of this project, still under construction in the adjacent photo. 4 inches of polyurethane insulation was sandwiched between a 2 inch outer skin and a 4 inch inner skin. The two concrete wythes were interconnected by concrete ribs to cause them to act as one unit in resisting transverse forces.

With the need to conserve energy and considering the new energy regulations being implemented at both the state and national levels, precast concrete sandwich panels offer an ideal building solution. With sandwich panels, a durable surface is provided on both sides of the panel. The surface may be smooth on both sides, or exposed aggregate, sandblasted, or patterned on the exterior surface. The concrete surface offers resistance to weather, fire, vandalism, and is easy to maintain. With sandwich panels the project construction time can be reduced, resulting in an economic savings to the owner. The "U" thermal transmission) value of a six inch solid normal weight concrete wall is 2.38. For a sandwich panel consisting of 2 inches of concrete for one wythe, 1½ inches of polyurethane, and five inches of concrete the U value is 0.10. Sandwich panels can be effectively used for commercial, educational, governmental, industrial and residential buildings. In designing with sandwich panels, the architect still has the range of form, pattern, surface texture, and color that he has with solid non-insulated precast concrete panels. Sandwich panels may be designed as either load bearing or non-load bearing elements. They may be designed to function as either composite or non-composite where the design temperatures are different on each side of the wall. A non-composite panel is one in which one wythe is non-structural and is usually from 1½ to 2½ inches thick. This wythe is supported from the structural wythe with flexible hangers and is a floating wythe free to react to temperature and other volumetric changes. Its main purpose is to protect the insulative core. A composite panel is one in which the two wythes act as a unit to resist transverse forces and the wythes are connected together with concrete ribs or steel shear connectors.

Various types of sandwich panels are as follows:

(A) DOUBLE TEE WALL PANEL

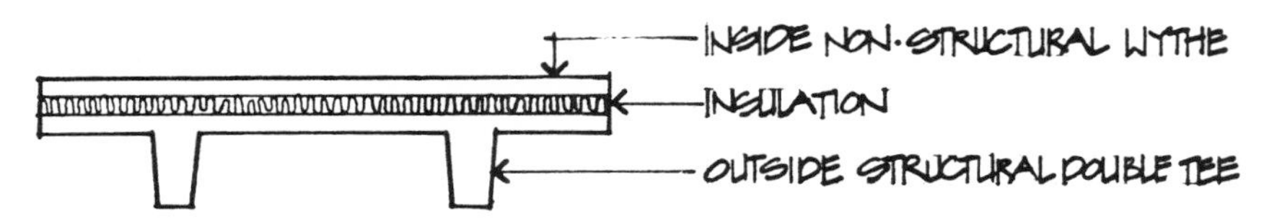

(B) SCULPTURED WALL PANEL

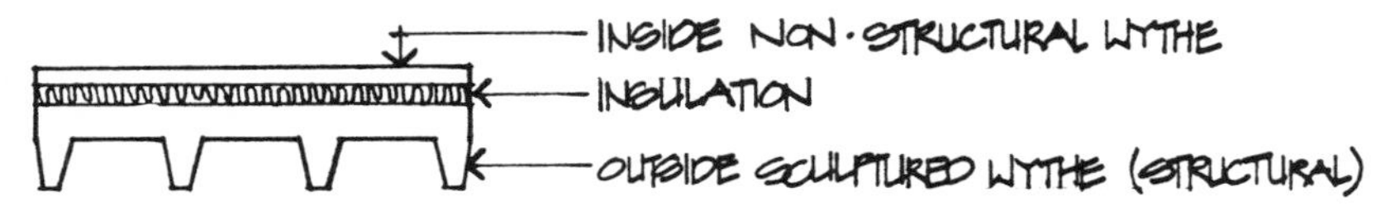

(C) HOLLOW CORE PLANK WALL PANEL

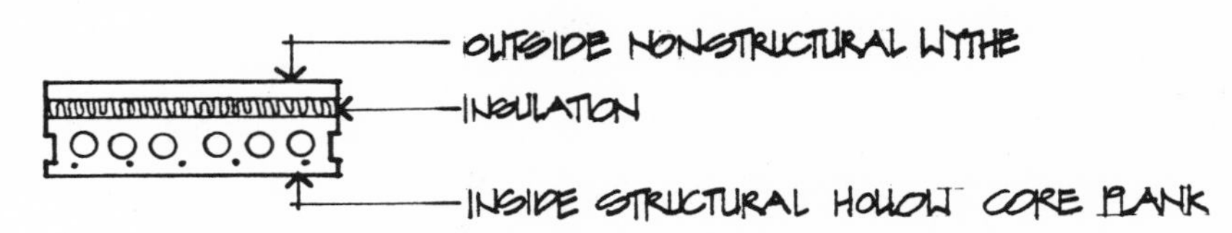

(D) PRESTRESSED FLAT WALL PANEL

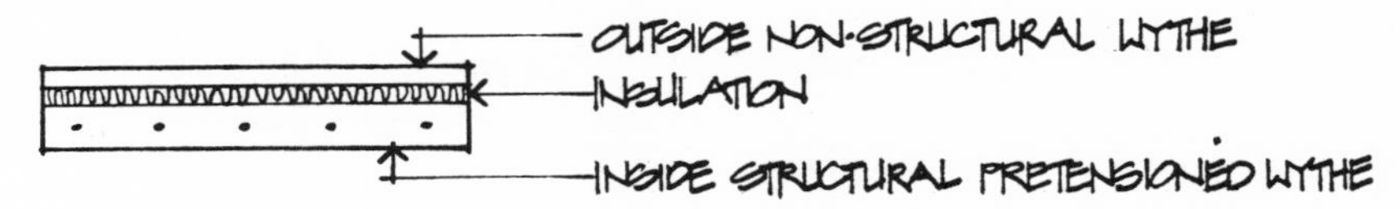

(E) NON-PRESTRESSED FLAT WALL PANEL

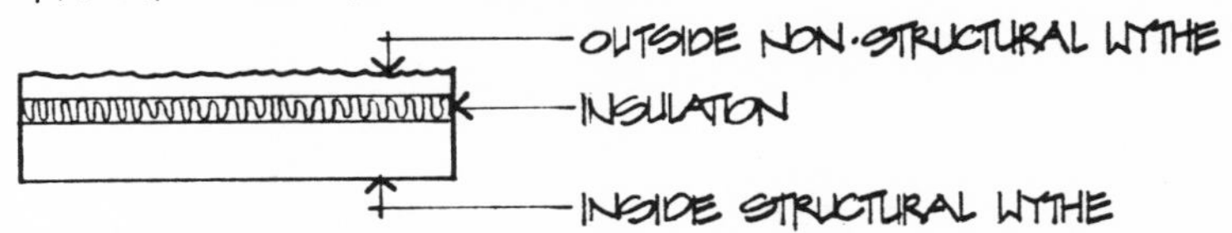

(F) COMPOSITE NON-LOAD BEARING PANEL

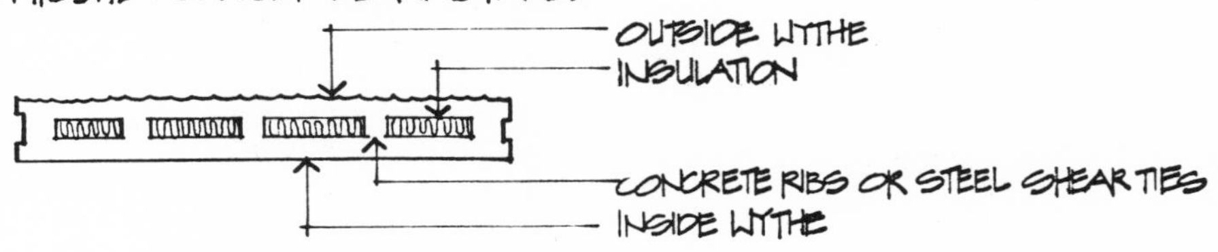

INSULATION

The type of insulation used is subject to the U value desired, wall thickness, cost and type of panel. Polystyrene has an R value (thermal resistance) of 4.00/inch of thickness, polyurethane has an R value of 6.25/inch of thickness and fiberglass board has an R value of 4.00/inch of thickness.

$$U = \frac{1}{R} = \frac{1}{R_{\text{Outside Air}} + R_{\text{Ext. Conc.}} + R_{\text{insulation}} + R_{\text{Int. Conc.}} + R_{\text{Inside Air}}}$$

To reduce the required wall thickness, polyurethane would be the better insulation. The cost is affected by the current economic market, the thickness and the R value desired. For composite panels, insulation with a high bond surface is desired but with a minimum water absorption potential from the fresh concrete. For non-composite panels a bond breaker sheet or a treated surface adjacent to the floating wythe is necessary to prevent bonding of the concrete to the insulation.

Non-Composite Sandwich Panels

With a Sandwich Panel the outside wythe is subject to volumetric changes which differ from those of the inside wythe. These volumetric changes result from differential temperature changes between the two wythes, differential temperature changes due to time, shrinkage, lateral load stresses, and creep. For non-composite panels, one wythe is made thin, from 1½ to 2½ inches thick with an optimum thickness of 2 inches. This wythe is attached to the other wythe with a combination of flexible and non-flexible ties. The flexible ties allow the floating wythe to accommodate movements from its different volumetric changes while the non-flexible tie anchors or fixes one portion of the floating wythe relative to the other wythe. The other wythe is the structural wythe and resists the forces placed on the panel due to stripping, plant handling, transportation, erection, and in-situ loadings. This structural wythe is the thicker wythe and its thickness is determined by the stresses placed upon it and building code requirements. Usually the thinner wythe is placed on the outside and the structural wythe on the inside. The exception to this is when the structural wythe is made with a pretensioned double tee section. In colder climates, attention should be given to avoidance of any pockets that could trap water and result in freezing water cracking the wythes.

The flexible tie may be one or several of the following types.:

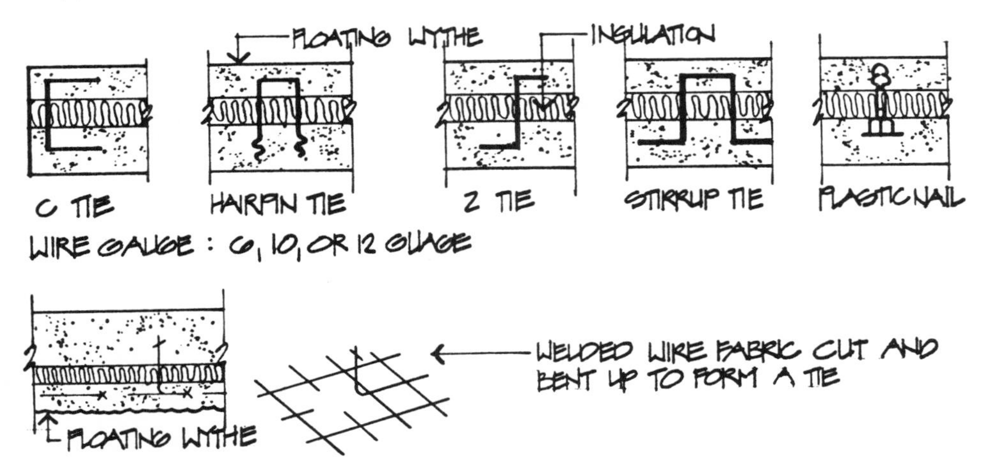

The non-flexible ties or hangers may be one of the following types:

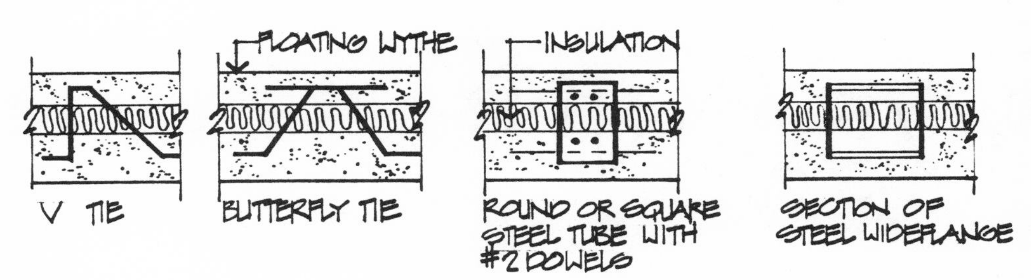

The ties or hangers are galvanized or stainless steel and are either #2 reinforcing steel or 6-gage wire. A 6-gage wire tie will provide a tensile strength of about 1400 lbs. The ties are usually spaced at 4 feet on centers each way. The insulative core runs out to the panel edges thus allowing the floating wythe to move independent of the structural wythe. The insulation is either covered with a bond breaking sheet, or is spray coated with a liquid bond breaking compound.

Lifting Stresses

Double tee and other structural ribbed panels are usually cast with the floating wythe on top and in a horizontal position. However, flat and sculptured architectural panels are usually cast with the floating wythe down. Since the floating wythe is non-structural, all of the lifting stresses are resisted by the structural wythe. The structural wythe is designed on the concept of an uncracked section with the extreme fiber in tension not exceeding $f_{t\ design} = \frac{7.5\sqrt{f'c}}{1.5}$, where 1.5 is a factor of safety and f'c is the concrete compressive strength at the time of lifting. It should also be large enough such that frictional resistance of the floating panel should not induce strain which would cause cracking of the floating panel where it may be in tension. Here the precast manufacturer's experience is an important factor. The value of $f_{t\ actual} = \frac{M\ C}{I}$, where I and C are for the structural wythe and M is based upon the total panel weight, including an allowance for impact.

Lifting, handling, and erection inserts and connections

The location and special detailing of these inserts depends upon which wythe of the panel is cast up, which wythe is placed to the outside, whether the panel is bottom or side supported, load bearing or non-load bearing, the panel story height, and the method of plant handling, transportation and erection. If the inserts pierce the floating wythe they should be isolated such that they do not bear against the floating wythe. Styrofoam is usually placed around the insert at the floating wythe.

However, one or several inserts may be bonded to both the floating wythe and the structural wythe where hanger ties are used, or at corbels. This is the point where the two wythes are fixed together. If prestressing strands are used as lifting bales, the strands can be burnt off after the panel is erected and the remaining holes grouted. A preformed pocket is usually provided at these points if they are in the exposed face of the panel or exposed to the weather.

Floating wythe reinforcement

The Uniform Building Code has no provisions for reinforcement in the floating wythes of sandwich wall panels. For walls in general the minimum requirements for welded wire fabric are 0.002 bt horizontal and 0.0012 bt vertical (UBC-79, Sec. 2614). For a 2 inch wythe the horizontal min. requirement is 0.002 x 12 x 2 = 0.048 sq. inch/ft. (6 x 6 - W4.0 x W4.0 provides 0.08 in^2/ft)

On the next 4 pages, several types of non-composite panel types and details are shown.

CONTRA COSTA COUNTY DETENTION CENTER - MARTINEZ

The precast concrete sandwich panels provide energy savings, security, and economy as well as an attractive appearance, allowing the building to blend harmoniously with the surrounding neighborhood.

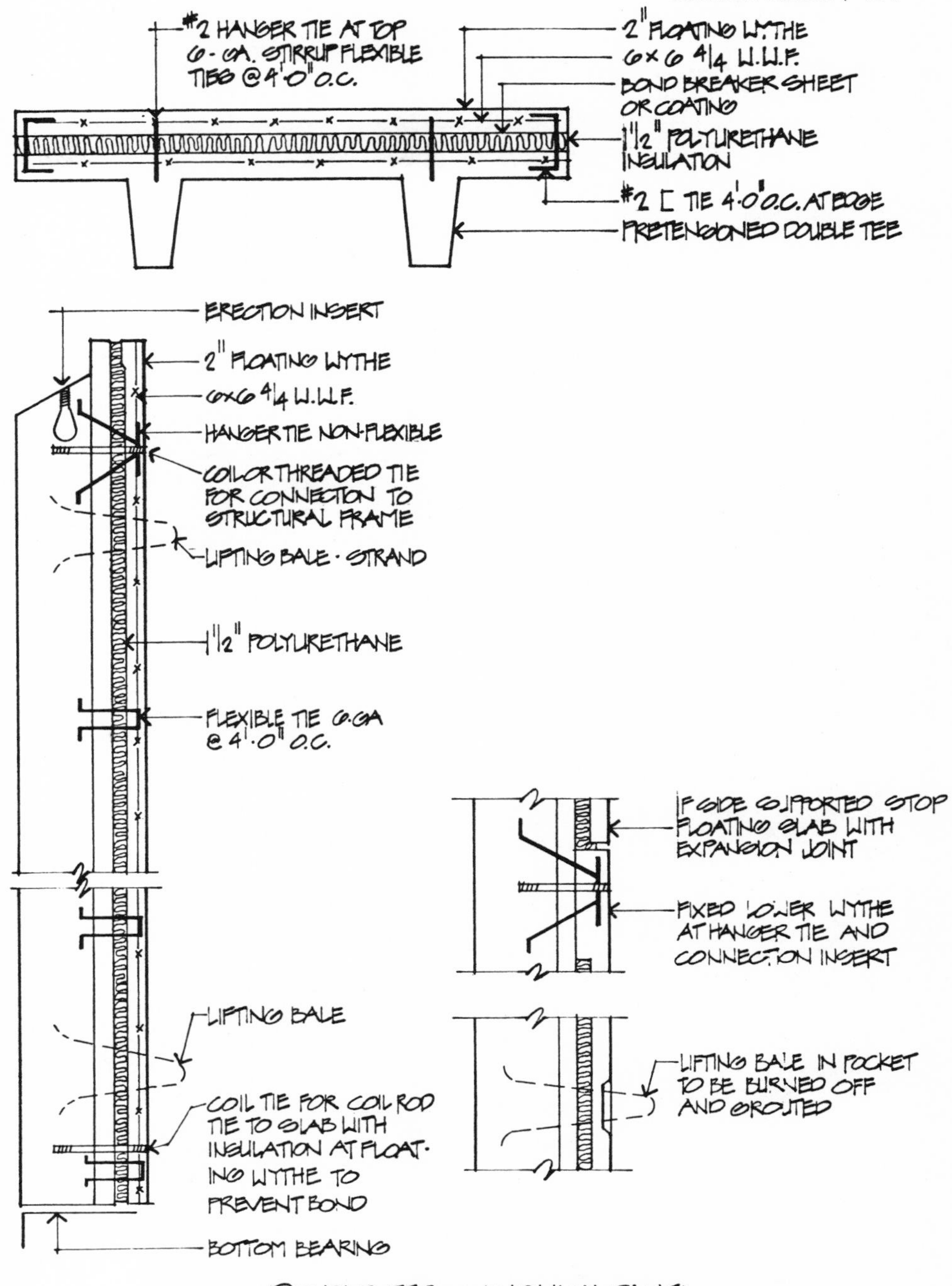

DOUBLE TEE SANDWICH PANEL

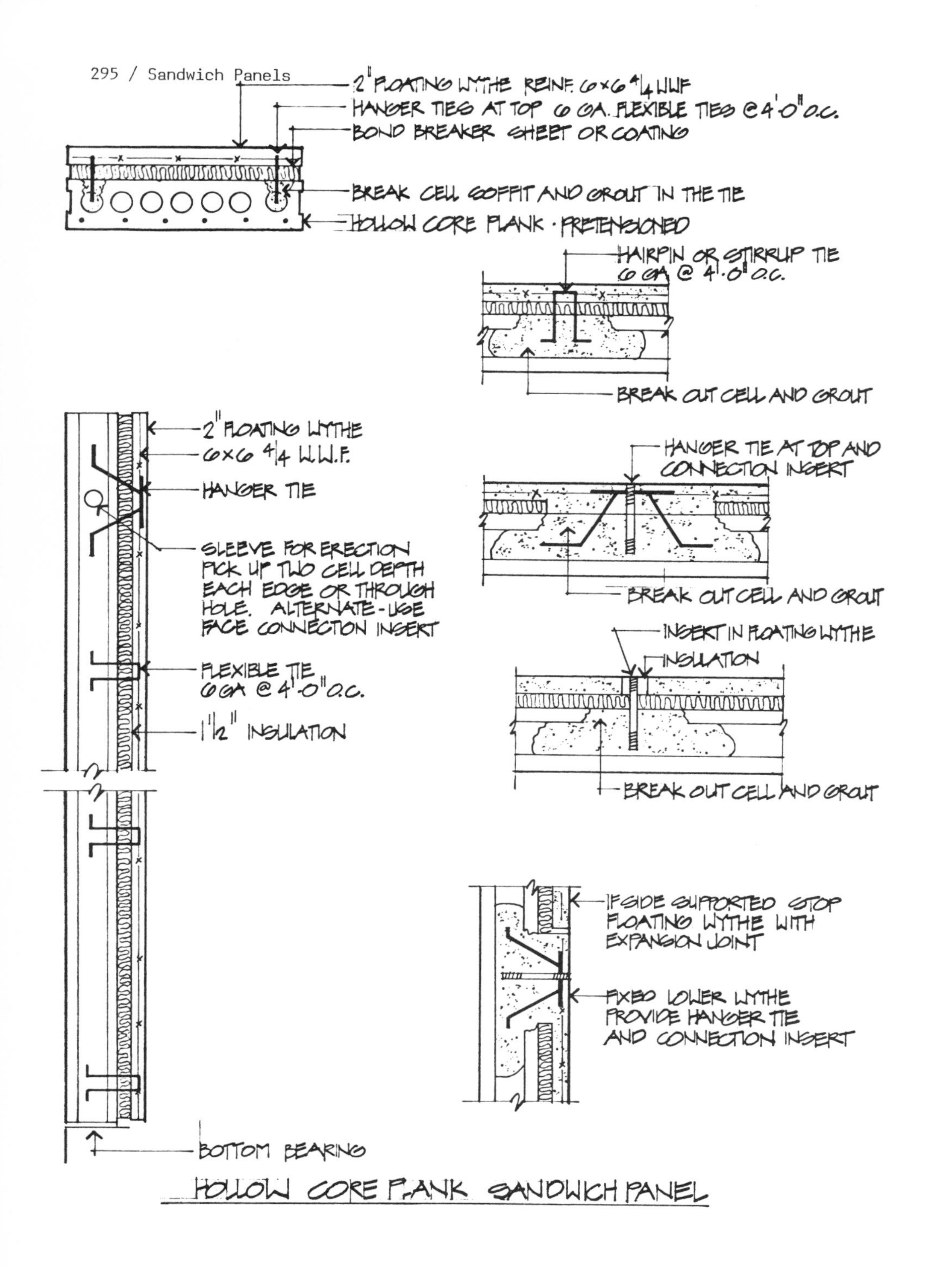

HOLLOW CORE PLANK SANDWICH PANEL

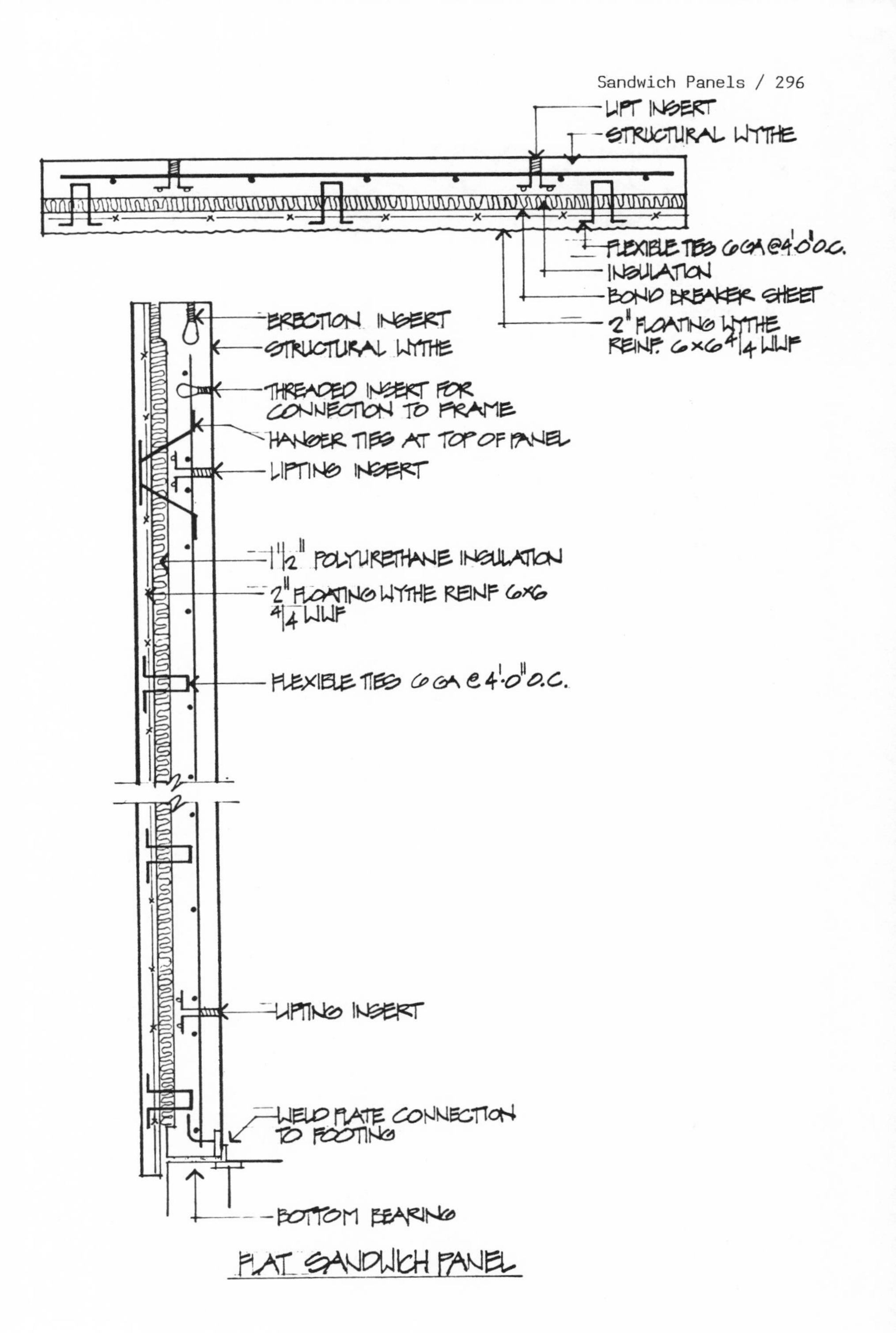

FLAT SANDWICH PANEL

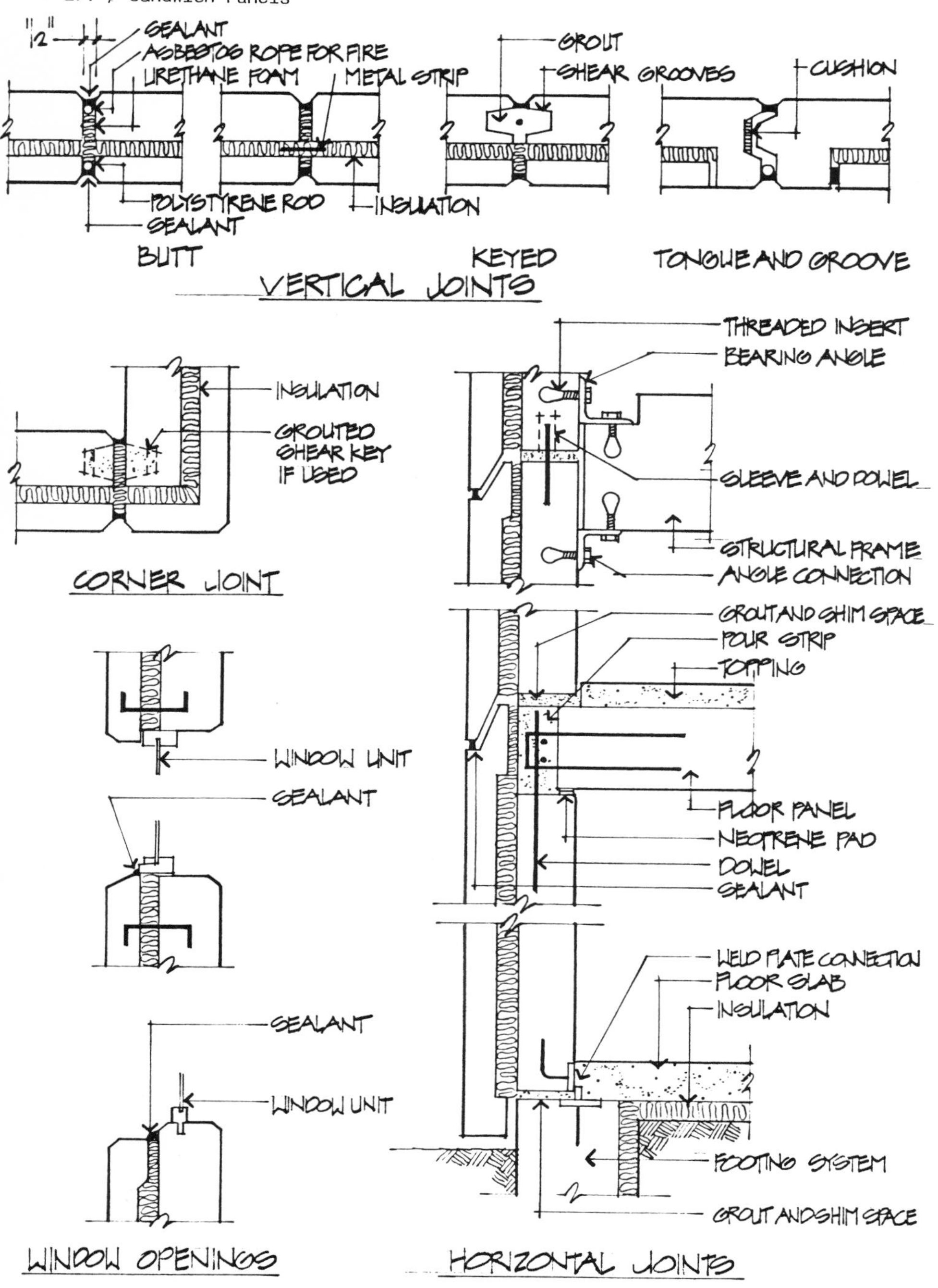
1/2"
SEALANT
ASBESTOS ROPE FOR FIRE
URETHANE FOAM
METAL STRIP
POLYSTYRENE ROD
SEALANT
INSULATION
GROUT
SHEAR GROOVES
CUSHION
BUTT
KEYED
TONGUE AND GROOVE
VERTICAL JOINTS
INSULATION
GROUTED SHEAR KEY IF USED
CORNER JOINT
WINDOW UNIT
SEALANT
SEALANT
WINDOW UNIT
WINDOW OPENINGS
THREADED INSERT
BEARING ANGLE
SLEEVE AND DOWEL
STRUCTURAL FRAME
ANGLE CONNECTION
GROUT AND SHIM SPACE
POUR STRIP
TOPPING
FLOOR PANEL
NEOPRENE PAD
DOWEL
SEALANT
WELD PLATE CONNECTION
FLOOR SLAB
INSULATION
FOOTING SYSTEM
GROUT AND SHIM SPACE
HORIZONTAL JOINTS

Composite Panels

A composite panel is one in which the two wythes are designed to act as a unit in resisting transverse or vertical forces. The edges are usually solid concrete and the wythes are connected together with solid concrete ribs, steel ribs, or a combination of concrete and steel. The ribs are designed to resist the imposed horizontal shear forces. If the temperature and humidity difference between the outside and the inside surface of the composite panel is zero or small then the volumetric changes due to both temperature, shrinkage and creep will be the same for both wythes and they will work together with little differential movement. However, as the volumetric changes become different between the two wythes, the resultant effect will be warping or bowing of the panel. The advantage of composite panels is that they have higher strength than equal concrete non-composite panels. Composite panels are usually non-load bearing and as such resist transverse forces only. The shear ties or connecting ribs may be concrete, trussed steel joint reinforcement as used in masonry construction, trussed steel studs of many different patterns, or expanded metal lath embedded in a concrete rib. If the panel is pretensioned then both wythes are pretensioned and the ribs are concrete. The panel edges are usually solid concrete; however, panels have been designed without concrete solid edges, but with metal stud or trussed reinforcement as shear ties.

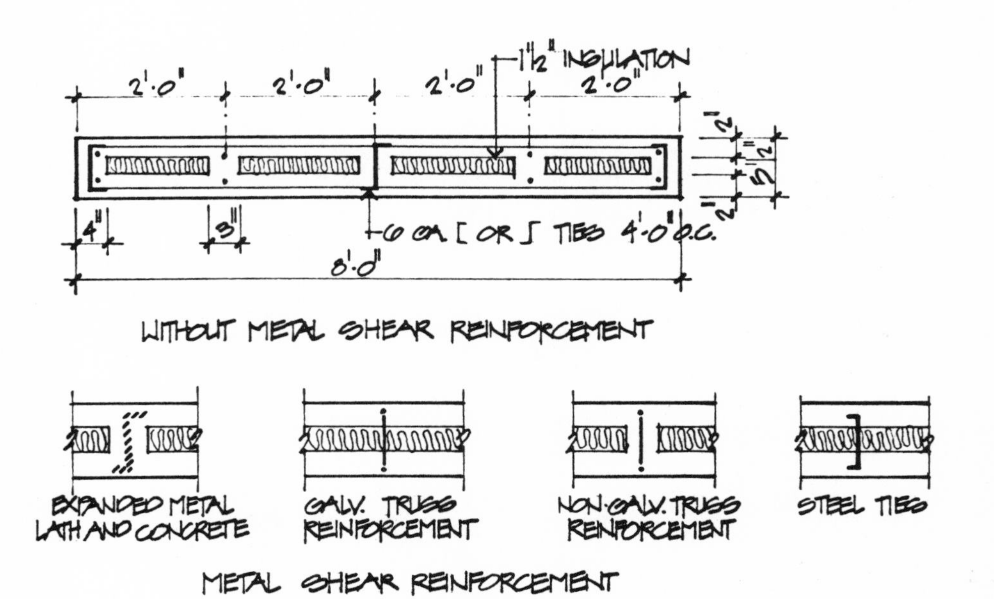

The spacing of the metal ties is often governed by the standard supplied widths of the insulation sheets.

Thermal bridge effect

The use of concrete ribs and edges and metal shear ties reduces the thermal efficiency of the panels. They produce thermal bridges (heat flow paths through the panel). The approximate width of the thermal bridge is as follows:

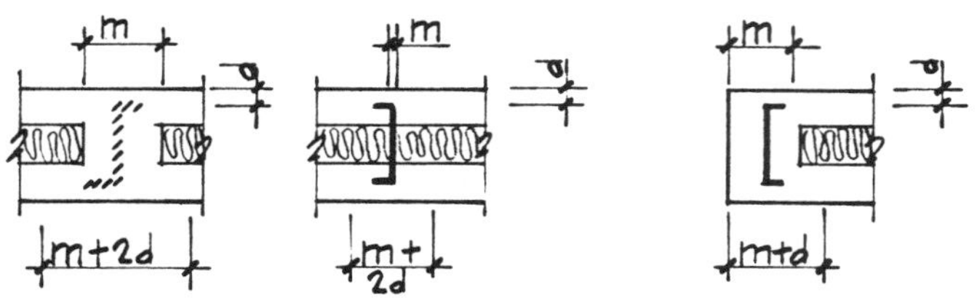

m = WIDTH OF THE CONCRETE RIB OR THE METAL SHEAR TIE
d = DISTANCE FROM THE PANEL SURFACE TO THE METAL TIE

$$U_{ave} = \frac{\text{U insulated area} + \text{U thermal bridge area}}{\text{Total panel area}} + \text{mass coefficient}$$

The mass coefficient is used to calculate the total heat loss due to the heat storage factor of the wall mass, but is omitted when comparing wall U values.

Insulation

The insulation used may be polystyrene, polyurethane, or fiberglass board. Again the type and thickness is determined by the desired R value, cost, and the panel thickness. For composite panels the insulation should not be coated nor have a bond breaker sheet since as much bond as possible is desired to be developed between the concrete and the insulation in order to improve the composite action.

Design Stresses

In calculating the value of $ft = \frac{MC}{I}$ for composite panels the value of I must be corrected due to the stiffness behavior of the shear ties or ribs. The theoretical value of I is:

$$I_t = \frac{b}{12} (d_o^3 - d_i^3).$$

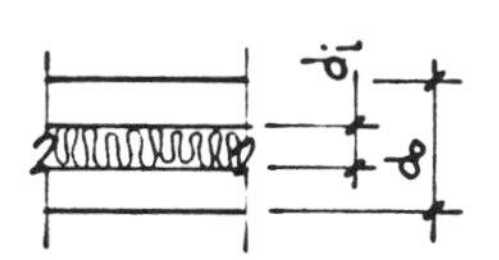

THE VALUE OF $I_{CORRECTED} = CI_t$
C = CORRECTION COEFFICIENT

Tests conducted by the Portland Cement Association have indicated the following percent of rigidity to that of I theoretical for composite panels. (correction coefficients)

22% - wythes connected by rigid board insulation or by metal connectors with no shear value

39% - panels with solid concrete periphery edges only

50% - panels with metal truss members, but no concrete edges

70% - panels with metal truss or steel stud shear connectors and concrete edges.

The above values are in the direction of loading of the metal shear connectors. If the panel has solid pick up points and solid edges and the width of the panel is small (8 feet or less) then composite action can be assumed in the transverse direction for the transverse band width without the use of metal shear ties in the transverse direction. Thus, the two wythes will work together to distribute the lifting stresses both ways. If the panel has no solid concrete edges then transverse shear connectors should be provided in the transverse band width at lifting points. The transverse band width is limited to the lesser of 10 times the total panel thickness, the distance from the pick up point to the closest edge, or half the spacing between the pick up points. The value of C is usually taken from 40 to 60%, subject to the panel design. The design value of $f_t = \frac{7.5\sqrt{f'c}}{F.S.}$, where F.S. (factor of safety) is usually 1.5. Handling design is done in the manner specified on pp. 165-176, taking into account the reduced value of "I" and the normal impact factors. If sand-light weight concrete is used, multiply f_t by 0.85.

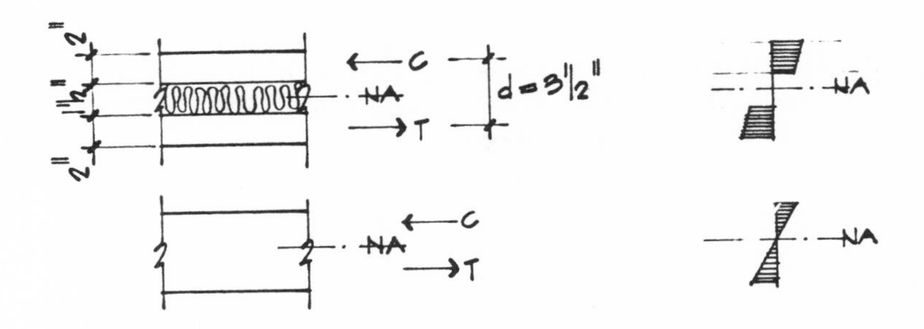

The concrete strength at pick-up is usually 2,000 psi and the 28 day strength is usually 5,000 psi or greater. The value of M (moment) is determined the same way as for non-sandwich panels and is dependent upon the location of the lifting and erection inserts.

Wythe reinforcement

The reinforcing steel in the wythe has little influence on the lifting and handling operations, hence the panels are designed as uncracked sections with the limiting factor being the tension stress in the concrete. However, the code minimum required reinforcing steel is provided to satisfy handling operations, service loads, and for volumetric changes. The code minimum is primarily to control volumetric changes; however, the lifting stresses are usually the most critical stresses to which the panels are ever subjected.

The code reinforcement requirement for welded wire fabric is 0.002 bt in the horizontal direction and 0.0012 bt in the vertical direction. For 40 ksi yield reinforcing bars the values are 0.0025 in the horizontal direction and 0.0015 in the vertical direction. With composite panels this can be divided proportionately between the two wythes. For pretensioned panels the strands are balanced and placed in the center of the wythes. In analyzing axial loads, the full section is used to determine the radius of gyration. However, composite panels are not recommended for use as load bearing panels due to volumetric strain differences between the wythes. Where they have been used as load bearing elements the axial load is usually carried by one wythe.

Pick-up points

Where pick-up points or inserts are provided the panel is usually solid. This solid area is made large enough to both enclose the insert and to resist the imposed shear forces.

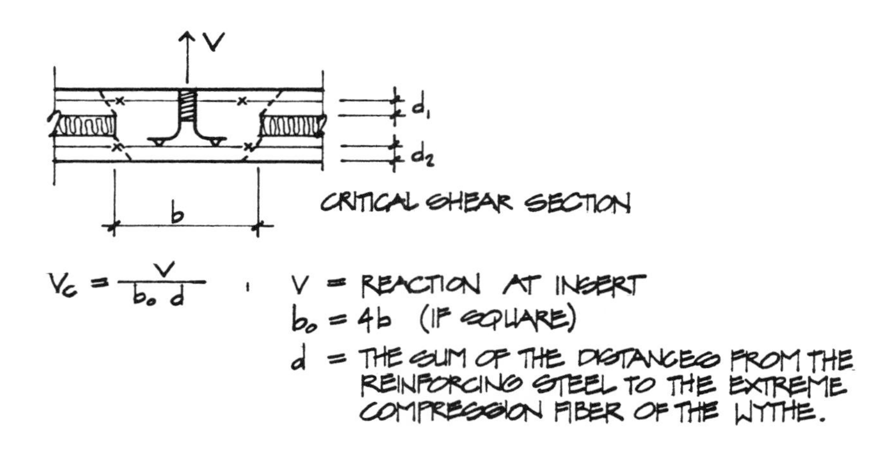

Waterproofing and Sealant Details

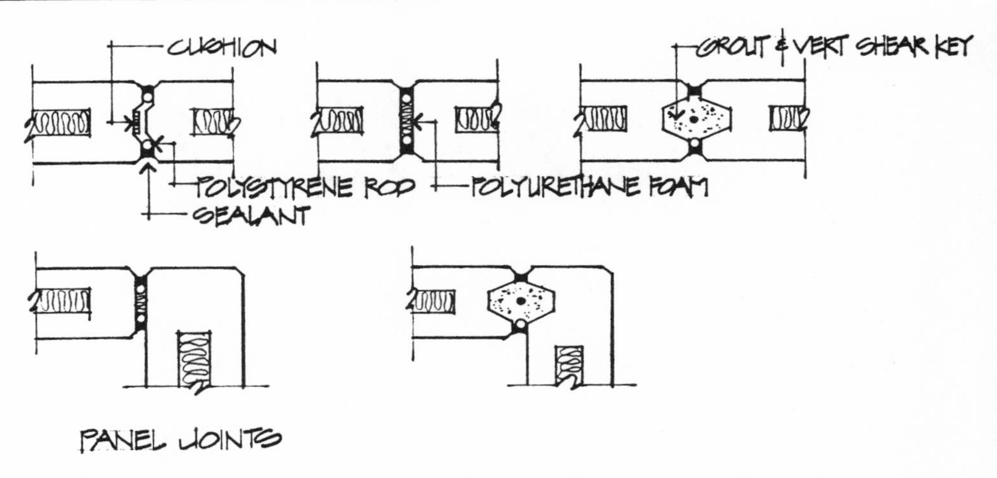

PRESTRESSED CONCRETE SOLID FLAT SLABS

BONAVENTURE HOTEL - LOS ANGELES

684,000 sf of prestressed concrete flat slabs were used for the floors in this 35 story multi-tower complex. The structural steel erector installed both the steel framing and the precast prestressed flat slabs in one continuous uninterrupted operation.

Solid prestressed concrete flat slabs are used for floor or roof members, load bearing or non load bearing wall panels, spandrel panels, or any situation where flat slabs in one direction may be used. They may be designed as either solid or insulated units. As deck members, they may be designed as simple spans or continuous spans with or without shoring and use a field poured cast-in-place concrete topping. They may also be designed as untopped deck elements, with due attention being given to the connections of the elements to ensure satisfactory performance of the diaphragm under seismic loading conditions. Since these elements are relatively thin sections, careful attention should also be given to the camber induced by prestressing as it relates to in-storage and in-situ conditions of slab dead weight and possible differential camber. These elements are used as floors in Motel and Hotel Construction, intermediate spanning slabs in spread tee or spread channel floor systems and offices and garages, stay in place deck panels in bridge construction and walls for industrial buildings.

BONAVENTURE HOTEL - LOS ANGELES

Shown here are floors in various stages of construction. The longest span was 24 feet with temporary midspan shoring. 3 1/2" thick prestressed concrete slabs with a 2 1/2" thick cast in place concrete topping formed the entire floor assembly with the underside of the slab being sprayed with acoustical textured paint to form the finished ceiling.

BONAVENTURE HOTEL - LOS ANGELES

Plant casting allowed close control over tolerances. Here we see slabs stockpiled in storage awaiting shipment. Openings for mechanical risers and vertical ventilator ducts were cast in the slabs during fabrication. The 16 gage metal edge former and uni-strut assembly eliminated perimeter edge forming and provided an instant attachment point for the window wall.

Solid prestressed concrete slabs are usually fabricated in depths ranging from 3" to 6" thick. For depths of 6" or greater it is usually more efficient to use hollow core slabs or shallow double tee units. The width of the units is restricted by shipping and the width of the precaster's casting beds. Optimum slab widths are from 8 to 12 feet. Casting beds in the factory may vary from 200 feet to as much as 800 feet in length. Many units are cast in a long line simultaneously, with bulkhead separators between units, or in a continuous pouring operation, with individual units be saw cut to length after the concrete has reached its required release strength. Untopped units are designed as simple spans only. Topped units are designed as simple spans or as continuous units with temporary midspan shoring. The minimum thickness of units is sometimes governed by non-structural considerations such as fire resistive requirements, or sound transmission criteria. The chart that follows indicates maximum spans for various slab thickness and design conditions:

PRELIMINARY SOLID SLAB DESIGN CHART (S_{LL} = 100 PSF)

	SLAB THICKNESS (INCHES)	MAXIMUM SIMPLE SPAN WITHOUT TOPPING (FEET)	MAXIMUM SIMPLE SPAN WITH 2½ IN. OF HARDROCK CONCRETE TOPPING (FEET)	MAXIMUM CONTINUOUS SPAN WITH 2½ IN. OF HARDROCK CONCRETE TOPPING & TEMPORARY MID-SPAN SHORING (FEET)**
LIGHTWEIGHT CONCRETE SLAB - 1 1/8" COVER OVER STRANDS	3½	*	14'-6	24'-0
	4	14'-0	16'-0	26'-0
	4½	15'-6	17'-6	28'-6
	5	17'-0	19'-0	31'-0
	5½	18'-6	20'-6	33'-6
	6	20'-0	22'-0	36'-0
HARDROCK CONCRETE SLAB - 1½" COVER OVER STRANDS	4	*	15'-0	25'-0
	4½	14'-6	16'-6	27'-6
	5	16'-0	18'-0	30'-0
	5½	17'-6	19'-6	32'-6
	6	19'-0	21'-0	35'-0

* Usually not permitted because of fire rating requirements.

** Longer spans may require third point shoring.

FIRE RATING REQUIREMENTS FOR SOLID PRESTRESSED CONCRETE FLAT SLABS (UBC-76)

	Item Description \ Fire Rating (Hours)	1	2	3	4
	Minimum Composite Floor Slab Thickness (inches Hardrock Concrete	3 1/2	4 1/2	5 1/2	6 1/2
	Minimum Composite Floor Slab Thickness (inches) Lightweight Concrete	3	4	4 1/2	5
UNRESTR RATING	Required Cover Over Strand (inches) Hardrock Concrete	1	1 1/2	2	2 1/2
UNRESTR RATING	Required Cover Over Strand (inches) Lightweight Concrete	3/4	1 1/8	1 1/2	1 7/8
RESTR RATING	Required Cover Over Strand (inches) Hardrock Concrete	3/4	3/4	1	1 1/4
RESTR RATING	Required Cover Over Strand (inches) Lightweight Concrete	3/4	3/4	3/4	1

Long Term Camber/Deflection

Initial camber and deflections are magnified with time due to several factors, among which are the loss of prestress force over time and the strength gain of concrete, as well as plastic deformation of the reinforced concrete complex with time, or creep. The second edition of the PCI Design Handbook on page 3-27 presents a chart with suggested multipliers to use in calculating long term cambers and deflections, taking into account the above factors. Calculate initial camber and member weight deflection using the initial concrete modulus of elasticity and initial prestress force, and always apply the multipliers listed to these values. Calculation of movements occurring due to loads applied in the final erected condition are naturally made using full values of modulus of elasticity. These factors will give the total approximate expected cambers and deflections for the condition being analyzed, and not just the additional effect, as is given in some codes and publications.

LOAD TABLES · SOLID PLS SLAB + 2 1/2" C.I.P. TOPPING
(PRESTRESSING SHOWN IS FOR 12'-0" WIDE SLAB

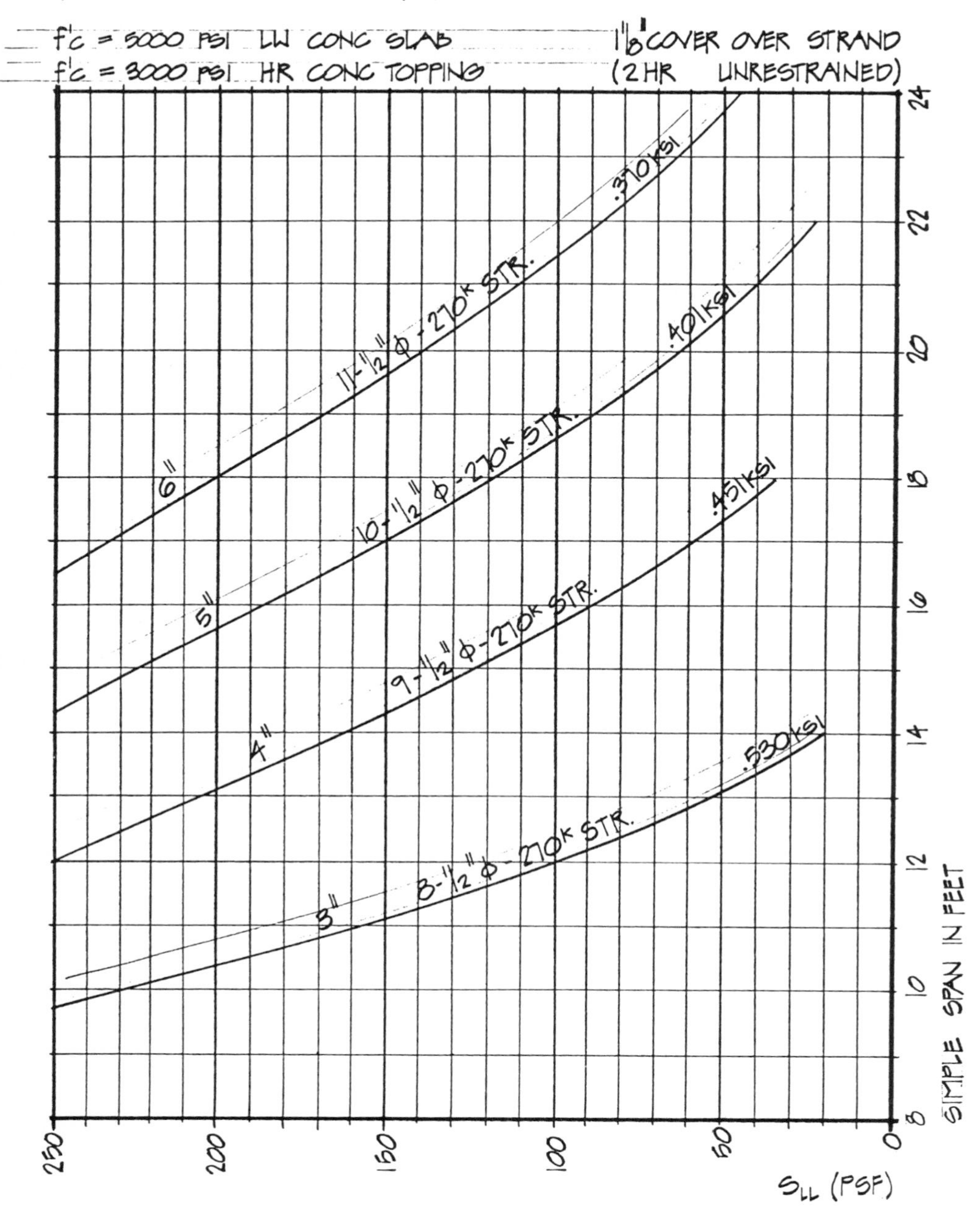

To satisfy energy requirements roof panels are usually covered with rigid insulation prior to applying the roofing material. Sandwich panels consisting of two flat slabs with an insulation core between may be used for both wall and roof panels. In this case one of the slabs is usually the structural slab and the other serves as a protective finish surface covering. This method also overcomes problems due to bowing caused by differential thermal expansion between the outer and inner surfaces of the panel.

ANALYSIS OF TOPPED PRESTRESSED CONCRETE FLAT SLABS

DESIGN EXAMPLE - FLAT FLOOR SLAB - UNSHORED

DESIGN A PRECAST PRETENSIONED FLOOR SLAB FOR A MIDRISE OFFICE BUILDING PROJECT. SLABS ARE TO BE 12 FEET WIDE AND SPAN 16 FEET. THEY ARE TO BE FABRICATED WITH SAND-LIGHTWEIGHT CONCRETE, AND MUST HAVE A 2 HOUR UNRESTRAINED FIRE RATING. SLAB CONCRETE STRENGTH AT RELEASE: 3500 PSI; AT 28 DAYS: 5000 PSI. TOPPING WILL BE NORMAL WEIGHT CONCRETE (f'_c = 3000 PSI) ½" φ - 270 K PRESTRESSING STRANDS WILL BE USED.

SOLUTION:

FROM PRELIMINARY DESIGN CHARTS A 4" THICK SECTION IS CHOSEN WITH 2½" OF CAST IN PLACE TOPPING.

$$\frac{E_{TPG}}{E_{P/C}} = \frac{3.3}{2.9} = 1.14$$

WT. SLAB = 38 PSF
WT. TPG. = 32 PSF
LL = 50 PSF

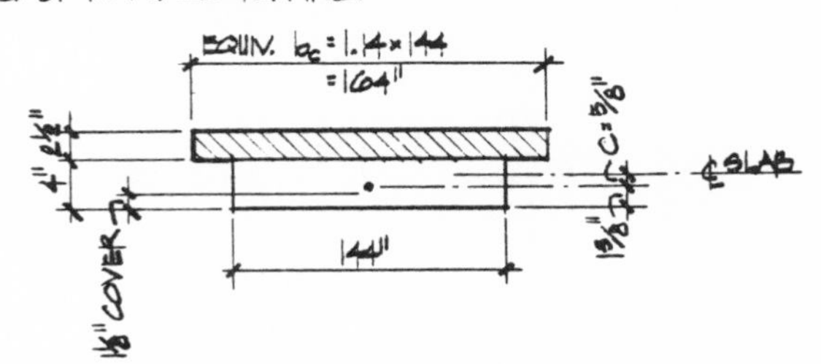

DETERMINE SECTION PROPERTIES:

BASIC

$A = (144)(4) = 576 \text{ IN}^2$

$I = (\frac{1}{12})(144)(4)^3 = 768 \text{ IN}^4$

$S_b = S'_t = (\frac{1}{2})(768) = 384 \text{ IN}^3$

COMPOSITE

PART	A	y_{TCC}	Ay_{TC}	I_o	Ad^2
BASIC	576	4.50	2592	768	1049
TOPPING	410	1.25	513	213	1480
TOTAL	986		3105	ΣI_c = 3510	

$$\bar{y}_{TCC} = \frac{3105}{986} = 3.15''$$

$y_{TC} = 0.65$ $\qquad S_{TC} = 5400 \text{ IN}^3$

$y_{BC} = 3.35$ $\qquad S_{BC} = 1048 \text{ IN}^3$

LOADS & MOMENTS - 12' WIDE SLAB:

SLAB WT. $M_S = (1/8)(12)(.038)(16)^2(12) = 175$ K-IN

TOPPING $M_C = (1/8)(12)(.032)(16)^2(12) = 147$ K-IN

LIVE LOAD $M_L = (1/8)(12)(.050)(16)^2(12) = 230$ K-IN

ALLOWABLE STRESSES

CONCRETE

@ TRANSFER ($f'_{ci} = 3500$ PSI)

COMPRESSION $0.6 \times 3500 = 2100$ PSI

TENSION $3 \times \sqrt{3500} = 177$ PSI

FINAL DESIGN ($f'_c = 5000$ PSI)

COMPRESSION $0.45 \times 5000 = 2250$ PSI

TENSION $6 \times \sqrt{5000} = 425$ PSI

STRAND

@ TRANSFER (ASSUME 10% LOSSES) $f_i = (0.9)(0.7)(270) = 170$ KSI

FINAL DESIGN (ASSUME 10% LOSSES) $f_i = (0.8)(0.7)(270) = 151$ KSI

CALCULATE PRESTRESS FORCE REQUIRED

$$F_f = \frac{\frac{M_D}{S_B} + \frac{M_{BC}}{S_{BC}} - f_t}{\frac{1}{A} + \frac{e}{S_B}} = \frac{\frac{175+147}{384} + \frac{230}{1048} - 0.425}{\frac{1}{576} + \frac{0.625}{384}} = 188 \text{ K}$$

$$N = \frac{188}{151 \times 0.1531} = 8.13$$

USE 8 - ½" Φ - 270 K STRANDS

USE INITIAL TENSION OF $\frac{188}{(8)(0.153)(270)(0.8)} = 0.71\ f_{pu}$

$f_i = (0.9)(0.71)(270) = 173$ PSI $\quad f_f = (0.8)(0.71)(270) = 153$ KSI

$F_i = (8)(173)(0.1531) = 212$ K $\quad F_f = (8)(153)(0.1531) = 187$ K

@ RELEASE $-\frac{F_i}{A} \pm \frac{F_i e}{S_B} \pm \frac{M_S}{S_B}$

@ END $f_t = -\frac{212}{576} + \frac{(212)(0.625)}{384} = -0.023$ (COMPRESSION)

$f_b = -0.368 - 0.345 = -0.713$ (COMPRESSION)

@ ℄ $f_t = -0.368 + 0.345 - \frac{175}{384} = -0.479$ (COMPRESSION)

$f_b = -0.368 - 0.345 + 0.456 = -0.257$ (COMPRESSION)

UNDER ALL LOADS - FINAL DESIGN CONDITION

@ ℄ OF SPAN

$$f_t = -\frac{187}{576} + \frac{(187)(.625)}{384} - \frac{322}{384} - \frac{230}{5400} = -0.902 < 0.45 f'_c \quad O.K.$$

$$f_b = -0.324 - 0.304 + 0.839 + \frac{230}{1048} = +0.430 \approx 6\sqrt{f'_c} \quad O.K$$

CHECK ULTIMATE STRENGTH

$$\rho_p = \frac{A_{ps}}{bd} = \frac{(3)(0.1531)}{(144)(5.13)} = 0.001658$$

$$f_{ps} = f_{pu}\left(1 - 0.5\rho_p \frac{f_{pu}}{f'_c}\right) = 270\left(1 - 0.5 \times 0.001658 \times \frac{270}{5}\right) = 257.9 \text{ KSI}$$

$$a/2 = \frac{A_{ps} f_{ps}}{1.7 f'_c b} = \frac{(3)(.1531)(257.9)}{(1.7)(3)(144)} = 0.48''$$

$$M_u = \phi A_{ps} f_{ps}(d - a/2) = (0.9)(3)(.1531)(257.9)(5.13 - 0.48) = \underline{1336 \text{ K-IN}}$$

$$UM = 1.4 M_D + 1.7 M_L = (1.4)(175 + 147) + (1.7)(230) = 842 \text{ K-IN} < M_u \quad O.K.$$

TRANSVERSE STEEL IN PRECAST SLAB (TEMPERATURE + HANDLING)

$$A_s = (0.002)(4)(12) = 0.096 \text{ IN}^2/\text{FT REQ'D.}$$

USE #3 @ 12" O.C. TRANSVERSE

CHECK SHEAR

$$v_u = \frac{UV}{bd} = \frac{(1.4)(0.07)(8)(12) + (1.7)(0.05)(8)(12)}{(144)(5.13)} = 0.024 \text{ KSI}$$

$$v_c = 0.85\,\phi\, 2\sqrt{f'_c} = \frac{(0.85)(0.85)(2)(\sqrt{5000})}{1000} = 0.102 \text{ KSI} \quad O.K.$$

CHECK DEFLECTIONS

INSTANTANEOUS CAMBER

$$\Delta_{P/S}\uparrow = \frac{F_i e L^2}{8EI} = \frac{(212)(.63)(16)^2(144)}{(8)(2400)(768)} = 0.33''\uparrow$$

DEFLECTION SLAB SELF WGT. (ASSUME SLAB IS BLOCKED IN STORAGE ONE FT. FROM EACH END)

$$\Delta_S\downarrow = \frac{5wL^4}{384EI} = \frac{(5)(.038)(12)(14)^4(1728)}{(384)(2400)(768)} = 0.21''\downarrow$$

USING A CREEP FACTOR OF 1.8, WHEN THE SLAB REACHES THE JOB SITE IT WILL EXHIBIT A NET UPWARD DEFLECTION OF:

$$1.8 \times (0.33 - 0.21) = 0.21''\uparrow$$

WHEN WET TOPPING IS POURED, THE SLAB WILL DEFLECT AN ADDITIONAL:

$$\Delta_{TPG}\downarrow = \frac{(5)(.032)(12)(16)^4(1728)}{(384)(2900)(768)} = 0.25''\downarrow$$

LONG TERM EFFECT OF TOPPING WEIGHT (ADDITIONAL CREEP DEFLECTION):

$$1.5\times\frac{(5)(.032)(12)(16)^4(1728)}{(384)(3100)(3510)} = 0.05''\downarrow$$

INSTANTANEOUS LIVE LOAD DEFLECTION:

$$\Delta_{LL} = \frac{(5)(0.05)(12)(16)^4(1728)}{(384)(3100)(3510)} = 0.08''\downarrow$$

MAXIMUM LL DEFLECTION ALLOWABLE FROM ACI 318-77 TABLE 9.5(b):

$$\frac{L}{480} = \frac{16\times12}{480} = 0.40'' > \Delta_{LL} \quad \text{O.K.}$$

NET CONDITION ALL LOADS

$$+\uparrow\Sigma\Delta = +0.21 - 0.25 - 0.05 - 0.08 = -0.17''\downarrow$$

CHECK HANDLING CONDITIONS

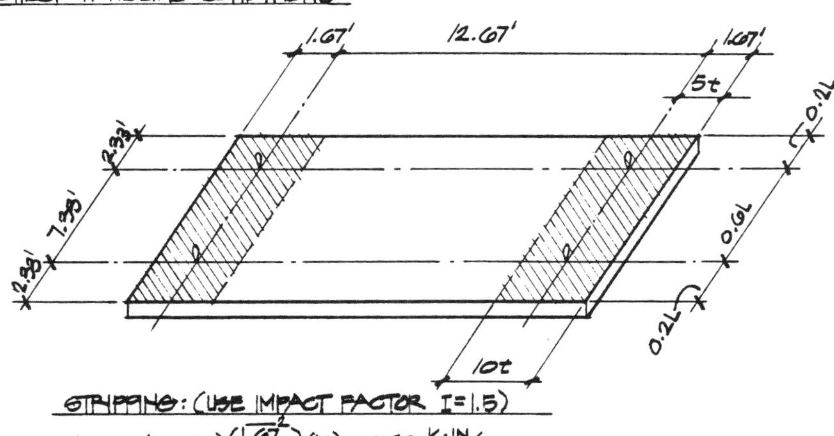

STRIPPING: (USE IMPACT FACTOR I=1.5)

$$M_{NEG} = (0.056)\left(\frac{1.67^2}{2}\right)(12) = 0.93\ \text{K·IN/FT.}$$

$$M_{POS} = (0.056)\left(\frac{12.67^2\times12}{8}\right) - 0.93 = 12.55\ \text{K·IN/FT.}$$

448# 448#

1.67' 12.67' 1.67'

$37\times1.5 = 56$ #/FT.

CHECK TENSILE STRESSES

@CANTILEVER

$$f_{top} = -\frac{P_i}{A} + \frac{P_i e}{S_T} + \frac{M_{STRIP}}{S_T} = -\frac{17.7}{48} + \frac{(17.7)(.63)}{32} + \frac{0.95}{32}$$

$$= -0.369 + 0.348 + 0.030 = +0.009 \text{ KSI O.K.}$$

@MIDSPAN

$$f_{bot.} = -\frac{17.7}{48} - \frac{(17.7)(.63)}{32} + \frac{12.55}{32} = -0.325 \text{ KSI O.K.}$$

ALLOWABLE TENSILE STRESS AT STRIPPING

$$F_T = \frac{f_{ri} \times 0.85}{F.S.} = \frac{7.5\sqrt{3500} \times 0.85}{2.0} = +0.189 \text{ KSI}$$

TRANSVERSE CONDITION (ASSUME TRANSVERSE BEAM 40" WIDE)

$$M^+ = M^- = \frac{(0.444)(2.33)^2}{2} = 14.46 \text{ K-IN}$$

$$f_t = \frac{M}{S} = \frac{14.46}{32 \times 3.33} = +0.136 \text{ KSI} < 0.189 \text{ O.K.}$$

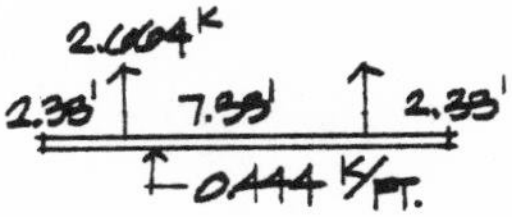

2.664^K

2.33' | 7.33' | 2.33'

0.444 K/FT.

SHIPPING:

BLOCK @ LIFTING POINTS WITH TRANSVERSE DUNNAGE; USE I = 2.0
BY INSPECTION FROM LONGITUDINAL STRIPPING CONDITION, THIS IS NOT CRITICAL.

PALO ALTO FINANCIAL CENTER

This totally precast concrete building features the use of exterior and deck elements acting compositely with cast in place concrete. Here we see 2 3/4" thick solid prestressed lightweight concrete slabs being erected on temporary midspan shoring. Final design span is 18 feet with shoring being removed after 2 3/4" of lightweight concrete topping is cured.

WESLEY TOWERS HOUSING FOR THE ELDERLY
CAMPBELL, CA.

4" thick solid pre-stressed concrete slab being erected on precast concrete "voided" bearing walls. Temporary midspan shoring is used until the cast in place topping is poured and cured, thus providing a 24' clear span floor.

DESIGN EXAMPLE - SHORED SOLID SLAB SYSTEM

DESIGN A SOLID PRESTRESSED CONCRETE SLAB TO SPAN 25'-0" BETWEEN MASONRY BEARING WALLS FOR A MOTEL PROJECT. TEMPORARY MIDSPAN SHORING WILL BE PROVIDED. A 2 HOUR FIRE RATING IS REQUIRED.

SOLUTION:

FROM UBC-76, TABLE 43-A, IN ORDER TO ACHIEVE A 2 HOUR RATING, 1½" OF COVER ARE REQUIRED FOR HARD ROCK CONCRETE, AND 1⅜" OF COVER ARE REQUIRED FOR LIGHTWEIGHT CONCRETE. BY REFERRING TO PRELIMINARY LOAD/SPAN DESIGN TABLES, A 4 INCH THICK HARDROCK CONCRETE SLAB WITH 2½ INCHES OF CAST IN PLACE CONCRETE TOPPING, SPANNING CONTINUOUSLY BETWEEN SUPPORTS IS SELECTED. (NOTE ONLY A SLIGHT INCREASE IN MAXIMUM SPAN IS AFFORDED BY USING LIGHTWEIGHT AGGREGATE CONCRETE.)

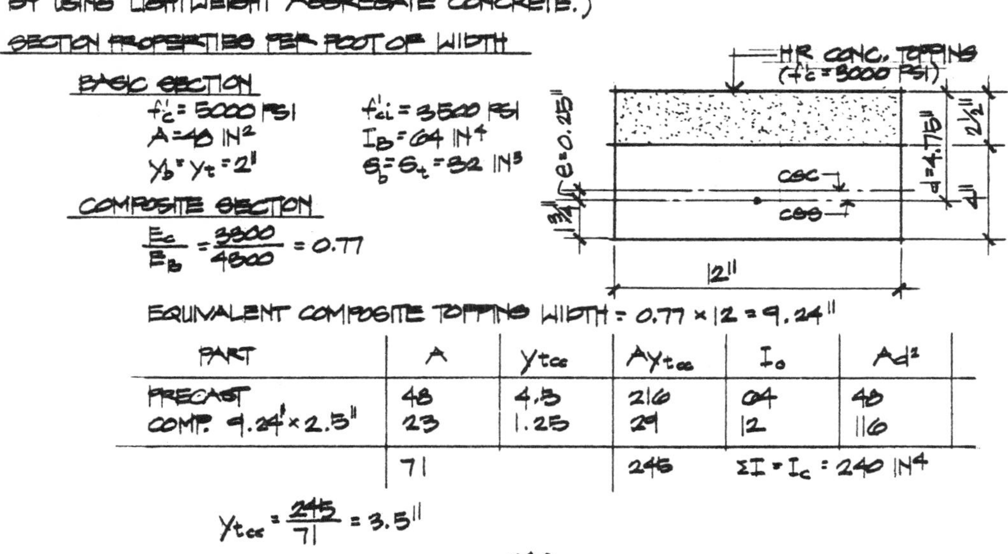

SECTION PROPERTIES PER FOOT OF WIDTH

BASIC SECTION

$f'_c = 5000$ PSI $\qquad f'_{ci} = 3500$ PSI

$A = 48$ IN^2 $\qquad I_B = 64$ IN^4

$y_b = y_t = 2"$ $\qquad S_b = S_t = 32$ IN^3

COMPOSITE SECTION

$$\frac{E_c}{E_B} = \frac{3300}{4300} = 0.77$$

EQUIVALENT COMPOSITE TOPPING WIDTH = 0.77 × 12 = 9.24"

PART	A	y_{tcc}	Ay_{tcc}	I_o	Ad^2
PRECAST	48	4.5	216	64	48
COMP. 9.24' × 2.5"	23	1.25	29	12	116
	71		245	$\Sigma I = I_c = 240$ IN^4	

$$y_{tcc} = \frac{245}{71} = 3.5"$$

$$y_{bc} = 3.0" \qquad S_{bc} = \frac{240}{3} = 80 \text{ IN}^3$$

SERVICE LOADS

4" PRESTRESSED SOLID SLAB	50 PSF
2½" C.I.P. CONC. TOPPING	32 PSF
PARTITIONS & MISCELLANEOUS	15 PSF
LIVE LOAD	40 PSF
TEMPORARY CONSTRUCTION LOAD (S_{D+L}) (CAL. OSHA SEC. 1717)	100 PSF

SOLID SLAB (CONT.)

I. ELASTIC ANALYSIS - SERVICE LOAD DESIGN

a) MAXIMUM POSITIVE MOMENT CONDITION

(1) WEIGHT OF P/S SLAB + WET TOPPING

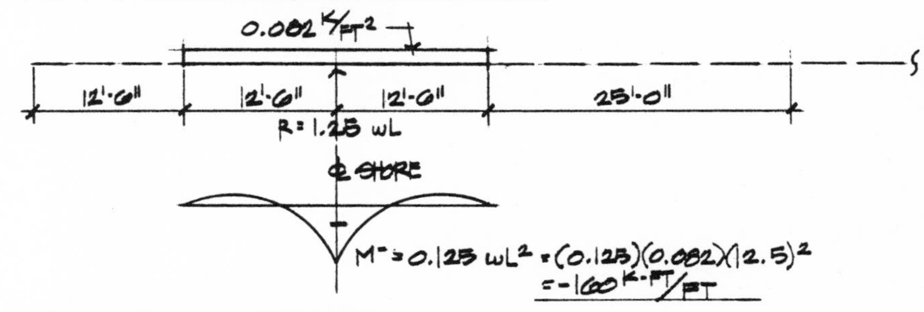

(2) EFFECT OF SHORE REMOVAL

PLACE TWO BEARING WALLS 12'-6" APART AT ENDS OF BUILDING AND AT EITHER SIDE OF EXPANSION JOINTS IN ORDER TO MINIMIZE POSITIVE MOMENT IN FIRST 25 FT. INTERIOR SPAN

$1.25\,wL = (1.25)(0.082)(12.5) = 1.28^{K}$

$I = 80$ CONSTANT

MOMENT SIGN CONVENTION (+ clockwise)

A — 12'-6" — B — 25'-0" — C — 25'-0" — D

	A	B	B	C	C	D	D
K	4.8	6.4	3.2	3.2	3.2	3.2	3.2
D.F. ($K/\Sigma K$)	1.0	.67	.33	.5	.5	.5	.5
F.E.M.	0	0	+4.00	-4.00	0	0	0
BAL.		-2.68	-1.32	+2.00	+2.00		
C.O.			+1.00	-0.66		+1.00	
BAL.		-0.67	-0.33	+0.33	+0.33	-0.50	-0.50
C.O.			+0.16	-0.16	-0.25	+0.16	
BAL.		-0.11	-0.05	+0.21	+0.20	-0.08	-0.08
ADJ. MOM.	0	-3.46	+3.46	-2.28	+2.28	+0.58	-0.58

$K_{AB} = \frac{80}{12.5} = 6.4$

$K_{BA} = \frac{3}{4} \times 6.4 = 4.8$

$K_{BC} = \frac{80}{25} = 3.2$

$F.E.M. = \frac{PL}{8} = \frac{(1.28)(25)}{8} = 4.0^{K\text{-}FT/FT}$

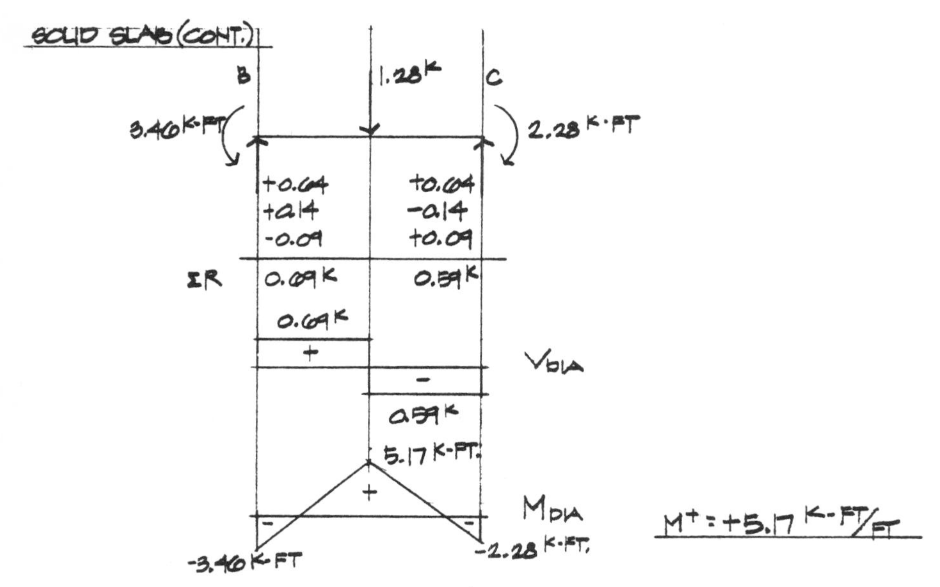

(3) SUPERIMPOSED DEAD & LIVE LOAD ON FIRST 25 FT. INTERIOR SPAN

0.055 K/FT²

A — 12'-6" — B — 25'-0" — C — 25'-0" — D

	A	B	B	C	C	D	D
K	4.8	6.4	3.2	3.2	3.2	3.2	3.2
D.F. (K/ΣK)	1.0	.67	.33	.5	.5	.5	.5
F.E.M.	0	0	+2.86	-2.86	0	0	0
BAL.		-1.92	-0.94	+1.43	+1.43		
C.O.			+0.72	-0.47		+0.72	
BAL.		-0.48	-0.24	+0.24	+0.23	-0.36	-0.36
C.O.			+0.12	-0.12	-0.18	+0.12	
BAL.		-0.08	-0.04	+0.15	+0.15	-0.06	-0.06
ADJ. MOM.	0	-2.48	+2.48	-1.63	+1.63	+0.42	-0.42

$$F.E.M. = \frac{wL^2}{12}$$

$$= 0.055 \times \frac{(25)^2}{12}$$

$$= 2.86 \text{ K-FT/FT}$$

SOLID SLAB (CONT.)

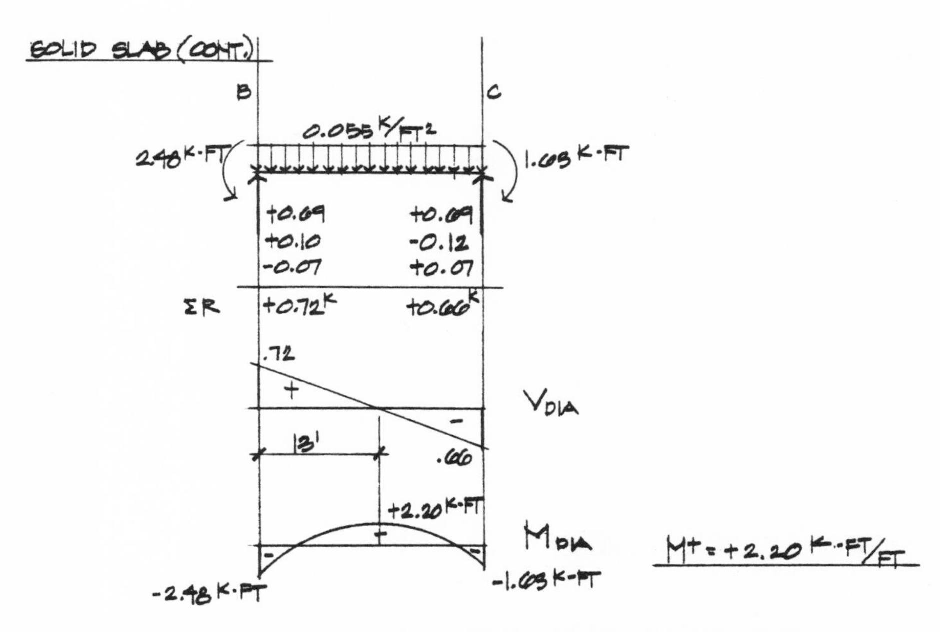

$M^+ = +2.20$ K·FT/FT

PRESTRESS FORCE REQ'D. TO DEVELOP POSITIVE MOMENT

(3-10)*

$$F_f = \frac{\frac{M_B}{S_b} + \frac{M_c}{S_{bc}} - f_t}{\frac{1}{A} + \frac{e}{S_b}} = \frac{-\frac{1.60 \times 12}{32} + \frac{(5.17+2.20)(12)}{80} - 0.425}{\frac{1}{48} + \frac{0.25}{32}}$$

$$= \frac{-0.600 + 1.106 - 0.425}{0.02083 + 0.00781} = \frac{0.081}{0.029} = 2.8 \text{ K/FT (SMALL)}$$

∴ ULTIMATE STRENGTH ANALYSIS GOVERNS STRAND REQUIREMENT

b) MAXIMUM NEGATIVE MOMENT CONDITION IN SLAB
CAL OSHA SEC 1717 GOVERNS

$M^- = (0.125)(0.100)(12.5)^2 = 1.95$ K·FT/FT

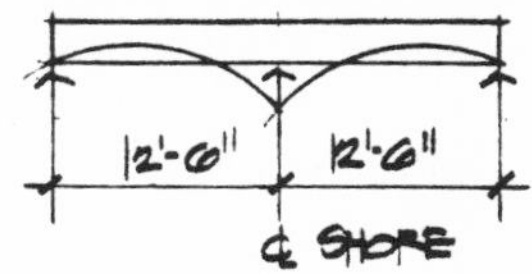

*( ) PAGE REFERENCE IN 2ND EDITION OF PCI DESIGN HANDBOOK

SOLID SLAB (CONT.)

PRESTRESS FORCE REQUIRED TO DEVELOP NEGATIVE MOMENT

(3-14) USE ALLOWABLE TENSION EQUAL TO MODULUS OF RUPTURE

$f_r = 7.5\sqrt{f'_c} = 0.530$ KSI

$f_t = -\frac{F_f}{A} + \frac{F_f \times e}{S_t} + \frac{M}{S_t}$

$F_f\left(\frac{1}{A} - \frac{e}{S_t}\right) = \frac{M}{S_t} - f_t$

$F_f = \frac{\frac{M}{S_t} - f_t}{\frac{1}{A} - \frac{e}{S_t}} = \frac{\frac{(1.95)(12)}{32} - 0.530}{\frac{1}{48} - \frac{0.25}{32}} = \frac{0.731 - 0.530}{0.02083 - 0.00781} = \frac{0.201}{0.013} = 15.5\ ^{K}/_{FT}$

(3-15) FOR A 12 FOOT WIDE SLAB, THE REQUIRED NUMBER OF $\frac{1}{2}''\phi$ - 270 K STRANDS IS:

(3-27) $A_{ps} = 0.1531$ $F_o = 28.9\ ^{K}/_{STRAND}$ $F_f = 0.78 \times 28.9 = 22.5\ ^{K}/_{STRAND}$

$N = \frac{15.5 \times 12}{22.5} = 8.3 =$ 9 STRANDS REQ'D.

2. ULTIMATE STRENGTH DESIGN

$UM^+ \cong (1.4)(5.17) + (1.4)(^{13}/_{55})(2.2) + (1.7)(^{42}/_{55})(2.2)$

$= 7.24 + 0.84 + 2.72$

$= 10.8\ ^{K\text{-}FT}/_{FT}$ OF SLAB

(3-50) DETERMINE PRESTRESSING REQUIREMENTS

$K_u = \frac{12000\ UM}{bd^2} = \frac{(12000)(10.8)}{(12)(4.75)^2} = 479$

FOR $f'_c = 3000$ & $K_u = 480$

$\overline{\omega}_p = 0.228 < 0.300$ O.K. - UBC 76 SEC 2618(i)

$A_{ps} = \overline{\omega}_p\, bd\, ^{f'_c}/_{f_{pu}}$

$= (0.228)(12)(4.75)(\frac{3}{270}) = 0.144\ ^{IN^2}/_{FT}$

(3-15) $N = \frac{0.144 \times 12}{0.1531} = 11.3$

USE 12 - $\frac{1}{2}''\phi$ - 270 K STRANDS

SOLID SLAB (CONT.)

(3-27) CHECK DEFLECTIONS

$$\Delta \cong 0.008\frac{PL^3}{EI} + 0.004\frac{wL^4}{EI}$$

WHERE $P = 1.28^K$ (SHORE)
$w = 0.015\ ^K/_{FT}\ (S_{DL})$
$$E = \frac{(4300)(4)+(3300)(2.5)}{6.5} = 3900\ \text{KSI}$$

$$= \left\{0.008\left[\frac{(1.28)(25)^3}{(3900)(240)}\right] + 0.004\left[\frac{(0.015)(25)^4}{(3900)(240)}\right]\right\}(1728)$$

$= (0.000171 + 0.000025)(1728)$
$= 0.34''$ $\quad \Delta_{LT} \cong 2.5 \times 0.34 = 0.85''$

$$\frac{L}{240} = \frac{25\times12}{240} = 1.25''\ \text{O.K.} \cdot \text{UBC-76 TABLE 23-D}$$

(3-21) CHECK HORIZONTAL SHEAR BETWEEN SLAB & TOPPING
(SEE SHEAR DIAGRAMS FOR CRITICAL LOADING CONDITIONS)

POSITIVE MOMENT REGION

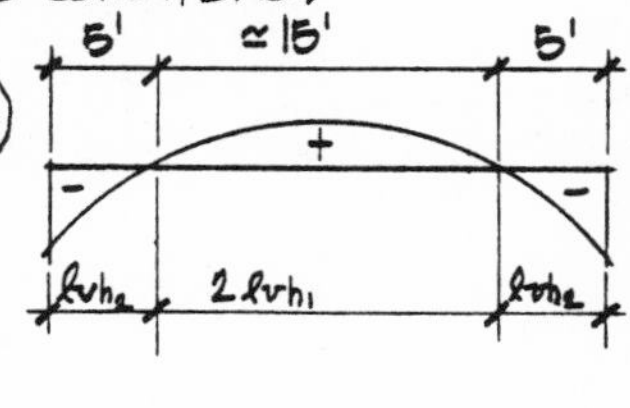

$$UV^+ = (1.4)(0.05) + (1.4)\left(\frac{15}{55}\right)\left(\frac{0.69}{2}\right) + (1.7)\left(\frac{40}{55}\right)\left(\frac{0.69}{2}\right)$$
$= 1.47^K$

$\phi V_{uh} = \phi \times 80\ b_v d$

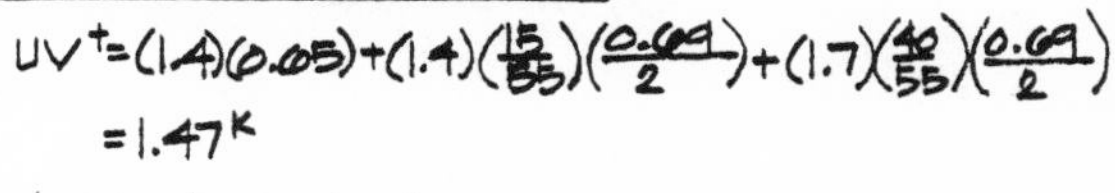

$$= \frac{(0.85)(80)(12)(4.75)}{1000} = 3.88^K > UV\ \text{O.K.}$$

NEGATIVE MOMENT REGION

$$UV^- = (1.4)(0.09) + (1.4)\left(\frac{15}{55}\right)(0.60) + (1.7)\left(\frac{40}{55}\right)(0.60) = 1.93^K > UV\ \text{O.K.}$$

∴ ROUGHEN TOP OF SLAB TO SATISFY HORIZ. SHEAR

CHECK VERTICAL SHEAR

$$v_{u\ MAX} = \frac{UV^-}{bd} = \frac{1.93}{(12)(4.75)} = 0.034\ \text{KSI}$$

$\frac{1}{2}\phi v_c = \frac{1}{2} \times 0.85 \times 2\sqrt{5000} = 0.060\ \text{KSI} > v_u$ O.K.

NO WEB REINFORCING REQUIRED

CHECK HANDLING - 25'-0" LONG SLABS SHIPPING GOVERNS

TRY 2 PT. SUPPORT TRANSVERSE

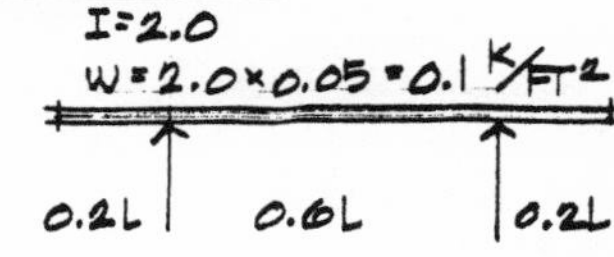

$M^- = M^+ = 0.025 \times 0.1 \times \overline{25}^2 = 1.56\ ^{K\cdot FT}/_{FT}$

$$f^+ = \frac{M}{S} = \frac{1.56\times12}{32} = 0.585\ ^K/_{IN^2}\ \text{EXCESSIVE}$$

SOLID SLAB (CONT.)

USE CONTINUOUS LONGITUDINAL DUNNAGE ON TRUCK

$M = 0.025 \times 0.1 \times 12^2 = 0.36 \ ^{K\cdot FT}/_{FT}$

$f^{TRANSVERSE} = \frac{0.36 \times 12}{32} = 0.135 \ KSI$

$f_T^{ALLOWABLE} = \frac{f_r}{1.5} = \frac{0.530}{1.5} = 0.350 \ KSI$ O.K.

DETERMINE NEGATIVE MOMENT REQUIREMENTS IN TOPPING OVER SUPPORTS

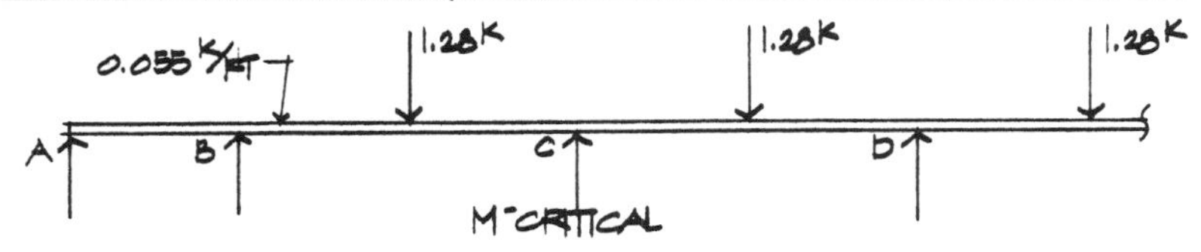

BY USING MOMENT DISTRIBUTION M^- MAXIMUM IS FOUND TO OCCUR AT SUPPORT "C"

$M_C^- = \frac{\text{SHORE LOAD}}{0.135\,PL} \frac{50+L}{+0.088\,wL^2} = \frac{\text{SHORE LOAD}}{4.33} + \frac{50+L}{3.04} = 7.37 \ ^{K\cdot FT}/_{FT}$

$UM^- = (1.4)(4.33) + (1.4)(\frac{15}{50})(3.04) + (1.7)(\frac{40}{50})(3.04) = 11.5 \ ^{K\cdot FT}/_{FT}$

USING ASTM A615-60 REINFORCEMENT

FROM ACI DESIGN HANDBOOK

$A_s = \frac{UM}{a_u d} = \frac{11.5}{4.18 \times 5.5} = 0.50 \ IN^2/FT$, LESS TOPPING MESH AREA

USE #6 @ 12" o/c ; 5000 PSI BETWEEN SLABS @ WALLS

TRANSVERSE TEMPERATURE STEEL $0.002 \times 12 \times 4 = 0.096$

USE #3 @ 12" o/c TRANSVERSE

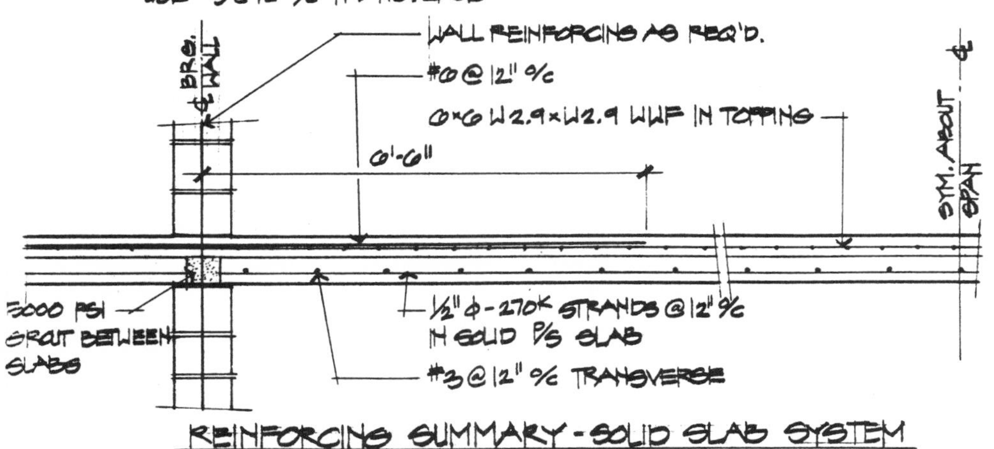

REINFORCING SUMMARY - SOLID SLAB SYSTEM

ANALYSIS OF UNTOPPED PRESTRESSED CONCRETE FLAT SLABS

Untopped prestressed concrete slabs are a very economical solution for floors in hotels, housing, or in conjunction with spread systems used in parking structures. Other types of structures such as offices use cast in place topping to surround cast in electrical conduit and communication services in addition to performing the structural function, thereby normally precluding the use of untopped slabs. When untopped slabs are used, the boundaries should be tied into supporting members or chord beams by poured in place concrete and bars or strand left exposed beyond the ends of the members. It is recommended that shear keys between adjacent slabs be wide enough to allow the grouted joint top surface to form a gradual change in elevation between adjacent slabs exhibiting a small degree of differential camber. This keyway should also be castellated to develop horizontal shear forces developed by the composite untopped slab diaphragm. A recommended joint detail and typical forces acting on the untopped slab diaphragm are shown below:

In analyzing these diaphragms, the following assumptions are made:

1. The slab diaphragm is assumed to be a rigid plane, fully supported laterally.

2. The diaphragm transfers lateral shears to stiffening elements (shear walls, frame elements, etc.) which provide overall structural stability.

3. A properly designed slab diaphragm resists both shear and flexural stresses.

The three types of loads shown above that are resisted by the untopped diaphragm and its joints are:

V_s Horizontal shear parallel to the grouted slab joint. V_s is calculated by the relationship

$$v_s = \frac{Vay}{Ib} \quad \text{(unit shear stress in K/in}^2\text{)}$$

with the slab diaphragm acting as a beam.

V_e Horizontal shear parallel to the end of the slab at the bearing. V_e is usually the end reaction of the slab diaphragm acting as a beam, where V_e is the drag force being taken into a lateral stiffening element. The force V_e is usually transferred from the untopped slab diaphragm by bars or strand protruding beyond the ends of the slabs providing a positive mechanical load transfer at the interface.

V_p Vertical shear perpendicular to the grouted slab joint. V_p is a result of load transfer from a point loading from one side of the slab joint.

The shear castellation may be conservatively estimated to have the following capacity:

Cross sectional area of castellation:

$$(2.5)\ (1.5)\ (0.5) = 1.875\ in^2$$

using 3000 psi grout, and an ultimate bearing capacity of 0.7 f'c = (0.7) (3000) or 2100 psi, the ultimate resisting load per key is:

$$(1.875)\ (2100) = 3939\#/key$$

or

$$1.938^k/key \times \frac{12}{10}$$

$= \underline{4.73^k/ft}$ ultimate horizontal shear resistance of castellated shear key

Preliminary load tables are shown for untopped slabs fabricated in both hardrock and lightweight concrete. In most applications, the as-built condition will permit these slabs to be considered as being restrained. Finally, in designing these slabs, the prestress force should be such that the instantaneous camber at release is 1.1 times the instantaneous dead load deflection produced by the slab self weight. The following example serves to demonstrate that proper design of untopped slabs is a result of a camber analysis, with the load carrying capacity being determined afterwards and compared with that required. This procedure will assure successful performance in the finished structure, minimizing the potential of problems resulting from differential camber and creep sag occurring with time.

In lieu of providing protruding end steel from the slabs and forming the castellated shear key, which may make smaller projects excessively costly, the diaphragm may be formed by using intermediate and boundary reinforcing calculated by the use of the shear-friction theory as permitted by UBC-79, Section 2611(p). The use of this type of reinforcing to maintain the integrity of diaphragms comprised of untopped slab elements subjected to seismic (cyclic) loading has been substantiated by testing. This shear-friction reinforcing is in addition to any reinforcing required to satisfy flexural loading conditions.

LOS ANGELES COUNTY PUBLIC SOCIAL SERVICES PARKING GARAGE - LONG BEACH, CA.

5" thick untopped prestressed concrete slabs span between spread double tee channels which span 63'-0 between supports. Shear castellations in the plank sides and a positive tie into the field poured concrete over the channel ensure that the completed floor acts as a rigid diaphragm to transmit lateral forces.

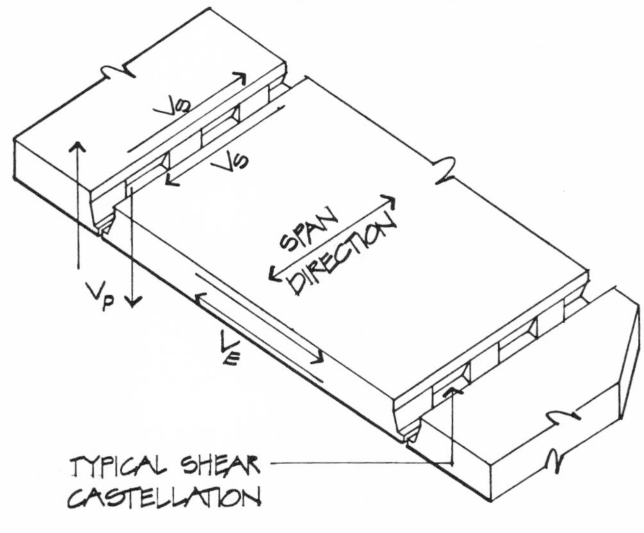

UNTOPPED DIAPHRAGM SLAB FORCES

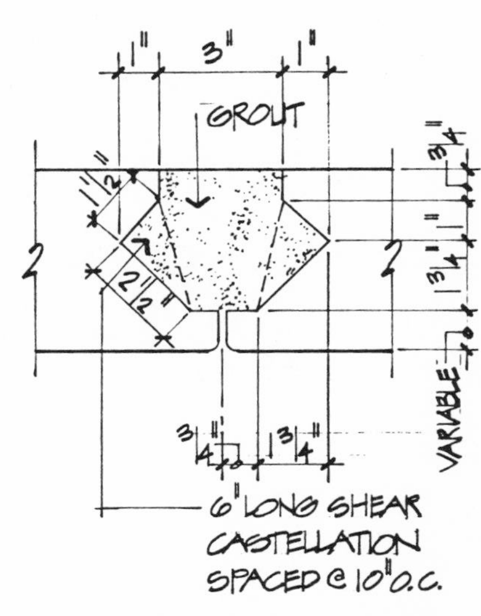

RECOMMENDED LATERAL SLAB JOINT DETAIL

LOS ANGELES COUNTY PUBLIC SOCIAL SERVICES
PARKING GARAGE - LONG BEACH, CA.

5" thick untopped slabs being erected. The slabs span 16'-0 between supports. The prestressing strands were allowed to extend out beyond the ends of the slabs to form a positive tie with the cast in place concrete over the prestressed channels.

UNTOPPED SOLID SLABS

f'_c = 5000 psi HR CONCRETE

1/2" COVER (2 HR UNRESTRAINED RATING)

12'-0" WIDE SLABS

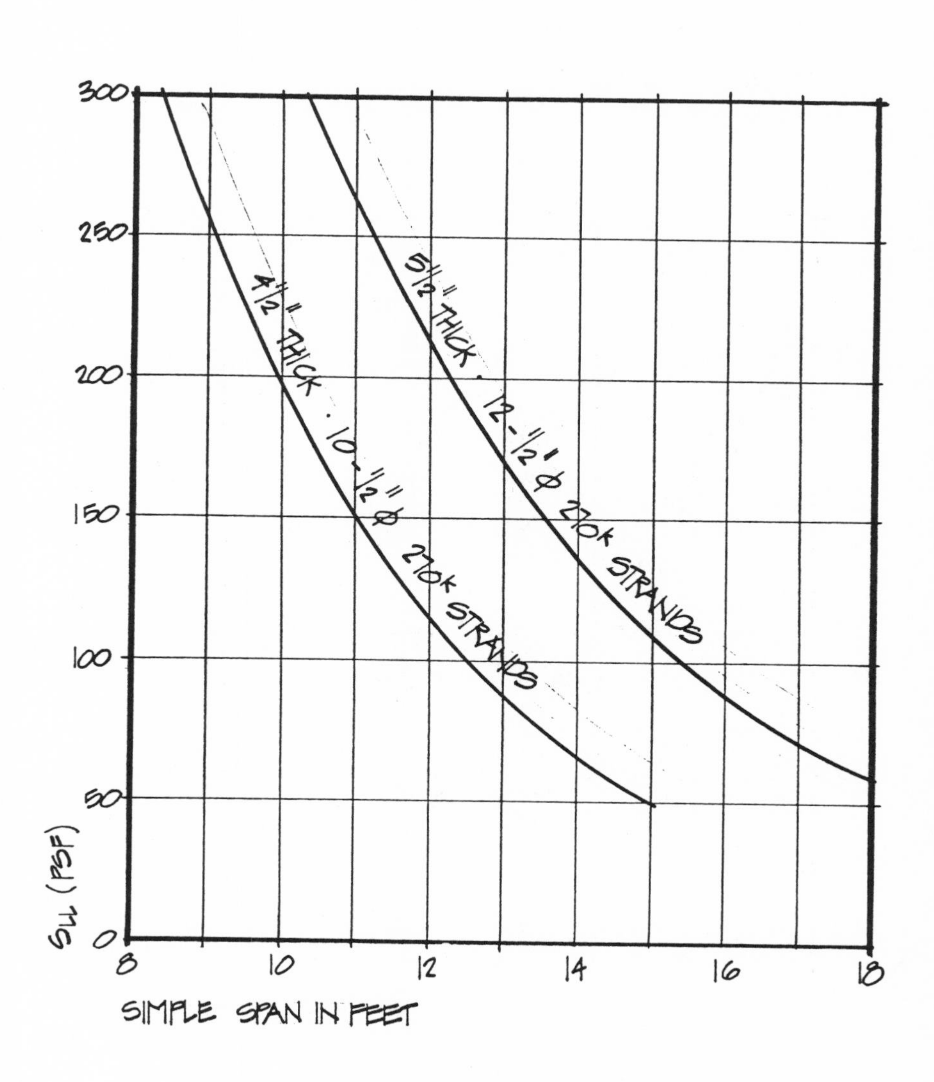

UNTOPPED SOLID SLABS
f'c = 5000 PSI LW CONCRETE
3/4" COVER (2 HR UNRESTRAINED RATING)
12'-0" WIDE SLABS

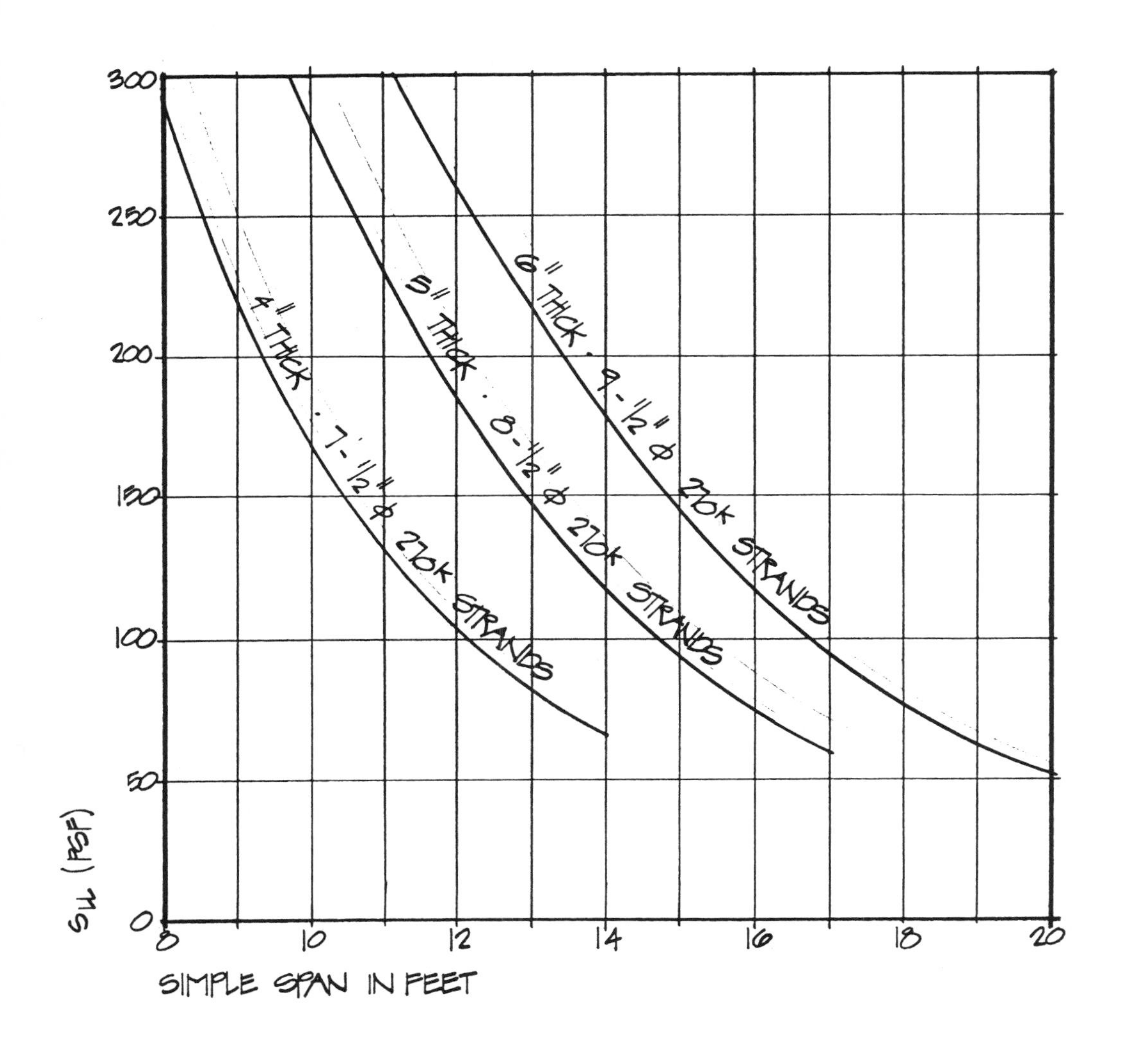

DESIGN EXAMPLE - UNTOPPED PRESTRESSED FLAT SLAB

DESIGN AN UNTOPPED SLAB FOR A PARKING STRUCTURE SPANNING 17'-0" BETWEEN SPREAD CHANNELS. THE AREA OVER THE CHANNELS WILL SUBSEQUENTLY BE FILLED WITH CAST-IN-PLACE TOPPING FLUSH WITH THE TOPS OF THE UNTOPPED SLABS, SO THAT THE SLABS MAY BE CONSIDERED TO BE RESTRAINED. A 2 HOUR RATING IS REQUIRED. USE 1/2" ϕ - 270^K STRANDS.

SOLUTION:

FROM PRELIMINARY DESIGN CHARTS, A 5" THICK SECTION IS SELECTED. 3/4" OF COVER ARE REQUIRED.

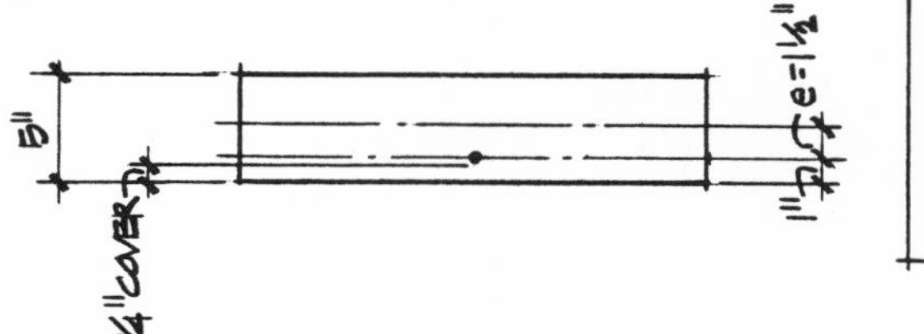

S = 50 IN^3	A = 60 IN^2/FT
I = 125 IN^4	WT = 63 PSF
f'_{ci} = 3500 PSI	E_{ci} = 3600
f'_c = 5000 PSI	E_c = 4300
S_{LL} = 50 PSF	

CAMBER CONTROL

$$\Delta i_{\text{SLAB WT.}} = \frac{5wL^4}{384 E_{ci} I} = \frac{(5)(.063)(17)^4(1728)}{(384)(3600)(125)} = -0.26'' \downarrow$$

$$\therefore \text{REQ'D } \Delta_{P/si} = 1.1 \times 0.26 = +0.29'' \uparrow$$

CALCULATE F_i REQ'D

$$\Delta_{P/si} = \frac{F_i e L^2}{8 E_{ci} I}$$

$$0.29 = \frac{F_i (1.5)(17)^2(144)}{(8)(3600)(125)}$$

$$F_i = 16.7 \text{ K/FT}$$

FOR A 12'-0" WIDE SLAB

$$N = \frac{16.7 \times 12}{0.9 \times 28.9} = 7.7$$

USE 8 - 1/2" ϕ - 270^K STRANDS

$$F_i = (0.9)(8)(28.9) = 208^K$$

$$F_f = (0.78)(8)(28.9) = 180^K$$

CHECK FINAL STRESSES (USE 22% LOSSES)

$M_{SLAB} = 1/8 \times .063 \times 12 \times 12 \times 17^2 = 328^{K\text{-}IN}$

$M_{SLL} = 1/8 \times .050 \times 12 \times 12 \times 17^2 = 260^{K\text{-}IN}$

$UM = (1.4)(328) + (1.7)(260) = 901^{K\text{-}IN}$

$f_{bot}^{℄} = -\frac{180}{720} - \frac{(180)(1.5)}{600} + \frac{328+260}{600}$

$= -0.250 - 0.450 + 0.980 = +0.280 \text{ KSI} < 6\sqrt{f'_c}$ O.K.

$(6\sqrt{f'_c} = +0.425 \text{ KSI})$

AT RELEASE

$f_{top}^{end} = -\frac{208}{720} + \frac{(208)(1.5)}{600}$

$= -0.289 + 0.520 = +0.231 < 6\sqrt{f'_{ci}}$

$(6\sqrt{f'_{ci}} = +0.354 \text{ KSI})$

CHECK ULTIMATE STRENGTH

$\rho_p = \frac{A_{ps}}{bd} = \frac{(8)(.1531)}{(144)(4)} = 0.002126$

$f_{ps} = f_{pu}\left(1 - 0.5\rho_p \frac{f_{pu}}{f'_c}\right) = 270\left(1 - 0.5 \times .002126 \times \frac{270}{5}\right) = 254.5 \text{ KSI}$

$a/2 = \frac{A_{ps} f_{ps}}{1.7 f'_c b} = \frac{(8)(.1531)(254.5)}{(1.7)(5)(144)} = 0.25''$

$M_u = \phi A_{ps} f_{ps}(d - a/2) = (0.9)(8)(.1531)(254.5)(4 - 0.25)$

$= 1052^{K\text{-}IN} > UM$ O.K.

CHECK CAMBER/DEFLECTION IN FINAL CONDITION - LT/CREEP

$\Delta_{P/S\,R} = \frac{(208)(1.5)(17)^2(144)}{(8)(3600)(1500)} \times \overset{C.F.}{2.45} = +0.74'' \uparrow$

$\Delta_{SLAB\,R} = 0.26'' \times 2.70 = -0.70'' \downarrow$

FINAL LONG TERM CONDITION WITH CREEP $+0.04'' \uparrow$ O.K.

$\Delta_{SLL} = \frac{50}{63} \times 0.26'' = 0.21'' \downarrow$

$\frac{L}{240} = \frac{12 \times 17}{240} = 0.85'' > 0.21$ O.K.

CONNECTION DETAILS

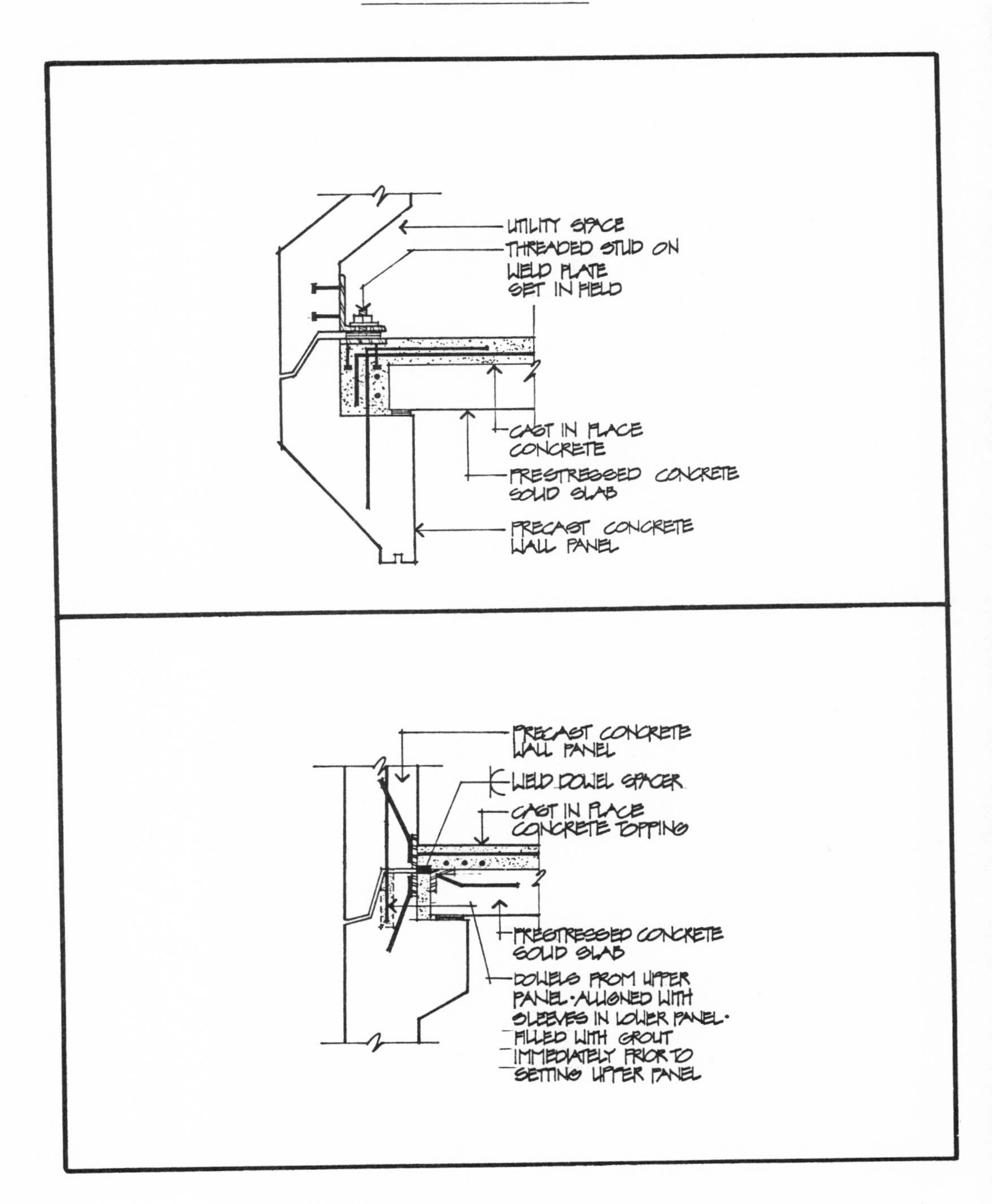

CONNECTION DETAILS

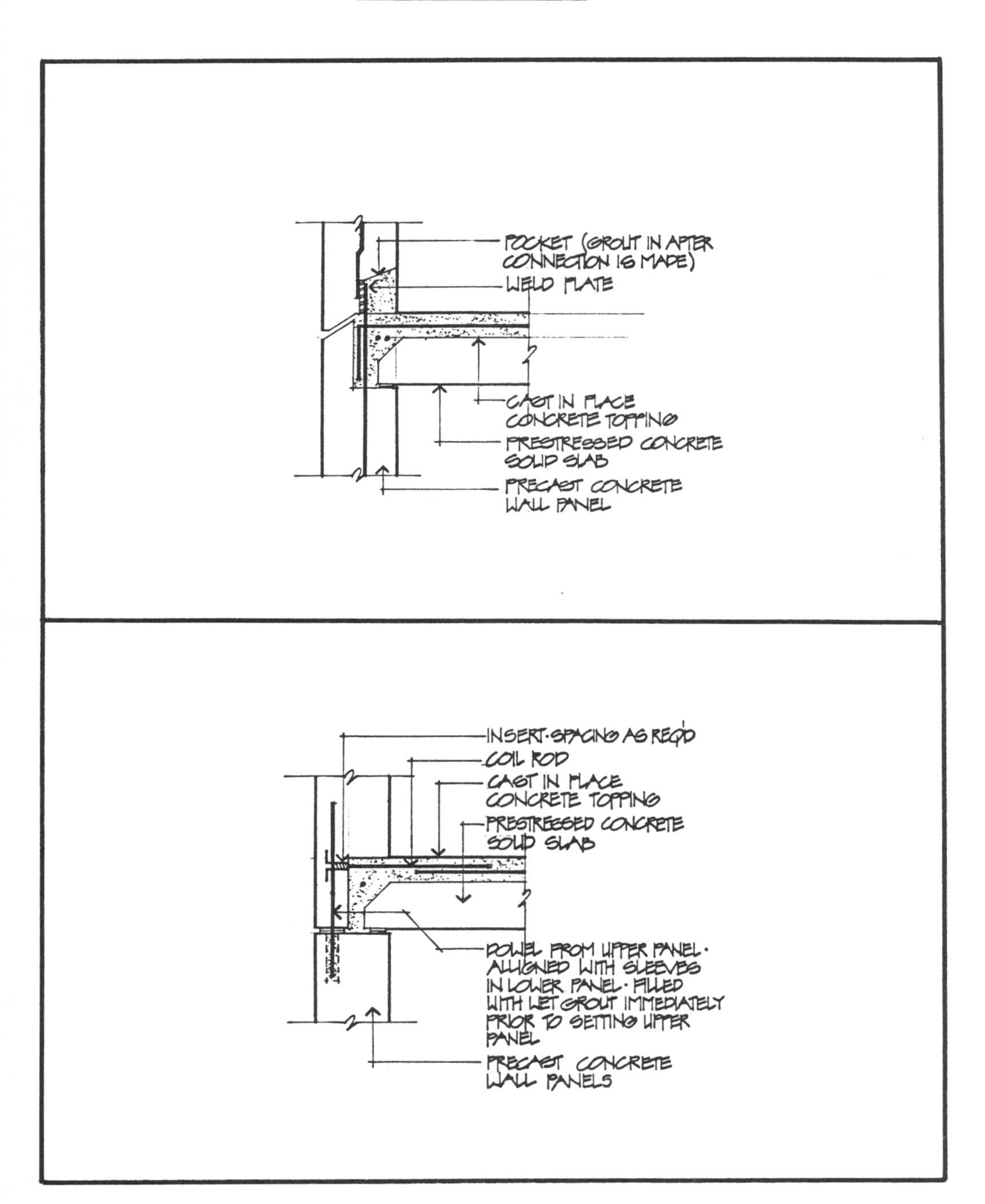

CONNECTION DETAILS

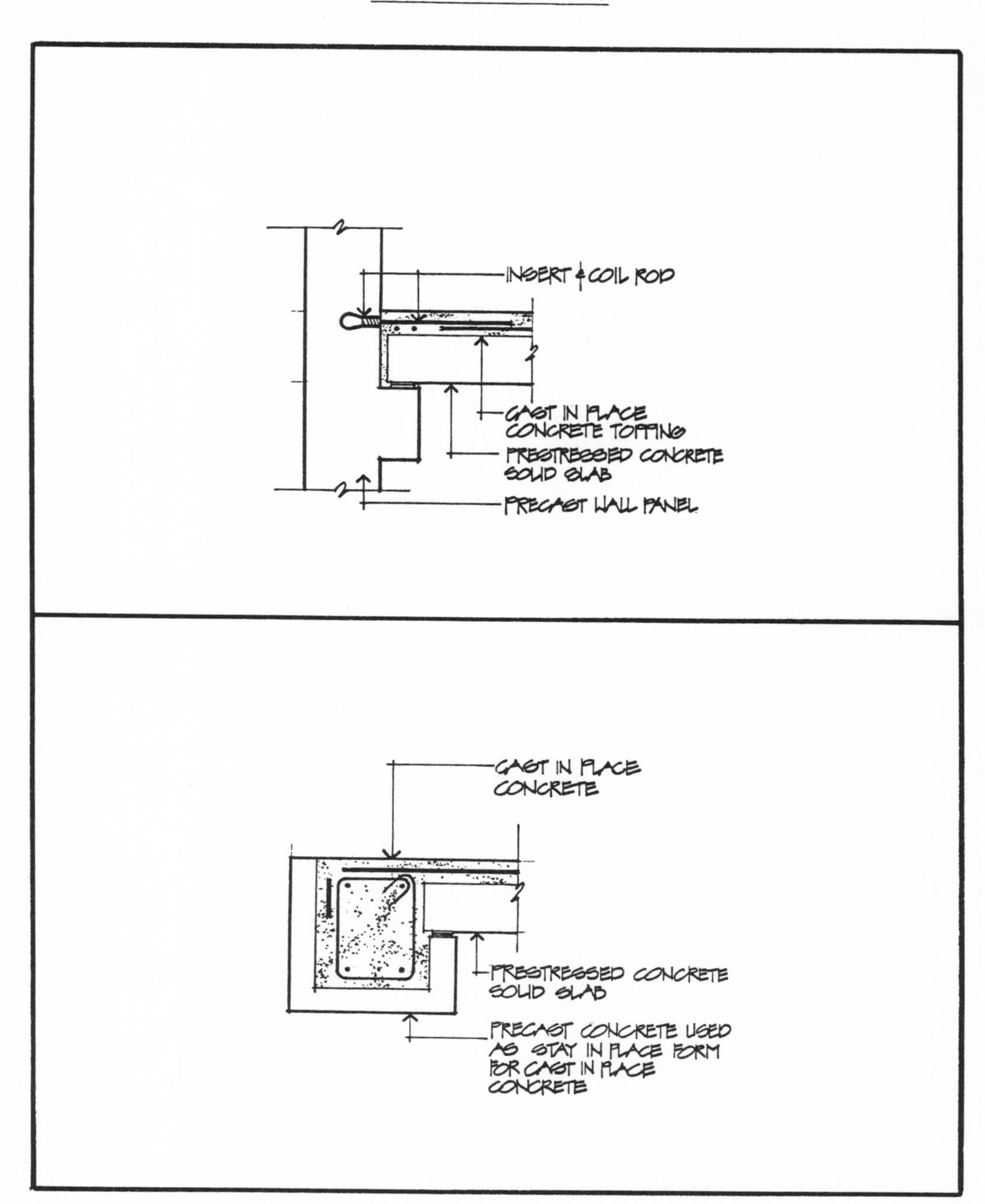

CONNECTION DETAILS

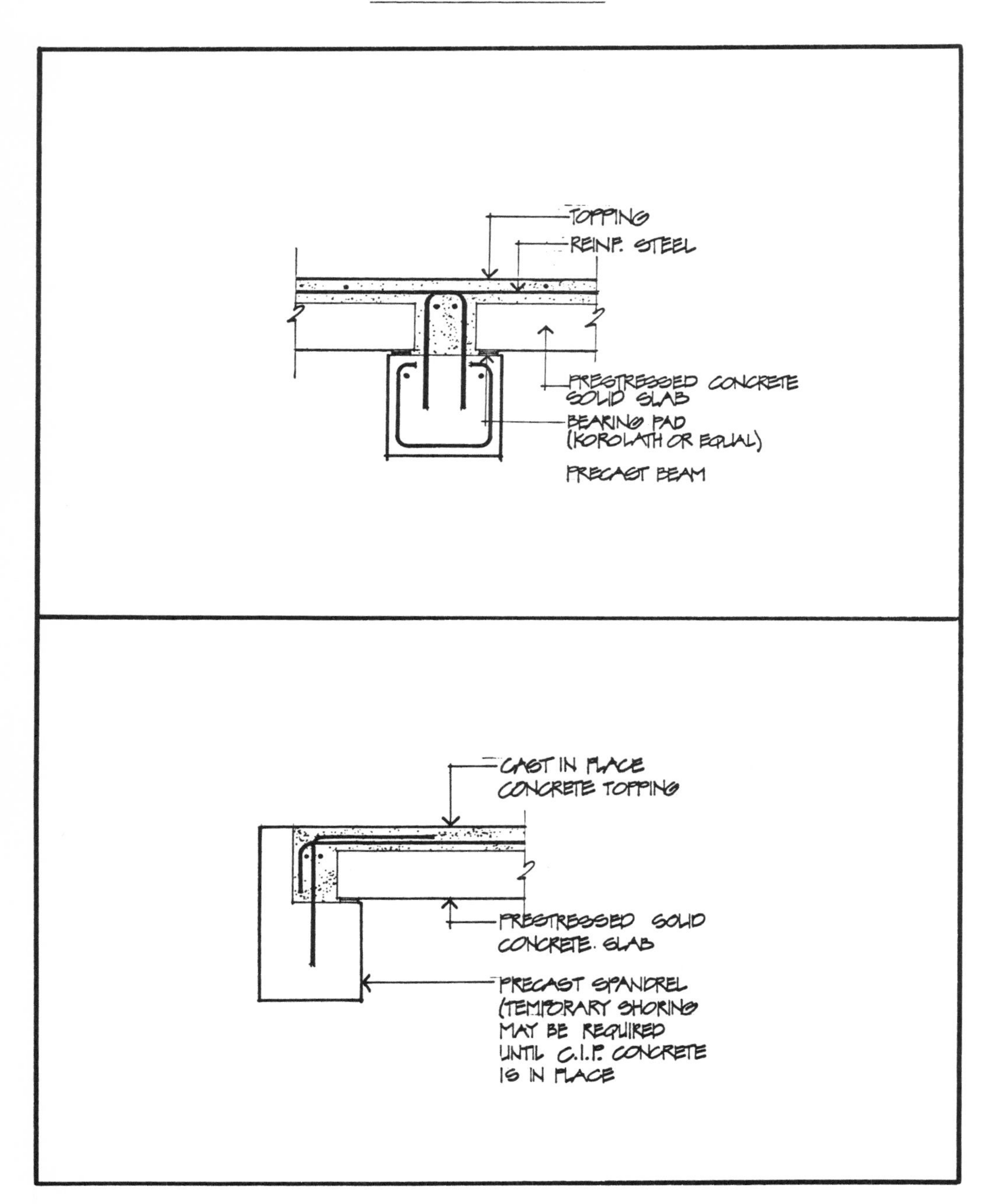

CONNECTION DETAILS

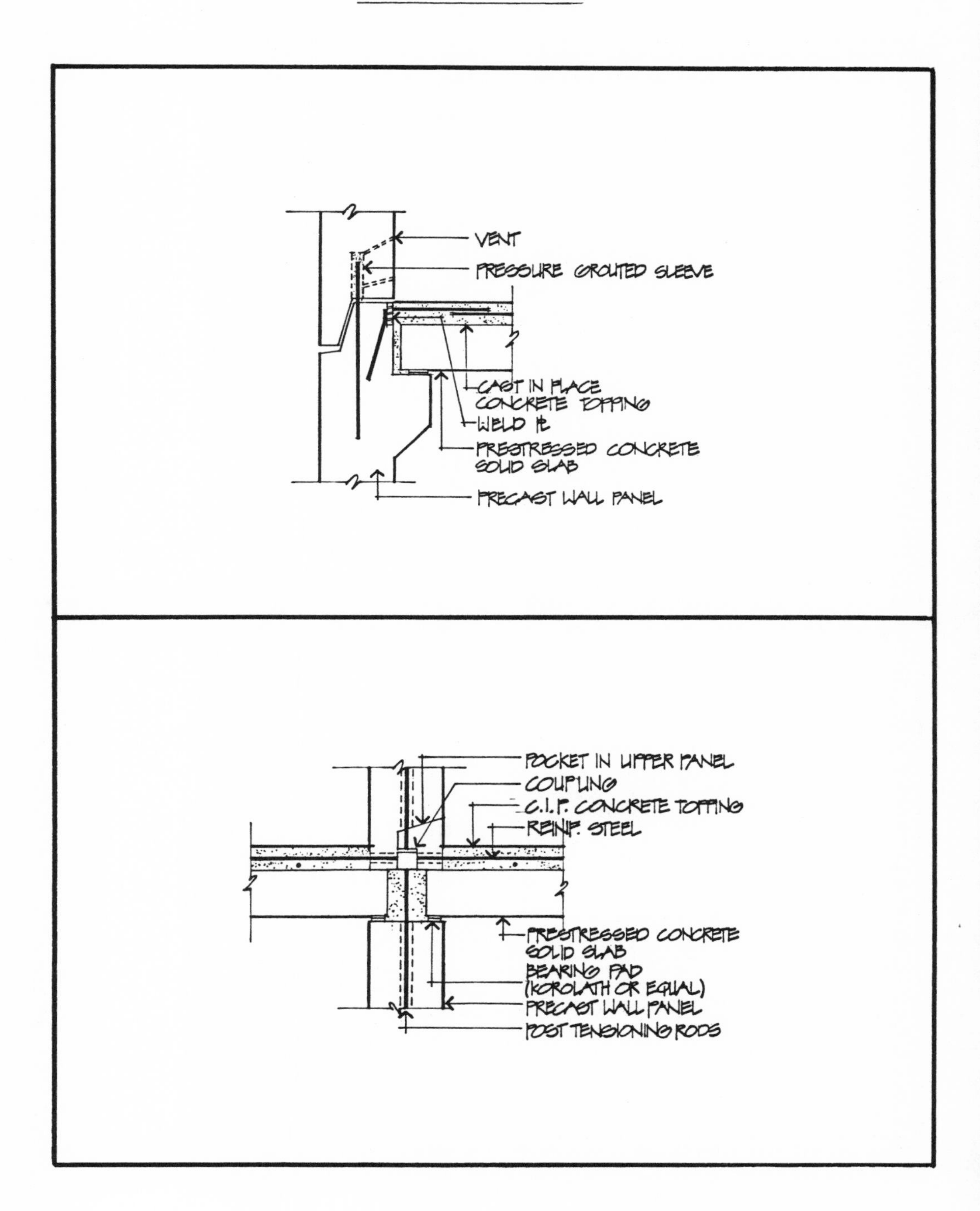

CONNECTION DETAILS

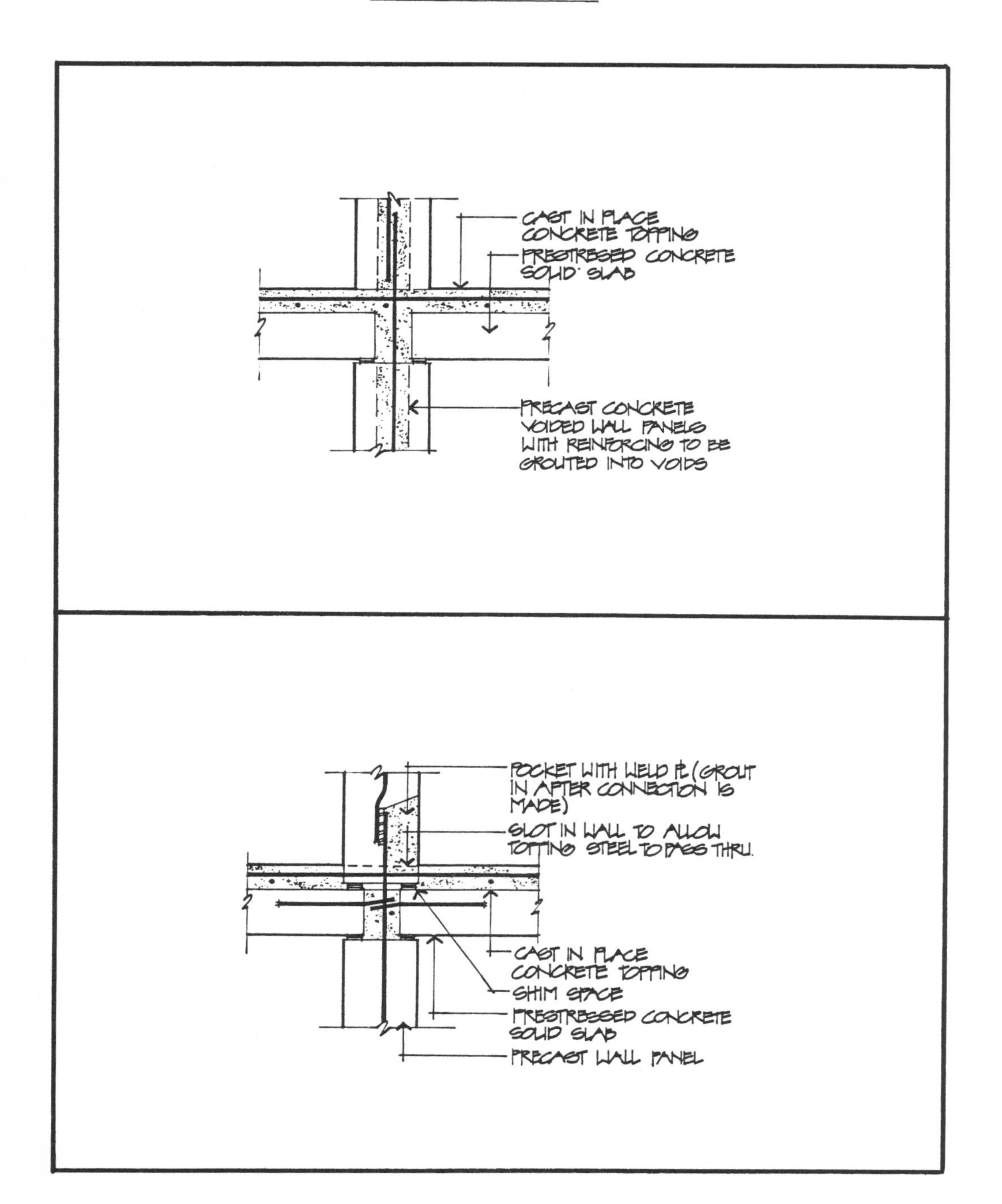

CONNECTION DETAILS

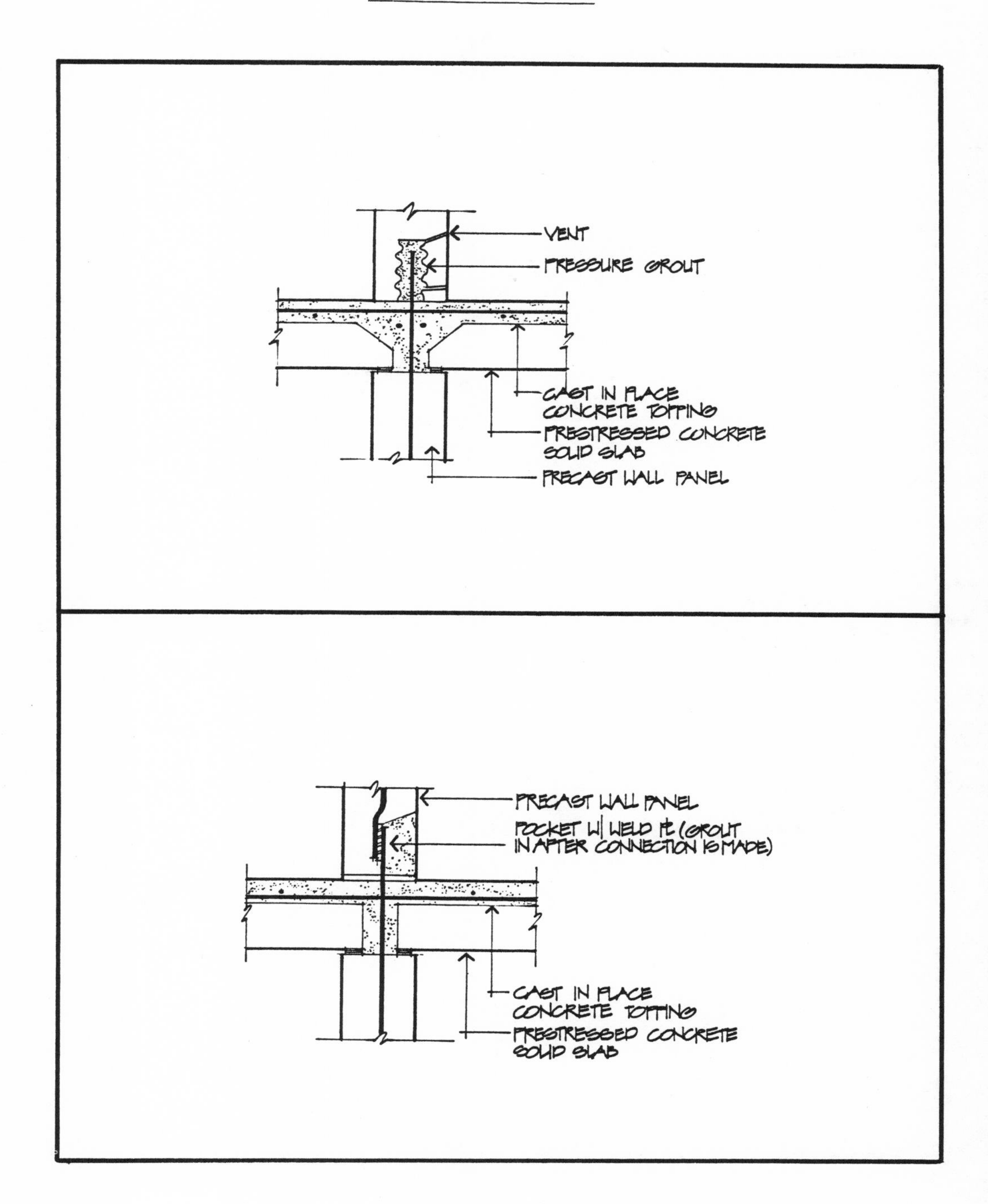

CONNECTION DETAILS

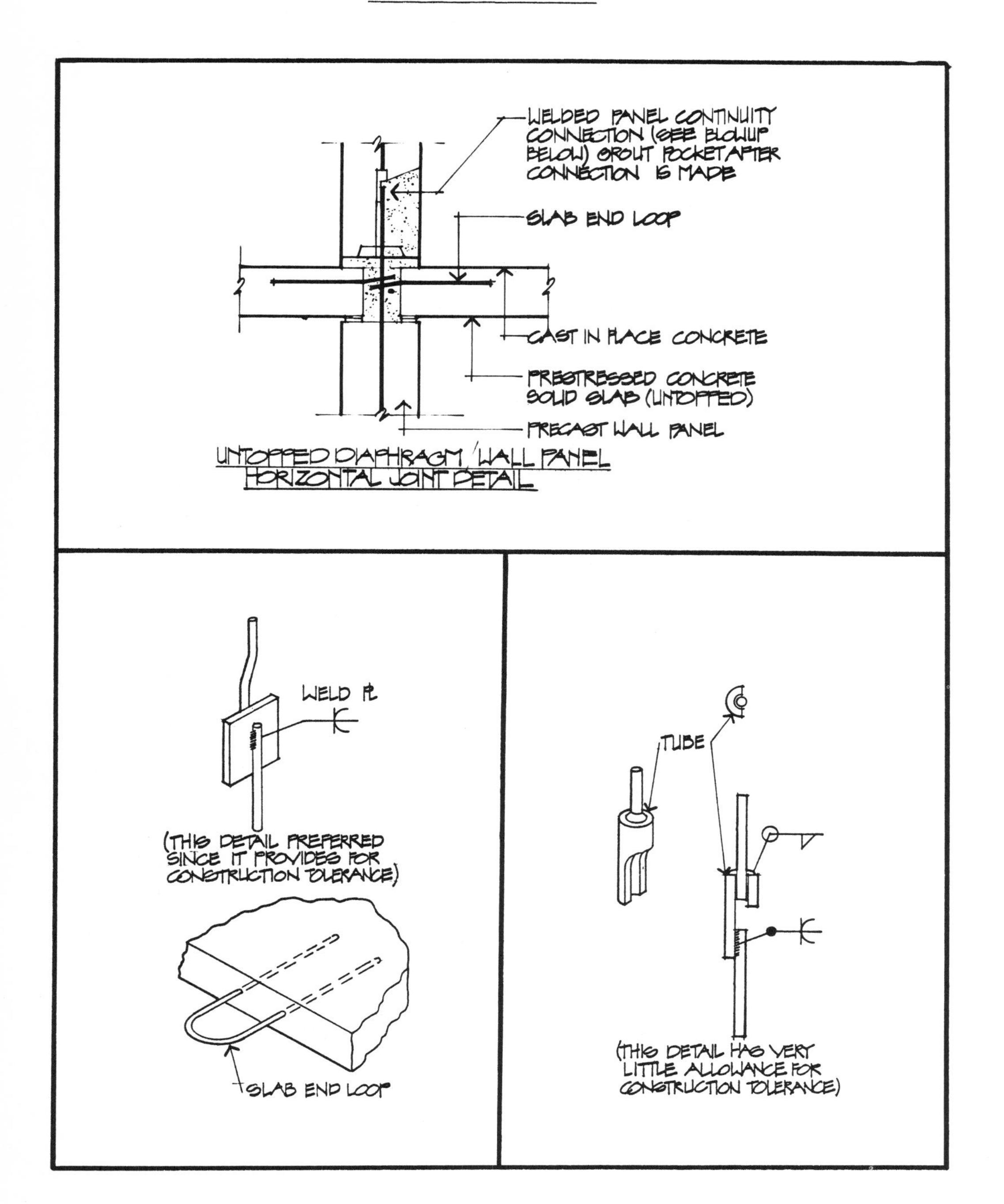

BONAVENTURE HOTEL - LOS ANGELES

Tower cranes were used to erect structural steel, prestressed concrete flat slabs and perform hoisting for other trades working on the building. Two levels of steel framing were erected, then two levels of prestressed slabs, etc.

HOLLOW CORE PLANKS

HERITAGE HOUSE - CONCORD

118,000 square feet of 8" Spancrete hollow core planks form the floors of this 8 story residence for senior citizens. One entire floor was erected each week on masonry bearing walls. This fast construction allowed the building to be completed 2 months ahead of schedule. The hollow core plank also provides inherent fire protection and prevents the passage of both impact noise and sound transmission through the floor.

Hollow core planks are voided precast, pretensioned concrete deck units used mainly to form floors and roofs in buildings. They can also be used as non-insulated or insulated wall panels in either load bearing or non load bearing structural functions. These units are made by commercially franchised processes using specialized forming machinery. The six principal processes produced in the United States are:

<u>DYNASPAN</u> made in 8' widths by a slip forming process with low-slump concrete. Each slab has 14 cores.

<u>FLEXICORE</u> a wet cast product cast in 2' widths in 60' long pans. Voids are formed with deflatable rubber tubes.

<u>SPAN-DECK</u> a wet cast product cast in two sequential operations, with the second operation slip forming 8' wide planks with rectangular voids.

<u>SPANCRETE</u> made in 40" wide units by tamping an extremely dry mix with 3 sequential sets of tampers compacting the mixes around slip forms.

<u>SPIROLL</u> an extruded product made in 4' wide units with round voids formed by augers which are a part of the casting machine.

<u>DY-CORE</u> an extruded 4 foot wide product made by compressing zero slump concrete into a solid mass by a set of screw conveyors in the extruder. High frequency vibration combined with compression around a set of dies in the forming chamber of the machine produces the plank with oblate, or octagonal shaped voids.

Hollow core slabs are made in 6", 8", 10", and 12" depths and used in spans from 18' to 42'. They are used as floors or roofs in hotels, motels, offices, shopping malls, department stores, schools, hospitals, multi-family housing, as decks in parking structures, and as infill systems spanning between spread channels. They may be erected side by side or spread apart a distance of 2 or 3 feet with the space in between spanned with metal deck. These spread systems are covered with cast in place concrete topping to form an integral diaphragm and achieve the required fire rating. In certain applications, such as in roofs or floors in multi-family housing, the units may be installed without a cast in place concrete topping provided that the completed diaphragm maintains its integrity in withstanding design seismic stresses.

Fire ratings of up to 4 hours are achieved by providing adequate cover over the strands, using cast in place concrete topping, and lightweight aggregate concrete. 8" thick untopped hollow core units meet HUD sound transmission criteria for floors in multi-family multi-story construction.

MANUFACTURING

The two hollow core plank processes currently being manufactured in California are Span-Deck and Spancrete. In the Span-Deck process, 2 separate casting machines are used. The first machine (bottom casting machine) lays down a 1 3/4" thick soffit of fluid workable concrete around the tensioned prestress strands, usually 1/2" diameter strands. Then, the second machine (top casting machine) deposits the balance of the concrete section around temporary rectangular voids of pea gravel deposited from inside a slip form. This second machine has two large hoppers - one for the pea gravel and one for the concrete. Then the freshly cast slabs are steam cured, sawed to length, and transported to a void material recovery area where the pea gravel is dumped out of the voids by tilting the planks on end. In the wet cast Span-Deck process, side rails with a shear key former keep the wet concrete inside the bed. These side rails pivot down prior to sawing and stripping. Large preformed openings may be made in this product, to accomodate mechanical risers or HVAC ducts. Insert plates and top reinforcing may be placed in the wet concrete. The side rails may be altered to form a castellated shear key in the plank (see page 326). The Span-Deck beds are cast on a daily cycle and are 400-600' long.

The Spancrete process uses a very dry mix which may vary in aggregate quality in each layer that is deposited. These dry mixes are placed in 3 layers by the successive compaction of tampers which are part of the machine. The top 2 layers are tamped around and above slip forms, leaving the voids in the product. One pass is made each day on 750' long beds. The product is stack cast (one layer cast directly on the preceding layer). The beds are cast up to 5 stacks high and the product is ambient cured for one week. Water is also run through the voids in the initial curing phase. The shear key is slipformed by the side of the machine. Large openings for shafts, HVAC ducts, or mechanical risers may be saw cut in the plant. Due to the nature of the dry mix process, it is not possible to cast plates or rebar in the slabs. Top strand may be placed subject to design limitations. Spancrete is tensioned with 1/4" diameter through 1/2" diameter strands.

In both processes, the top finish may be roughened sufficient to satisfy section 2617(h) of UBC-79. If the planks are to be used untopped, as in roof construction, they should have smooth top finishes. Other advantages of hollow core planks are the use of the voids as electrical raceways, and in specially designed instances, as HVAC ducts. Hangers for suspended ceilings or pipes may be placed by the respective trades in the field in the plank joints prior to grouting. Small openings - less than 10" x 10" - should be core drilled by the respective trades in the field.

ATTACHMENT OF PARTITIONS

When hollow core planks are used in housing or motel applications where the underside of the plank forms the finished ceiling, partition attachment details should allow for movement of the plank due to temperature changes and creep and shrinkage effects on camber and plank self weight deflection. The use of proper attachment details will ensure that unsightly gaps do not form at the tops of the partitions, or conversely, prevent cracking of partitions due to long term downward movement of the floor system. Some suggested details are shown at the end of this chapter.

HOLLOW CORE PLANK PRODUCTION

Spancrete production beds showing casting machine and long line stack casting method used.

Spandeck line and casting machines. Bottom casting machine shown in foreground places soffit concrete. Top casting machine in background follows, slipforming balance of section containing temporary aggregate void forming material.

DESIGN

Below are preliminary design charts for topped and untopped hollow core plank. Preliminary design information for hollow core plank can also be found in the 2nd edition of the PCI Design Handbook, pp 2-25 to 2-32. Following these preliminary charts is a sample of a detailed load table for a specific manufacturer's hollow core plank system.

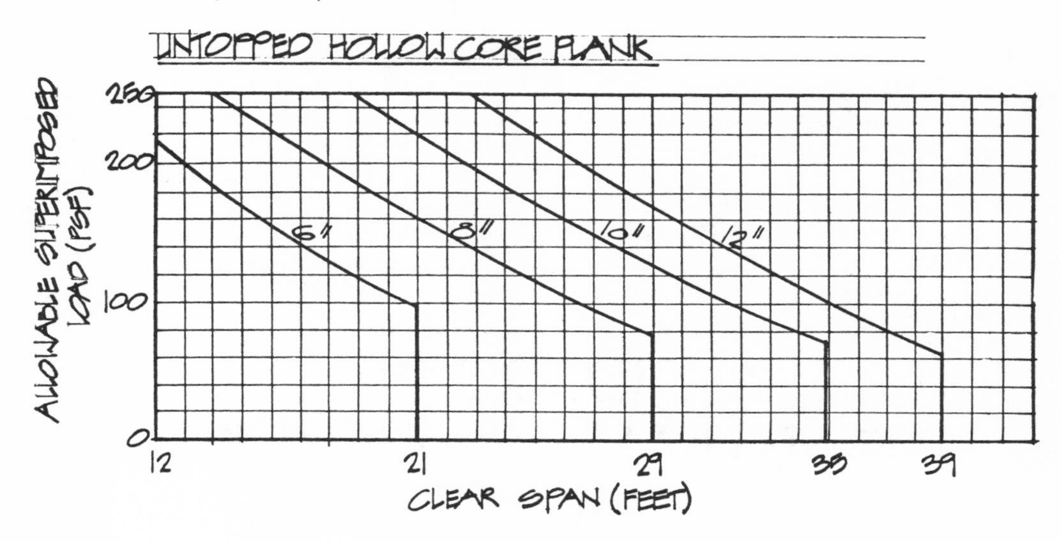

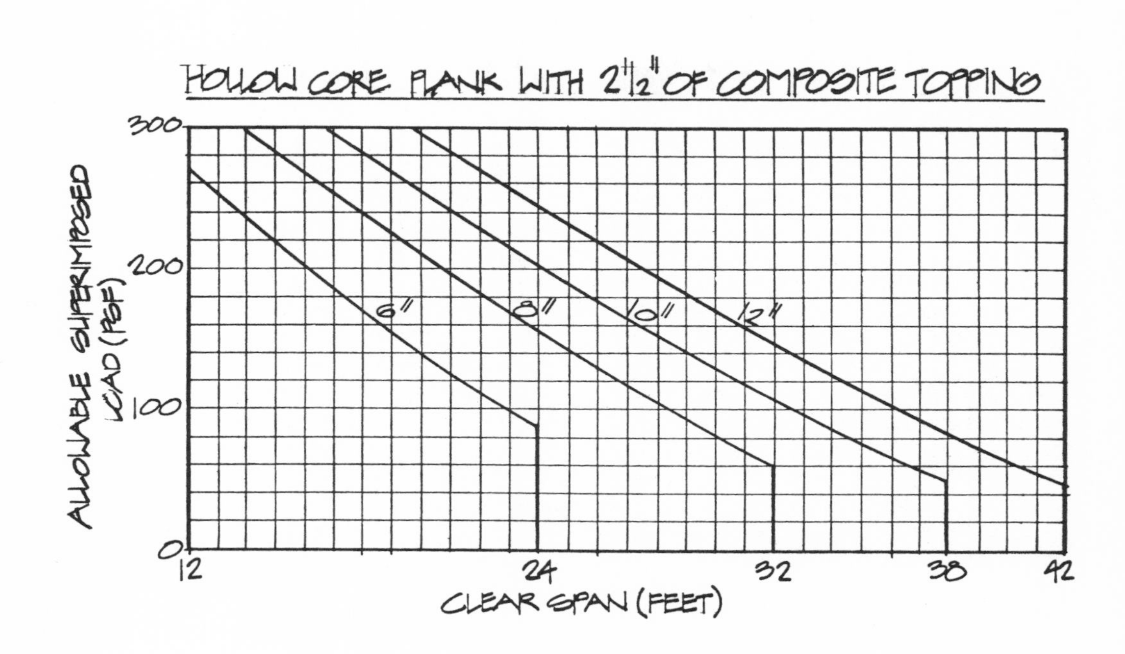

8"x8'-0" SPAN-DECK W/2" COMPOSITE TOPPING

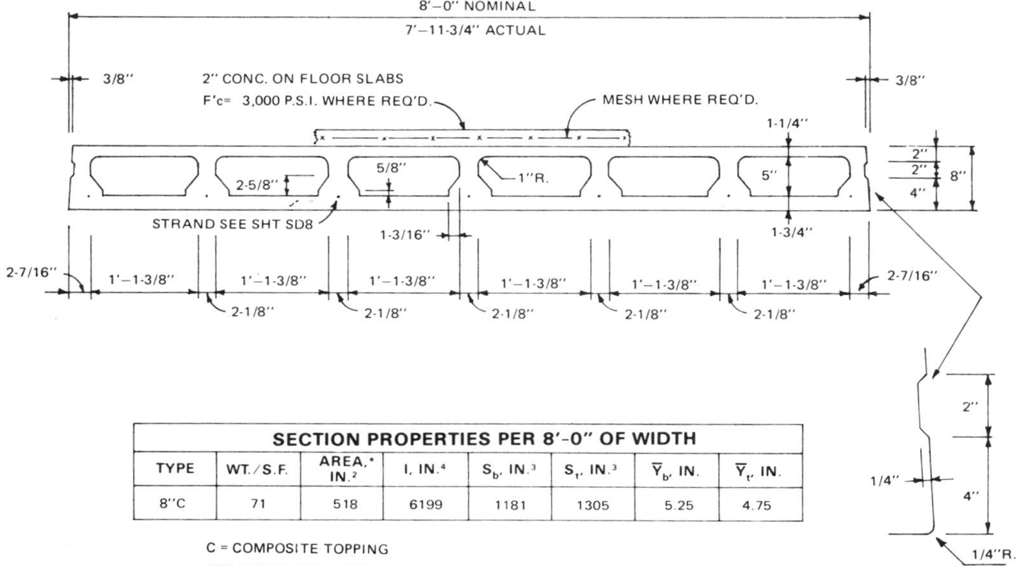

SECTION PROPERTIES PER 8'-0" OF WIDTH							
TYPE	WT./S.F.	AREA,* IN.²	I, IN.⁴	S_b, IN.³	S_t, IN.³	$\bar{Y}_b$, IN.	$\bar{Y}_t$, IN.
8"C	71	518	6199	1181	1305	5.25	4.75

C = COMPOSITE TOPPING

*TRANSFORMED AREA,
WIDTH OF TOPPING TRANSFORMED
ACCORDING TO RATIO

$\frac{E3000}{E5000}$ (E_c= $W^{1.5}$ 33 $\sqrt{f'c}$, W = 145

ALLOWABLE SUPERIMPOSED UNIFORM LOADING (P.S.F.) FOR SIMPLE SPANS																					
DESIGN NO.	SPAN IN FEET																				
	17	18	19	20	21	22	23	24	25	26	27	28	29	30	31	32	33	34	35	36	37
8C76	127	107	90	76	63	52	43														
8C5627	147	125	106	90	76	64	54	44													
8C3647		140	120	102	87	74	63	53	44												
8C77			140	121	104	90	77	66	56	47	40										
8C5728				136	118	102	88	76	66	56	48	40									
8C3748					131	114	99	87	75	65	56	48	41								
8C78						132	116	101	89	78	68	59	51	44							
8C88							137	121	107	94	83	73	64	56	49	42					
8C98								140	124	110	98	87	77	68	60	53	46				
8C108									141	126	112	100	89	80	71	63	56	49			
8C118										140	126	113	101	91	81	73	65	58	50		
8C128											139	125	112	101	91	82	73	66	59	51	43

HOLLOW CORE PLANK DESIGN

DESIGN A HOLLOW CORE PLANK WITH 3½" OF HARDROCK COMPOSITE TOPPING TO SPAN 28'-8" IN A DEPARTMENT STORE FLOOR. A FIRE RATING OF 2 HOURS IS REQUIRED. USE HARDROCK CONCRETE FOR THE HOLLOW CORE PLANK.
f'_c (PRECAST) = 5000 PSI $\quad f'_c$ (TOPPING) = 3000 PSI

SOLUTION:

FROM A LOCAL MANUFACTURER'S LOAD SPAN TABLES, AN 8'-0" WIDE × 8" DEEP SECTION IS SELECTED. NOTE THAT THE LOAD TABLES INDICATE THAT 8" DEEP SPAN-DECK HOLLOW CORE PLANK WITH 2" OF HARDROCK CONCRETE COMPOSITE TOPPING CARIES A SUPERIMPOSED LOAD OF 112 PSF FOR A 29 FOOT SPAN. WITH THE 3½" THICK TOPPING, AS SPECIFIED IN THE DESIGN DESCRIPTION, THIS SYSTEM SHOULD CARRY THE SUPERIMPOSED LOADING OF 75 PSF LIVE LOAD AND 25 PSF OF PARTITIONS AND MECHANICAL.

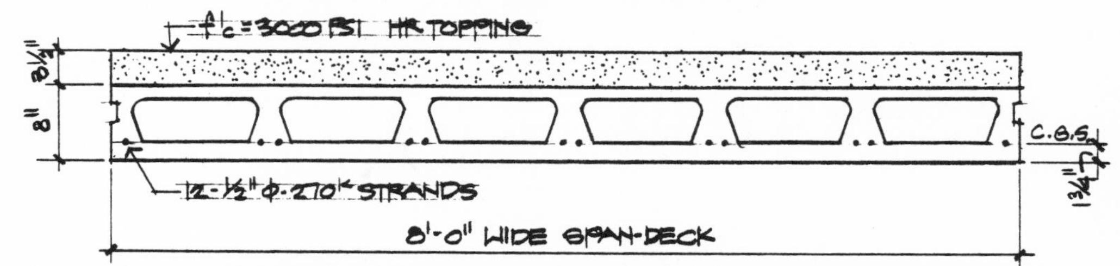

NOTE ALSO THAT BECAUSE THE MANUFACTURING PROCESS USED FOR FORMING THE VOIDS RESULTS IN RIBS SLIGHTLY THICKER THAN THE THEORETICAL DIMENSIONS SHOWN, THIS IS TAKEN INTO ACCOUNT IN THE DESIGN IN ASSIGNING A SLIGHTLY HIGHER VALUE TO THE AREA OF THE CROSS SECTION, BASED UPON ACTUAL PLANT MEASUREMENTS.

BASIC SECTION (CONC. WT. 150 PCF)

ACTUAL AREA: $A = 390$ IN² $\quad$ WT. = 51 PSF
$I_B = 3217$ IN⁴
$y_t = 4.26 \quad S_t = 755$ IN³
$y_b = 3.74 \quad S_b = 860$ IN³
$e = 3.74 - 1.75 = 1.99"$

COMPOSITE SECTION W/3½" C.I.P. H.R. TOPPING

PART	A	y_{tcc}	Ay_{tcc}	I_o	Ad^2
8" SPAN-DECK	390	7.76	3026	3217	2246
3½" C.I.P. 0.77×3.5×96 =	259	1.75	453	264	3375
COMPOSITE SECTION	649		3479	$\Sigma I = I_c = 9102$ IN⁴	

$\bar{y}_{tcc} = \frac{3479}{649} = 5.36"$ $\quad y_{tc} = 1.86" \quad S_{tc} = 4894$ IN³

$y_{bc} = 6.14" \quad S_{bc} = 1482$ IN³

LOADS AND MOMENTS PER 8'-0" WIDE SLAB

TO BASIC SECTION	K/FT	$M_{℄}$	× L.F.	= $UM_{℄}$
8" SPAN-DECK	$1.5 \times 0.408 \times \overline{28.67}^2$	= 503	1.4	705
3½" C.I.P. TOPPING	$1.5 \times 0.350 \times \overline{28.67}^2$	= 432	1.4	605
		M_B = 935 K-IN		

TO COMPOSITE SECTION				
S_{DL} (25 PSF)	$1.5 \times 0.200 \times \overline{28.67}^2$	= 246	1.4	345
S_{LL} (75 PSF)	$1.5 \times 0.600 \times \overline{28.67}^2$	= 740	1.7	1258
		M_C = 986 K-IN		UM = 2913 K-IN

(3-10)* PRESTRESS FORCE REQUIRED

$$F_f = \frac{\frac{M_B}{S_b} + \frac{M_C}{S_{bc}} - f_t}{\frac{1}{A} + \frac{e}{S_b}} = \frac{\frac{935}{860} + \frac{986}{1482} - 6\sqrt{5000}}{\frac{1}{390} + \frac{1.99}{860}}$$

$$= \frac{1.087 + 0.665 - 0.425}{0.002564 + 0.002314} = \frac{1.327}{0.004878} = 272^K$$

USING ½"φ - 270K STRANDS

$A_{ps} = 0.1531 \text{ IN}^2$

$F_o = 28.9$ K/STR (0.7 f_{pu})

$F_i = 0.9 \times 28.9 = 26.0$ K/STR

$F_f = 0.78 \times 28.9 = 22.5$ K/STR

$$N = \frac{272}{22.5} = 12.09$$

USE 12 - ½"φ - 270K STRANDS

CHECK END STRESSES AT RELEASE

$$f^{TOP} = -\frac{(12)(26)}{390} + \frac{(12)(26)(1.99)}{755} = -0.800 + 0.822 = +0.022 \text{ O.K.}$$

$$f_{BOT} = -\frac{(12)(26)}{390} - \frac{(12)(26)(1.99)}{860} = -0.800 - 0.722 = -1.522 \text{ O.K.}$$

$< 0.6 f'_{ci}$ ∴ O.K.

CHECK FINAL STRESSES AT ℄ SPAN

$$f^{TOP} = -\frac{(12)(22.5)}{390} + \frac{(12)(22.5)(1.99)}{755} - \frac{935}{755} - \frac{986}{4894}$$

$$= -0.692 + 0.712 - 1.238 - 0.201 = 1.419 < 0.45 f'_c \text{ O.K.}$$

$$f_{BOT} = -\frac{(12)(22.5)}{390} - \frac{(12)(22.5)(1.99)}{860} + \frac{935}{860} + \frac{986}{1482}$$

$$= -0.692 - 0.625 + 1.087 + 0.665 = +0.435 \simeq 6\sqrt{f'_c} \text{ O.K.}$$

*( ) PAGE REFERENCE 2ND EDITION OF PCI DESIGN HANDBOOK.

CHECK ULTIMATE STRENGTH

(3-5) (3-53) $f_{se} = 147$ KSI $\quad \omega_p = c\frac{A_{ps}}{bd}\cdot\frac{f_{pu}}{f'_c} = 1.0\left(\frac{12\times.1531}{96\times9.75}\right)\left(\frac{270}{3}\right) = 0.18$

FROM CHART READ $f_{ps}/f_{pu} = 0.96$

$\therefore f_{ps} = 0.96\times270 = 259.2$ KSI

$a/2 = \frac{A_{ps}f_{ps}}{1.7f'_c b} = \frac{(1.837)(259.2)}{(1.7)(3)(96)} = 0.97''$

$M_u = \phi A_{ps}f_{ps}(d - a/2) = (0.9)(1.837)(259.2)(9.75-0.97) = 3763$ K-IN $> UM = 2913$ K-IN O.K.

CHECK CAMBER AND DEFLECTION

(3-66) $e = 1.99''$ $\quad E_{ci} = 3600$ KSI $\quad E_c = 4300$ KSI

CAMBER

$\Delta_{P/S}\uparrow = \frac{F_i e l^2}{8E_{ci}I_b} = \frac{(312)(1.99)(28.67)^2(144)}{(8)(3600)(3217)} = 0.79''$

SLAB SELF WEIGHT DEFLECTION

$\Delta_{SLAB}\downarrow = \frac{5wl^4}{384E_{ci}I_b} = \frac{(5)(0.408)(28.67)^4(1728)}{(386)(3600)(3217)} = 0.54''$

TOPPING WEIGHT DEFLECTION

$\Delta_{TPG}\downarrow = \frac{5wl^4}{384E_cI_b} = \frac{(5)(0.350)(28.67)^4(1728)}{(386)(4300)(3217)} = 0.38''$

SUPERIMPOSED LOADS

$\Delta_{SD+L}\downarrow = \frac{5wl^4}{384E_cI_c} = \frac{(5)(0.800)(28.67)^4(1728)}{(384)(4300)(9102)} = 0.31''$

(3-27) DETERMINE CREEP EFFECTS:

AT ERECTION	Δ_i	×	C.F.	=	$\Delta_{L.T.}$
CAMBER	0.79	×	1.80	=	+1.42
SLAB WT.	0.54	×	1.85	=	−1.00
Σ BASIC SECTION					+0.42'' ↑
TOPPING WEIGHT					−0.38 ↓
Σ AT TOPPING POUR					+0.04'' ↑

LONG TERM AS-BUILT CONDITION				
CAMBER	0.79	×	2.20	+1.74
SLAB WT.	0.54	×	2.40	−1.30
TOPPING	0.38	×	2.30	−0.87
Σ LONG TERM COMPOSITE				−0.43'' ↓

(3-23) CHECK Δ_{SD+L} ALLOWABLE (UBC-79 TABLE 23-D)

$\frac{L}{360} = \frac{28.67\times12}{360} = 0.96'' > 0.31''$ O.K.

CHECK HORIZONTAL SHEAR BETWEEN PLANK & TOPPING

(3-21) $UV = [(1.4)(0.408 + 0.350 + 0.200) + (1.7)(0.600)] \frac{28.67}{2} = 33.8^K$

$\phi V_{nh} = \phi \times 80 b_v d = (0.85)(80)(96)(9.75)/1000 = 63.6^K$

∴ ROUGHEN TOP OF PLANK TO SATISFY HORIZONTAL SHEAR REQUIREMENT.

CHECK VERTICAL SHEAR

(3-59) $b_w d \cong (21.1)(9.75) = 206 \text{ IN}^2$

v_u SUPPORT $= \frac{UV}{\phi b_w d} = \frac{33.8}{(0.85)(206)} = 0.19$ KSI

$\frac{L}{D} = \frac{28.67 \times 12}{9.75} = 35$

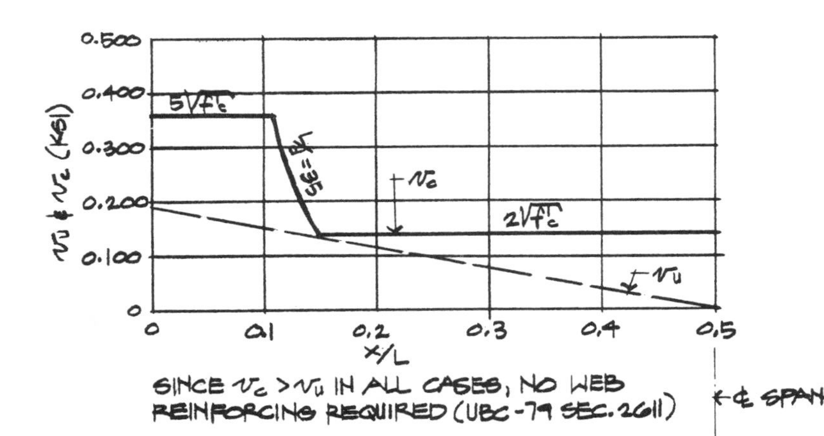

SINCE $v_c > v_u$ IN ALL CASES, NO WEB REINFORCING REQUIRED (UBC-79 SEC. 2611)

HOLLOW CORE PLANK ERECTION

SEARS - EASTRIDGE (SAN JOSE)

8" spancrete spanning 28 feet. A 2½" thick cast-in-place concrete topping is subsequently placed to form the completed diaphragm and take lateral loads to shearwalls. Precast concrete beams and columns complete the building frame.

SEARS - EASTRIDGE (SAN JOSE)

8" spancrete being erected 2 at a time to maximize crane efficiency.

HOLLOW CORE PLANK DESIGN

DESIGN A HOLLOW CORE PLANK WITH 2½" OF HR COMPOSITE TOPPING TO SPAN 37'-0" %C BEARINGS OVER PARKING, AND TO SUPPORT MOTEL CONSTRUCTION; 3 HOUR RATING IS REQUIRED.

(2-30)* USE LIGHTWEIGHT CONCRETE WITH 1½" OF COVER TO ACHIEVE THE DESIRED FIRE RATING, FROM PCI HANDBOOK CHARTS ON P. 2-30, A 10" SECTION SHOULD WORK.

$f'_c = 4000$ PSI , $f'_{ci} = 3500$ PSI

$f'_c = 3000$ PSI HR TPG. — 2½" — 10" — 1 1/16" — 3'-4" — SPANCRETE

(2-39) BASIC SECTION

$A = 272\ IN^2$ $\quad I_b = 2970\ IN^4$

$y_t = 5.09"$ $\quad S_t = 585\ IN^3$

$y_b = 4.91"$ $\quad S_b = 604\ IN^3$

$e = 4.91 - 1.69 = 3.22"$ $\quad$ WT = 65 PSF

COMPOSITE SECTION

$I_c = 6611\ IN^4$ $\quad y_{tc} = 3.02"$ $\quad S_{tc} = 2186\ IN^3$

$y_{bc} = 6.98"$ $\quad S_{bc} = 947\ IN^3$

LOADS AND MOMENTS

TO BASIC SECTION (K/FT)		M_ϕ K-IN ×	L.F.	= UM_ϕ
10" %C (LW)	$1.5 \times 0.217 \times \overline{37}^2$	= 446	1.4	624
C.I.P. TOPPING	$1.5 \times 0.104 \times \overline{37}^2$	= 214	1.4	299
		660 K-IN		
TO COMPOSITE SECTION				
S_{DL} (15 PSF)	$1.5 \times 0.050 \times \overline{37}^2$	= 103	1.4	144
S_{LL} (40 PSF)	$1.5 \times 0.133 \times \overline{37}^2$	= 273	1.7	464
		376 K-IN		$\Sigma UM_\phi = 1531$ K-IN

(3-10) PRESTRESS FORCE REQUIRED

$$F_f = \frac{\frac{M_s}{S_b} + \frac{M_c}{S_{bc}} - f_t}{\frac{1}{A} + \frac{e}{S_b}} = \frac{\frac{660}{604} + \frac{376}{947} - 0.425}{\frac{1}{272} + \frac{3.22}{604}} = \frac{1.063}{0.00901} = 118.2^K$$

(3-15) USING 3/8" φ - 270K STRANDS:

(3-29) $A_{ps} = 0.085$ $\quad F_o = 16.1$ K/STRAND $\quad F_i = (0.9)(16.1) = 14.5$ K/STR.

$F_f = (0.75)(16.1) = 12.1$ K/STR.

$$N = \frac{118.2}{12.1} = 9.77$$

∴ USE 10 - 3/8" φ 270K STRANDS

* ○ PAGE REFERENCE IN 2ND EDITION OF PCI DESIGN HANDBOOK

(3-14) CHECK END STRESSES AT RELEASE

(-) COMPRESSION
(+) TENSION

$f^{TOP} = -\frac{145}{272} + \frac{(145)(3.22)}{585} = -0.533 + 0.798 = +0.265 < 6\sqrt{f'_{ci}}$ O.K.

$f_{BOT} = -\frac{145}{272} - \frac{(145)(3.22)}{604} = -0.533 - 0.773 = -1.306 < 0.6\sqrt{f'_{ci}}$ O.K.

CHECK SERVICE LOAD STRESSES @ ℄ SPAN

$f^{TOP} = -\frac{121}{272} + \frac{(121)(3.22)}{585} - \frac{660}{585} - \frac{376}{2186} = -0.445 + 0.666 - 1.138 - 0.172$

$= -1.089 < 0.45 f'_c$ O.K.

$f_{BOT} = -\frac{121}{272} - \frac{(121)(3.22)}{604} + \frac{660}{604} + \frac{376}{947} = -0.445 - 0.645 + 1.093 + 0.397$

$= +0.400 < 6\sqrt{f'_c}$ O.K.

(3-53) CHECK ULTIMATE STRENGTH

(3-5)

$f_{se} = 142$ KSI $\quad c\overline{\omega}_p = (1.0)\left(\frac{0.85}{40 \times 10.81}\right)\left(\frac{270}{3}\right) = 0.177$

FROM CHART READ $f_{ps}/f_{pu} = 0.96 \quad \therefore f_{ps} = 259.2$ KSI

$a/2 = \frac{A_{ps} f_{ps}}{1.7 f'_c b} = \frac{(0.85)(259.2)}{(1.7)(3)(40)} = 1.08''$

$M_u = \phi A_{ps} f_{ps} (d - a/2) = (0.9)(0.85)(259.2)(10.81 - 1.08) = 1929^{K\cdot IN} > UM$ O.K.

(3-21) CHECK HORIZONTAL SHEAR BETWEEN PLANK & TOPPING

$UV \left[(1.4)(0.217 + 0.104 + 0.050) + (1.7)(0.133)\right]\frac{37}{2} = 13.8^K$

$\phi V_{nh} = \phi \times 80 b_v d = (0.85)(80)(40)(10.81)/1000 = 29.4^K > UV$

$\therefore$ ROUGHEN TOP OF PLANK TO ASSURE SHEAR TRANSFER

CAMBER AND DEFLECTION

$\Delta_i\, P/S = +1.61''\uparrow \qquad \Delta_i\, S/C = -1.28''\downarrow$

$\Delta_{TPG.} = -0.57''\downarrow \qquad \Delta_{SDL} = -0.03''\downarrow$

(3-27) (3-23) $\Delta_{SLL} = 0.09'' < \frac{L}{480} = 0.93''$ OK

AT ERECTION

$\Delta = +(1.80)(1.61) - (1.85)(1.28) = \underline{+0.53'\uparrow}$

ALL SUSTAINED LOADS

$\Delta = +(2.20)(1.61) - (2.40)(1.28) - (2.3)(0.57) - (3.0)(0.03) = -0.93''\downarrow$

(3-59) CHECK VERTICAL SHEAR

$b_w d = (13)(10.81) = 141$

v_u SUPPORT $= \frac{UV}{\phi b_w d} = \frac{13.8}{(0.85)(141)} = 0.115$ KSI

$L/d = \frac{37 \times 12}{10.81} = 40$

0.500
0.400
0.300
0.200
0.100
0.0
0
0.1
0.2
0.3
0.4
0.5
$5\lambda\sqrt{f'_c}$
$L/d = 40$
v_c
$2\lambda\sqrt{f'_c}$
v_u
X/L
℄ SPAN

∴ SINCE $v_c > v_u$ IN ALL CASES, NO WEB REINFORCING IS REQUIRED [UBC-79 SEC. 2611(b)]

REINFORCING SUMMARY - 10" LW SPANCRETE

10-3/8"φ-270K STRANDS
10"
1 3/4"
3'-4"

SPREAD PLANK SYSTEMS

Sometimes it is economical to spread hollow core planks apart and span the resulting gap with metal deck. Following the photos shown below of a typical installation during construction is a design example using 10" Spancrete hollow core planks.

SEARS - MODESTO

10" spancrete spanning 28'-8" and spread 34 3/4" apart. 2 1/2" of cast in place concrete topping is placed over the planks and the metal deck spanning the gaps is covered with 4 1/2" of cast in place concrete. The spread system also makes vertical penetrations for mechanical and electrical systems vastly simplified.

SPREAD HOLLOW CORE PLANK SYSTEM

DESIGN A SPREAD PLANK SYSTEM FOR A DEPARTMENT STORE FLOOR. REQUIRED FIRE RATING OF 2 HR. DESIGN SPAN = 26.5'. PLANKS ARE SPREAD APART 2'-10".

DESIGN DATA FOR 10" SPREAD SPANCRETE SYSTEM $\quad S_{LL}$ = 75 PSF $\quad S_{DL}$ = 25 PSF

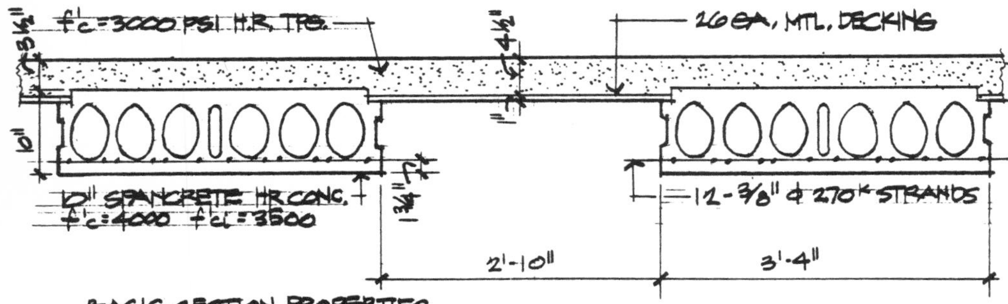

BASIC SECTION PROPERTIES

$A = 272\ IN^2$ $\quad I_b = 2970\ IN^4$ $\quad WT. = 85$ PSF

$y_t = 5.09''$ $\quad S_t = 585\ IN^3$ $\quad e = 3.16''$

$y_b = 4.91''$ $\quad S_b = 604\ IN^3$ $\quad \frac{E_{3000}}{E_{4000}} = \frac{3.3}{3.8} = 0.87$

COMPOSITE SECTION PROPERTIES

PART	A	y_{tcc}	Ay_{tcc}	I_o	Ad^2
10" S/C	272	8.59	2336	2970	2768
4.5" C.I.P. 0.87×A4.5 =	141	2.25	317	238	1399
3.5" C.I.P. 0.87×A.35 =	116	1.75	203	118	1545
COMPOSITE SECTION	529		2856	$\Sigma I = I_c$ =	9038 IN⁴

$y_{tcc} = 5.4''$ $\quad y_{tc} = 1.9''$ $\quad S_{tc} = 4757\ IN^3$

$y_{bc} = 8.1''$ $\quad S_{bc} = 1116\ IN^3$

LOADS & MOMENTS

TO BASIC SECTION

10" S/C	$1.5 \times 0.283 \times \overline{26.5}^2$ = 298	1.4	417
C.I.P. TPG.	$1.5 \times 0.326 \times \overline{26.5}^2$ = 343	1.4	481
	641 K-IN		

TO COMPOSITE SECTION

S_{DL}	$1.5 \times 0.154 \times \overline{26.5}^2$ = 162	1.4	227
S_{LL}	$1.5 \times 0.463 \times \overline{26.5}^2$ = 488	1.7	830
	650 K-IN		1935 K-IN

(3-10)* PRESTRESS FORCE REQUIRED

$$F_f = \frac{\frac{M_B}{S_b} + \frac{M_c}{S_{bc}} - f_t}{\frac{1}{A} + \frac{e}{S_b}} = \frac{\frac{641}{604} + \frac{650}{1116} - 0.425}{\frac{1}{272} + \frac{3.16}{604}} = \frac{1.219}{0.00891} = 136.8^K$$

(3-15) USING 3/8" ϕ-270K STRANDS

(3-29) $A_{ps} = 0.085$ IN² $F_o = 16.1$ K/STR. $F_i = 14.5^K$ $F_f = (.78)(16.1) = 12.6^K$

$N = \frac{136.8}{12.6} = 10.9$ USE 12 - 3/8" ϕ - 270K STRANDS

(3-14) CHECK END STRESSES AT RELEASE (-) COMPRESSION (+) TENSION

$$f^{TOP} = -\frac{174}{272} + \frac{(174)(3.16)}{585} = -0.640 + 0.940 = +0.300 < 6\sqrt{f'_{ci}} \text{ O.K.}$$

$$f_{BOT.} = -\frac{174}{272} - \frac{(174)(3.16)}{604} = -0.640 - 0.910 = -1.550 < 0.6f'_{ci} \text{ O.K.}$$

CHECK FINAL STRESSES AT ℄ SPAN

$$f^{TOP} = -\frac{151.2}{272} + \frac{(151.2)(3.16)}{585} - \frac{641}{585} - \frac{650}{4757}$$
$$= -0.556 + 0.817 - 1.096 - 0.137 = -0.972 < 0.45f'_c \text{ O.K.}$$

$$f_{BOT} = -\frac{151.2}{272} - \frac{(151.2)(3.16)}{604} + \frac{641}{604} + \frac{650}{1116}$$
$$= -0.556 - 0.791 + 1.061 + 0.582 = +0.296 < 6\sqrt{f'_c} \text{ O.K.}$$

(3-53) CHECK ULTIMATE STRENGTH

(3-5) $f_{se} = 147.4$ KSI $c\bar{\omega}_p = (1.0)\left(\frac{1.02}{74 \times 11.75}\right)\left(\frac{270}{3}\right) = 0.106$

FROM CHART READ $f_{ps}/f_{pu} = 0.975$ ∴ $f_{ps} = 263.3$ KSI

$$a/2 = \frac{A_{ps}f_{ps}}{1.7f'_c b} = \frac{(1.02)(263.3)}{(1.7)(3)(74)} = 0.71''$$

$$M_u = \phi A_{ps} f_{ps} (d - a/2)$$
$$= (0.9)(1.02)(263.3)(11.75 - 0.71) = 2668^{K\text{-}IN} > UM \text{ O.K.}$$

(3-27) DEFLECTIONS:

(3-23) AT ERRECTION: $\Delta = 0.63''$ ↑

ALL SUST. LOADS $\Delta_{LT} = 0.18''$ ↓

$\Delta_{LL} = 0.15''$ ↓ O.K.

*( ) PAGE REFERENCE IN 2ND EDITION OF PCI DESIGN HANDBOOK

(3-21) CHECK HORIZONTAL SHEAR BETWEEN PLANK AND TOPPING

$$UV = [(1.4)(0.283+0.326+0.154)+(1.7)(0.463)]\frac{26.5}{2} = 24.6^{K}$$

$$\phi V_{nh} = \phi \times 80\, b_v d = (0.85)(80)(38)(11.75)/1000 = 30.4^{K} \quad \text{O.K.}$$

∴ ROUGHEN TOP OF PLANK ONLY TO SATISFY SHEAR

(3-59) CHECK VERTICAL SHEAR

$$b_w d = (13)(11.75) = 153$$

$$v_{u\ SUP'T} = \frac{UV}{\phi b_w d} = \frac{24.6}{(0.85)(153)} = 0.189 \text{ KSI}$$

$$L/d = \frac{26.5 \times 12}{11.75} = 27$$

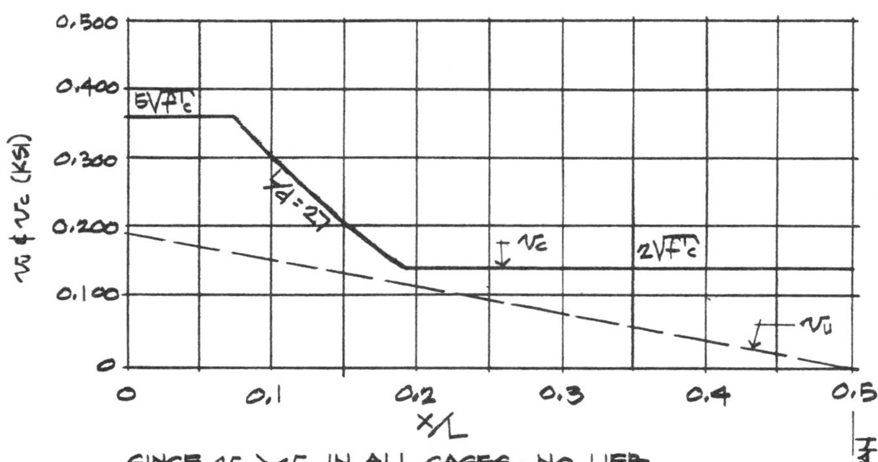

SINCE $v_c > v_u$ IN ALL CASES, NO WEB REINFORCING IS REQUIRED. [UBC-79 SEC. 2611(b)]

ADDITIONAL DESIGN CONSIDERATIONS

In designing untopped hollow core plank, as was discussed on page 325 for solid slabs, the prestress force should be such that the instantaneous camber at release is 1.1 times the instantaneous dead load deflection produced by the slab self weight. Proper design of untopped hollow core systems is a result of camber analysis with the load carrying capacity being determined afterwards and being compared with that required. This will assure that differential camber (the difference in in-situ elevations of adjacent units) will be minimized. In situations where heavy superimposed loads are intended to be carried on untopped hollow core slabs, several field expedients serve to level adjacent planks until grouting can be performed. These are:

1. Use of temporary wooden wedges driven into shear keys
2. 4" x 4" fish plates top and bottom connected by bolts through the shear keys
3. Placement of weights or water filled drums on high planks.

Untopped diaphragm action may be achieved by the use of edge weld plates cast into wet cast hollow core plank products. Dry mix hollow core plank products may use intermediate and boundary reinforcing calculated by the use of the shear friction theory to form the diaphragm. The use of this concept as permitted by UBC-79, Section 2611(p) has been substantiated by testing.

The manufacturing process also effects the design of cantilever reinforcing. Wet cast products can have top bars cast into the section. Dry cast products use top strand, and auxiliary mild steel cast in shear keys for situations where heavy point loadings occur on the cantilever ends.

After curing, hollow core planks are saw cut to specified lengths and placed into storage. Odd width planks are cut lengthwise. In the Span-Deck process shown here, the side rails rotate down and out of the way allowing full access for the saw.(Photo by Ted Gutt)

Hollow core planks in storage awaiting shipment. Plant fabrication is performed at the same time site work is done, thereby telescoping the total project construction time. Note the preformed opening in the plank in the foreground. (Photo by Ted Gutt)

CONNECTION DETAILS

A common situation in the use of hollow core plank is bearing them on masonry walls which also serve as the lateral force resisting system in the building. Close attention must be paid to assure that the transfer of lateral forces at the floor zone is not impaired by poor detailing at this critical interface. The details shown below should be used:

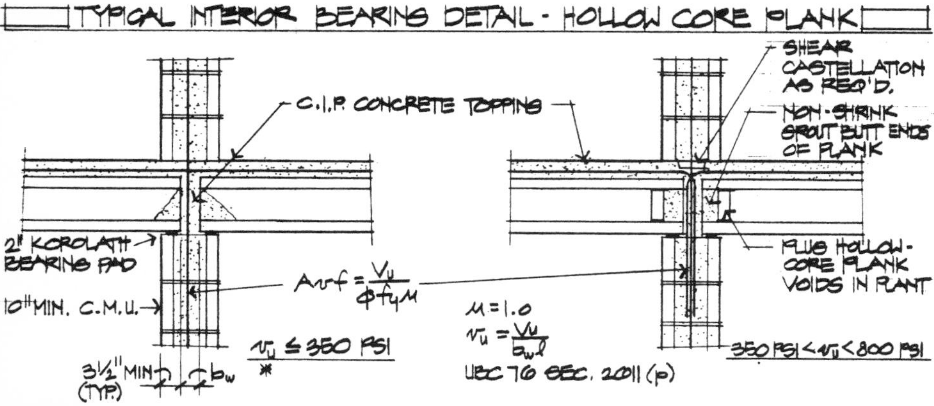

* NOTE THAT THE REDUCED EFFECTIVE WIDTH "b_w" MAY SERIOUSLY LIMIT THE LATERAL CAPACITY OF THE WALL IN RESISTING SEISMIC FORCES.

CONNECTION DETAILS

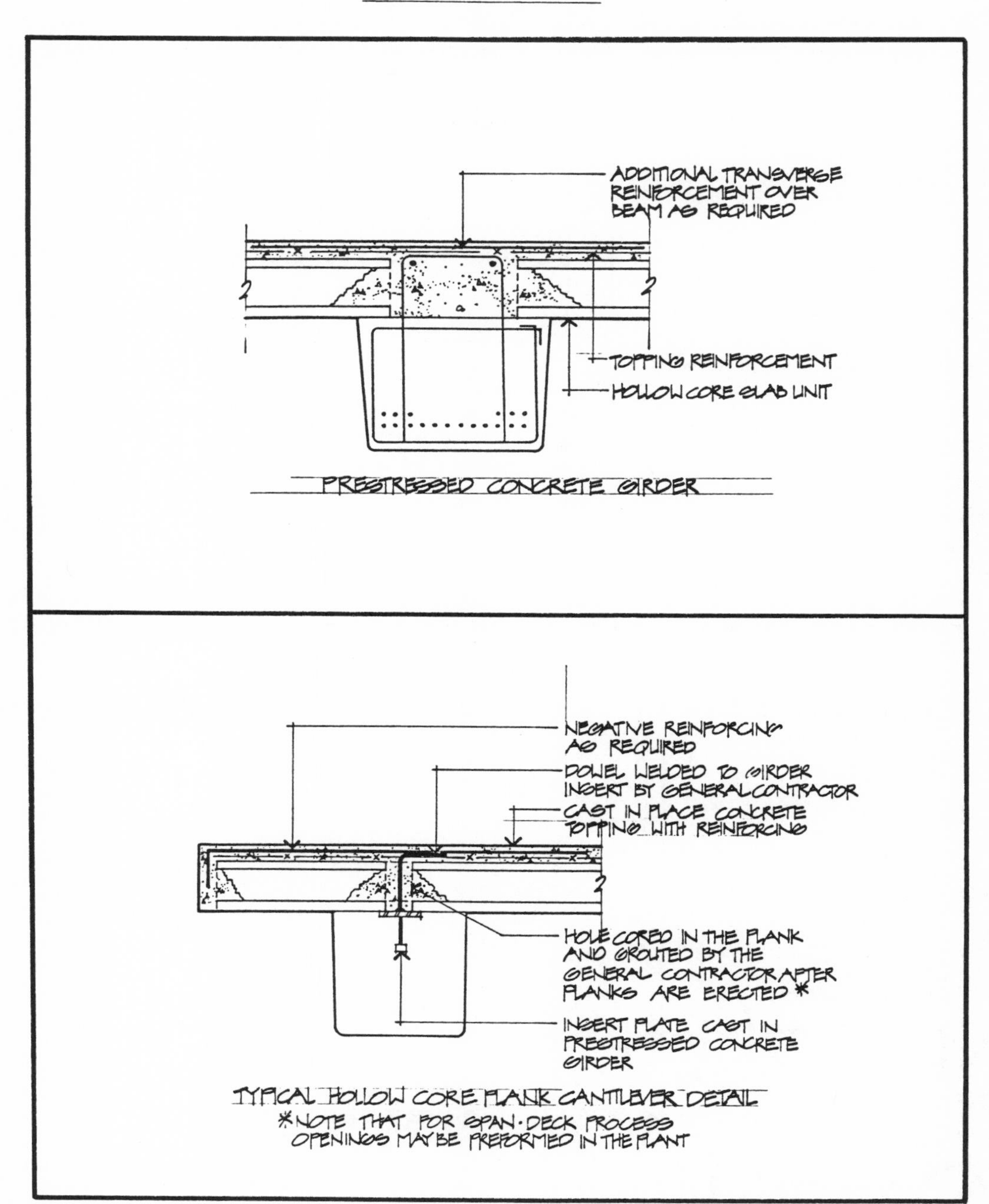

CONNECTION DETAILS

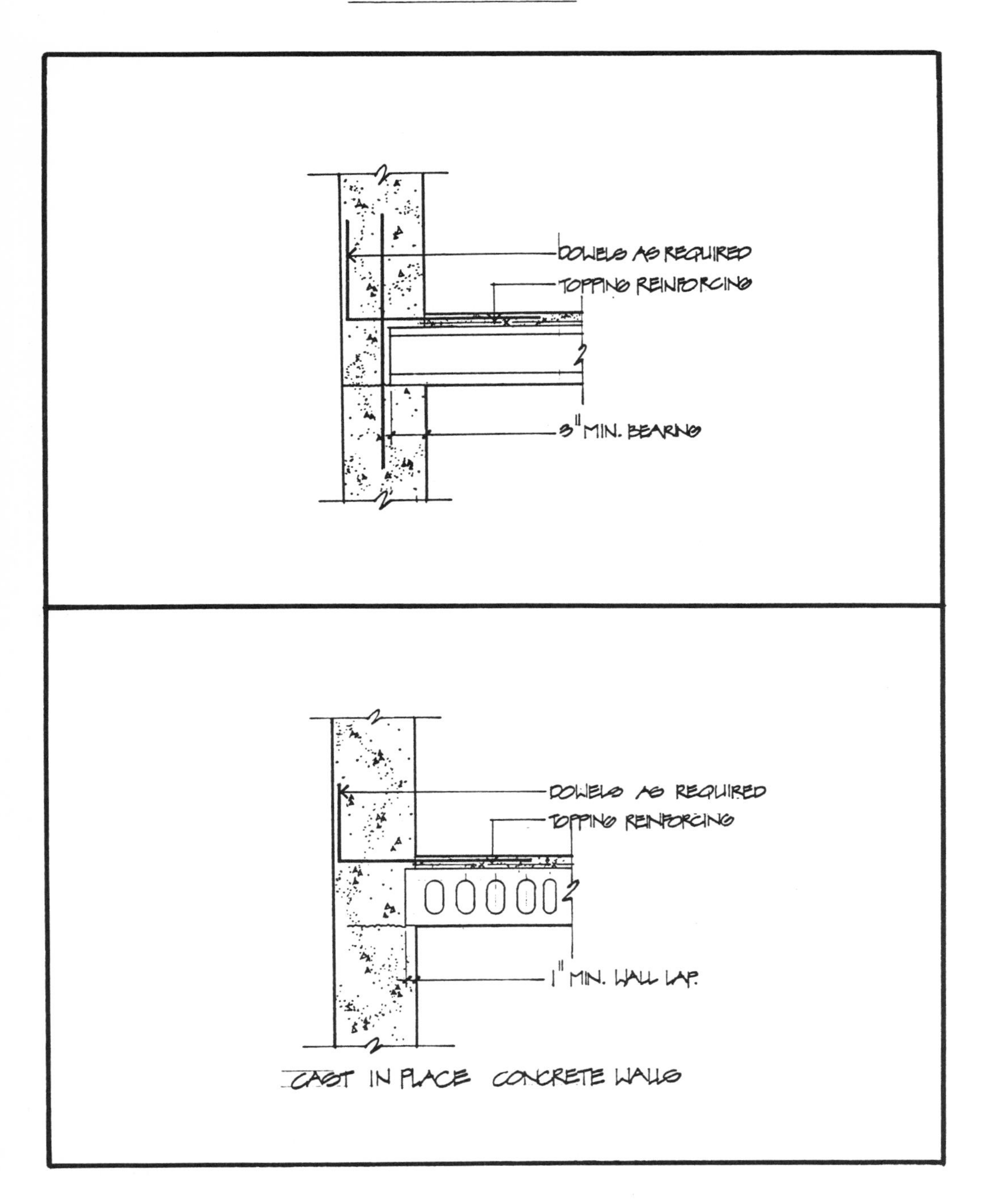

CONNECTION DETAILS

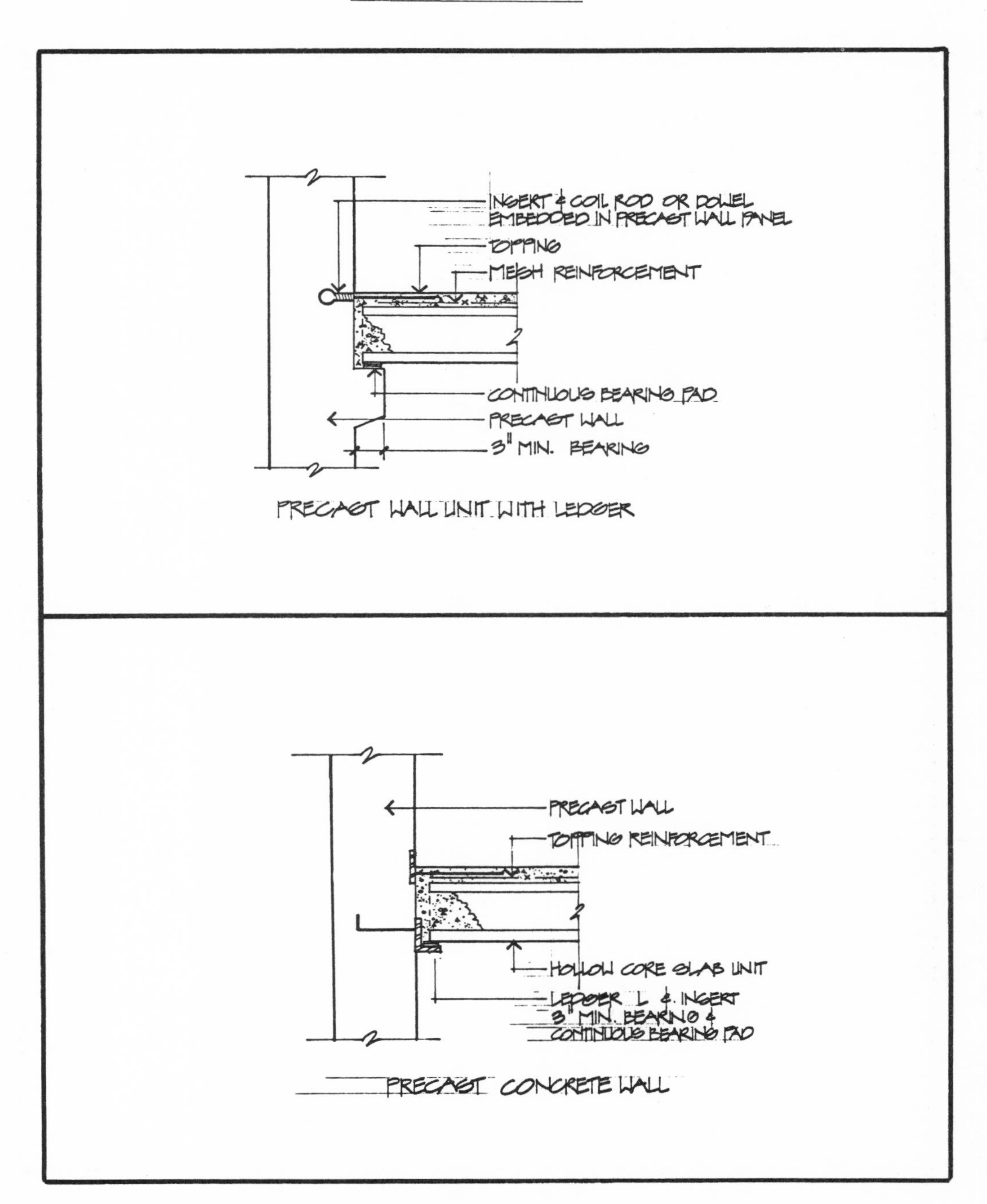

PRECAST WALL UNIT WITH LEDGER

PRECAST CONCRETE WALL

CONNECTION DETAILS

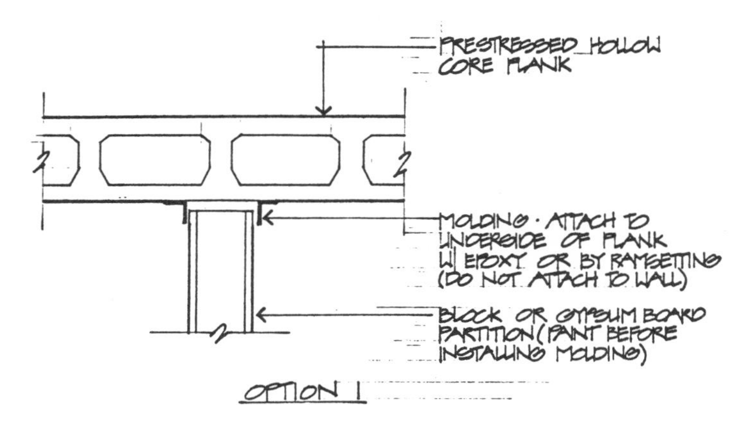

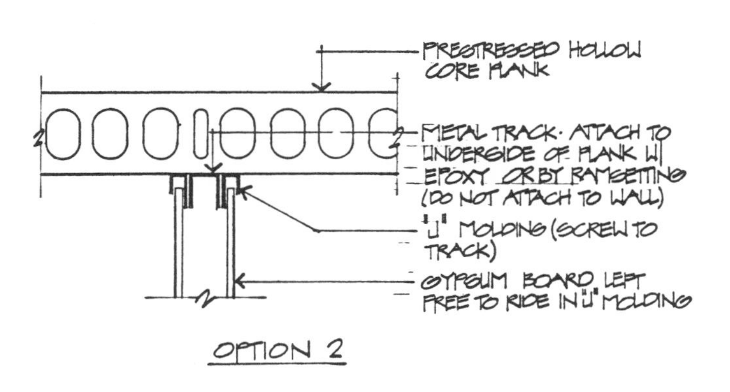

CONNECTION DETAILS

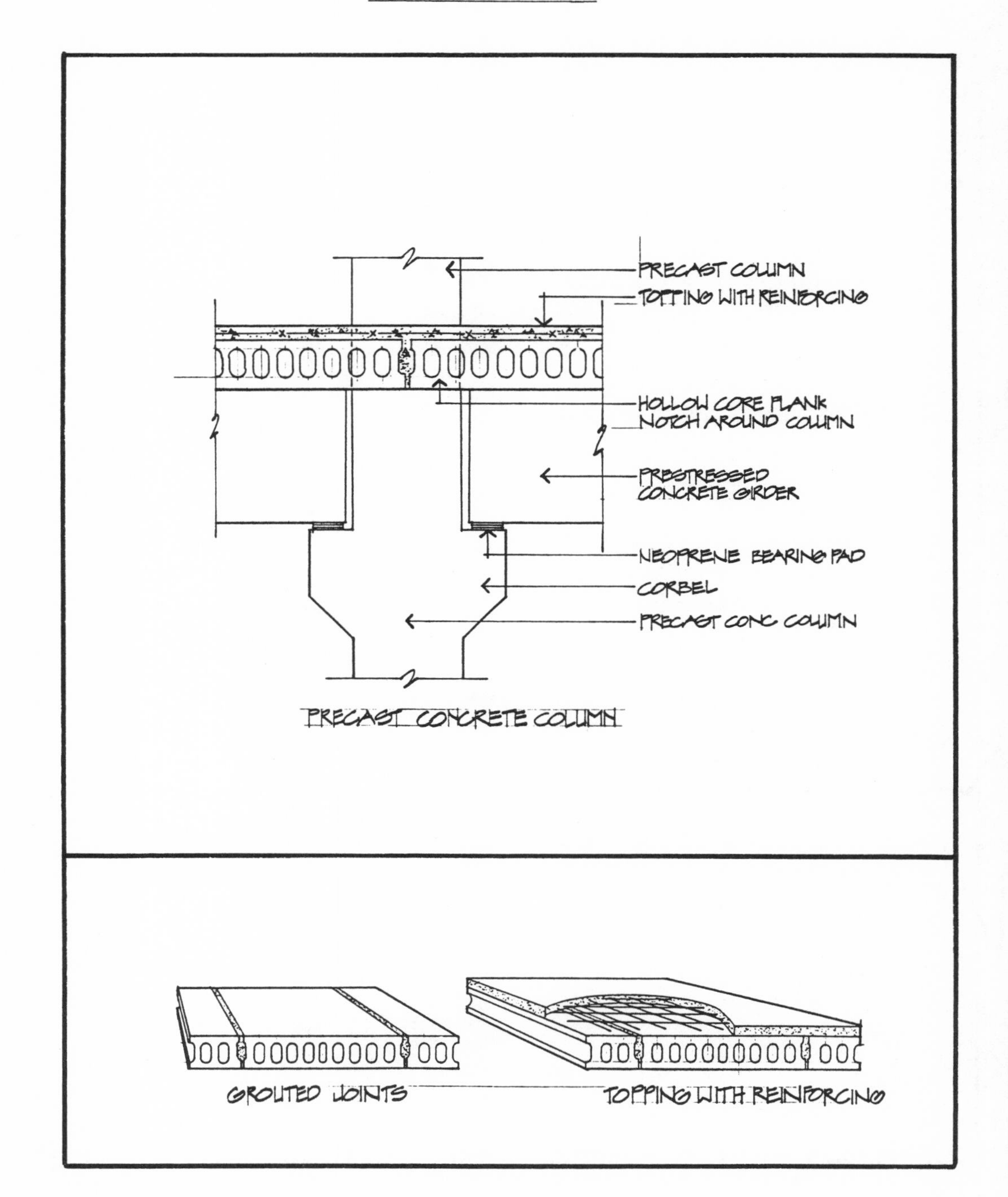

CONNECTION DETAILS

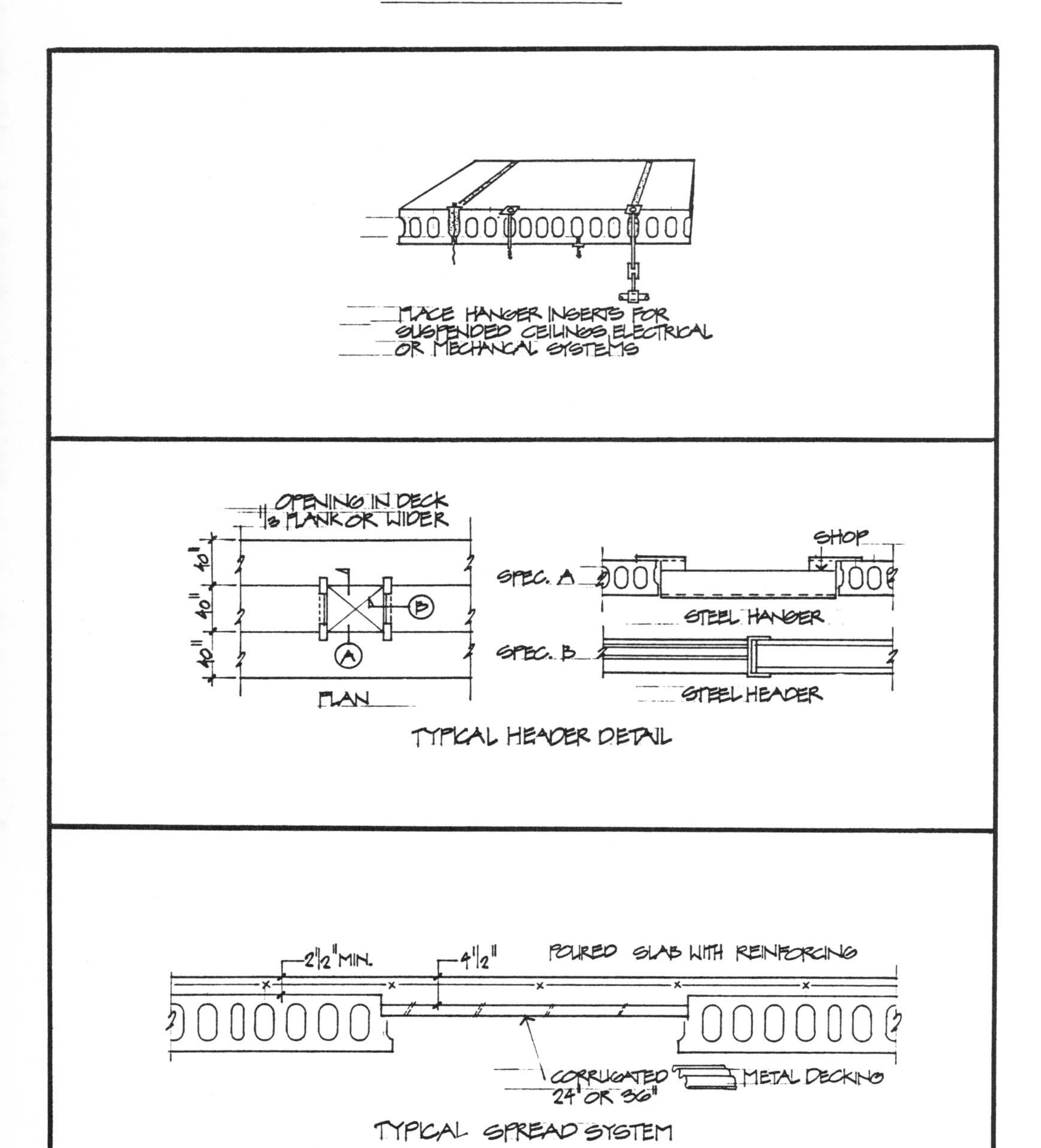

TYPICAL HEADER DETAIL

TYPICAL SPREAD SYSTEM

SPANDECK PRODUCTION OPERATION

The upper photograph shows the bottom (soffit) casting machine receiving concrete from a specialized yard delivery truck called a sidewinder. Below, the bottom casting machine as seen from the rear with the 1 3/4" thick soffit placed and ready for the subsequent top casting operation.(Photos by Ted Gutt)

PRESTRESSED CONCRETE DOUBLE TEES

ALVARADO SEWAGE TREATMENT PLANT - UNION CITY

Double tees up to 70 feet in length were provided for this complex made up of many buildings. Inverted tee girders and vertical and lateral load resisting wall panels were also plant precast and installed, quickly enclosing these buildings.

The double tee has become the mainstay of the prestressed concrete industry. It is used as a deck member in floors and roofs of offices, shopping malls, and department stores, schools, and in parking structures. Deeper sections are used to form clear span roofs over gymnasiums and swimming pools. In roof construction, rigid insulation can be incorporated into the flanges in the plant. The double tee can also be used as a wall panel in commercial and industrial buildings. It can be used to resist vertical as well as lateral loads in these applications. Wall panel units can be solid as well as insulated sandwich panel units. Windows and door openings may be placed in the flange areas between the stems.

In this chapter, we shall deal with double tees used as flexural deck members. As floor units, double tees usually receive a field poured cast-in-place concrete topping which provides a smooth level floor surface and serves as the horizontal diaphragm. Roof units may be designed without cast-in-place concrete topping, in which case the units are connected together by flange weld plates, thus forming the diaphragm.

AVAILABILITY

Double tees are made in various widths, with 8'-0 and 12'-0 wide units being standard in California. They are made in depths from a minimum of 10" to a maximum of 41". Standard 2" flange thickness may be increased by altering the side rail height. Double tee spans of up to 90' are possible in the deeper sections.

DOUBLE TEES IN STORAGE

8 foot wide - 32" deep "heavy stemmed" double tees in storage in Rockwin plant in Santa Fe Springs. Prestressed concrete double tees are fabricated and stockpiled while site work is done, thereby telescoping total project time. Note proper dunnage, clean access and well ordered rows of product.

UNION BANK - OCEANGATE PARKING STRUCTURE - LONG BEACH

12 foot wide - 36" deep double tees spread apart 4 feet provide the clear span deck for this 3 level parking structure. The area between the spread tees was formed and poured at the same time the cast-in-place concrete topping was poured. Long span column free parking bays are afforded by the 64 foot span double tees.

MANUFACTURING

Double tees are cast specifically for individual projects, and are not stockpiled prior to order. They are cast in long lines ranging from 300 to 600 feet in length. Many units are in the same bed with bulkhead separators being placed in between the individual units. They are cast in steel molds which have variable stem widths since the form side walls slope approximately 1/2" per foot of depth. Various depths of tees are fabricated by casting in temporary concrete soffit liners or by using preformed special metal soffit liners. Odd width units can be made by blocking out portions of the overhanging flange. Block outs can be placed on the bed to create openings in the unit, eliminate one flange, stop the stem short of a flange, stop the flange short of a stem or create single tees. Openings can also be placed in the stem of the member up to one third the total member depth with no decrease in flexural strength. These openings should be located to clear the prestressing strands, optimally in the upper portion of the web. For topped designs, the top surface of the tee should be roughened sufficiently to develop horizontal shear in accordance with the requirements of the Uniform Building Code. Untopped roof tees should have their top surfaces steel trowel finished. Concrete strengths for hardrock concrete are 5000 psi and 6000 psi. 5000 psi sand light weight concrete is also used, when the additional cost of the light weight aggregate is offset by savings in some other area, such as shipping or erection. Prestressing strand is usually 1/2" diameter, 270 ksi grade. Economical designs usually specify harped strand. This is accomplished by either depressing down with special devices from the top of the bed after tensioning the strand, or by using hold down devices fastened to anchors built-in under the bed. In the latter case, the hold down devices are positioned prior to tensioning the strand. Welded wire fabric is used for flange reinforcing. Either welded wire fabric or mild steel stirrups, fabricated in advance into cages, are used as stem reinforcing.

ALVARADO SEWAGE TREATMENT PLANT - UNION CITY

4 foot wide single tees were made in the double tee mold by installing a temporary continuous flat bar former down the centerline of the bed. This was done to accommodate the 4 foot module used to lay out the bays in the building.

ERECTION CONSIDERATIONS

For plant handling and erection, lifting bales consisting of loops made up from prestress strand are embedded in each end of each stem. After the double tees are erected, these may be burned off for untopped units, or left cast in the concrete topping. Access for trucks into each bay of the structure is essential when erecting these large units, which can weigh 15 tons or more. Often times, prudent architects and their engineers consult with precast concrete operations personnel early on in a project to determine the best way to design the long span members and not overlook practical details that could save money in the installation of the prestressed concrete components.

J. C. PENNEY STORE - SANTA ROSA

Prestressed concrete shallow double tees, inverted tee girders, and precast concrete columns form the basic building frame for the floors and roof of this department store, with lateral forces being transmitted to the masonry infill walls. Quick erection saves construction time and the precast decks form an instant working platform for other trades.

KAISER PERMANENTE PLANT
EXPANSION - CUPERTINO

Double tees form the walls, floors and roof of this complicated industrial building. Interior precast wall panels, columns, and beams complete the range of precast concrete products used in this totally precast concrete structure. Note bracing used to stabilize structure during erection. In the photo to the left, the large openings were preplanned and formed in the plant in these 80 foot span double tees.

DESIGN CONSIDERATIONS

The span of the double tees is dependent upon the architectural considerations, building module, transportation and economics. The architectural considerations are: use of the enclosed space, building shape and design effect (aesthetics). Certain applications such as parking structures have set column spacings and clear span requirements which give optimum parking arrangements and maximum number of spaces. For hotels and apartment buildings the size and arrangement of the room layouts will dictate the location of structural supports which in turn will dictate the span of the double tee. Office and industrial occupancies require large clear spans. For industrial or engineering research occupancies, the design live loads also have an effect on the span.

The module to be selected related to the tee span is such as to give the maximum number of units of the same length as well as to be compatible with the wall panel size or module. The effect of cantilevers is also taken into consideration. Usually the building is designed around an 8 ft. module in both directions, being a multiple of the tee width or bay spacing. It is desirable but not essential that all of the tee units be of the same length.

To reduce costs it is desirable to keep the spans within the truck and load lengths permitted without special permits. For double tees this results in a tee length of approximately 60 feet. But, of course, longer lengths may be transported at a small premium in shipping cost. Since economics is always a factor, a cost study is often made to determine which system as well as span length module would be the most desirable, i.e. hollow core plank with shorter spans vs long span double tees, solid double tee system vs spread channel system with pretensioned flat panels.

Untopped diaphragms may be formed in two ways. The first is with the use of flange weld plates spaced as required by design to develop diaphragm integrity. Care must be taken at boundaries and interfaces with lateral load resisting systems to adequately transfer seismic diaphragm forces and provide structural redundancy at these points. The second method is by the use of cast-in-place concrete edge beams and intermediate and boundary reinforcing calculated by the use of the shear friction theory to form the diaphragm. The use of this concept as permitted by UBC-79, Section 2611(p) has been substantiated by testing. Untopped diaphragms are formed with double tee units on roofs, and in certain specialized instances, in parking structures. The flanges are thickened to 4 or 4½" and are interconnected with special welded connections shown in the back of this chapter in the connections section. Other connection conditions are also shown for these untopped structures.

Preliminary design charts for double tee units commonly found in California are shown on the following two pages. The charts in the Second Edition of the PCI Design Handbook, pp 2-6 through 2-21 are also useful. Designers should always contact the prestressed concrete manufacturers in the general vicinity of the project to ascertain the specific details and sections available. A sample of a specific manufacturer's cross-section is shown following the load tables, after which an example double tee design calculation is given.

PRELIMINARY LOAD TABLES FOR "HEAVY STEMMED" DOUBLE TEES

(BOTTOM STEM WIDTH ≥ 6") ALL DOUBLE TEES ARE FABRICATED WITH HARD ROCK CONCRETE (f'c = 5000 PSI) 2 1/2" OF HARDROCK CONCRETE TOPPING CAST IN PLACE.

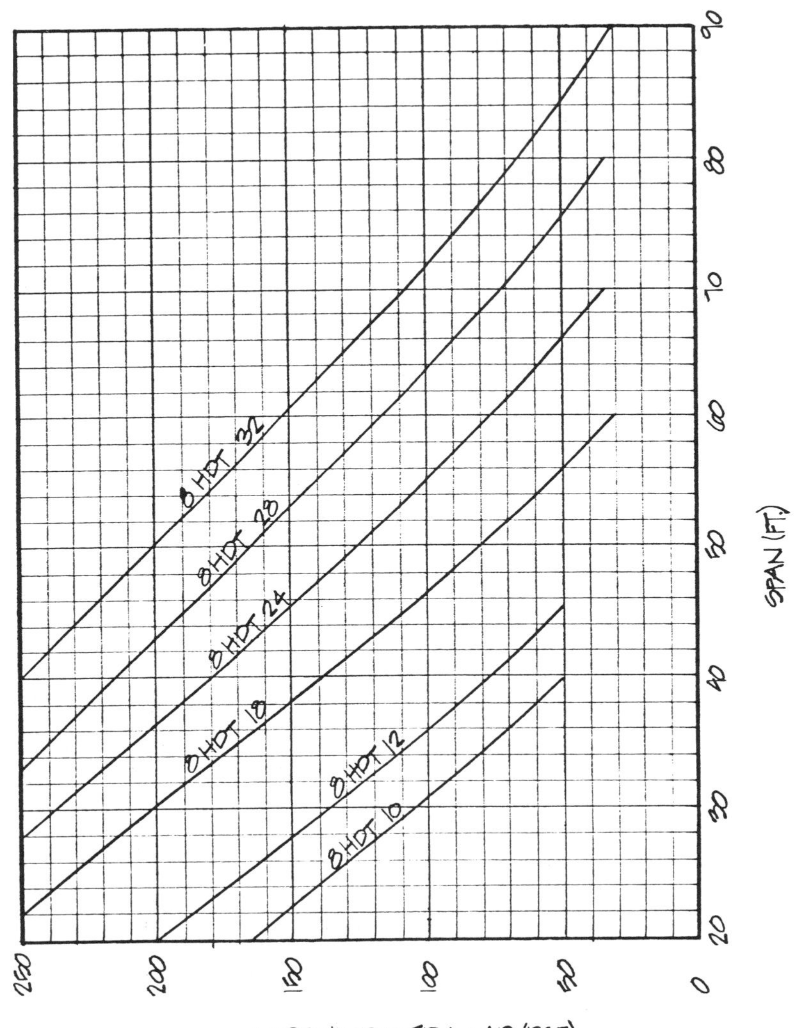

(TEE DESIGNATION 8HDT 24 MEANS 8' WIDE HEAVY STEMMED DOUBLE TEE 24" DEEP)

PRELIMINARY LOAD TABLES FOR "NORMAL STEM WIDTH" DOUBLE TEES

(BOTTOM STEM WIDTH < 4 3/4" FOR 8' WIDE TEES; 7 1/4" FOR 12' WIDE TEES) ALL DOUBLE TEES ARE FABRICATED WITH HARDROCK CONCRETE (f'_c = 5000 PSI) 2 1/2" OF HARDROCK CONCRETE TOPPING CAST IN FIELD.

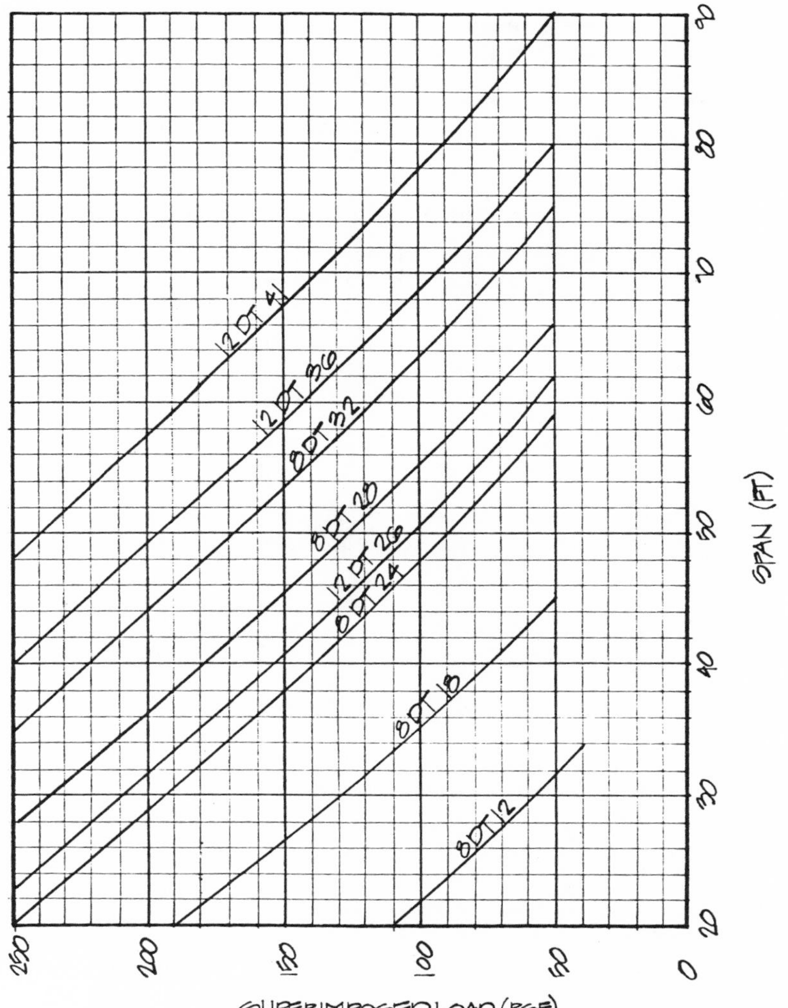

(TEE DESIGNATION 12DT36 MEANS 12' WIDE DOUBLE TEE 36" DEEP)

8′ LIGHT WEIGHT DOUBLE TEE x 24″

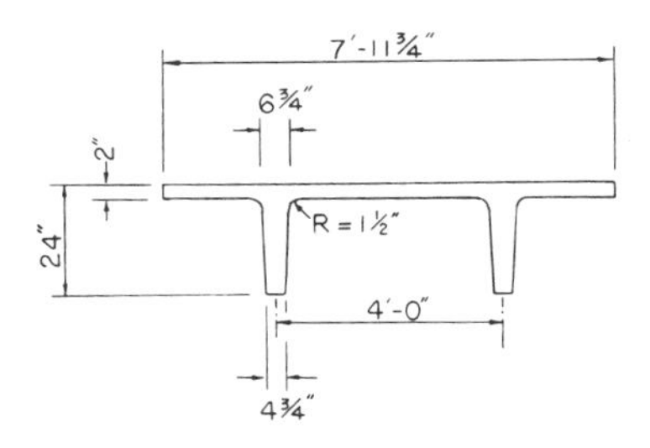

Section Properties:

A = 444.5 in.^{2} **Weight: 44.4 p.s.f. (L.W. @ 115 p.c.f.)**
57.9 p.s.f. (H.R. @ 150 p.c.f.)

No Topping: I = 24,237 in.^{4}

C_t = 7.47″	K_t = 3.29″	S_t = 3245 in.^{3}
C_b = 16.53″	K_b = 7.30″	S_b = 1466 in.^{3}

Composite with 2″ Topping Slab: I = 33,914 in.^{4}

C_t = 6.91″	S_t = 4905 in.^{3}
C_c = 4.91″	S_c = 6902 in.^{3}
C_b = 19.09″	S_b = 1776 in.^{3}

Composite with 2½″ Topping Slab: I = 36,205 in.^{4}

C_t = 6.91″	S_t = 5238 in.^{3}
C_c = 4.41″	S_c = 8208 in.^{3}
C_b = 19.59″	S_b = 1848 in.^{3}

Remarks: Fire Rating

2 Hour with 2″ Light Weight Topping.
2 Hour with 2½″ Hard Rock Topping.

Recommended Uses:

Parking Structures (with 2½″ H.R. Topping.)
Spans to 62′ with 50 p.s.f. Un-reduced L.L.
Spans to 64′ with 50 p.s.f. Reduced to 40 p.s.f. L.L.

Roof:
Spans to 64′ with or without Topping.

Floor:
Spans to 58′ with 2″ Topping.
Spans to 53′ with 2½″ Topping.

Load Assumptions:

Roof: 18 p.s.f. Super-imposed D.L.
20 p.s.f. L.L. to 50′ Span.
16 p.s.f. L.L. over 50′ Span.

Floor: 30 p.s.f. Super-imposed D.L.
50 p.s.f. L.L. Reduced for Area.

(Manufacturer's product literature courtesy of Kabo-Karr Corporation, Visalia, California)

DOUBLE TEE DESIGN

DESIGN AN 8'-0" WIDE DOUBLE TEE FOR A PARKING STRUCTURE

SPAN = 58.0 FT.
LL = 50 PSF

DOUBLE TEE CONC.
HARD ROCK
$f'_c = 5000$ PSI
$f'_{ci} = 3500$ PSI

CIP 3000 PSI HR TOPPING

2½" C, 2", 6¾", 24", 10.5", 6.03", e_e, CGC, 14.28", $e_¢$, 2.25", ½" CLEAR, 4¾", 2'-0", 4'-0", 2'-0", 8'-0" NOMINAL

(2-16)* FROM PCI DESIGN HANDBOOK, P. 2-16, A 24" DEEP SECTION IS SELECTED. FROM MANUFACTURER'S PRODUCT LITERATURE, SPECIFIC SECTION PROPERTIES FOR A 24" DT ARE NOTED.

BASIC SECTION

A = 444 IN.^{2} WT. = 0.480 K/FT.
I_b = 24,237 IN.^{4}
y_t = 7.47" S_t = 3245 IN^3
y_b = 16.53" S_b = 1466 IN^3

COMPOSITE SECTION

I_c = 36,205 IN.^{4}
y_{tc} = 4.41" S_{tc} = 8208 IN^3
y_{bc} = 19.59" S_{bc} = 1848 IN^3

(8-15) STRAND DATA FOR ½" φ - 270^K STRAND

f_{pu} = 270 KSI A_{ps} = 0.1531 IN^2/STRAND

INITIAL JACKING FORCE $F_o = 0.7 A_{ps} f_{pu}$ = 28.9 K/STRAND
AT RELEASE $F_i = 0.9 \times F_o$ = 26.0 K/STRAND
(3-29) FINAL DESIGN (22% LOSSES) $F_f = 0.78 \times F_o$ = 22.5 K/STRAND

LOADS AND MOMENTS

TO BASIC SECTION	$M_¢$ K-IN	× L.F.	= UM ¢
DT $\frac{1}{8} \times 0.480 \times \overline{58}^2 \times 12$ =	2422	1.4	3391
C.I.P. $\frac{1}{8} \times 0.250 \times \overline{58}^2 \times 12$ =	1262	1.4	1766
	3684		
TO COMPOSITE SECTION			
LL $\frac{1}{8} \times 0.400 \times \overline{58}^2 \times 12$ =	2018 K-IN	1.7	3431
			ΣUM ¢ = 8588 K-IN

* ◯ PAGE REFERENCE IN 2ND EDITION OF PCI DESIGN HANDBOOK

DETERMINE IF STRANDS MUST BE DEPRESSED

$$f_b = \frac{M_B}{S_b} + \frac{M_C}{S_{bc}} = \frac{3684}{1466} + \frac{2018}{1848} = 3.6 \text{ KSI}$$

$$f_t + \frac{f'_c}{2} = 6\sqrt{5000} + \frac{3500}{2} = 2.2 \text{ KSI} < f_b$$

∴ USE DEPRESSED STRANDS (SHALLOW DRAPE)

$M_{0.4L} = 0.96\,M_{℄}$

$e_{0.4L} \cong 0.93\,e_{℄}$

REQUIRED PRESTRESS FORCE

$$F_f = \frac{\frac{M_B}{S_b} + \frac{M_C}{S_{bc}} - f_t}{\frac{1}{A} + \frac{e}{S_b}} = \frac{\frac{(0.96)(3684)}{1466} + \frac{(0.96)(2018)}{1848} - 6\sqrt{5000}}{\frac{1}{444} + \frac{(0.93)(14.28)}{1466}}$$

$$= \frac{2.412 + 1.048 - 0.425}{0.00225 + 0.00906} = \frac{3.035}{0.01131} = 268\text{ K}$$

$$N = \frac{268}{22.5} = 11.9 \qquad \therefore \text{ USE } 12 - \tfrac{1}{2}''\phi - 270^K \text{ STRANDS}$$

(3-10) (8-14) CHECK STRESSES AT END AT RELEASE

$$f^{TOP} = -\frac{(12)(26)}{444} + \frac{(312)(6.03)}{3245} = -0.703 + 0.580 = -0.123 \text{ KSI COMP.}$$

$$f_{BOT} = -0.703 - \frac{(312)(6.03)}{1466} = -0.703 - 1.283 = -1.986 < 0.6f'_{ci} \quad \text{OK.}$$

CHECK BOTTOM TENSION UNDER ALL LOADS @ 0.4 L

$$e_{0.4L} = (0.8)(14.28 - 6.03) + 6.03 = 12.63''$$

$$f_{BOT} = -\frac{(12)(22.5)}{444} - \frac{(270)(12.63)}{1466} + \frac{(0.96)(3684)}{1466} + \frac{(0.96)(2018)}{1848}$$

$$= -0.608 - 2.326 + 2.412 + 1.048$$

$$= +0.526 \text{ KSI}$$

$$7.5\sqrt{f'_c} = +0.530 \text{ KSI} \quad \text{O.K.}$$

(3-5) (3-5X) CHECK ULTIMATE STRENGTH

1. APPROXIMATE CODE METHOD (UBC-79, SEC. 2618)

$$\rho_p = \frac{A_{ps}}{bd} = \frac{(12)(.1531)}{(96)(22.6)} = 0.000847$$

$$f_{ps} = f_{pu}\left(1 - 0.5\rho_p \frac{f_{pu}}{f'_c}\right) = 270\left(1 - 0.5 \times .000847 \times \frac{270}{5}\right) = 259.7$$

$$a/2 = \frac{A_{ps}\,f_{ps}}{1.7 f'_c\, b} = \frac{(1.837)(259.7)}{(1.7)(3)(96)} = 0.97''$$

$$M_u^{0.4L} = \phi A_{ps} f_{ps}(d - a/2) = (0.9)(1.837)(259.7)(22.6 - 0.97) = \underline{9287^{K\text{-}IN}}$$

$UM^{0.4L} = 0.96 \times 8588 = 8244$ K-IN < Mu O.K.

(3-8) (3-53) 2. STRAIN COMPATIBILITY

(3-8X)

USING FIGURE 3.9.5 :

$f_{sc} = 147$ KSI

$$c\bar{\omega}_p = c\frac{A_{ps}}{bd} \times \frac{f_{pu}}{f'_c} = (1.0)\left(\frac{1.837}{96 \times 22.6}\right)\left(\frac{270}{3}\right) = 0.076$$

FROM CHART:

$$\frac{f_{ps}}{f_{pu}} = 0.98 \qquad \therefore\ f_{ps} = (0.98)(270) = 264.6 \text{ KSI}$$

$$\frac{a}{2} = \frac{(1.837)(264.6)}{(1.7)(3)(96)} = 0.99''$$

$$M_u^{0.4L} = (0.9)(1.837)(264.6)(22.6 - 0.99) = \underline{9454 \text{ K-IN}}$$

(3-21) CHECK HORIZONTAL SHEAR BETWEEN PRECAST TEE & TOPPING

$$UV = \left[(1.4)(0.480 + 0.250) + (1.7)(0.400)\right]\frac{58}{2} = \underline{49.4\text{K}}$$

$$\phi V_{nh} = \phi(80 b_v d) = (0.85)(80)(96)(22.6)/1000 = \underline{147.5\text{K}}$$

∴ ROUGHEN TOP OF TEE; NO TIES REQUIRED

(THIS IS ANALOGOUS TO UBC-79 SEC. 2617 WHERE: $v_{dh} = \frac{UV}{\phi b_v d} \leq 80$ PSI ∴ ROUGHEN TOP ONLY)

CHECK SHEAR; PROPORTION WEB REINFORCEMENT

NOTE CERTAIN PROVISIONS OF UBC-79:

a. WEB REINFORCING MAY BE OMITTED WHERE $v_u < 0.5 v_c$

b. MAXIMUM SPACING SHALL BE THE LESSER OF

$0.75h = (0.75)(24) = 18''$

OR

$24''$

OR

$$\frac{A_v f_y}{50 b_w} = \frac{(0.22)(40{,}000)}{(50)(11.5)} = \underline{15.3''} \text{ GOVERNS}$$

(3-60) CONSTRUCT SHEAR DIAGRAM

$d = 0.8h = 19.2''$

$$v_{u\ \text{SUPPORT}} = \frac{UV}{\phi bd} = \frac{49.4}{(0.85)(11.5)(19.2)} = 0.263 \text{ KSI}$$

$$L/D = \frac{58 \times 12}{19.2} = 36$$

$5\sqrt{f'_c} = 0.350$

$2\sqrt{f'_c} = 0.140$

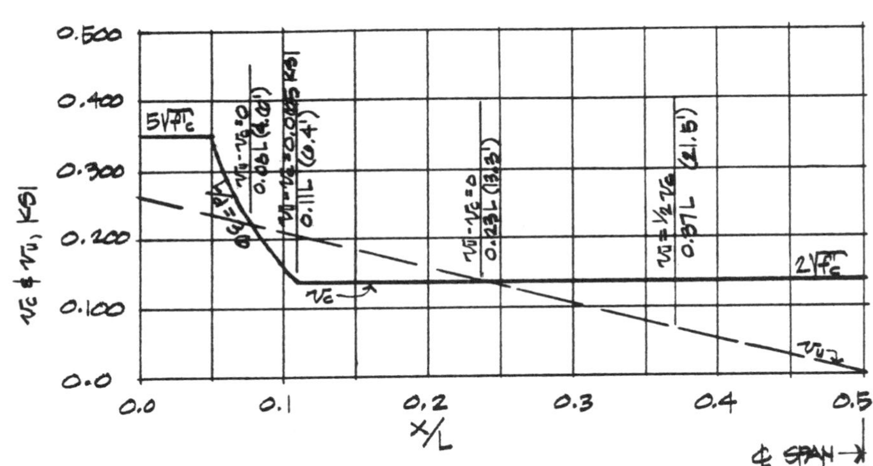

(3-64) DETERMINE MINIMUM WEB REINFORCEMENT REQUIREMENTS

USING #3 STIRRUPS (ONE ROW OF BARS - EACH TEE STEM)

$b_w d = (11.5)(19.2) = 220$ $\quad A_V = 0.22$ IN² $\quad A_{ps} = 1.837$ IN²

FROM CHART : $A_{v_{min}} = 0.12$ IN²/FT

$A_{v_{PROVIDED}}$ #3 @ 15" %c $= 0.22 \times \frac{12}{15} = 0.176$ IN² O.K.

(#3 STIRRUP - ONE ROW PER TEE STEM)

(3-65) DETERMINE WEB REINFORCING REQUIRED AT MAXIMUM VALUE OF

$(v_u - v_c) b_w = (0.065)(11.5) = 0.748$

(8-21) FOR #3 @ 12 ($A_V = 0.22$) $\quad (v_u - v_c) b_w = 0.733$ K/IN O.K.

FOR 4×4 - W5.5×W5.5 $\quad (v_u - v_c) b_w = 0.810$ K/IN O.K.

SELECT EITHER OPTION

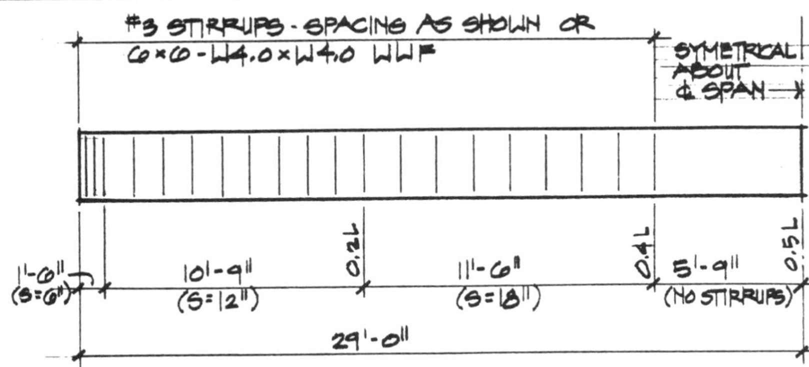

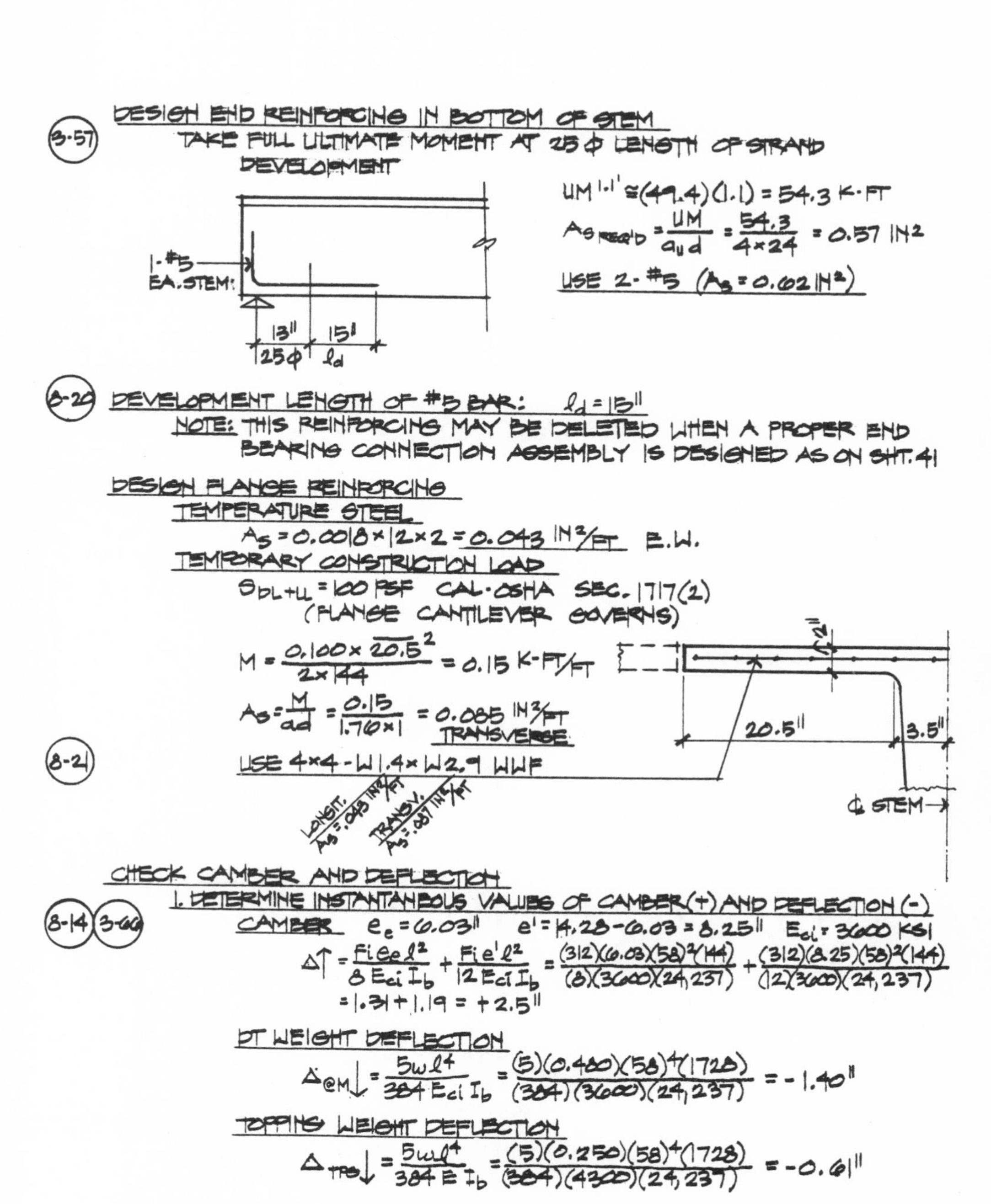
DESIGN END REINFORCING IN BOTTOM OF STEM
3-57
TAKE FULL ULTIMATE MOMENT AT 25 Φ LENGTH OF STRAND DEVELOPMENT
UM 1.1' ≅ (49.4)(1.1) = 54.3 K-FT
As REQ'D = UM / au d = 54.3 / 4×24 = 0.57 IN²
USE 2-#5 (As = 0.62 IN²)
1-#5 EA. STEM
13" 15"
25Φ ld
8-20
DEVELOPMENT LENGTH OF #5 BAR: ld = 15"
NOTE: THIS REINFORCING MAY BE DELETED WHEN A PROPER END BEARING CONNECTION ASSEMBLY IS DESIGNED AS ON SHT. 41
DESIGN FLANGE REINFORCING
TEMPERATURE STEEL
As = 0.0018 × 12 × 2 = 0.043 IN²/FT E.W.
TEMPORARY CONSTRUCTION LOAD
SDL+LL = 100 PSF CAL-OSHA SEC. 1717(2)
(FLANGE CANTILEVER GOVERNS)
M = 0.100 × 20.5² / 2× 144 = 0.15 K-FT/FT
As = M / ad = 0.15 / 1.76×1 = 0.085 IN²/FT TRANSVERSE
8-21
USE 4×4-W1.4×W2.9 WWF
LONGIT. As = .043 IN²/FT
TRANSV. As = .087 IN²/FT
20.5"
3.5"
℄ STEM
CHECK CAMBER AND DEFLECTION
8-14 3-66
1. DETERMINE INSTANTANEOUS VALUES OF CAMBER (+) AND DEFLECTION (-)
CAMBER ee = 6.03" e' = 14.28 - 6.03 = 8.25" Eci = 3600 KSI
Δ↑ = Fi ee l² / 8 Eci Ib + Fi e' l² / 12 Eci Ib = (312)(6.03)(58)²(144) / (8)(3600)(24,237) + (312)(8.25)(58)²(144) / (12)(3600)(24,237)
= 1.31 + 1.19 = +2.5"
DT WEIGHT DEFLECTION
Δ@M↓ = 5wl⁴ / 384 Eci Ib = (5)(0.480)(58)⁴(1728) / (384)(3600)(24,237) = -1.40"
TOPPING WEIGHT DEFLECTION
ΔTPG↓ = 5wl⁴ / 384 E Ib = (5)(0.250)(58)⁴(1728) / (384)(4300)(24,237) = -0.61"

LL DEFLECTION

$$\Delta_{LL}\downarrow = \frac{5w\ell^4}{384EI_c} = \frac{(5)(0.400)(58)^4(1728)}{(384)(4300)(36,205)} = -0.65''$$

(3-22) 2. DETERMINE LONG TERM EFFECTS OF CREEP

AT ERECTION	Δ_i	×	C.F.	$=\Delta_{LT}$
CAMBER	+2.50	×	1.80	= +4.50
DEFLECTION (DT WT)	-1.40	×	1.85	= -2.60
Σ				+1.90″ ↑

FINAL CONDITION	Δ_i		C.F.	Δ_{LT}
CAMBER	+2.50	×	2.20	= +6.00
DEFLECTION (DT WT)	-1.40	×	2.40	= -3.36
DEFLECTION (TOPPING)	-0.61	×	2.30	= -1.40
Σ				+1.24″ ↑
DEFLECTION (LL)				-0.65
				+0.59″ ↑

(3-23) CHECK Δ_{LL} ALLOWABLE - ACI 318-77 TABLE 9.5(b)
(SEE ALSO UBC-79 TABLE 23-D)

$$\frac{L}{360} = \frac{58 \times 12}{360} = 1.93'' \text{ O.K.}$$

REINFORCING SUMMARY - BDT 24

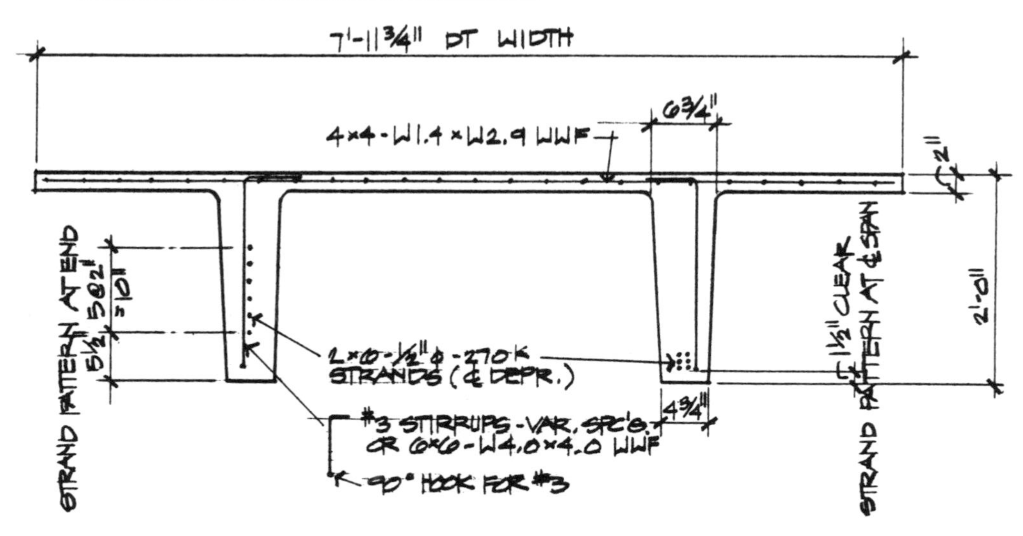

CONNECTION DESIGN

It can be stated axiomatically that the bottoms of double tees and other stemmed prestressed concrete members are never welded at their supports; they are left to "float" free on neoprene to allow long term creep and shrinkage movements to be accommodated without causing distress in the prestressed concrete elements and their connections.

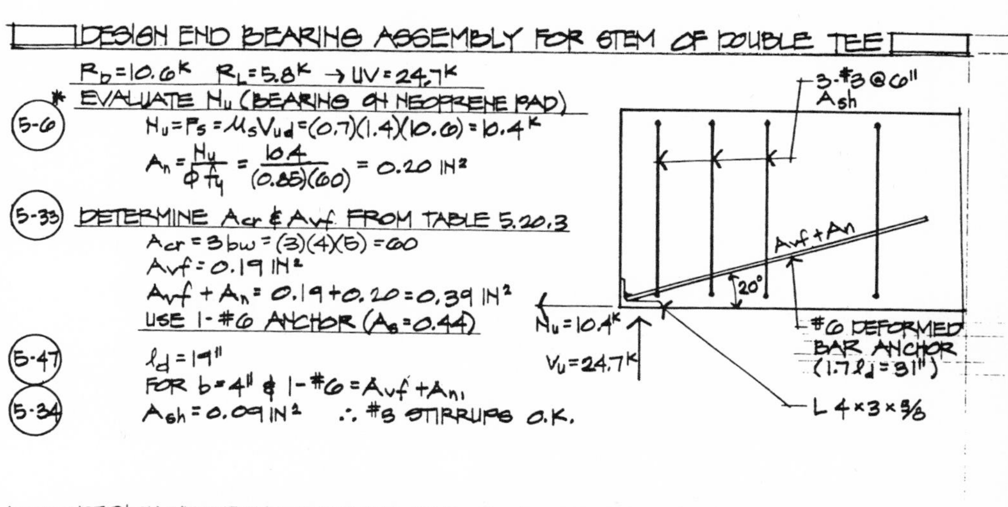

DESIGN NEOPRENE BEARING PAD TO ACCOMODATE LONG TERM MOVEMENT AND PRESSURE OF TEE STEM OF DOUBLE TEE. DOUBLE TEES ARE ERECTED 45 DAYS AFTER CASTING.

$R_{D+L} = 10.6 + 5.8 = 16.4$ K/STEM

BEARING PRESSURE $f = \frac{V}{wb} = \frac{16.4}{(4)(5)} = 0.820$ KSI

(4-13) CALCULATE LONG TERM MOVEMENT DUE TO CREEP AND SHRINKAGE

$f'_{ci} = 3500$ PSI AVG. $P/_S = 1000$ PSI $V/_S = \frac{444}{288} = 1.5$

	LT	− 45 DYS	= NET
CREEP STRAIN	524	259	265
SHRINKAGE STRAIN	560	251	309
TOTAL			574×10^{-6} IN/IN

1'' 4'' 1''
w
(b=5'')
DT
½'' SETBACK
IT GIRDER

$\Delta = 574 \times 10^{-6} \times 29 \times 12 = 0.20''$ MAX. STRAIN IN PAD

RECOMMENDED $t = 2\Delta = 2 \times 0.20 = 0.40''$

(5-4) TRY 3/8'' PAD

SHAPE FACTOR $= \frac{wb}{2(w+b)t} = \frac{(4)(5)}{(2)(4+5)(3/8)} = 3$

FROM FIG. 5.4.1 FOR S.F. = 3 & f = 820 PSI

USE PAD 4½ × 5 × 3/8 − 60 DUROMETER

*( ) PAGE REFERENCE IN 2ND EDITION OF PCI DESIGN HANDBOOK

CONNECTION DETAILS

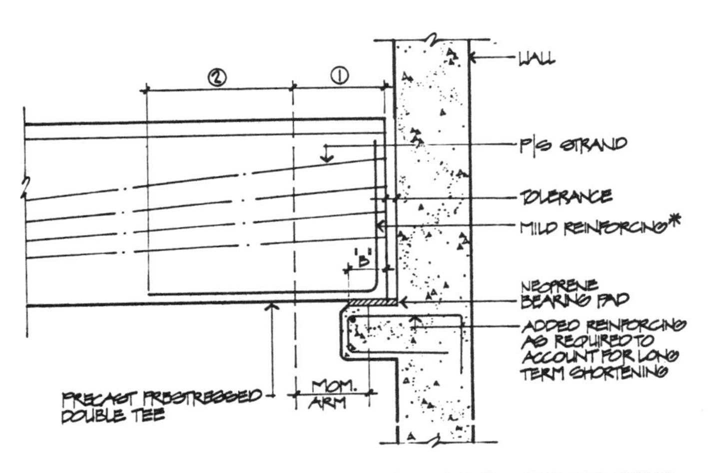

① ZONE TO BE DESIGNED AS REINF. CONC. SINCE PRESTRESSING FORCE NOT FULLY TRANSFERRED TO CONCRETE.

② ANCHORAGE DISTANCE FOR MILD REINF. IN FULLY PRESTRESSED CONCRETE.

'B' BEARING LENGTH DETERMED BY ALLOWABLE:
BEARING LOAD
P/S MEMBER WIDTH
TOLERANCES

* MAY BE DELETED WHEN END REINFORCING IS DESIGNED IN ACCORDANCE WITH CHAPTER 5 OF THE SECOND EDITION OF THE PCI DESIGN HANDBOOK

END ZONE DETAIL

CONNECTION DETAILS

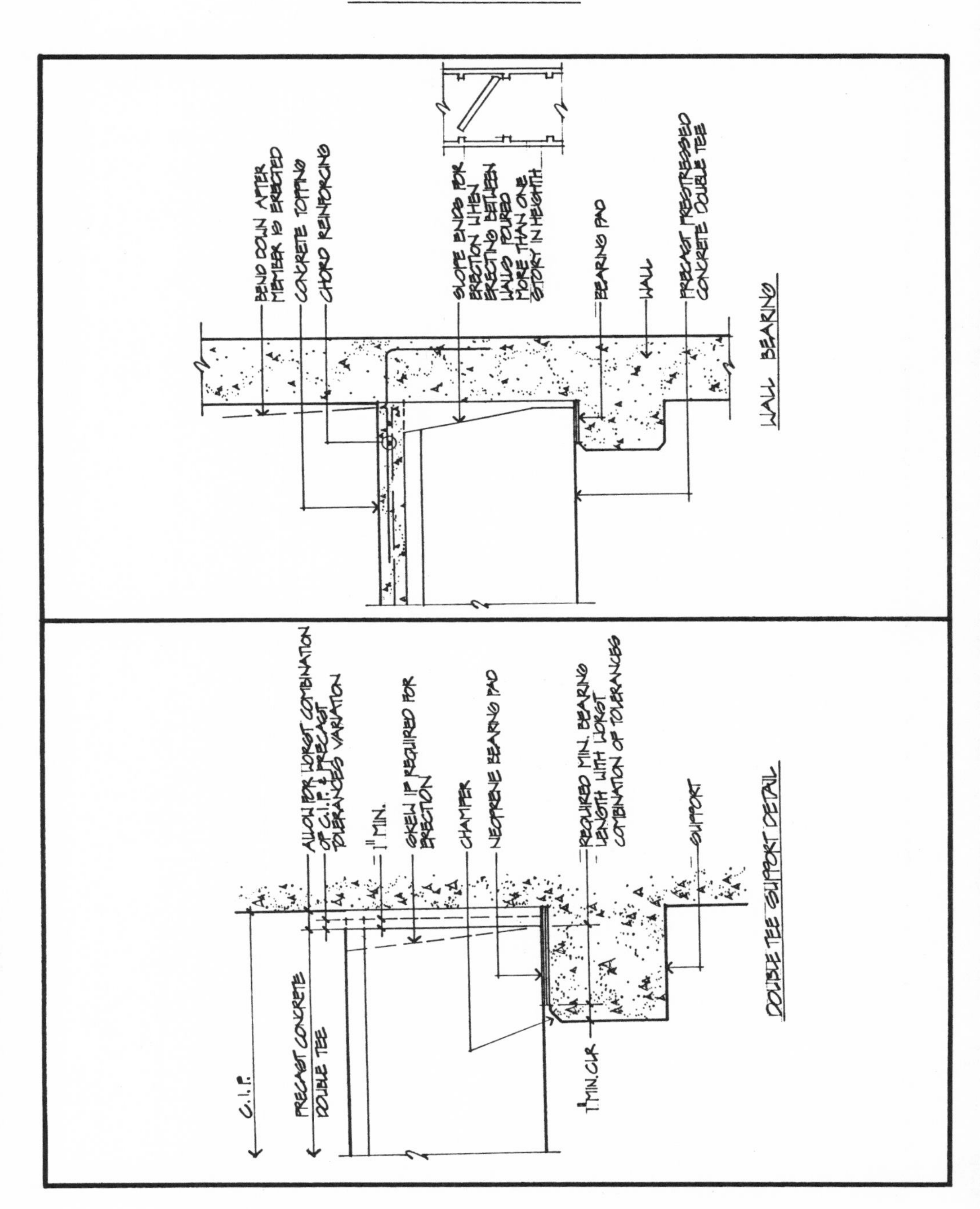

CONNECTION DETAILS

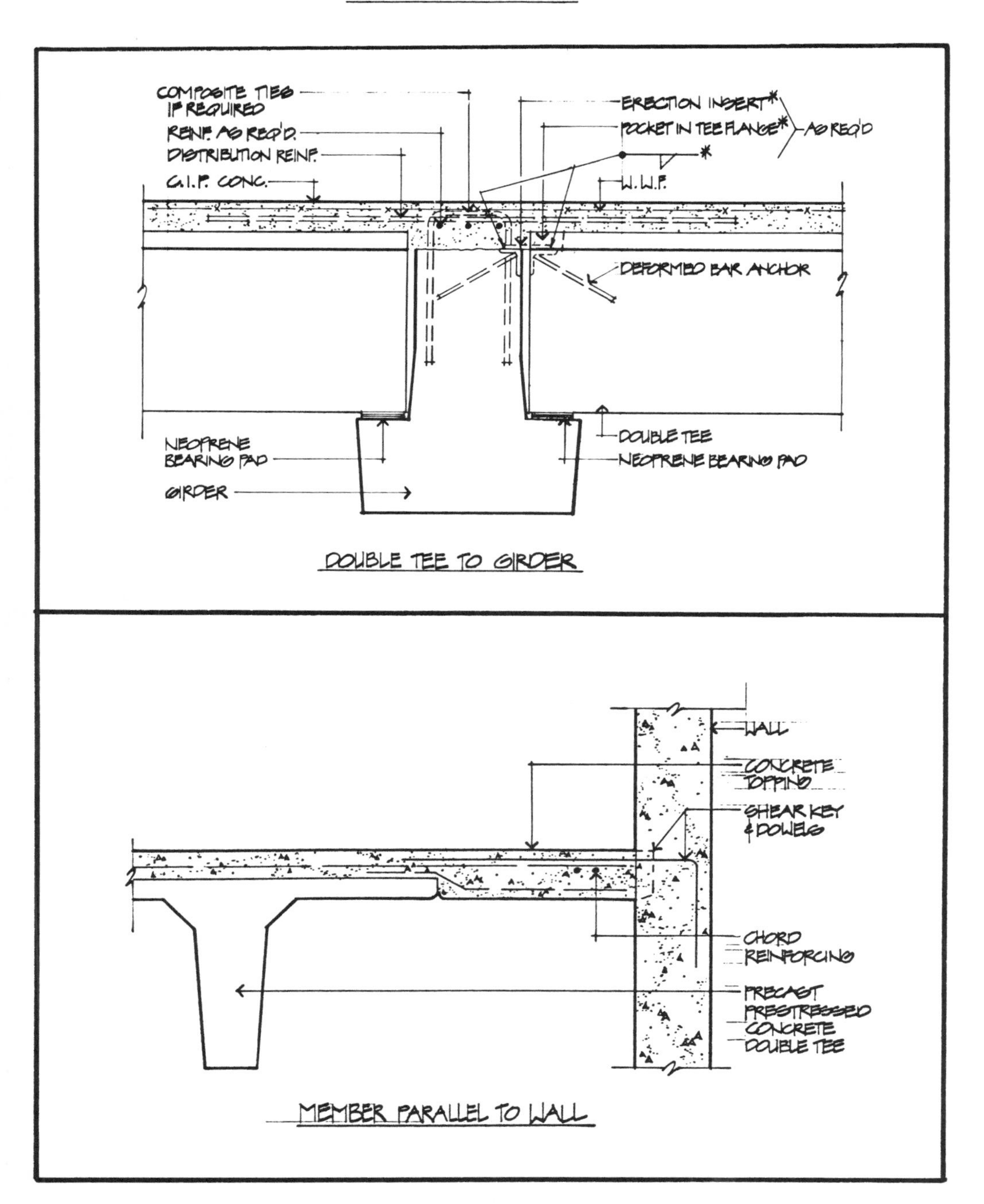

CONNECTION DETAILS

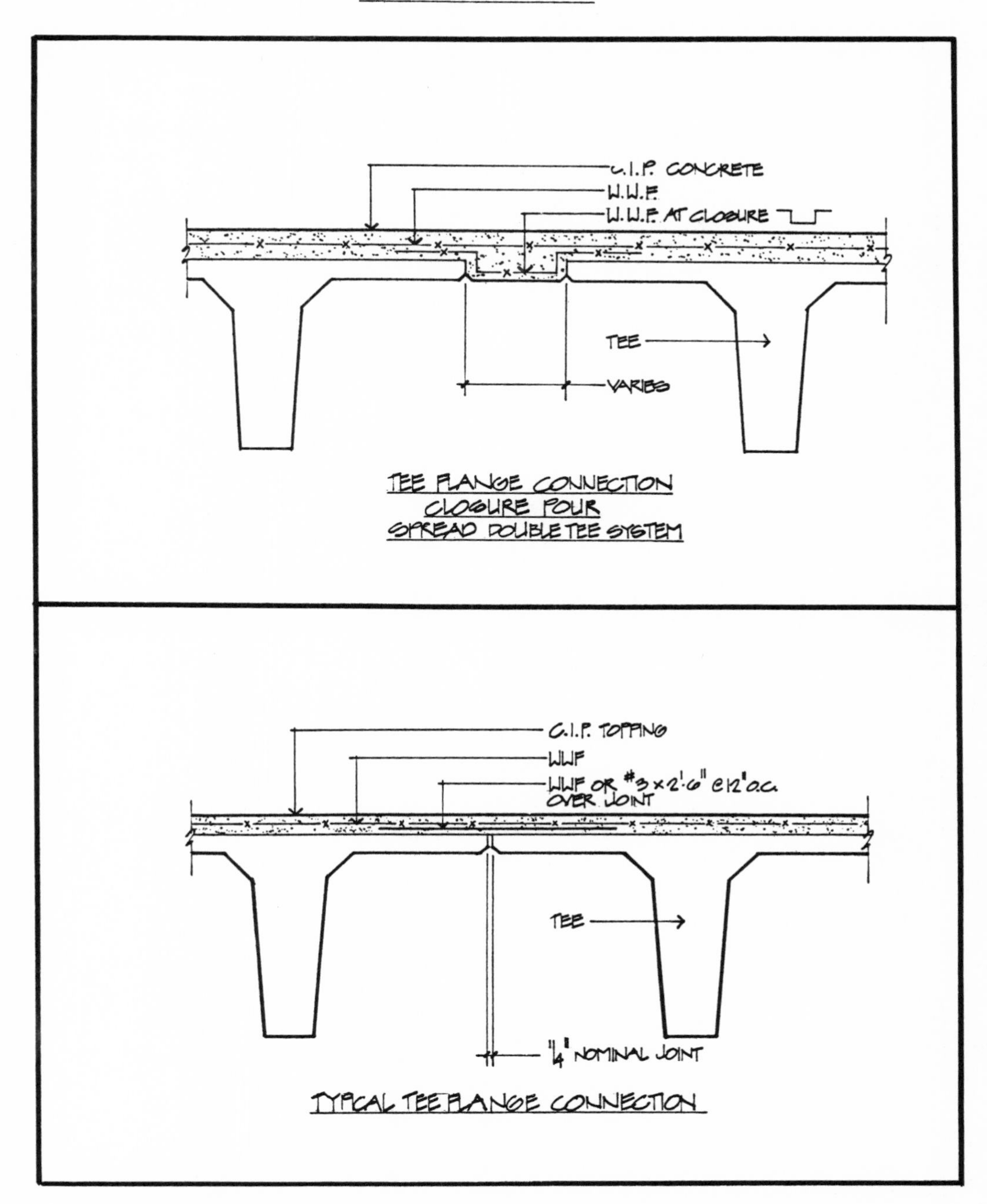

CONNECTION DETAILS

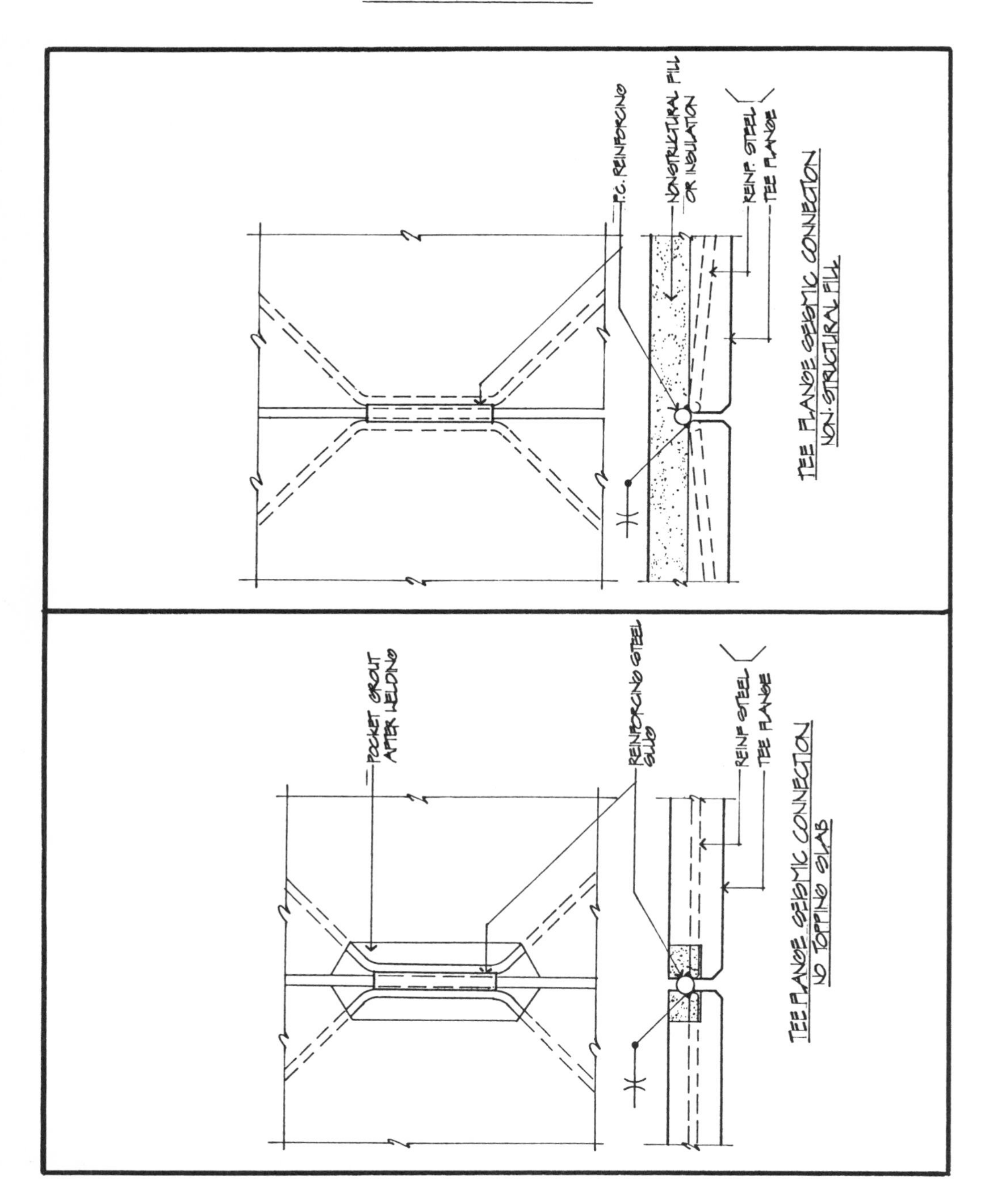

CONNECTION DETAILS

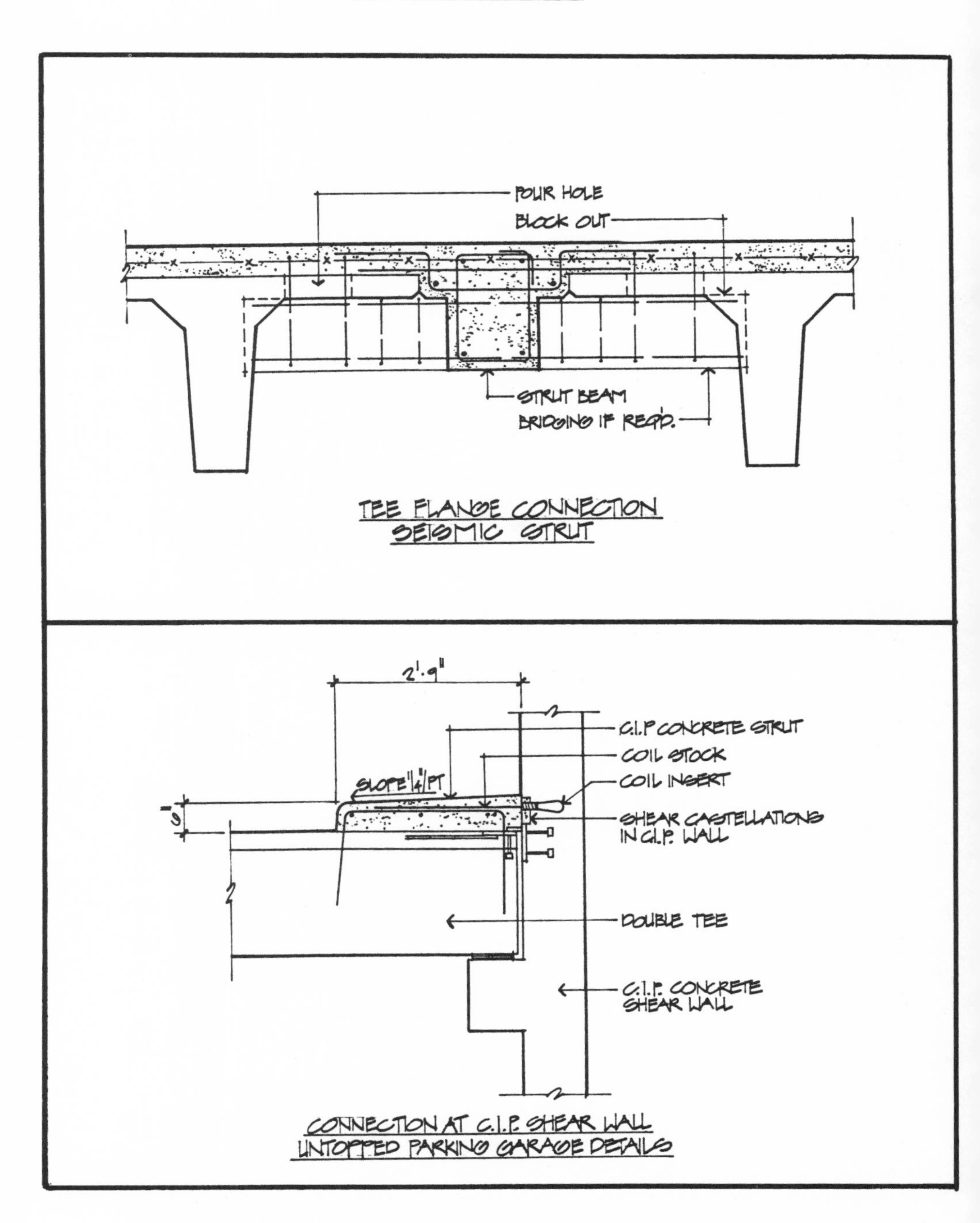

CONNECTION DETAILS

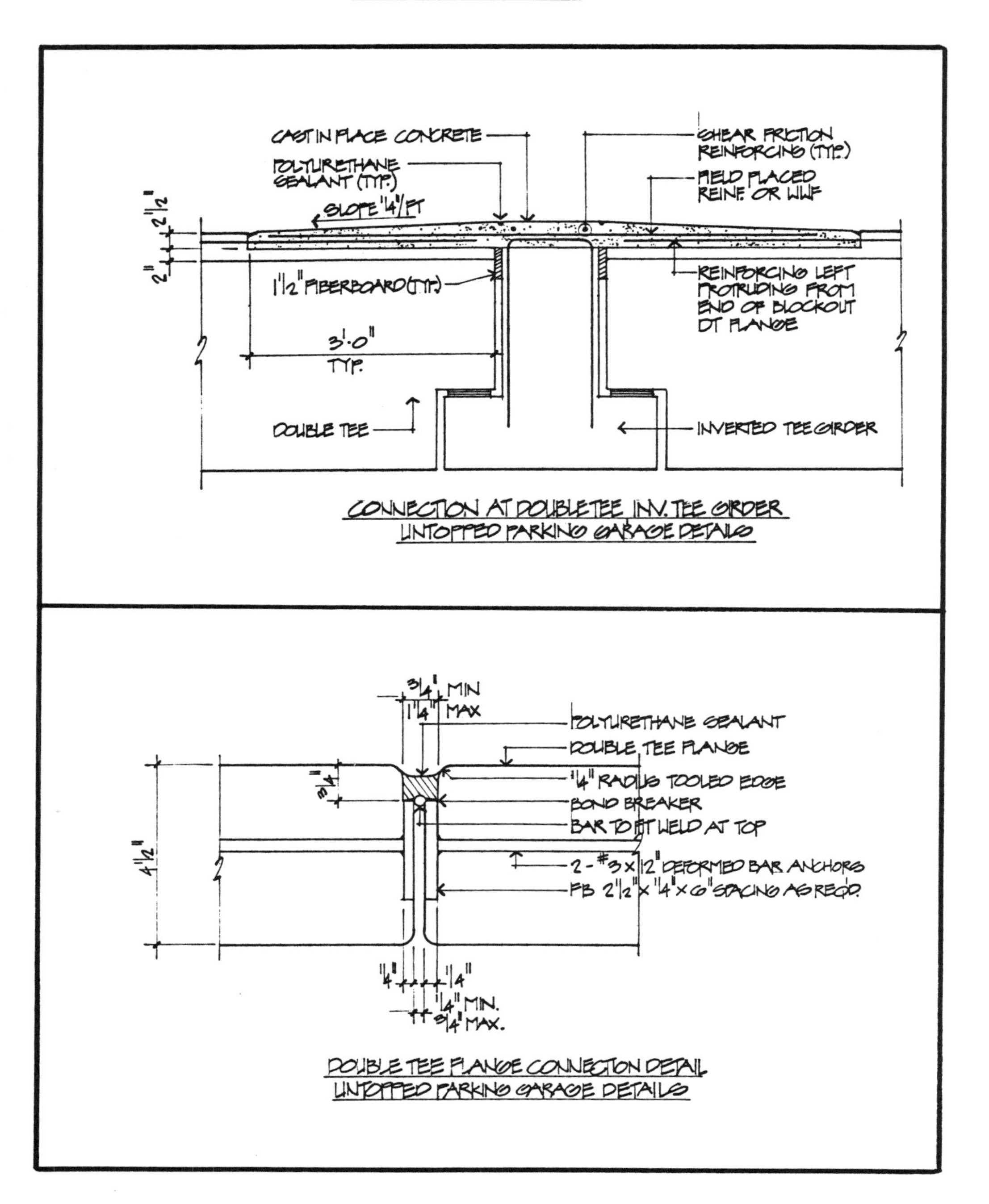

CONNECTION AT DOUBLE TEE INV. TEE GIRDER
UNTOPPED PARKING GARAGE DETAILS

DOUBLE TEE FLANGE CONNECTION DETAIL
UNTOPPED PARKING GARAGE DETAILS

ALVARADO SEWAGE TREATMENT PLANT - UNION CITY

8 foot wide by 24" deep double tee being erected. The specifications called for special paint on the underside of the tee. The painting contractor found it economically feasible to do this work in the Basalt Rock plant while the tees were in storage, eliminating on-site painting and saving time.

PRESTRESSED CONCRETE GIRDERS

THE BROADWAY - LA JOLLA

Prestressed concrete inverted tee girders being erected on 2 story precast concrete columns. The iron worker on the ladder is positioning a neoprene bearing pad on the column corbel in preparation for the next girder. The bottoms of these girders are left "floating" on elastomeric bearing pads so long term creep and shrinkage movements may be accommodated.

Plant cast prestressed concrete girders are normally used to support other prestressed concrete deck members, such as solid slabs, hollow core planks, or double tees, and deliver vertical loads to precast concrete columns or wall systems. They are classified in 3 general categories: rectangular, ledger, and inverted tee. They are made in depths up to 48" deep, although deeper sections can be precast subject to handling and erection limitations. The larger sizes may be voided to reduce weight. Some manufacturers also have forms for "I" shapes which have more efficient section properties than the rectangular shape. Voids, where used, are formed by hollow tubes, cardboard box forms, plywood liners, polystyrene blocks, or by using removable steel mandrels.

Standard sizes and cross-sections, subject to individual manufacturer mold inventories, are as follows:

Rectangular Girder:

from 12" wide by 18" deep
to 36" wide by 48" deep

Ledger Girder:

from 12" wide by 18" deep
to 30" wide by 48" deep

Inverted Tee Girder:

from 12" wide by 18" deep
to 24" wide by 48" deep

Widths and depths within the min-max ranges usually vary in 2" increments.

The widths given for ledger girders and inverted tee girders are the stem width at the top; standard ledge dimensions are 6" wide by 12" deep. Sometimes, girders are made with pockets sized and located to correspond with double tee framing. These are often used in bakeries and food processing plants where it is mandatory that no ledges be present to catch and collect dust. They are also used as parking garage transfer girders where the area between the tee stem and the pocket is subsequently dry packed to gain the additional web width and consequently increased load carrying capacity. Precast prestressed concrete girders are used as framing components in parking structures, shopping malls, department stores, office buildings, commercial buildings, subterranean parking decks, reservoir covers, schools, hospitals - anywhere where precast concrete construction is feasible.

MANUFACTURING

Girders are cast either in permanent type beds where the forms are fixed in place, or in "universal beds," where transportable molds are set up for the specific project being manufactured. The stressing lines are usually 200 to 400' long. Some manufacturers that have pile lines cast girders in these facilities which feature beds 800' long or more. Where universal beds are used with transportable forms, strand depressing capability usually does not exist. Plants with fixed girder forms often have strand hold down anchors cast below the pallet of the girder form as part of the bed. Fixed side forms exhibit a side slope (draft) which facilitates stripping. The draft is usually 3/8" or 1/2" in one foot. Girder forms with moveable sides obviously need not have draft. These forms are also the most efficient from the standpoint of initial costs involved in bed set up for individual project cross sections. Concrete strengths range to 6000 psi for hardrock concrete and 5000 psi for lightweight. 1/2" diameter 270 ksi strand is usually used for prestressing.

GIRDER CASTING BED

Shown here is a 430 foot long girder line in the Rockwin plant in Santa Fe Springs. One side is moveable in 2" increments to permit casting girders in widths up to 24". The side forms shown can produce girders to 3 feet in depth. Side rail extensions increase the maximum girder depth to 4 feet. Note the stressing abutment in the foreground with drillings for 1/2" diameter strand spaced 2" apart each way.

RECTANGULAR GIRDERS

14" x 31" prestressed concrete girders in storage for the Bedford Parking structure in Beverly Hills. Girders are 65 feet long.

LEDGER GIRDERS

Girders in storage in Kabo-Karr plant in Visalia.

INVERTED TEE GIRDERS

Girders awaiting shipment - 64' span girders destined for service in a parking structure.

ERECTION CONSIDERATIONS

In planning projects with large prestressed concrete girders, anticipated job site conditions and erection requirements should be investigated early. In general, any project containing precast concrete components weighing in excess of 10 tons should be analyzed on an individual basis, often by consulting with prestressed concrete manufacturing operations and erection personnel. Another practical consideration that often causes problems in the field is the introduction of torsional stresses in ledger and inverted tee girders as deck members are erected on them. Often, temporary bracing and shoring is required to resist these forces which can sometimes cause extremely large movements in the partially erected building frame. Sometimes it is prudent to provide for temporary bolted or welded connections between tee and beam or beam and column or wall to resist the problems caused by erection torsional forces. These temporary connections can usually be removed once all the members are erected and permanent connections are made and cast-in-place concrete topping and closures are poured.

THE BROADWAY - LA JOLLA

Photograph of completed precast concrete frame consisting of columns, ledger girders, inverted tee girders, and double tees. Double nut connection to column base plate was used to plumb the columns. The base plates were then grouted, giving sufficient rigidity to the two story column to negate any requirement for additional bracing. Top connections were made at the ledger-column juncture to prevent torsional rotation when the tees were erected.

THE BROADWAY - LA JOLLA

Inverted tee girder with roughened top and extended ties to develop horizontal shear with the cast-in-place concrete topping in accordance with UBC-79 Sections 2617 and 2611. Note also the double tees framing into the girder.

DESIGN CONSIDERATIONS

Due to the manufacturing conditions mentioned above, it is usually wise to design girders in straight strands, while allowing precast manufacturers with strand depressing capabilities to re-design the girders to take advantage of the inherent economy in their particular operation. When designing with straight strands, end stresses often become critical, and how this situation is handled in design is demonstrated in the following design example. Preliminary load tables for non-composite girders are contained in the second edition of the PCI Design Handbook, pp 2-50 through 2-52. However, it is often just as convenient to use the approximate formulae given in the beginning of the design example on the following page. Another option often overlooked in design, especially in large projects, is the material savings gained by utilizing temporary midpoint or thirdpoint shoring, especially for longer spans. This allows the bulk of the load to be carried by the composite section, as opposed to unshored designs where the basic section uses the major portion of its capacity in carrying supported dead loads from precast concrete deck members and cast-in-place concrete topping.

Introducing continuity at intermediate supports results in increased positive moment capacity; the resulting trade-off in negative reinforcing however often negates any economic gains in the prestressed concrete girder design. One last item that is often overlooked in design is the use of transformed girder section properties in calculating deflections and verifying service load performance. Deep, heavily prestressed girders with top reinforcing both in the basic section and the composite pour exhibit transformed section properties considerably higher than those indicated by the bare concrete cross sections.

GIRDER DESIGN

DESIGN AN INVERTED TEE GIRDER TO SUPPORT THE DOUBLE TEES IN THE PARKING STRUCTURE REFERRED TO IN THE EXAMPLE ON PAGE 380.

SELECT A GIRDER SPAN COMPATIBLE WITH THE PARKING LAYOUT AND THE DOUBLE TEE WIDTH

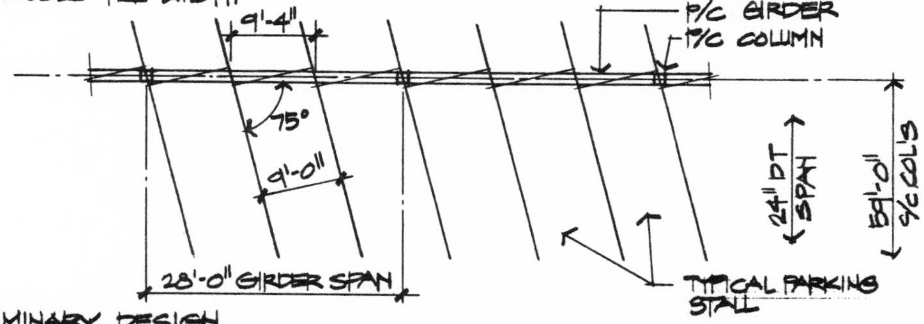

P/C GIRDER
P/C COLUMN
9'-4"
75°
9'-0"
24" DT SPAN
59'-0" C/C COL'S
28'-0" GIRDER SPAN
TYPICAL PARKING STALL

PRELIMINARY DESIGN

TRY A 28'-0" C/C COLUMN SPACING (APPROX. 27.0' GIRDER SPAN)

APPROX. LOADS TO BEAM

REDUCED LL	0.6 × 0.050 × 59	= 1.77 K/FT.
DT WT	0.060 × 58	= 3.48
TOPPING WT	0.031 × 59	= 1.83
BEAM WT (EST.)		= 0.60
Σw		= 7.68 K/FT

WHEN PRELIMINARY LOAD SPAN TABLES ARE NOT AVAILABLE, THE FOLLOWING FORMULAE ARE USEFUL FOR SELECTING TRIAL SIZES:

NON COMPOSITE GIRDER

$$h = \sqrt{\frac{2.5 M_T}{b}}$$

COMPOSITE GIRDER

$$h = \sqrt{\frac{1.7 M_T}{b}}$$

WHERE: h = PRECAST BEAM DEPTH, INCHES
M_T = TOTAL MOMENT, IN-KIPS
b = BEAM STEM WIDTH, INCHES

FOR $b = 12''$, $h = \sqrt{\frac{(1.7)(1.5)(7.68)(27)^2}{12}} = 34.5''$

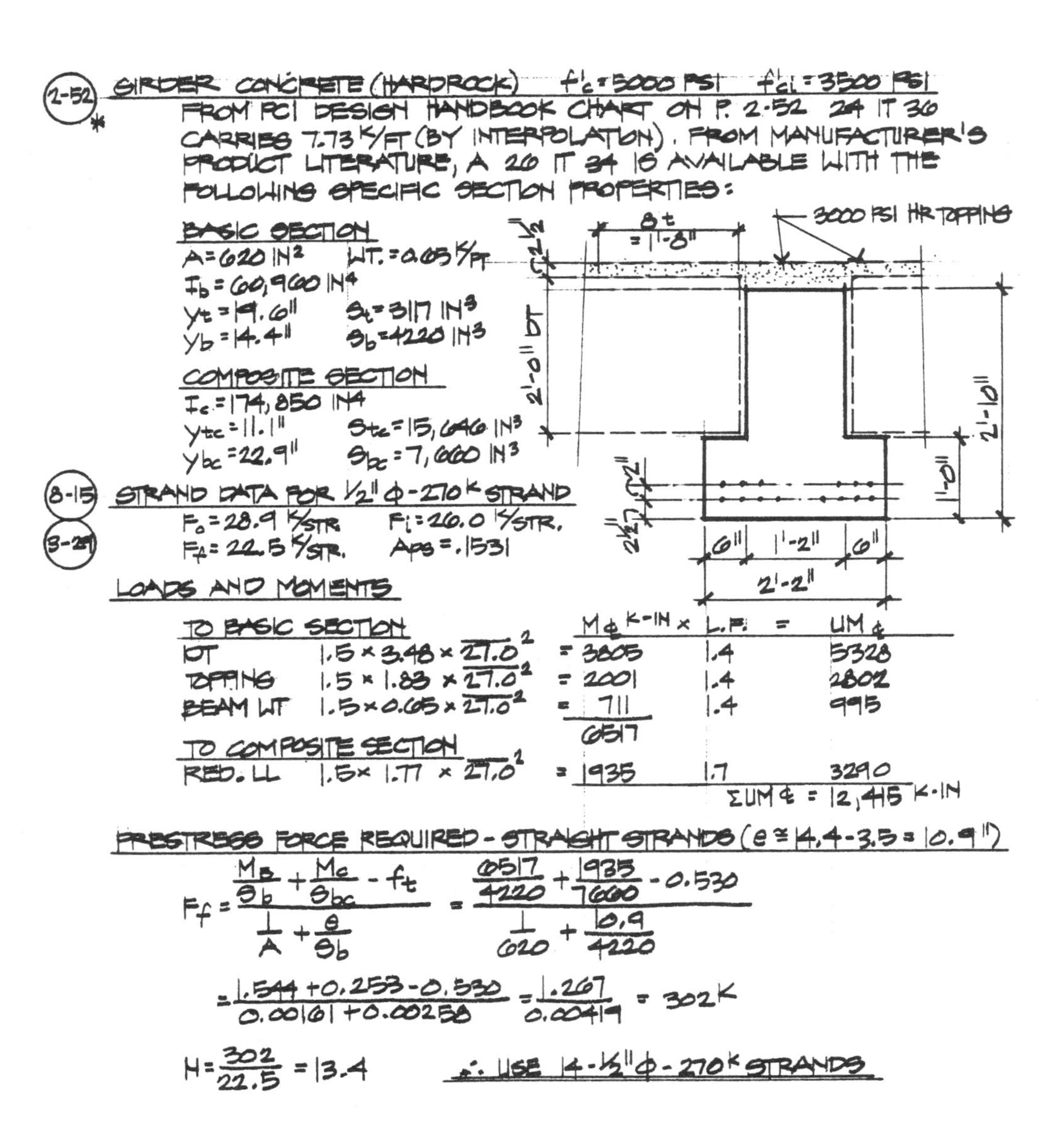

(2-52)* GIRDER CONCRETE (HARDROCK) $f'_c = 5000$ PSI $f'_{ci} = 3500$ PSI

FROM PCI DESIGN HANDBOOK CHART ON P. 2-52 24 IT 36 CARRIES 7.73 K/FT (BY INTERPOLATION). FROM MANUFACTURER'S PRODUCT LITERATURE, A 26 IT 34 IS AVAILABLE WITH THE FOLLOWING SPECIFIC SECTION PROPERTIES:

BASIC SECTION

$A = 620$ IN² WT. = 0.65 K/FT

$I_b = 60,960$ IN⁴

$y_t = 19.6''$ $S_t = 3117$ IN³

$y_b = 14.4''$ $S_b = 4220$ IN³

COMPOSITE SECTION

$I_c = 174,850$ IN⁴

$y_{tc} = 11.1''$ $S_{tc} = 15,646$ IN³

$y_{bc} = 22.9''$ $S_{bc} = 7,660$ IN³

(8-15) (3-29) STRAND DATA FOR ½" φ - 270 K STRAND

$F_o = 28.9$ K/STR. $F_i = 26.0$ K/STR.

$F_f = 22.5$ K/STR. $A_{ps} = .1531$

LOADS AND MOMENTS

TO BASIC SECTION		$M_{℄}$ K-IN	× L.F. =	UM ℄
DT	$1.5 \times 3.48 \times \overline{27.0}^2$	= 3805	1.4	5328
TOPPING	$1.5 \times 1.83 \times \overline{27.0}^2$	= 2001	1.4	2802
BEAM WT	$1.5 \times 0.65 \times \overline{27.0}^2$	= 711	1.4	995
		6517		
TO COMPOSITE SECTION				
RED. LL	$1.5 \times 1.77 \times \overline{27.0}^2$	= 1935	1.7	3290
				ΣUM ℄ = 12,415 K-IN

PRESTRESS FORCE REQUIRED - STRAIGHT STRANDS ($e \cong 14.4 - 3.5 = 10.9''$)

$$F_f = \frac{\frac{M_B}{S_b} + \frac{M_C}{S_{bc}} - f_t}{\frac{1}{A} + \frac{e}{S_b}} = \frac{\frac{6517}{4220} + \frac{1935}{7660} - 0.530}{\frac{1}{620} + \frac{10.9}{4220}}$$

$$= \frac{1.544 + 0.253 - 0.530}{0.00161 + 0.00258} = \frac{1.267}{0.00419} = 302^K$$

$$N = \frac{302}{22.5} = 13.4$$

∴ USE 14 - ½" φ - 270 K STRANDS

* ◯ PAGE REFERENCE IN 2ND EDITION OF PCI DESIGN HANDBOOK

(3-10) CHECK STRESSES $F_i = (14)(26.0) = 364^K$

(3-14) END AT RELEASE

$f^{TOP} = -\frac{364}{620} + \frac{(364)(10.9)}{3117} = -0.587 + 1.272 = +0.685 \text{ KSI} > 6\sqrt{f'_{ci}}$

$f_{BOT} = -\frac{364}{620} + \frac{(364)(10.9)}{4220} = -0.587 - 0.940 = -1.527 \text{ KSI} < 0.6 f'_{ci} \text{ O.K.}$

(3-56) TOP END STEEL IS REQUIRED TO RESIST FULL TENSION FORCE

+0.685; T = 50.3K; 10.5" x; 34.0"; 23.5" 34-x; +; −; −1.527

FROM FIG. 3.9.8
bh = 476
$f_b = 1.527$
READ T = 48 K

$\frac{x}{0.685} = \frac{34}{0.685 + 1.527}$ $\therefore x = 10.5''$

$T = (\tfrac{1}{2})(0.685)(10.5)(14) = 50.3K$

$A_s = \frac{50.3}{1.33 \times 24} = 1.58 \text{ IN}^2$

USE 2-#8 IN TOP END

$A_s = 1.58 \text{ IN}^2$

DETERMINE WHERE END TOP STEEL IS NO LONGER REQUIRED

∴ WHERE BEAM WEIGHT MOMENT REDUCES TOP TENSION TO $6\sqrt{f'_{ci}}$

COMPRESSION REQ'D FROM BEAM WT.

$= 0.685 - 6\sqrt{3500} = 0.685 - 0.355 = 0.330 \text{ K/IN}^2$

$\frac{M_{BM}}{S_t} = 0.330$

0.65 K/FT; 9K; x; 28'; 9K

@ ℄ SPAN

$\frac{711}{3117} = 0.228$ ∴ RUN TOP STEEL CONTINUOUSLY SINCE TOP TENSION IS ALWAYS $> 6\sqrt{f'_{ci}}$

CHECK BOTTOM TENSION UNDER ALL LOADS $F_f = (14)(22.5) = 315$

$f^{℄}_{BOT} = -\frac{315}{620} - \frac{(315)(10.9)}{4220} + \frac{6517}{4220} + \frac{1935}{7660}$

$= -0.508 - 0.814 + 1.544 + 0.253 = +0.475 \text{ K/IN}^2 < 7.5\sqrt{f'_c}$

(3-53) CHECK ULTIMATE STRENGTH (FIG. 3.9.5)

(3-5) $f_{se} = 147 \text{ KSI}$ $A_{ps} = (14)(.1531) = 2.143 \text{ IN}^2$

$c\bar{\omega}_p = c\frac{A_{ps}}{bd} \times \frac{f_{pu}}{f'_c} = (1.0)\left(\frac{2.143}{54 \times 35}\right)\left(\frac{270}{3}\right) = 0.10$

FROM CHART, $\frac{f_{ps}}{f_{pu}} = 0.97$ $\therefore f_{ps} = (0.97)(270) = 261.9 \text{ KSI}$

$$\frac{a}{2} = \frac{A_{ps} f_{ps}}{1.7 f'_c b} = \frac{(2.143)(261.9)}{(1.7)(3)(54)} = 2.04''$$

(3-9) ASSUME TOPPING AND TEE FLANGE ACT COMPOSITELY SO COMPRESSION BLOCK DEPTH "a" IS WITHIN FLANGE

$$\therefore M_u = \phi A_{ps} f_{ps}(d - a/2) = (0.90)(2.143)(261.9)(35.0 - 2.04)$$

$$= 16,649^{K\text{-}IN} > UM = 12,415^{K\text{-}IN} \quad O.K.$$

(3-21) DETERMINE TIE REQUIREMENTS TO TRANSFER HORIZONTAL SHEAR FORCE BETWEEN PRECAST AND CAST IN PLACE CONCRETE

$$T = A_{ps} f_{ps} = (2.143)(261.9) = 561^K \qquad l_{vh} = \frac{27 \times 12}{2} = 162'' \quad f_y = 40 \text{ KSI}$$

$$C = (0.85) f'_{cc} A_{TOP} = (0.85)(3)(2.5 \times 40 + 4.5 \times 14) = 416^K = F_h$$

MAXIMUM ALLOWABLE W/O TIES

$$= 40\,\phi\, b_h\, l_{vh} = (40)(0.85)(14)\left(\frac{162}{1000}\right) = 77^K < F_h$$

$\therefore$ TIES ARE REQ'D.

$$\mu_e = \frac{1000\lambda^2 b_v l_{vh}}{F_h} = \frac{(1000)(1.0)^2(14)(162)}{416,000} = 5.45$$

$$A_{cs} = \frac{F_h}{\phi \mu_e f_y} = \frac{416}{(0.85)(5.45)(40)} = 2.25 \text{ IN}^2$$

MINIMUM REQUIREMENTS

1. $A_{cs\,(MIN)} = \dfrac{120\, b_v l_{vh}}{f_y} = \dfrac{(120)(14)(162)}{40,000} = 6.8 \text{ IN}^2$

   OR $1.33 \times 2.25 = 2.99 \text{ IN}^2$ (CONTROLS)

2. $A_{cs\,(MIN)} = \dfrac{50\, b_v l_{vh}}{f_y} = \dfrac{(50)(14)(162)}{40,000} = 2.84 \text{ IN}^2$

USE #4 TIES $\qquad S = \dfrac{(13.5)(12)(0.40)}{2.99} = 21.7''$ $\quad$ USE 18'' MAX

(MAXIMUM SPACING $4 \times 4.5 = 18''$

CALCULATE HORIZONTAL SHEAR STEEL REQUIREMENTS USING UBC-79

$$v_{dh} = \frac{V_u}{\phi b_v d} = \frac{153}{(0.85)(14)(35)} = 0.367 \text{ KSI} > 0.350 \quad \text{(SEC. 2617)}$$

$$A_{vf} = \frac{V_u}{\phi f_y \mu} = \frac{153}{(0.85)(40)(1.0)} = 4.50 \text{ IN}^2 \quad \text{(SEC. 2611-P)}$$

$$S = \frac{162 \times 0.40}{4.50} = 14.4'' \qquad \text{USE } 15''$$

(3-59) DETERMINE WEB REINFORCING REQUIREMENTS $\qquad b_w d = (14)(35) = 490 \text{ IN}^2$

$$v_{u\,\text{SUPPORT}} = \frac{UV}{\phi b d} = \frac{153}{(0.85)(490)} = 0.367 \text{ KSI}$$

$$L/d = \frac{27 \times 12}{35} = 9.5'' \qquad A_{ps} = 2.143 \text{ IN}^2$$

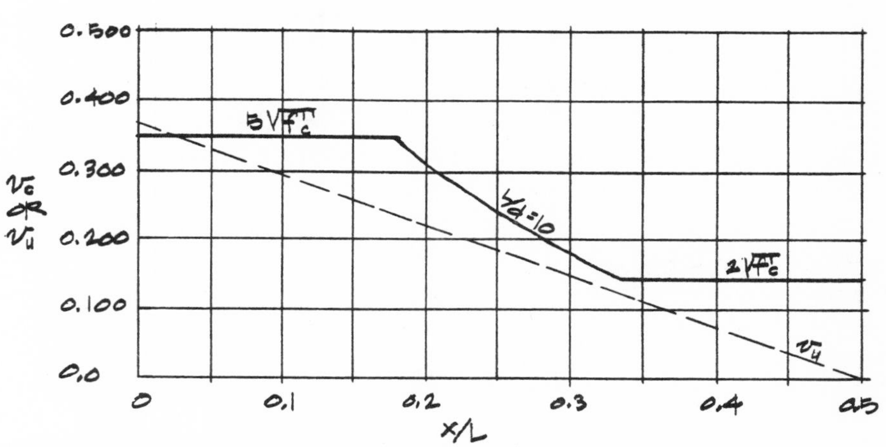

BY INSPECTION, MINIMUM SHEAR REINFORCEMENT SUFFICES.

(3-64) FROM CHART $A_{V\,MIN} = 0.10\ IN^2/FT.$

USING #4 STIRRUPS,

$$S_{MAX} = \frac{A_v f_y}{50\, b_w} = \frac{(0.40)(40,000)}{(50)(14)} = 22.9''$$

USE #4 @ 15" o/c ; EXTEND INTO TOPPING

(3-66) CHECK CAMBER AND DEFLECTION

(8-14)

CAMBER $\Delta_{P/S\,i} = \frac{F_i e l^2}{8EI_b} = \frac{(364)(10.9)(27)^2(144)}{(8)(3600)(60,960)} = +0.24''$

BM. WT. $\Delta_{BM\,i} = \frac{5wl^4}{384EI} = \frac{(5)(0.65)(27)^4(1728)}{(384)(3600)(60,960)} = -0.04''$

DT WT. $\Delta_{DT\,i} = \frac{3.48}{0.65} \times 0.035 = -0.19''$

TOPPING $\Delta_{TPG\,i} = \frac{1.83}{0.65} \times 0.035 = -0.10''$

FINAL CONDITION	Δ_i	×	C.F.	= Δ_{LT}
CAMBER	+0.24	×	2.2	= +0.53
DT + BM	-0.23	×	2.4	= -0.55
TOPPING	-0.10	×	2.3	= -0.23
Σ LONG TERM				-0.23" ↓

$$\Delta_{LL} = \frac{(5)(1.77)(27)^4(1728)}{(384)(4300)(174,850)} = 0.03''$$

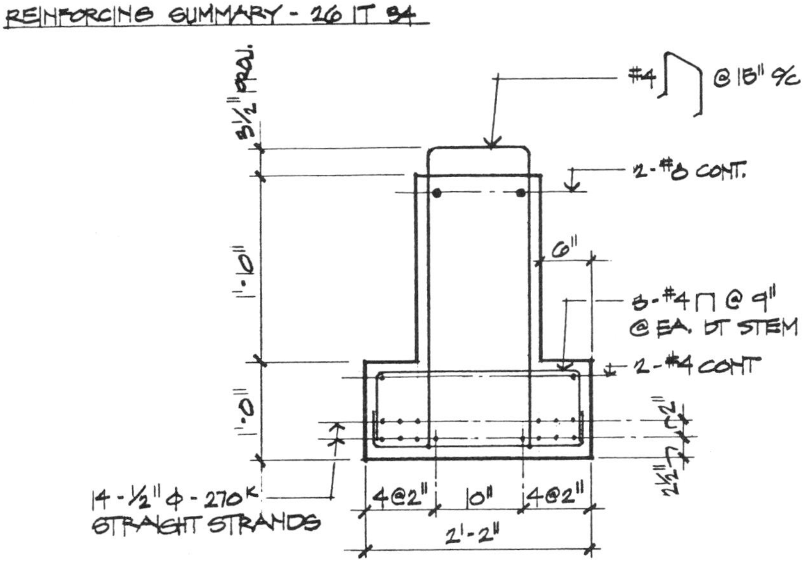

REINFORCING SUMMARY - 26 IT 34

ALVARADO SEWAGE TREATMENT PLANT - UNION CITY

Inverted tee girders supported on cast-in-place concrete bents and supporting double tees up to 70 feet long. Notice the clean lines and close dimensional control afforded by the plant cast precast concrete members.

Special attention should be paid to the design and detailing of corbels and ledges, due to the high stress concentrations at these critical interfaces. Floating neoprene bearings under girders assure relief of creep and shrinkage movement stresses.

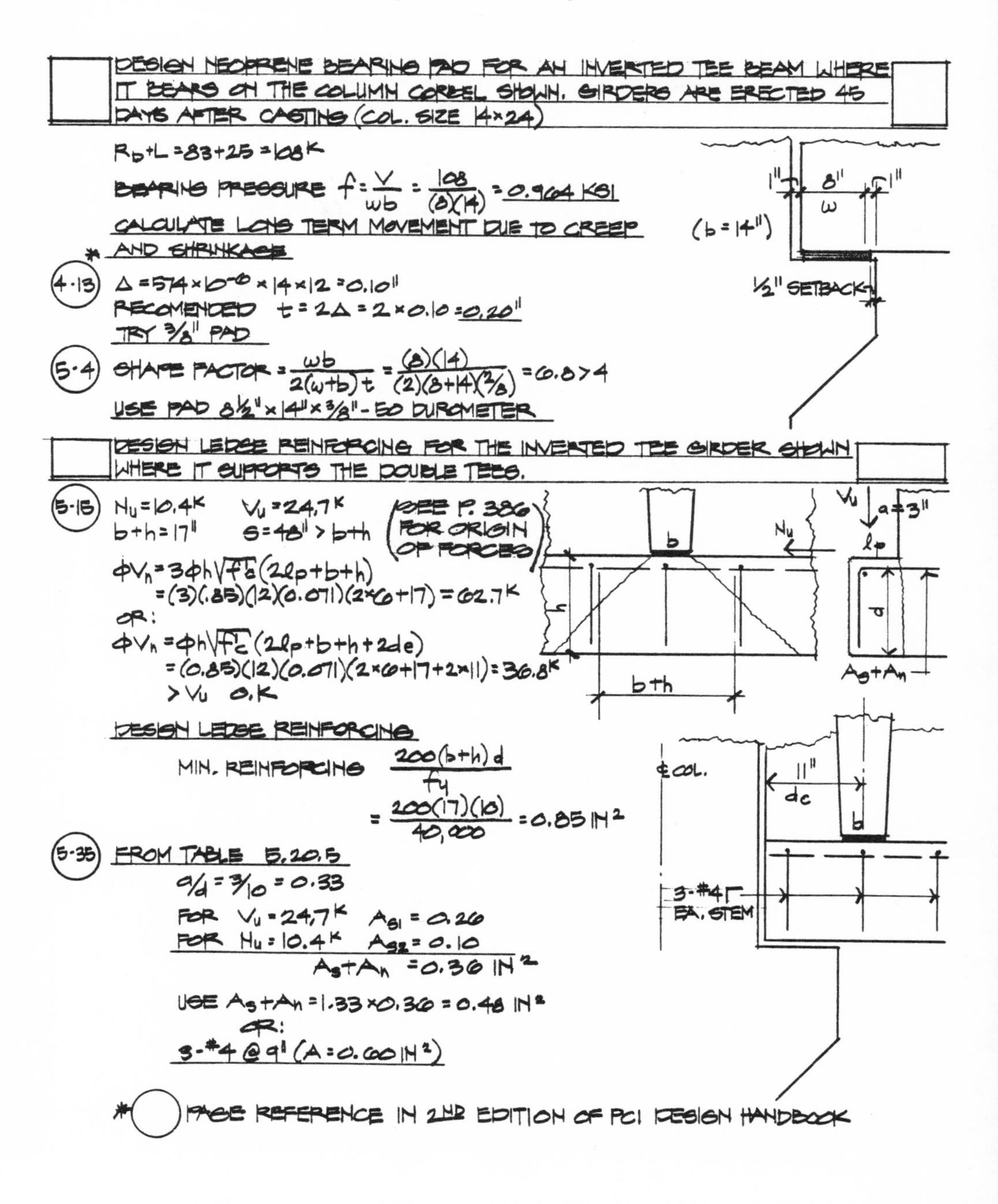

CONNECTION DETAILS

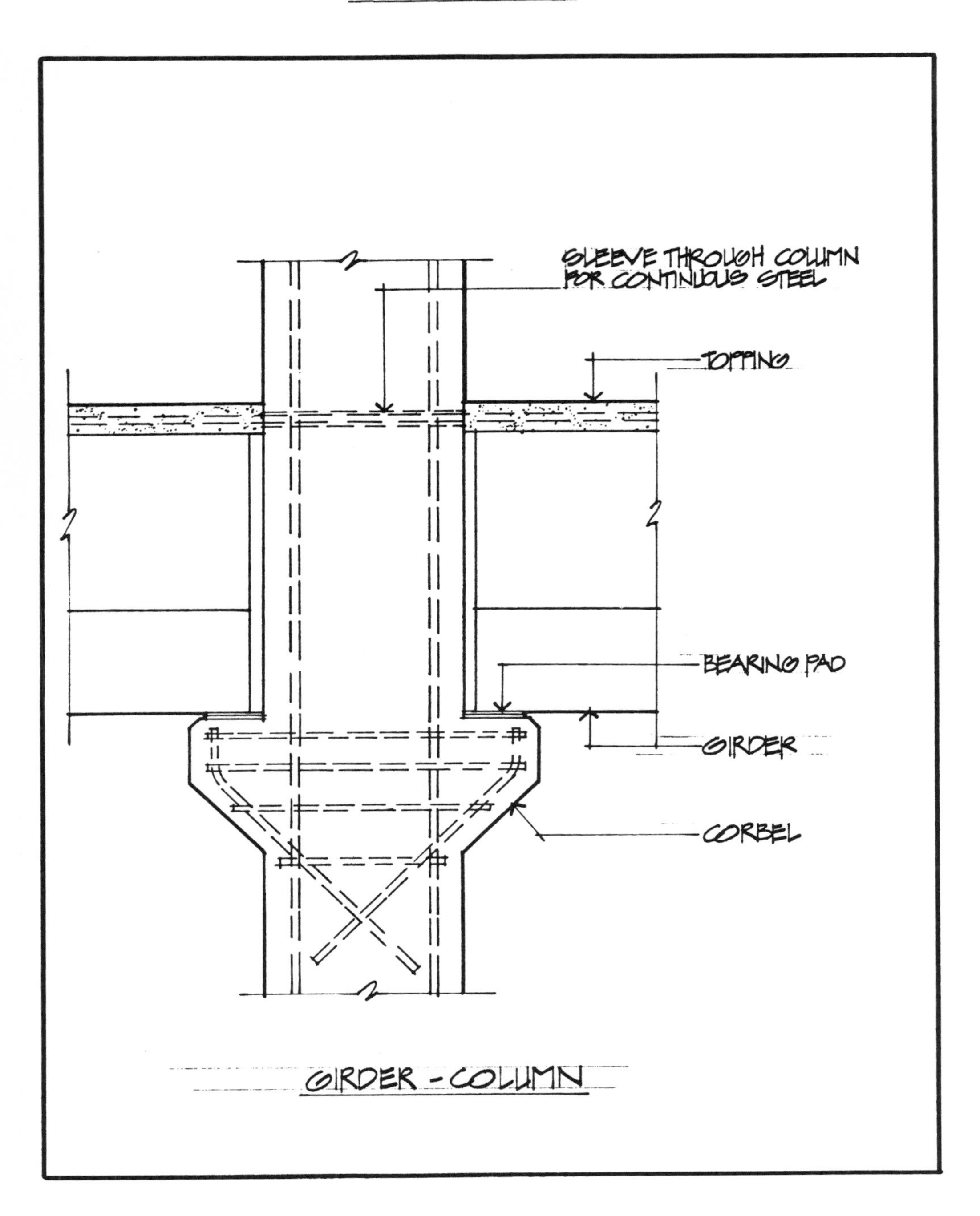

GIRDER - COLUMN

CONNECTION DETAILS

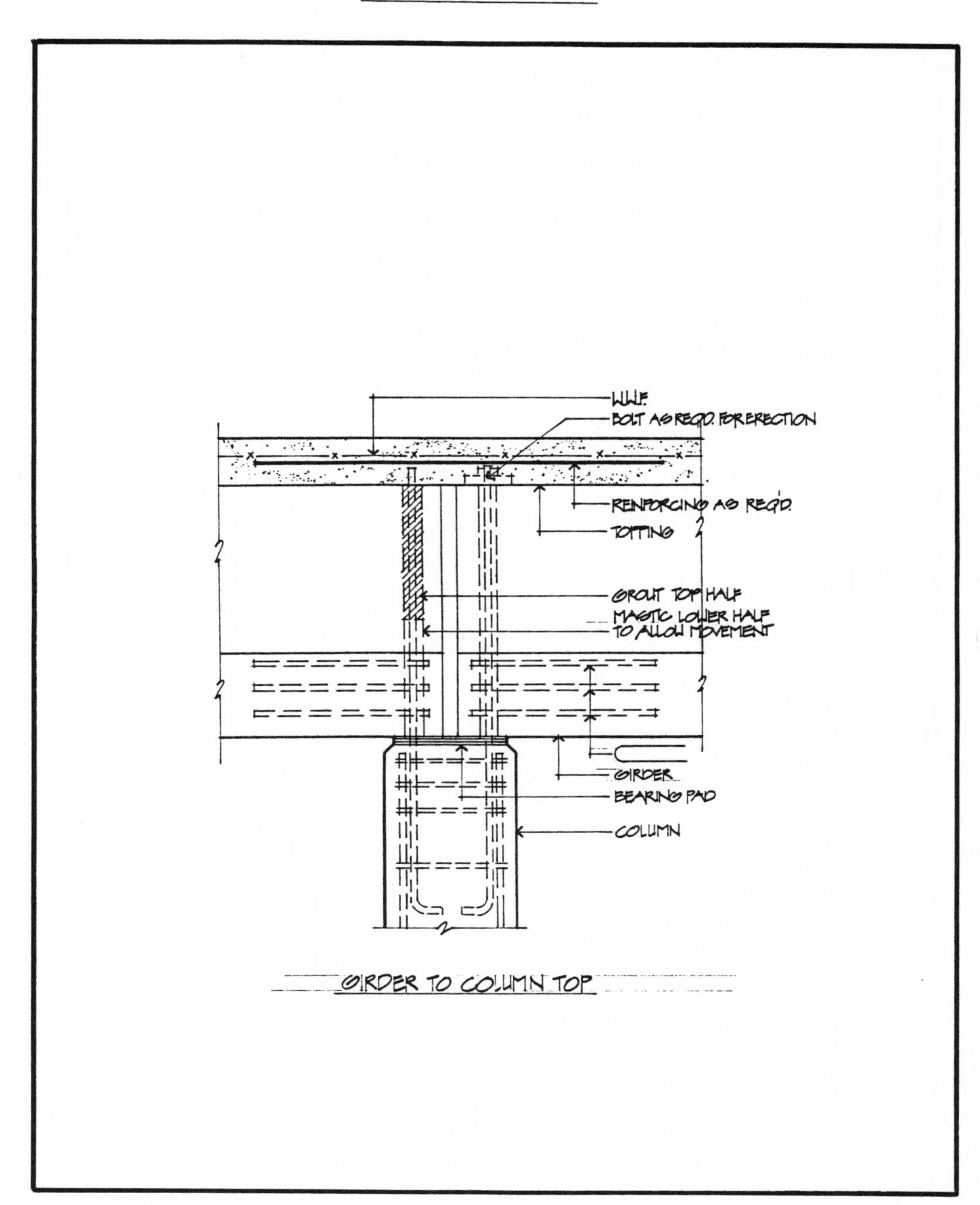

GIRDER TO COLUMN TOP

CONNECTION DETAILS

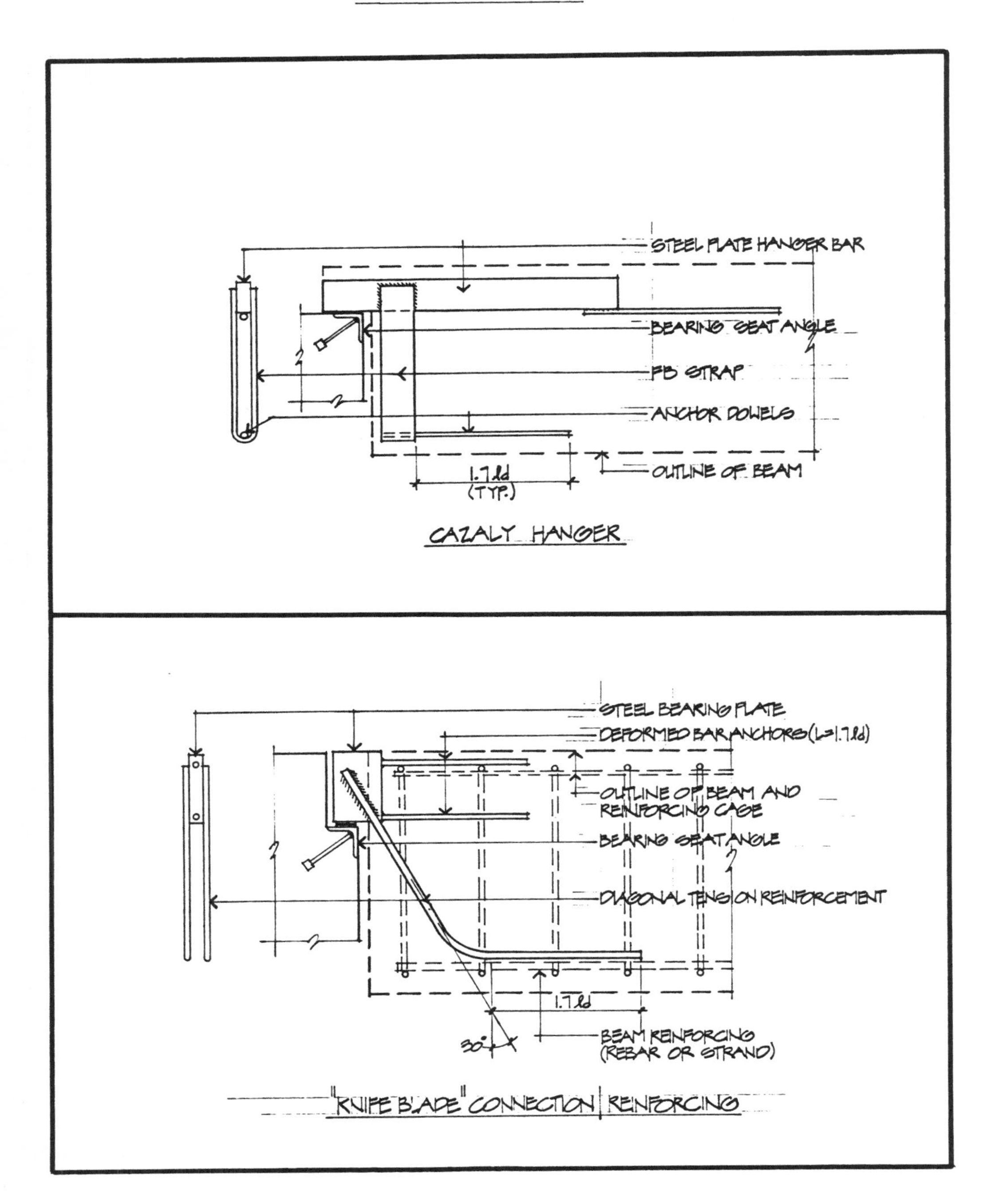

CONNECTION DETAILS

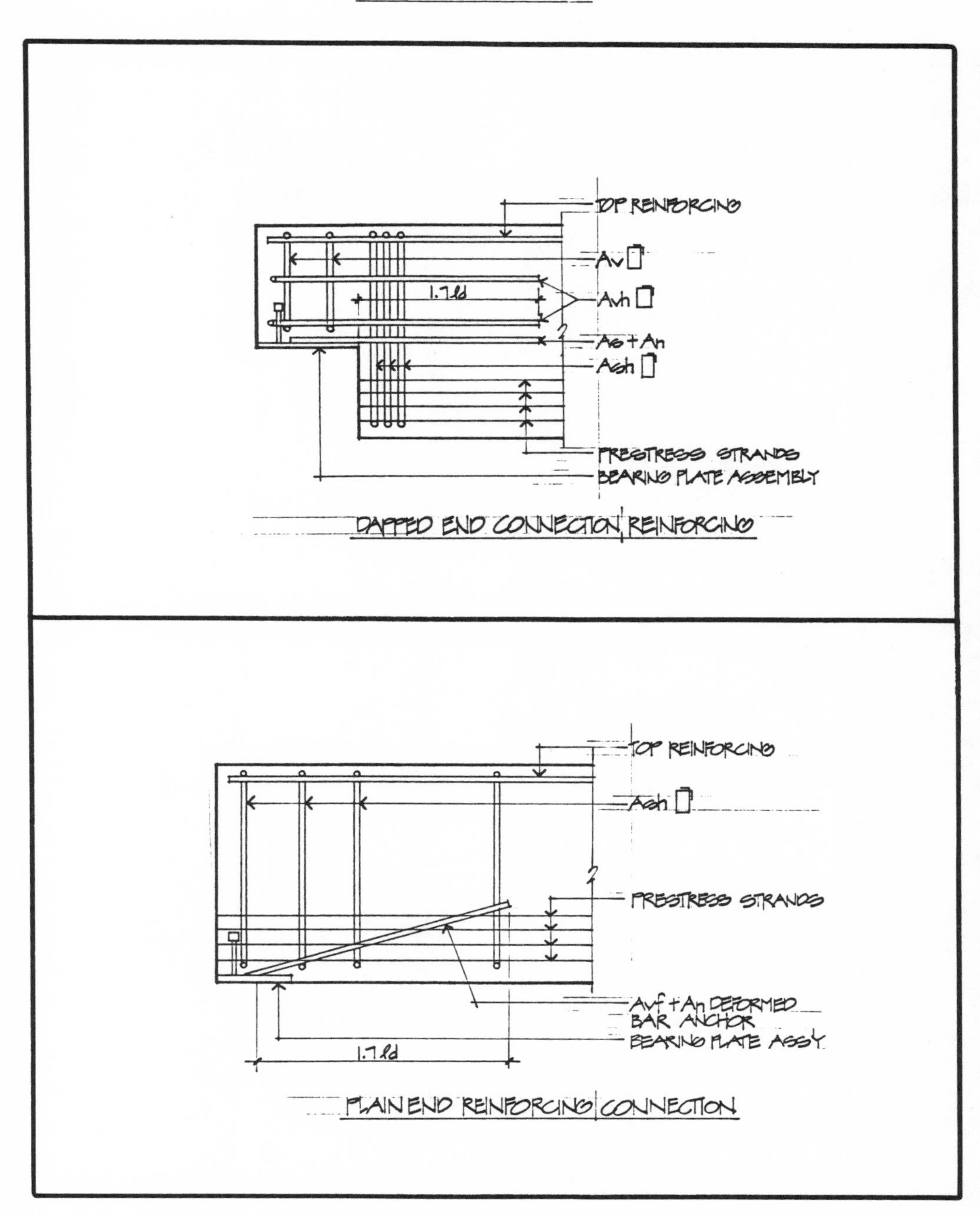

PRECAST CONCRETE COLUMNS

CLASSROOM BUILDING - UNIVERSITY OF CALIFORNIA AT DAVIS

87 foot long prestressed concrete columns form the vertical load carrying frame for this 9 story building. Other precast components are exposed aggregate wall panel units, precast channel floor units and single tee roof members seen cantilevering out above.

Precast concrete columns may be made in many different cross-sections, but square or rectangular cross-sections are standard because precast concrete manufacturers have these types of forms available in their inventories. Standard column sizes range from a minimum of 10" by 10" square to a maximum of 24" by 24" square. Rectangular columns are available consistent with the size of pallet, or soffit form standard in a particular manufacturer's mold inventory. For larger quantities of the same type of column, say over 1,000 lineal feet of product on a given project, the selection of available mold cross-section is not as critical as for smaller quantities where standard column form sizes should definitely be ascertained in the planning stages of the structure. Precast concrete columns can be designed in lengths consistent with handling requirements at the plant and the jobsite. Shipping will usually limit the maximum length of the column since only two points of support are normally feasible. However, special rigs are available with double pivot supports that will permit handling long pieces. These rigs are used to haul piling in excess of 100 feet in length. Splices may be used to cut down the length of individual pieces to be handled, yet satisfy final design situations in multi-story structures. Another way to keep handling stresses within allowable limits is by prestressing. This will allow transportation of longer pieces yet reduce tension under lateral loads induced by trucking as well as by service loads in the completed structure.

UNION BANK - OCEANGATE BUILDING - LONG BEACH

Precast concrete columns form the exterior architecture as well as performing a load bearing function in this structure. In addition, the beam-column connection was made using details conforming to code provisions for cast-in-place concrete moment resisting ductile frames. The column is designed and detailed to conform to these code requirements.

PRODUCTION

Standard column cross sectional dimensions are determined by the location of the standard corbel in the side forms. Standard column heights (measured in the corbel width dimension) are 12", 16", 20" and 24". Standard column widths (plan dimension of the mold soffit, or pallet) are from 10" to 24" varying in 2" increments. To make a corbel on the third side, a box is built across the top of the mold, and the corbel is cast at the same time the balance of the column is cast. For corbels on the fourth side of the column (bottom side as cast) a second cast is normally performed around a steel bracket welded to an insert plate cast in the bottom of the column. In performing a second cast operation in a precast plant, the primary cast shape, in this case a precast column, is stripped, moved to a finishing area, and turned over 180° so that the bottom side as cast is then on top. Then, a form is placed around the corbel location, and the corbel on the fourth side is cast. For very large projects, it is more economical to cut a hole in the soffit liner, and introduce a bottom form for the corbel on the fourth side of the column, so that the entire member is cast in one operation, thereby eliminating double handling in the plant.

PRECAST COLUMN PRODUCTION

Parking garage columns in storage at Kabo-Karr plant in Visalia. Note corbels on 3 sides, and keyed base detail for socketed connection in footing in lieu of customary steel base plate design.

DESIGN

The state of the art of the use of precast and prestressed concrete columns is such that these elements are normally only used as vertical load resisting elements in buildings, unless they are built into part of the lateral force resisting system of the building, such as a boundary element at the end of a cast-in-place concrete shear wall. As such, due consideration should be given to the column's ability to accomodate seismic drift in the connections at the base and other points where it is tied into the structure. Moreover, the column should not be inadvertantly built into the structure in such a manner that it contributes stiffness not accounted for in the lateral analysis of the building resulting in undesirable cracking, or failure. The proper use of bolted base plate connections (as opposed to socketed type footings) and free bottom neoprene bearings at beam/corbel interfaces will usually allow sufficient drift movements to be accommodated without distress, especially in relatively stiff shear wall buildings.

The 1979 edition of the Uniform Building Code gives some basic requirements for the reinforcing of columns. Some of these are listed here as follows:

- 2610(j)1. Limits of vertical reinforcement (minimum & maximum)
- 2618(o)1. Waiver of minimum vertical reinforcement for prestressed columns with effective prestress exceeding 225 psi
- 2618(o)3.; 2607(m)3. Lateral reinforcing - mild steel ties
- 2618(o)3.; 2607(m)2; 2610(j)2. Lateral reinforcing - mild steel spirals

UNION BANK BUILDING
LONG BEACH

Precast concrete columns form a part of a ductile moment resisting frame in this 16 story building. Main reinforcing was cast into the column and extended out into "U" shaped precast spandrels around the perimeter at each floor. Cast-in-place concrete ductile moment resistant frame provisions of the Uniform Building Code were used in detailing the rebar and cast-in-place concrete closure pours. The floor system was formed with spread single tees and hollow core plank infill.

Both prestressed and mild steel reinforced columns may be designed with mild steel ties as lateral reinforcement. In designing and producing large quantities of columns, it is more economical to prestress these members than to fabricate them in all mild steel units. The principle savings is due to material efficiency resulting from the use of prestress strand over reinforcing bars and savings in plant labor. For a given column cross-section and desired ultimate capacity at the balance point (axial load capacity at simultaneous ultimate strain of concrete and yielding of tension steel) it is considerably more advantageous to use strand in lieu of costly reinforcing bars. At the time of this editing, this cost advantage was more than 400%. In terms of plant labor, long line prestressing effects optimum efficiencies in preliminary and casting labor. Column ties are "hung" in bundles off tensioned strand in the beds, and wired in position as compared with the secondary operation required for fabricating cages and subsequent double handling of these reinforcing units.

In reinforced concrete column design, for a given nominal superimposed load, the greatest economy is achieved with the cross-section resulting from the use of the highest strength concrete attainable under plant conditions and with the minimum allowable vertical reinforcing (As = 0.01 Ag). In California where 6000 psi hard rock concrete is readily available, and combined with the fact that most applications for pin ended columns result in designs where slenderness does not govern the design of the column, preliminary designs using this strength of concrete and 2% of vertical steel with minimum column eccentricity usually result in optimum selections. Slenderness effects are covered in Section 2610(1) of the Uniform Building Code. Sometimes, such as in exterior parking garage columns, an appreciable bending moment is induced due to the summation of eccentric loadings delivered to the column corbels at each parking level. Table 4.6.5 on page 4-31 of the Second Edition of the PCI Design Handbook is helpful in quickly determining the sum of the induced moment from this unbalanced loading. The resulting maximum moment condition may also be determined by the use of moment distribution.

An unfortunate aspect of current code provisions makes the use of spiral reinforcing for "pin-ended" prestressed concrete columns uneconomical in that relatively large spiral bar diameters and closely spaced pitches result from the use of the code equation 10 - 3. For this reason, the use of mild steel ties is recommended in all cases for plant cast precast and prestressed concrete columns. Also, the use of cold drawn wire spirals, while offering promise as a lateral reinforcing material for precast concrete columns, is not recognized in the code for columns, as it is for prestressed concrete bearing piles.

The following design example demonstrates the procedure for selecting an economical column shape and shows how to construct the interaction diagram for the design chosen.

COLUMN DESIGN

DESIGN A PRECAST CONCRETE COLUMN SUPPORTING 4 LEVELS OF INTERIOR GIRDERS AS SHOWN FOR THE PARKING STRUCTURE ON PAGE 402.

CALCULATE LOADS TO COLUMN

		P^K	×	L.F.	=	UP^K
DOUBLE TEES	$0.060^{K/FT^2} \times 58 \times 28 \times 4$ =	390	×	1.4	=	546
C.I.P. TOPPING	$0.031 \times 59 \times 28 \times 4$ =	205	×	1.4	=	287
GIRDER WT.	$0.65 \times 27 \times 4$ =	70	×	1.4	=	98
REDUCED L.L.	$0.40 \times 0.050 \times 59 \times 28 \times 4$ =	132	×	1.7	=	225
				ΣUP	=	1156^K

PRELIMINARY DESIGN

(2.57)* WHEN PRELIMINARY COLUMN LOAD TABLES, OR INTERACTION CHARTS SUCH AS THOSE IN THE CRSI HANDBOOK ARE NOT AVAILABLE, THE FOLLOWING FORMULAE ARE USEFUL IN SELECTING TRIAL SIZES:

$f'_c = 5000$ PSI
$A_{ST} = 0.02 A_g$
$e = 0.1\,t$
$A_g = \frac{UP}{2.9}$

$f_y = 60$

$f'_c = 6000$ PSI
$A_{ST} = 0.02 A_g$
$e = 0.1\,t$
$A_g = \frac{UP}{3.4}$

THE MOST ECONOMICAL CROSS-SECTION WILL RESULT FROM USING THE HIGHEST STRENGTH CONCRETE ATTAINABLE FOR A GIVEN STEEL PERCENTAGE.

FOR OUR PROBLEM USE $f'_c = 6000$ PSI $\quad f_y = 60$ KSI

$\therefore A_g = \frac{1156}{3.4} = 340 \text{ IN}^2 \quad E_s = 29{,}000$

TRY 14×24 COLUMN WITH 4-#11 BARS (2 E.F.)

CONSTRUCT INTERACTION DIAGRAM

DETERMINE THE FOLLOWING PARAMETERS:

$\beta_1 = 0.85 - 0.10 = 0.75$
$d = 24 - 2.5 = 21.5''$ $\quad d' = 2.5''$
$y_t = 12''$ $\quad 0.85 f'_c = 5.1$ KSI
$A_g = 24 \times 14 = 336 \text{ IN}^2$
$A_s = A'_s = 3.14 \text{ IN}^2$

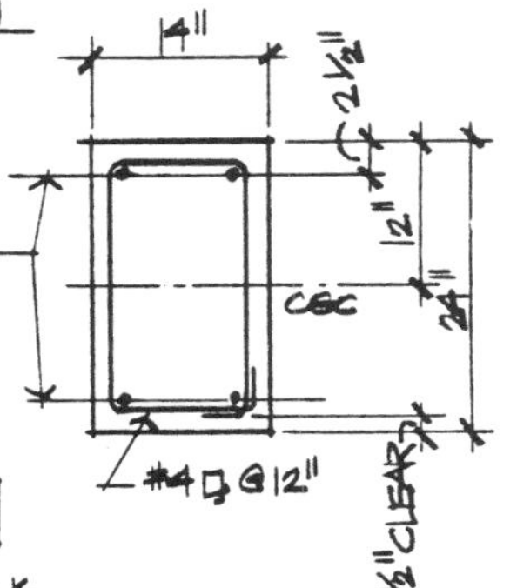

(3-32) STEP 1 DETERMINE P_o FROM FIG. 3.6.1 (c)

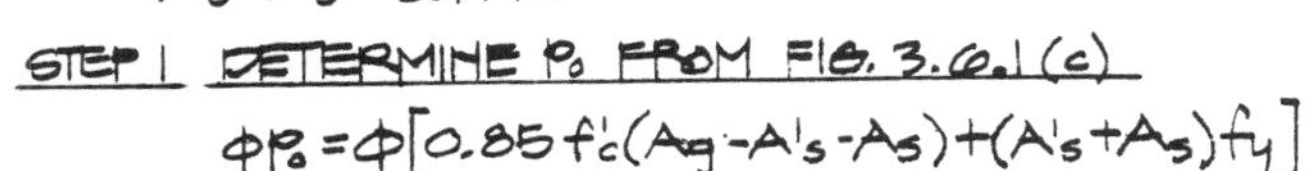

$$\phi P_o = \phi[0.85 f'_c (A_g - A'_s - A_s) + (A'_s + A_s) f_y]$$

$$= 0.70[5.1(336 - 6.28) + (6.28)(60)] = 1441^K$$

 *( ) PAGE REFERENCE IN 2ND EDITION OF PCI DESIGN HANDBOOK

STEP 2 DETERMINE P_{nb} AND M_{nb} FROM FIG. 3.6.1.(d)

$$c = \frac{0.003d}{0.003 + f_y/E_s} = \frac{(0.003)(21.5)}{0.003 + \frac{60}{29000}} = 12.72''$$

$$f'_s = E_s\left[\frac{0.003}{c}(c - d')\right] \le 60$$

$$= 29000\left[\frac{0.003}{12.72}(12.72 - 2.5)\right] = 69.9 \quad \therefore f'_s = f_y = 60 \text{ KSI}$$

$$A_{COMP.} = ab = \beta cb = (0.75)(12.72)(14) = 133.6 \text{ IN}^2$$

$$y' = a/2 = \frac{(0.75)(12.72)}{2} = 4.77''$$

$$\phi P_{nb} = \phi\left[(A_{comp} - A'_s)0.85f'_c + A'_s f'_s + A_s f_y\right]$$

$$= 0.70\left[(133.6 - 3.14(5.1) + (3.14)(60) - (3.14)(60)\right] = \underline{460^K}$$

$$\phi M_{nb} = \phi P_{nb}e = (0.70)\left[(665)(12 - 4.77) + (188.4)(21.5 - 12) + (188.4)(12 - 2.5)\right]$$

$$= 0.70(4808 + 1790 + 1790) = \underline{5872^{K\text{-}IN}} \quad (490^{K\text{-}FT})$$

STEP 3 DETERMINE M_o (NEGLECT COMPRESSIVE REINFORCEMENT)

$$a = \frac{A_s f_y}{0.85 f'_c b} = \frac{(3.14)(60)}{(5.1)(14)} = 2.64''$$

$$M_o = A_s f_y(d - a/2) = (3.14)(60)\left(21.5 - \frac{2.64}{2}\right) = 3802^{K\text{-}IN}$$

$\phi = 0.9$ @ $\phi P_n = 0$ $\quad \phi M_o = (0.9)\left(\frac{3802}{12}\right) = 285^{K\text{-}FT}$

($\phi = 0.7 \quad \therefore \phi M_o = 222^{K\text{-}FT}$)

@ $\phi = 0.7 \quad \phi P_n = 0.1 f'_c A_g = (0.1)(6)(336) = 202^K$

MAXIMUM DESIGN LOAD $0.8\phi P_o = (0.8)(1441) = 1153^K$

GIVES CURVE POINT @ $e = 0.1t$

$$\therefore \phi M_n = (1153)\left(\frac{2.4}{12}\right) = \underline{231^{K\text{-}FT}}$$

STEP 4 CALCULATE AN INTERMEDIATE CURVE POINT IN THE COMPRESSION RANGE ($a > a_{bal} < .75d$) FIG. 3.6.1 (a)

(3-33)

SET $a = 16''$ $\quad \therefore c = 16/.75 = 21.3''$ $\quad A_{COMP} = ab = (16)(14) = 224 \text{ IN}^2$

$$f'_s = E_s\left[\frac{0.003}{c}(c - d')\right] = 29000\left[\frac{.003}{21.3}(21.3 - 2.5)\right] = 76.7^K \text{ USE } 60^K$$

$$f_s = E_s\left[\frac{0.003}{c}(d - c)\right] = 29000\left[\frac{.003}{21.3}(21.5 - 21.3)\right] = 1^K \quad (y' = a/2 = 8'')$$

$$\phi P_n = 0.7\left[(224 - 3.14)5.1 + (3.14)(60) - 3\right] = 0.7(1126 + 188 - 3) = \underline{918^K}$$

$$\phi M_n = 0.7\left[(1126)(12 - 8) + (3.14)(1)(21.5 - 12) + (3.14)(60)(12 - 2.5)\right]$$

$$= 4427^{K\text{-}IN} \ (369^{K\text{-}FT})$$

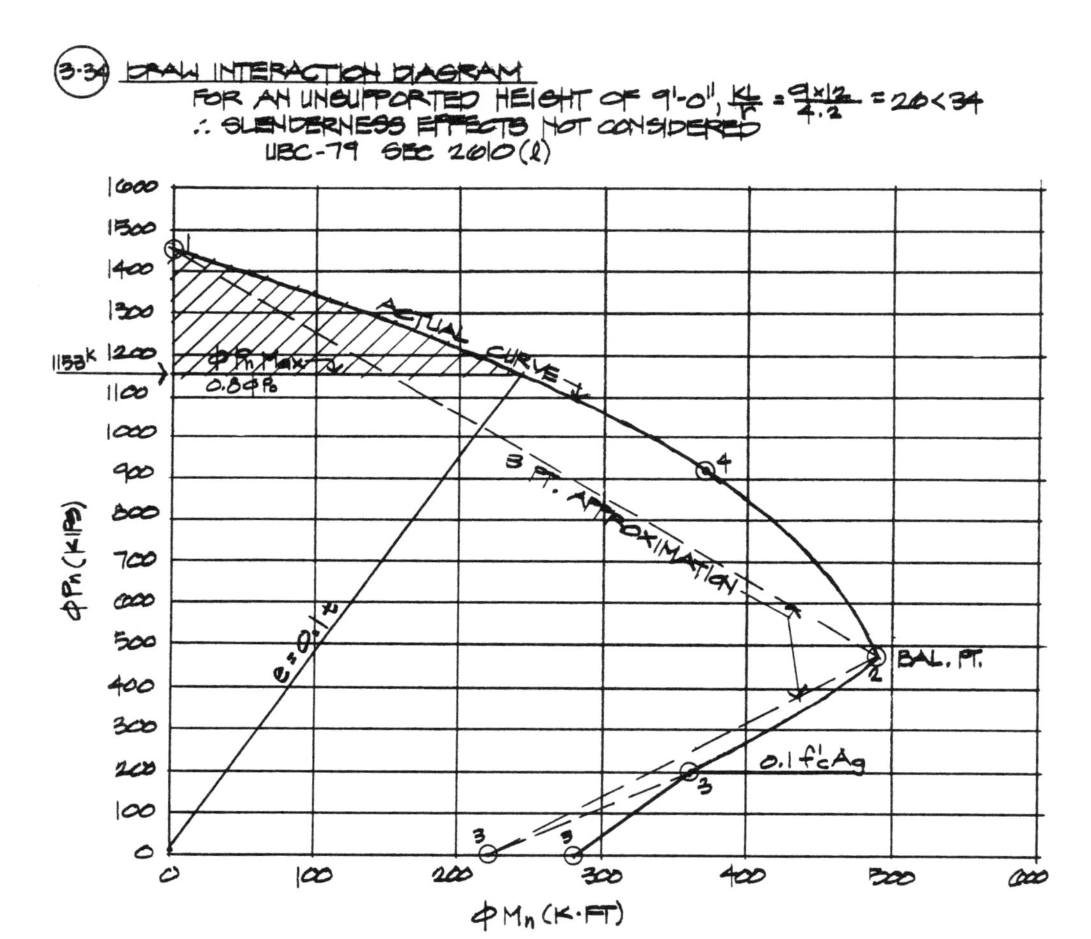

CONNECTION DESIGN

On the next page are shown example calculations for a typical column corbel, and for the design of a steel column base plate. In designing precast column base plates, 2 conditions of loading should be investigated. A temporary erection loading exists where the ungrouted base plate is subjected to the column weight plus erection loads. The final condition, shown on the next page, is the dead plus live loading on the grouted base plate. Page 5-27 of the 2nd Edition of the PCI Design Handbook gives additional information on base plate design.

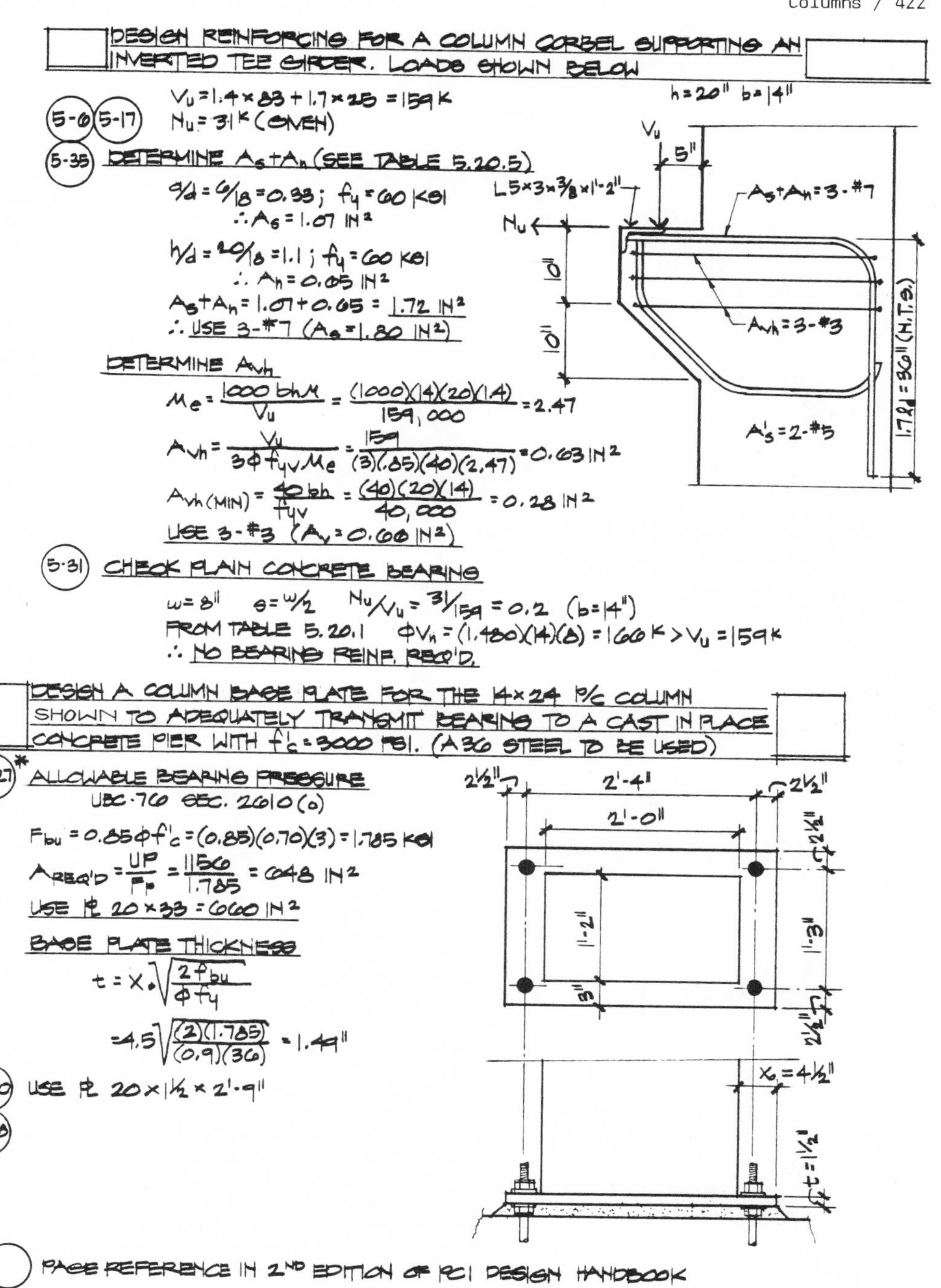

DESIGN REINFORCING FOR A COLUMN CORBEL SUPPORTING AN INVERTED TEE GIRDER. LOADS SHOWN BELOW

(5-6) (5-17)

$V_u = 1.4 \times 83 + 1.7 \times 25 = 159\,K$

$N_u = 31^K$ (GIVEN)

(5-35) DETERMINE $A_s + A_n$ (SEE TABLE 5.20.5)

$a/d = 6/18 = 0.33$; $f_y = 60$ KSI

$\therefore A_s = 1.07$ IN²

$h/d = 20/18 = 1.1$; $f_y = 60$ KSI

$\therefore A_n = 0.65$ IN²

$A_s + A_n = 1.07 + 0.65 = 1.72$ IN²

$\therefore$ USE 3-#7 ($A_s = 1.80$ IN²)

DETERMINE A_{vh}

$$M_e = \frac{1000\,bh\mu}{V_u} = \frac{(1000)(14)(20)(1.4)}{159{,}000} = 2.47$$

$$A_{vh} = \frac{V_u}{3\phi f_{yv} M_e} = \frac{159}{(3)(.85)(40)(2.47)} = 0.63\ \text{IN}^2$$

$$A_{vh\,(MIN)} = \frac{40\,bh}{f_{yv}} = \frac{(40)(20)(14)}{40{,}000} = 0.28\ \text{IN}^2$$

USE 3-#3 ($A_v = 0.66$ IN²)

(5-31) CHECK PLAIN CONCRETE BEARING

$w = 8''$ $s = w/2$ $N_u/V_u = 31/159 = 0.2$ ($b = 14''$)

FROM TABLE 5.20.1 $\phi V_n = (1.480)(14)(8) = 166^K > V_u = 159^K$

$\therefore$ NO BEARING REINF. REQ'D.

DESIGN A COLUMN BASE PLATE FOR THE 14×24 P/C COLUMN SHOWN TO ADEQUATELY TRANSMIT BEARING TO A CAST IN PLACE CONCRETE PIER WITH $f'_c = 3000$ PSI. (A36 STEEL TO BE USED)

(5-27)* ALLOWABLE BEARING PRESSURE
UBC-76 SEC. 2610 (o)

$F_{bu} = 0.85\phi f'_c = (0.85)(0.70)(3) = 1.785$ KSI

$$A_{REQ'D} = \frac{UP}{F_p} = \frac{1156}{1.785} = 648\ \text{IN}^2$$

USE PL 20×33 = 660 IN²

BASE PLATE THICKNESS

$$t = X_o\sqrt{\frac{2f_{bu}}{\phi f_y}}$$

$$= 4.5\sqrt{\frac{(2)(1.785)}{(0.9)(36)}} = 1.49''$$

(5-50) USE PL 20×1½×2'-9"

(5-28)

* ( ) PAGE REFERENCE IN 2ND EDITION OF PCI DESIGN HANDBOOK

CLASSROOM BUILDING - U. C. DAVIS

The top photo shows the bracing scheme used to stabilize the 9-story prestressed concrete columns during construction. The 26 columns subsequently received precast concrete infill panels, single tees at the roof, prestressed channel deck elements, and precast sunshades to complete the structure.

CONNECTION DETAILS

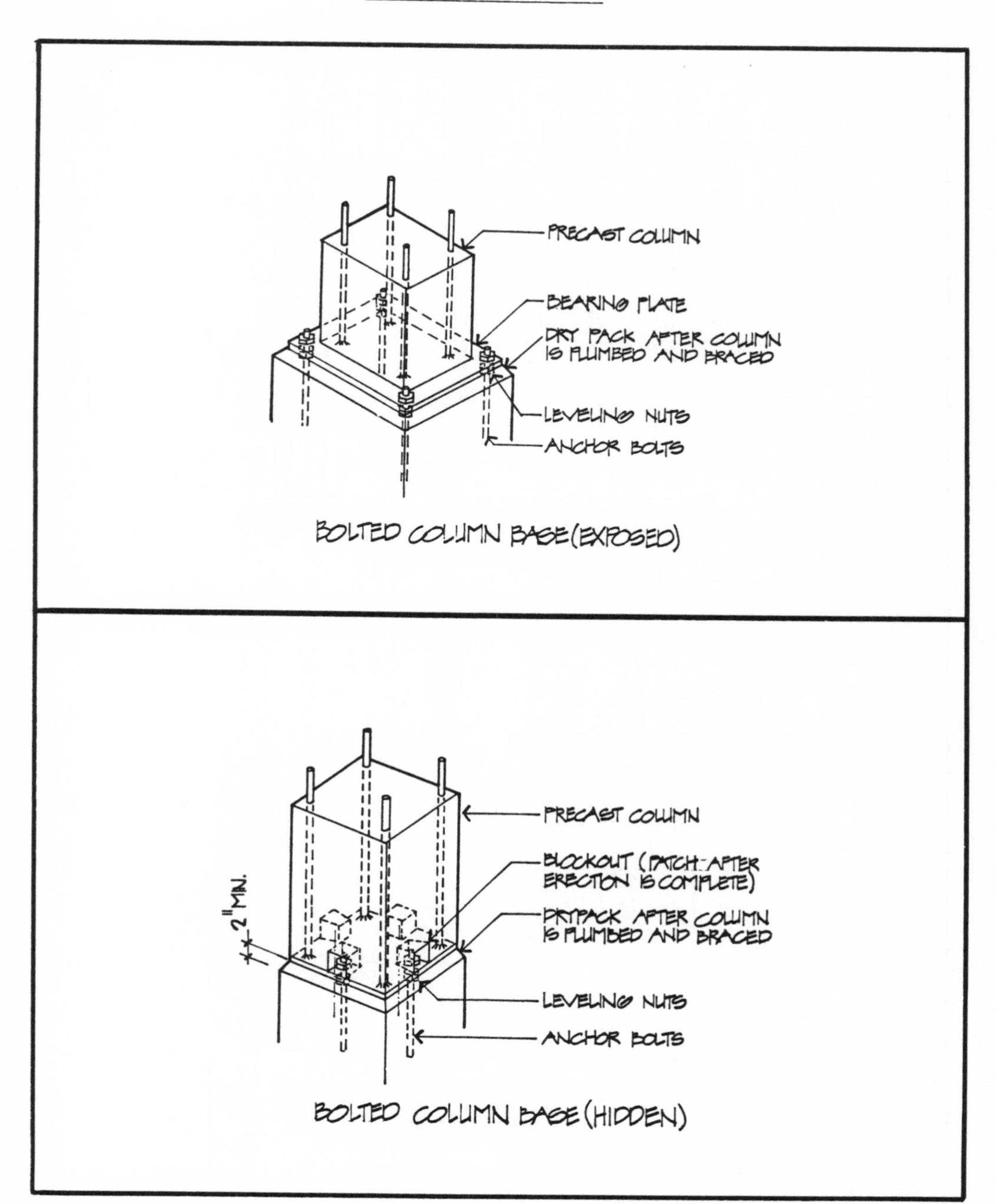

CONNECTION DETAILS

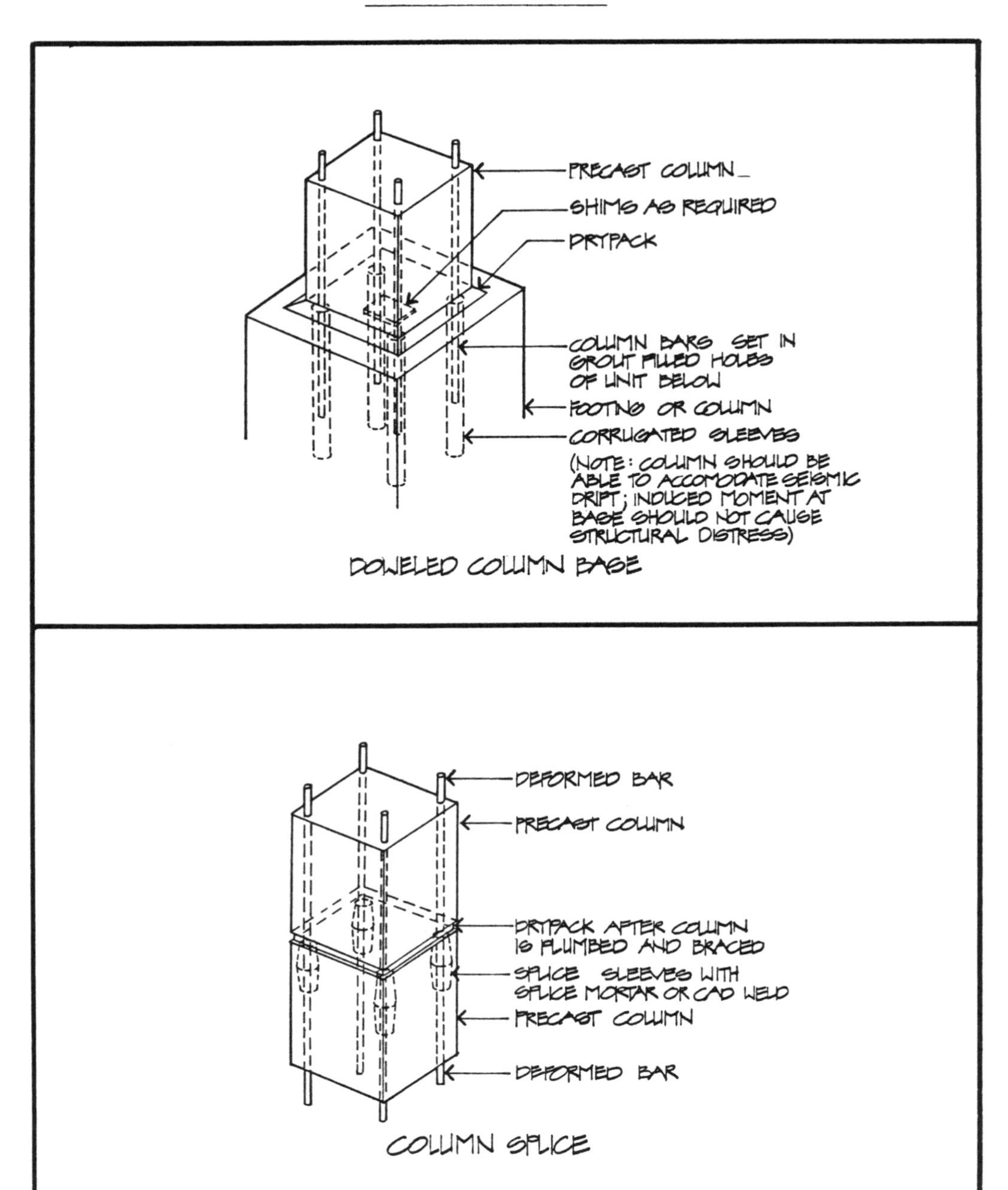

CONNECTION DETAILS

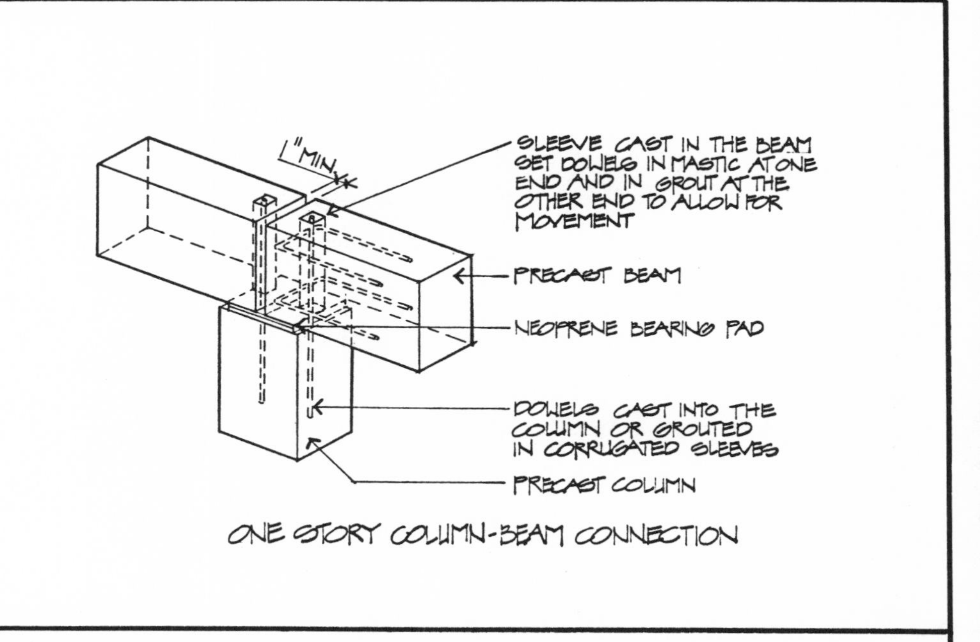

ONE STORY COLUMN-BEAM CONNECTION

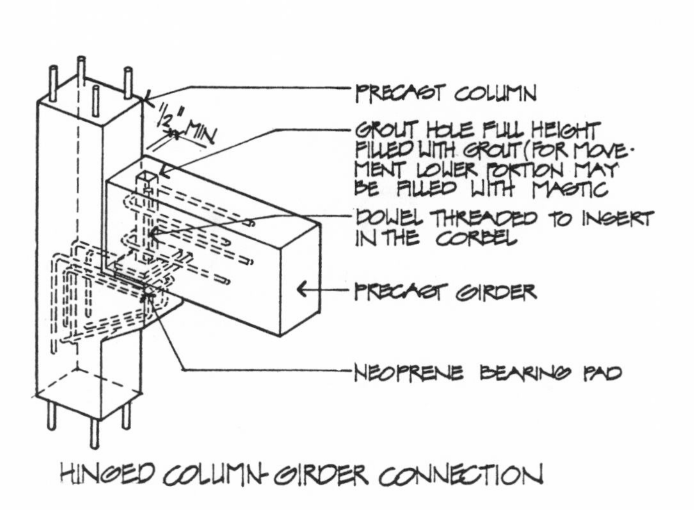

HINGED COLUMN-GIRDER CONNECTION

CONNECTION DETAILS

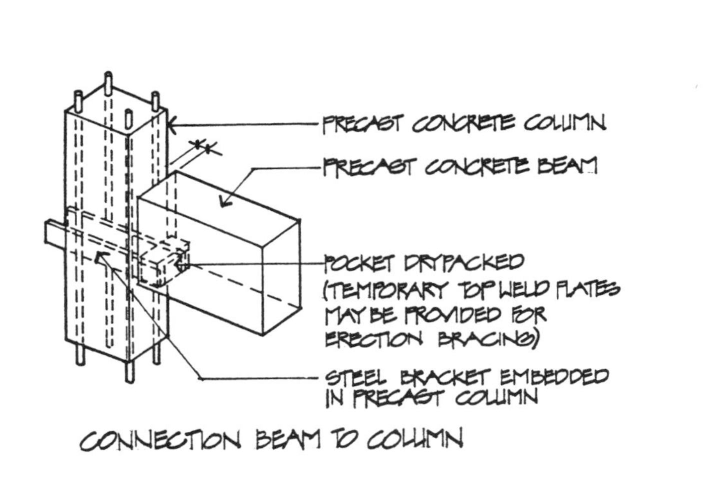

CONNECTION BEAM TO COLUMN

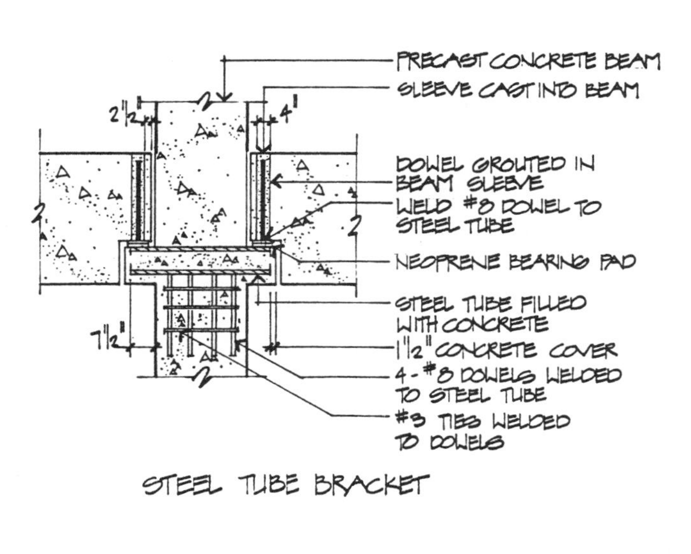

STEEL TUBE BRACKET

CONNECTION DETAILS

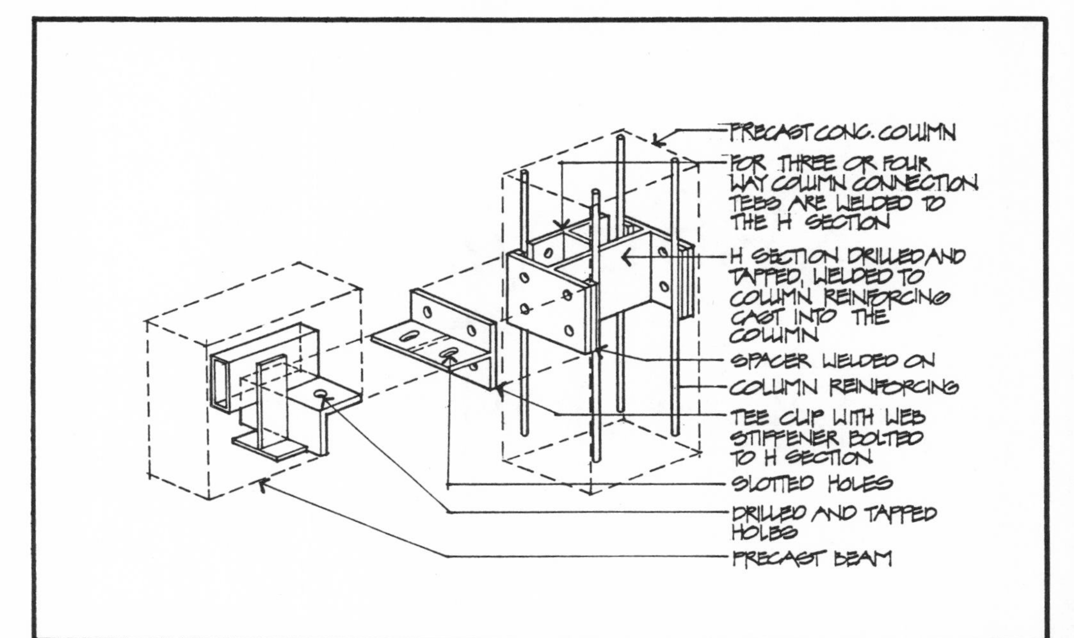

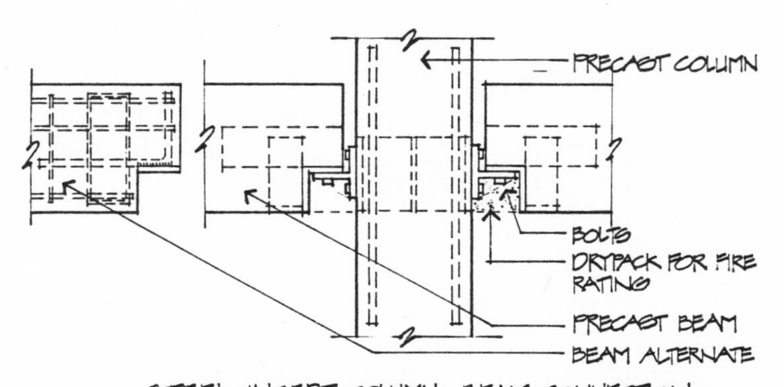

STEEL INSERT COLUMN - BEAM CONNECTION
WITH THIS CONNECTION THE COLUMN CAN BE CAST IN A CONTINUOUS BEAM OR PILING FORM.

PRESTRESSED CONCRETE BEARING PILES

GOLDEN GATEWAY III - SAN FRANCISCO

14" square prestressed concrete piling being driven in downtown San Francisco. Note the cast-in plate in the piling units in the foreground, which form a part of the DYN-A-SPLICE mechanical pile splice. This eliminated the problem with transporting long units on mid city streets. First, the lower 80' section was driven, then the upper 40' section was spliced and driving continued immediately.

Where the vertical loads delivered to the base of a building or bridge structure cause allowable soil pressures to be exceeded under reasonably sized spread footings or mat foundations, a bearing pile foundation system is used. Sometimes, the proximity of bedrock to the ground surface will result in economy over spread footings, even when soil bearing capacities are satisfied. Where end bearing conditions are not attainable, piles are driven until sufficient skin friction is developed to safely support the applied loads. Prestressed concrete piles, having a large constant surface area, develop frictional resistance in a shorter distance than other types of piling. Friction piles are most commonly used in silty sand, clay-silt, and clay-sand types of soils. The history of the use of piling dates back to before the Roman era, and evidence of their use exists worldwide. Prestressed concrete piling are also immune from fungus attack, marine borers, termites, corrosion, and fire. When considering the proven splicing methods available, prestressed concrete piling can be driven to any length. For these reasons, the majority of all piling used for foundations is pretensioned concrete piling. The advantage of prestressing is that compressive stresses are introduced which counteract tension stresses resulting from handling, stress waves produced by driving, or tension stresses due to buckling or eccentricity during driving. A complete overview of the use of concrete piling in general is given in ACI Committee 543 report entitled RECOMMENDATIONS FOR DESIGN, MANUFACTURE, AND INSTALLATION OF CONCRETE PILES. A wealth of practical information on the use of prestressed concrete bearing piles is given in Professor Ben C. Gerwick Jr.'s book entitled CONSTRUCTION OF PRESTRESSED CONCRETE STRUCTURES.

The principle competing types of piling systems are structural steel, timber, and concrete filled tapered metal casing piles. In addition to cost and construction time savings, the following advantages are also offered by prestressed concrete piling:

HIGH LOAD CARRYING CAPACITY

Because of their superior strength, high axial loads, well within the allowable limits of regulating codes, can be carried by prestressed piles to the extent allowed by soil conditions. Higher loads per pile mean fewer piles, smaller footings and, consequently, in most instances, lower costs per ton carried.

DURABILITY

The dense, high-quality concrete, maintained permanently under stress is relatively free from shrinkage and other cracks, and is impervious to water. Experience plus accelerated tests have proven pretensioned concrete piles to be exceptionally durable under even the most severe conditions of exposure.

EASE OF HANDLING

Fewer picking points and the great strength of pretensioned piles greatly facilitate their transportation and handling and contribute to lower driving costs.

ABILITY TO TAKE HARD DRIVING

The ability to withstand driving allows the use of heavier hammers for driving through denser soils, thus permitting the development of adequate bearing in the soil.

GREATER COLUMN STRENGTH

The long-column behavior of prestressed concrete piling is excellent, and thus permits high axial loads even when the piles are also subject to eccentricities or lateral forces and moments. When bending resistance is critical the pile capacity can be increased by increasing the effective prestress up to the recommended maximum.

RESISTANCE TO UPLIFT

Pretensioned piles can be utilized very effectively in tension where uplift must be resisted, and the transfer of tension to the pile can be easily accomplished.

QUALITY CONTROL

Close supervision by experienced personnel of materials and workmanship in a centrally controlled plant insures a high quality product with 6000 psi concrete. Guaranteed strengths to 8000 psi may be provided at slight additional cost.

ECONOMY

Prestressed concrete piles, when properly used, have shown substantial economic advantages for pile foundations.

AVAILABILITY

Piling sizes can obviously vary subject to design selection and local availability of standard sizes. Some of the standard sizes available in California are:

Solid Square	12", 14", 16", 20"
Solid Octagonal	18", 24"

These sizes are available from several manufacturers in both Southern and Northern California. Some of the manufacturers make a larger number of different types and sizes, including hollow sections and large diameter hollow cylinder piles. On the following page, product information on a more complete range of prestressed concrete piling is shown on the Santa Fe Pomeroy piling property sheet.

Santa Fe-Pomeroy, Inc.
Engineering and Construction Subsidiary of Santa Fe International Corporation
500 Hopper Street Extension Petaluma, California 94952 Telephone (707) 763-1918

PROPERTIES OF PRETENSIONED PRESTRESSED CONCRETE PILES

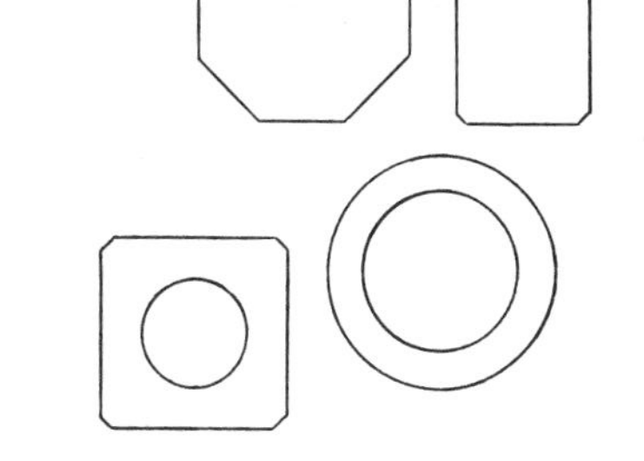

PILE SIZE DIAMETER	SHAPE	SOLID OR HOLLOW	Ac	WEIGHT PLF	NUMBER OF STRANDS	EFFECTIVE PRESTRESS (TO NEAREST 5 PSI)	I	I/C	r	PERIMETER	ALLOWABLE MOMENT 300 PSI TENSION	ALLOWABLE MOMENT 600 PSI TENSION	ALLOWABLE LOADS BASED ON f'c 6000 PSI	ALLOWABLE LOADS BASED ON f'c 7000 PSI
INCHES (1)		(2)	SQ. IN. (3)	LBS. (4)	PER PILE (5)	PSI (6)	IN.^{4}	IN.^{3}	IN.	INCHES	KIP INCHES (7)	KIP INCHES (7)	TONS (8)	TONS (8)
10''	SQUARE	SOLID	98	105	4—7/16"	760*	790	158	2.84	38	167	215	87	103
12''	SQUARE	SOLID	142	152	6—7/16"	785*	1,664	277	3.42	46	300	384	125	149
14''	SQUARE	SOLID	194	209	6—1/2"	730	3,112	445	4.00	54	458	592	172	204
15''	OCTAGONAL	SOLID	186	196	6—1/2"	760	2,765	368	3.86	50	390	500	165	195
16''	SQUARE	SOLID	254	273	9—1/2"	835	5,344	668	4.59	62	758	958	222	264
18''	OCTAGONAL	SOLID	268	288	11—7/16"	725	5,705	634	4.61	60	650	840	239	283
18''	SQUARE	SOLID	322	346	11—1/2"	805	8,597	955	5.17	70	1,055	1,342	283	336
20''	SQUARE	SOLID	398	428	13—1/2"	770	13,146	1,315	5.75	78	1,407	1,801	353	418
20''	SQUARE	11'' H. C.	303	326	10—1/2"	775	12,427	1,243	6.40	78	1,336	1,709	268	318
24''	SQUARE	14'' H. C.	418	450	13—1/2"	730	25,490	2,124	7.81	94	2,188	2,825	372	440
36''	ROUND	26'' H. C.	487	524	17—1/2"	820	60,016	3,334	11.10	113	3,734	4,735	428	508
48''	ROUND	38'' H. C.	675	726	24—1/2"	835	158,222	6,593	15.31	151	7,483	9,460	592	703
54''	ROUND	44'' H. C.	770	829	28—1/2"	855	233,409	8,645	17.41	170	9,985	12,578	673	800

(1) Nominal pile-size.
(2) Holes for hollow core piles are circular.
(3) Reduction in area for chamfers on square piles has been taken into account.
(4) Tables are based on regular concrete of 155 lb./cu. ft. density. The use of high strength lightweight concrete in piles for certain specific applications, such as fender piles, should be considered, when available, because its lower E value gives greater deflection and energy absorption characteristics. With lightweight concrete an f'c of 5000 psi should be used, and the values in the table should be adjusted accordingly.
(5) Based on 1/2'' and 7/16'' diameter ASTM A416 Grade 270 strands with ultimate strengths of 41,300 lbs. and 31,000 lbs., respectively. If different diameter strand is used, the number of strands per pile should be increased or decreased, in accordance with strand manufacturer's tables, to provide approximately the same minimum effective prestress shown in the table.
(6) Effective prestress assumes a uniform distribution of strands resulting in a uniform prestress. Piles marked with * have effective prestress based on 60% of ultimate strand strength. All other piles have effective prestress based on initial prestress in strand of 70% of ultimate minus losses of 35,000 psi.
(7) Allowable bending moments listed are for a permissable concrete tensile stress of 300 psi with an effective prestress as given in the table, no external axial load, f'c=6000 psi and assuming a modulus of rupture of 600 psi. Allowable moments for earthquake or similar transient loads are based on a tension of 600 psi. Piles with both axial load and bending should be analyzed considering the effect of the sustained external load. When bending resistance is critical, the allowable moment may be increased by using more strands to raise the effective prestress to a maximum of 0.2 f'c psi.
(8) Allowable design loads are based on the accepted formula of N=Ac (0.33 f'c - 0.27 fpe), and are computed for f'c =6000 psi and 7000 psi. For concrete strength in excess of 7000 psi, consult our Plant Engineering Department for information on practicability and economics.

FABRICATION

The advantage of pretensioning is that the piling can be quickly and efficiently produced in continuous steel forms with steam curing on a daily cycle. The piling are cast in long lines often 800 feet or more in length. The square sizes are often cast in "gang" molds, with several piling being cast side by side to achieve more efficient utilization of bed labor. Octagonal or round shapes are made by using molds with the upper portions being hingeJ to swing out of the way during strand placement and positioning of spiral reinforcement. Hollow sections are formed by using hollow tubes which are cast in, or by using a moving mandrel which forms the inner void. As with other precast concrete products, vertical sides of molds will have a slight draft on the order of ½" per foot to facilitate stripping. 6,000 psi concrete is normally used for piling for more cost efficient design, even though required release strengths rarely exceed 3,000 psi. Piling to be used in extremely corrosive environments, such as in tidal areas with heavy concentrations of salts are made with cement rich concrete mixes containing 7½ or 8 sacks of cement per cubic yard of concrete. In addition, as for all structural precast or prestressed concrete products, the water cement ratio should be kept below 0.40. Corrosion resistant piling also exhibit increased cover over the reinforcing assembly, on the order of 2½ to 3 inches. In some instances a 3 to 7 day moist or water cure is required to assure durability.

16" SQUARE PILE FORMS

A view of 16" square gang forms in the Santa Fe Pomeroy Plant in Petaluma. Seen in the foreground is the multiple strand stressing jack assembly used in this plant. The vertical wide flange beams are abutments designed to withstand the stressing force of almost 15 tons per strand.

CYLINDER PILE PRODUCTION

Shown is the production line for 54" diameter hollow cylinder piling being produced for the Dumbarton Bridge project. The spiral reinforcing is being spaced as per design; the top portion of the form is being placed in the background. 246 piles, each 90 feet long, were produced for this project and shipped to the bridge site by barge.

SQUARE PILE PRODUCTION

These 800 foot long lines in the Peter Kiewit plant in Richmond contain 16" square piling for the Embarcadero Center 4 project. Note lift loops cast in for stripping and handling. Gaps shown in foreground are bulkheads between units with space to permit the strands to be burnt to release the prestress force into the member.

HANDLING

Due to the extreme length of prestressed concrete piling (lengths in excess of 120 feet are not uncommon) particular care and planning are required to assure that allowable tensile stresses are not exceeded during stripping, transportation, and "tripping up" into the vertical position in the field. The usual impact factors are applied to working load moments, and actual calculated tensile stresses are compared with allowable stresses, usually the modulus of rupture of the concrete divided by 1.5. See pp 165-176 for various picking points and specifics on product handling.

ANHEUSER BUSCH PLANT EXPANSION - FAIRFIELD

Special 4 point pivot support rigs are used to ship these 110 foot long sections of 14" square piling. Note special spreader beams used to offload and handle sections at the site.

INSTALLATION

When driving into hard strata or rock, a steel tip consisting of a wide flange or "H" section is cast into the end of the prestressed concrete pile. The standard tip is the flat concrete surface with chamfered edges. High capacity requires that the final seating be done with a large hammer. Hammers delivering an energy of 30,000 to 60,000 ft. lbs. per blow are generally employed for developing design load capacities of 150 to 250 tons. These piles must be seated to or driven into a satisfactory bearing stratum as determined by the soils engineer. To ensure penetration and to prevent the excessive absorption of driving energy from upper strata, special methods must sometimes be employed, depending on the characteristics of the overlying soil. Jetting, including pilot jetting, internal jetting, and external jetting during driving, are often effective and practical. Pre-drilling with a wet drill, with or without the aid of a bentonite slurry to hold the hole, is also done.

STONERIDGE OVERCROSSING
PLEASANTON

70 foot long sections of 12" square prestressed concrete piling shown here being driven on a 20 degree batter.

DESIGN CONSIDERATION FOR PILE PLACEMENT AND DRIVING

The minimum center to center spacing of piles not driven in rock (friction piles) should not be less than twice the average diameter of a round or octagonal pile nor less than 1.75 times the diagonal dimension of a square pile nor less than 2'-6". For sand or clay or clay-sand soil, the spacing should be increased by 10% for each interior piling in a group. This increase need not exceed 40%. Piling driven to rock (bearing piles) should have a minimum center to center spacing of not less than twice the average diameter of round piles, nor less than 1.7 times the diagonal dimension of rectangular piles, nor less than 2'-0". A column or pier supported by piles, unless connected to permanent construction which provides adequate lateral support, should rest on not less than 3 piles. When the supporting capacity of a single row of piles is adequate for a wall, effective measure should be taken to provide for eccentricity and lateral force, or the piles should be driven alternately in lines spaced at least one foot apart and located symmetrically under the center of gravity of the carried loads. The normal tolerance for placing prestressed concrete piling is within ±6" from the theoretical location shown on the drawings. It has been found from experience that heavier rams and shorter strokes reduce driving stresses. The use of soft wood cushion blocks 6" to 12" in thickness will reduce the driving stresses by one half or more. These compress during driving and thus do not adversely affect the transmission of energy during hard driving. New blocks should be used for each pile. Recently, tests conducted in Sweden show that new type hammers with special impact absorbing caps have even better, more uniform force delivering characteristics. These hammers reduce the impact force in the pile and impart more energy than conventional wood caps. This means fewer blows to install the pile. In soft or irregular driving it is desirable to reduce the velocity of the blow to minimize the magnitude of the rebounding tension wave.

ANHEUSER BUSCH PLANT
EXPANSION - FAIRFIELD

Three point pick is used to lift a 110 foot long pile into the vertical position and place into the leads of the pile driving rig for driving.

SPLICING

Often it is advantageous to making piling in shorter lengths and splice them together in the field. Splices are used when:

- The ordered length of a prestressed concrete pile is insufficient to obtain the specified bearing value.

- The estimated length of a prestressed concrete pile cannot be economically or feasibly transported or safely handled in the driver.

- Engineering opinions vary on estimated pile lengths derived from analysis of widely spaced borings.

- Non-uniform underground conditions.

- Experience in local areas indicate erratic lengths which could not be forecast with any accuracy.

When splicing is used, the individual pile lengths are kept below 60' to 80'. A splice must be equal to, or greater than the piles it connects in load carrying capacity, moment resistance, shear, and uplift, if applicable. The splice must be durable, economical, and quickly effective so as to allow driving to be continued as soon as possible. Two types of splices have proven to be very effective. Both have had their capacities verified by actual full scale testing.

1. Cement Dowelled Splice

   This splice is made with the upper pile section being cast with protruding rebar dowels, and the lower section having sleeves cast in the top of the pile. after the lower section is driven down, the upper section is installed but separated from the lower section by a ½" shim stack at the center of the pile. Then a steel splice boot with pouring pockets is clamped around the sections to be spliced. Plasticized cement is poured into the pockets to seal the joint. In 10 to 15 minutes, the plasticized cement has developed a compressive strength of 5,000 psi, and the boot is removed and driving can continue. (Epoxy can also be used. In this case the epoxy is poured into the sleeves prior to lowering the upper section which displaces epoxy into the joint between sections, where it is retained by a form clamped around the perimeter of the piles.) A practical note: the dowel lengths should vary to facilitate slabbing during mating of the pile sections. This splice is sometimes specified where corrosive conditions prohibit the bolted splice.

2. DYN-A-SPLICE (Bolted Splice)

This splice is made up of upper and lower steel plates attached to threaded rebar cast in the pile sections. The threaded rebar is fastened to the plate section with special lock bolts which fix the plates in exact mated positions. The splice is completed in the field by hammering in wedges which interlock the bolts at each corner of the pile. This splice is preferred over the cement dowelled splice since less time is lost in making the connection.

DYN-A-SPLICE MECHANICAL PILE SPLICE

Note how the upper and lower sections are quickly mated in the field. The center aiding pin automatically centers the upper unit over the driven lower section. (Photograph courtesy A-Joint Corp., Campbell, CA.)

CONNECTION OF PILE TO CAP/FOOTING

The design varies in individual cases. The connection may be designed as a hinge to eliminate moment or a moment connection may be required. Frequently, a satisfactory connection is obtained by extending the prestressing strands into the footing or capping beam. The length of strand required may be exposed during cut-off of the pile to grade, or by allowing for the extra length during manufacture. An effective moment connection can be made by embedment of the pile head into the cap by about two feet. At this depth the full moment resistance of the pile can be developed.

Where the thickness of the cap will not permit embedment, the connection is made by mild steel dowels. Where the length of piles can be predetermined within one or two feet the dowels can be either: (a) fully cast into the head of the pile and exposed during cut-off to grade; or (b) left projecting at the head and a specially notched head or follower used during driving; or (c) inserted after driving and grouted into holes formed in the head during manufacture or drilled in after driving. Often, where the pile length is variable it is more economical and practical to drill the holes after cut-off. However, during manufacture ties must be cast into the pile along a sufficient length to be effective on the dowels in their eventual position. The preferred method for connecting piles to pile cap is with the use of pile anchor dowels anchored to sleeves (or holes drilled into the pile) with neat cement paste. The diameter of the holes shall be 1/8" to 1/4" larger than the outside diameter of the pile anchor dowels. The cement paste is placed into the holes before the dowels are inserted so no voids remain. The dowels are left undisturbed until the paste has hardened.

TYPICAL REINFORCING AND JET PIPE DETAILS

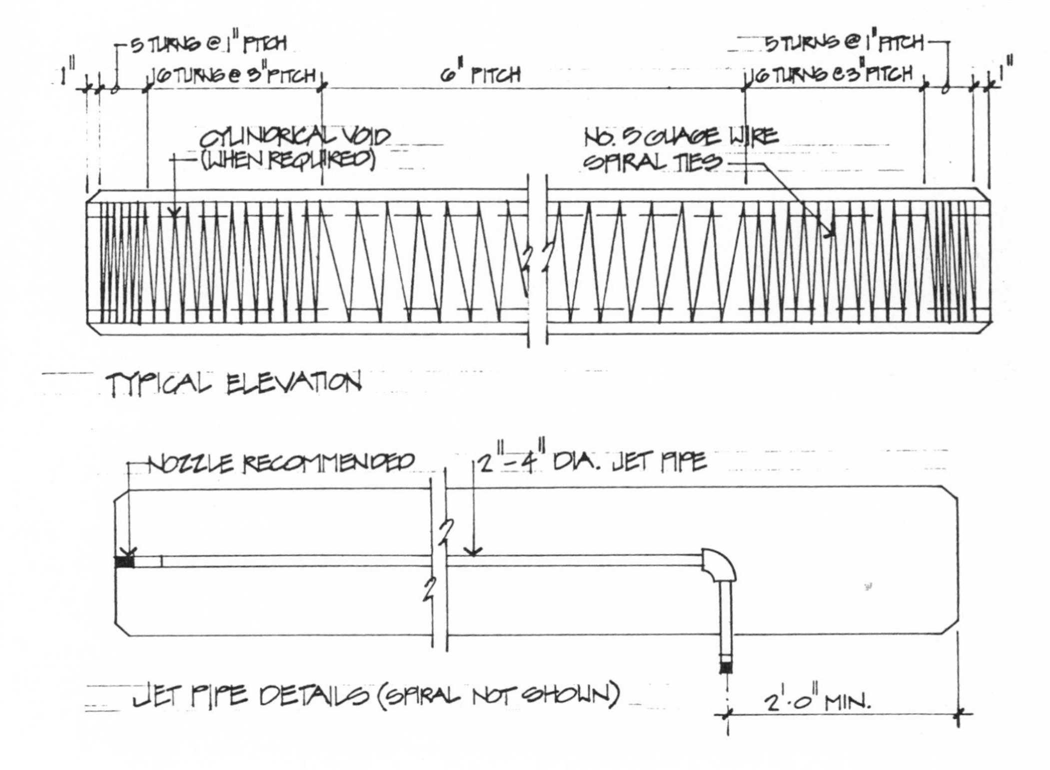

DESIGN CONSIDERATIONS

The soils engineer plays an important part in the design of prestressed concrete bearing piles. He will recommend expected lengths of pile that will be required to develop required frictional resistance, minimum pile stiffness required to develop fixity at a specified depth below the surface, and any imposed curvatures resulting from layered soil mass movements occurring during seismic disturbances.

Building code requirements for prestressed concrete piling are given in UBC 79 Section 2909, where a formula is given for the allowable compressive stress in the concrete due to externally applied axial loads. (fc = 0.33 f'c - 0.27 fpc) This formula, being somewhat arbitrary, should be checked against the specific stress conditions imposed by axial and bending loads and compared with allowable stresses given in UBC 79, Section 2618. Also, load combinations with seismic or wind are permitted a 33% increase in these allowable stresses. The code also indicates minimum effective prestress required for piling, - 700 psi for piles greater than 50 feet in length. The Santa Fe Pomeroy Pile Property Chart gives allowable concentric loads for various pile sizes and concrete strengths based upon the empirical formula from Section 2909 of the Uniform Building Code. These values are helpful in preliminary design. Page 2-62 of the Second Edition, PCI Design Handbook also gives similar information. In final design, the interaction diagram for the specific section being considered is drawn based upon design criteria outlined in the PCI Design Handbook on pages 3-34 through 3-36. The actual ultimate load combinations are plotted on the resulting diagram to see if they fall within the required envelope. Following the design example presented on the next page, another Santa Fe Pomeroy chart is presented, which has interaction diagrams already drawn for several standard pile sizes.

Another area where the reinforcement requirements for piling differ is the required lateral reinforcing - UBC-79, Section 2909 indicates the requirements for steel wire spiral lateral reinforcing varying from 5 ga for 16" diameter piles and smaller to 3 ga (¼" diameter) for piles larger than 20" in diameter. Spacing requirements are also given for this reinforcing. Recent studies and tests on piling to develop data on reinforcing required to enable prestressed concrete piling to withstand imposed curvatures from extreme (catastrophic) seismic conditions indicate that the diameter of the spiral reinforcing should be larger, and the spacing (pitch) of the spiral be smaller, similar to the lateral reinforcement provisions for spiral reinforcing in concrete columns. In other words, spiral reinforcing for these extreme design situations should be about 0.7% where the normal pile lateral reinforcing is about 0.1%. Additional mild steel may also be required to develop additional ultimate moment capacity and ductility at the pile/pile cap interface.

Previously, the requirements recommended for piling to be designed in corrosive or extremely saline environments were covered. These included additional cover, cement rich mixes coupled with a low water cement ratio, and additional curing requirements. Allowable tensions in these areas should be held to zero, or the modulus of rupture divided by a factor of safety of 2.5 at the most. This item is often overlooked in some codes.

PRESTRESSED CONCRETE BEARING PILE DESIGN

DESIGN PRESTRESSED CONCRETE PILING TO SUPPORT A PARKING GARAGE SHEAR WALL. COLUMN FOUNDATION. THE SKETCH BELOW SHOWS THE STRUCTURAL CONFIGURATION AND LOADS DELIVERED TO THE PILE CAP. THE SOILS ENGINEERS REPORT RECOMMENDS USING PILING WITH A MINIMUM OF STIFFNESS (EI) OR 7×10^9 LB.-IN.^{2}. BASED ON THIS RECOMMENDATION, THE PILE WILL DEVELOP A POINT OF FIXITY 12 FEET BELOW THE BOTTOM OF THE PILE CAP. IT IS ESTIMATED THAT 70' OF PILING WILL BE REQUIRED TO DEVELOP THE VERTICAL LOAD CAPACITY WHICH IS SUFFICIENT TO DEVELOP THE REQUIRED HORIZONTAL LATERAL RESISTANCE. PILE/PILE CAP CONN. IS TO PROVIDE FIXITY.

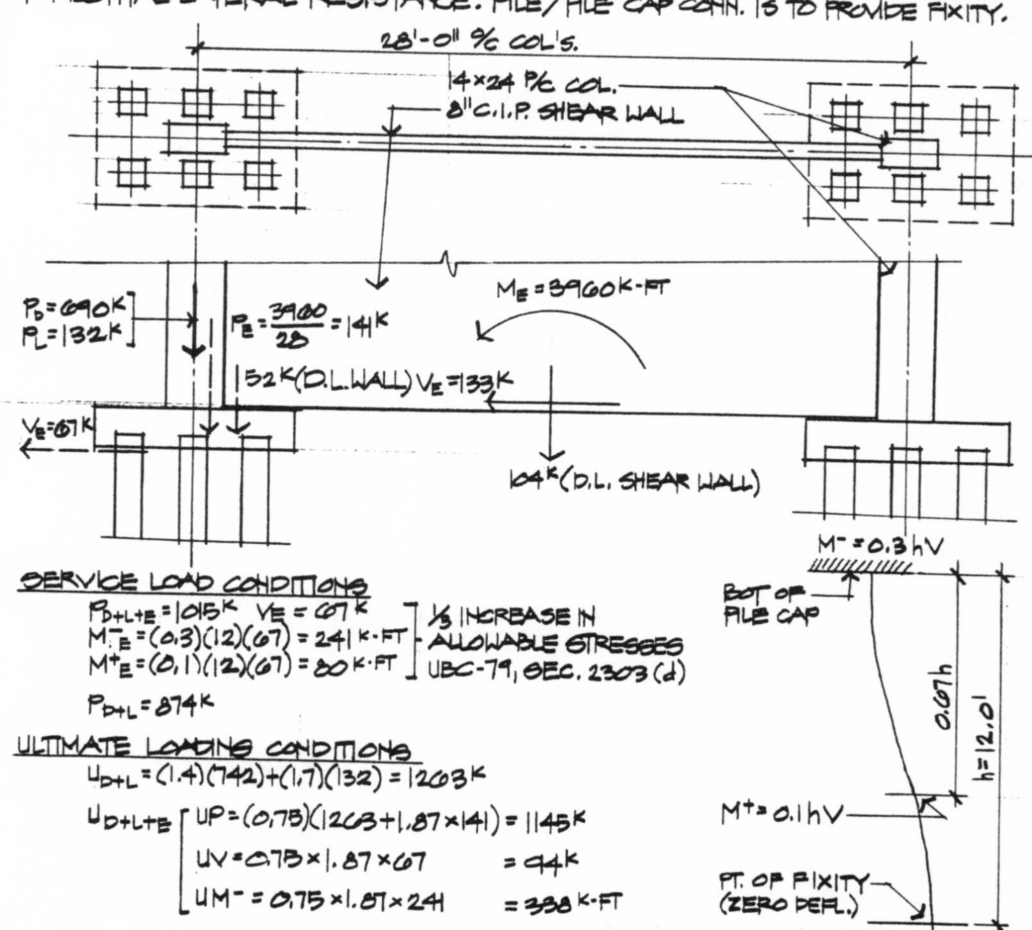

(2-62)* FOR PRELIMINARY DESIGN, USE THE VERTICAL LOAD CONDITION TO PICK A TRIAL PILE GROUP

$\frac{D+L}{4} = \frac{874}{4} = 219^K/\text{PILE}$ (110 TONS)

(−) COMPRESSION
(+) TENSION

TRY 4·12" SQ. PILES ($f'_c = 6000$ PSI)

INVESTIGATE SERVICE LOAD CONDITION FOR SEISMIC CONDITION

$f = -\frac{P}{A} \pm \frac{M_{EXT}}{S} - \frac{P_{EXT}}{A}$ (WORK WITH SANTE FE POMEROY PILE DESIGN CHARTS)

$f^+ = -0.785 + \frac{241 \times 12}{277 \times 4} - \frac{1015}{142 \times 4} = -0.785 + 2.610 - 1.787 = +0.038$ O.K.

(3-14) $f^- = -0.785 - 2.610 - 1.787 = -5.182$ KSI N.G. ($1.33 \times 4.5 f'_c = 3.59$ KSI)

TRY 6-12" SQ. PILES W/ 6-7/16" ϕ - 270^K STRANDS

(3-15)(3-29) $P_f = (6)(21.7)(0.78) = 101.6^K$

$f^- = -\frac{101.6}{142} - \frac{241 \times 12}{277 \times 6} - \frac{1013}{142 \times 6} = -0.715 - 1.740 - 1.191 = -3.646$

f'_c REQ'D. $= \frac{3.646}{1.33 \times 0.45} = 6.078$ SPECIFY 6100 PSI

(2-34) CHECK ULTIMATE STRENGTH FOR EACH PILE

1. $UP_{D+L} = \frac{1263}{6} = 211^K$

2. SEISMIC $\begin{cases} UP_E = \frac{1145}{6} = 191^K \\ UM_E = \frac{338}{6} = 56^{K\cdot FT} \end{cases}$

BOTH CONDITIONS FALL WITHIN ENVELOPE OF INTERACTION DIAGRAM

(DIAGRAM FROM OF POMEROY CHART)

COMPRESSION
400
200
0
200
$\phi P_n{}^K$
e=0.05t
1
2
12" SQ. PILE $f'_c = 6000$
6-7/16" ϕ 270^K STRANDS
25
50
75
100
$\phi M_n{}^{K\cdot FT}$
TENSION

*( ) PAGE REFERENCE IN 2ND EDITION OF PCI DESIGN HANDBOOK

CHECK HANDLING STRESSES FOR 70' LONG PILE

STRIPPING

(← IMPACT FACTOR)

$W = 1.5 \times 0.150 = 0.225\ ^K/_{FT}$

$M = (0.0214)(0.225)(70)^2(12) = 283\ ^{K\text{-}IN}$

@ RELEASE $F_i = (0.9)(6)(21.7) = 117.2^K$

$f_t = \frac{117.2}{14.2} + \frac{283}{277} = -0.825 + 1.020 = +0.195$ KSI

$F_{t\ ALLOW} = \frac{F_r}{F.S.} = \frac{7.5\sqrt{3500}}{1.65} = +0.269$ KSI O.K

2 PT. STRIP

0.2L | 0.6L | 0.2L

$M^+ = M^- = 0.0214\,wL^2$

STORAGE

BLOCK AT 4 POINTS

4 PT. STORE & SHIP

0.1L | 0.3L | 0.2L | 0.3L | 0.1L

$M^+ = M^- = 0.0056\,wL^2$

SHIPPING

(← IMPACT FACTOR)

$w = 2.0 \times 0.150 = 0.300\ ^K/_{FT}$

TRY SUPPORTING AT 2 PTS

$M = (0.0214)(0.3)(70)^2(12) = 377\ ^{K\text{-}IN}$

$f_t = -0.715 + \frac{377}{277} = +0.645$ KSI N.G.

SHIP ON SPECIAL RIGS SUPPORTED AT 4 PTS

ERECTION

$w = 1.25 \times 0.150 = 0.188\ ^K/_{FT}$

TRY 3 POINT PICK

$M = (0.02)(0.188)(70)^2(12) = 221\ ^{K\text{-}IN}$

$f_t = -0.715 + \frac{221}{277} = +0.082$ KSI O.K. · ERECT AS SHOWN

0.42wL | 0.42wL | 0.16wL

0.2L | 0.4L | 0.4L

$M = 0.02\,wL^2$

MILD STEEL REQ'D AT END OF PILE TO DEVELOP MOMENT AT PILE CAP

$A_s = \frac{UM}{a_u d} = \frac{56}{4 \times 6} = 2.33\ IN^2$ USE 4 - #7 ($A_s = 2.40\ IN^2$)

USE NEAT CEMENT GROUT IN SLEEVES CAST IN PILE

REINFORCING SUMMARY - 12" SQ P/S CONC. PILE

$f'_{ci} = 3500$

$f'_c = 6100$

2½"

6-7/16" φ - 270K STRANDS

5 GA. SPIRAL
5 TURNS @ 1" PITCH
16 TURNS @ 3"
BAL. @ 6" EA. END

1¼" φ FLEXITUBE DOWEL HOLE × 6'-0" LONG

12" SQUARE

SANTA FE - POMEROY, INC.

Engineering and Construction Subsidiary of Santa Fe International Corp
500 Hopper Street Extension, Petaluma, California 94952
Telephone (415) 982-7500, (707) 763-1918

PRETENSIONED PRESTRESSED CONCRETE PILES
PILE INTERACTION CURVES

SOLID SQUARE SHAPE PILES -
10'', 12'', 14'', 16'', and 18''

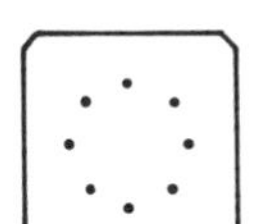

The interaction curves shown hereon are presented as an aid to the Engineer when designing structures using the ultimate strength method.

The curves are derived from a computer solution in accordance with ACI's Building Code. A rectangular stress block was assumed in the concrete with concrete strain limited to .003 in/in. Steel stresses are based on average values of stress strain curves for 1/2 inch and 7/16 inch diameter ASTM A416 grade 270 strands. Concrete Strength is 6,000 psi in 28 days.

A strength reduction factor ϕ of 0.7 has already been included in the diagrams for axial loads in the range from 100% to 10% of ultimate axial load. Below 10%, the ϕ factor increases linearly to a maximum of 0.9 for the case of pure flexure and combined tension and bending.

For section properties, effective prestress, allowable working design loads and other technical information for these piles see Santa Fe's "Properties of Pretensioned Prestressed Concrete Piles" sheet.

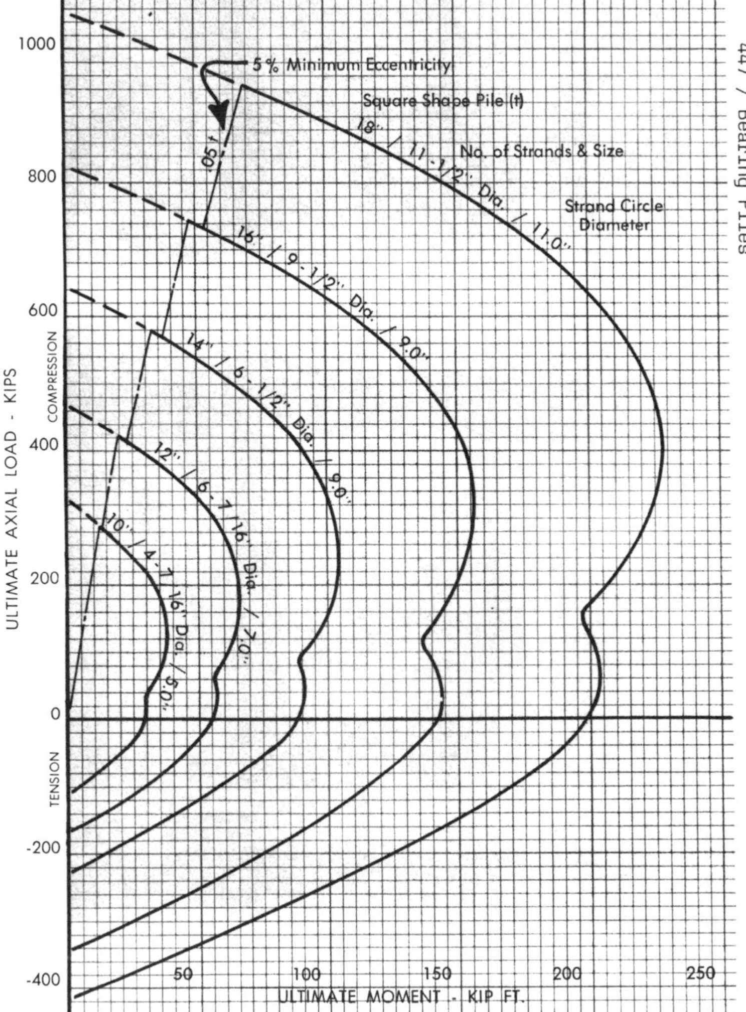

SPLICE DETAILS

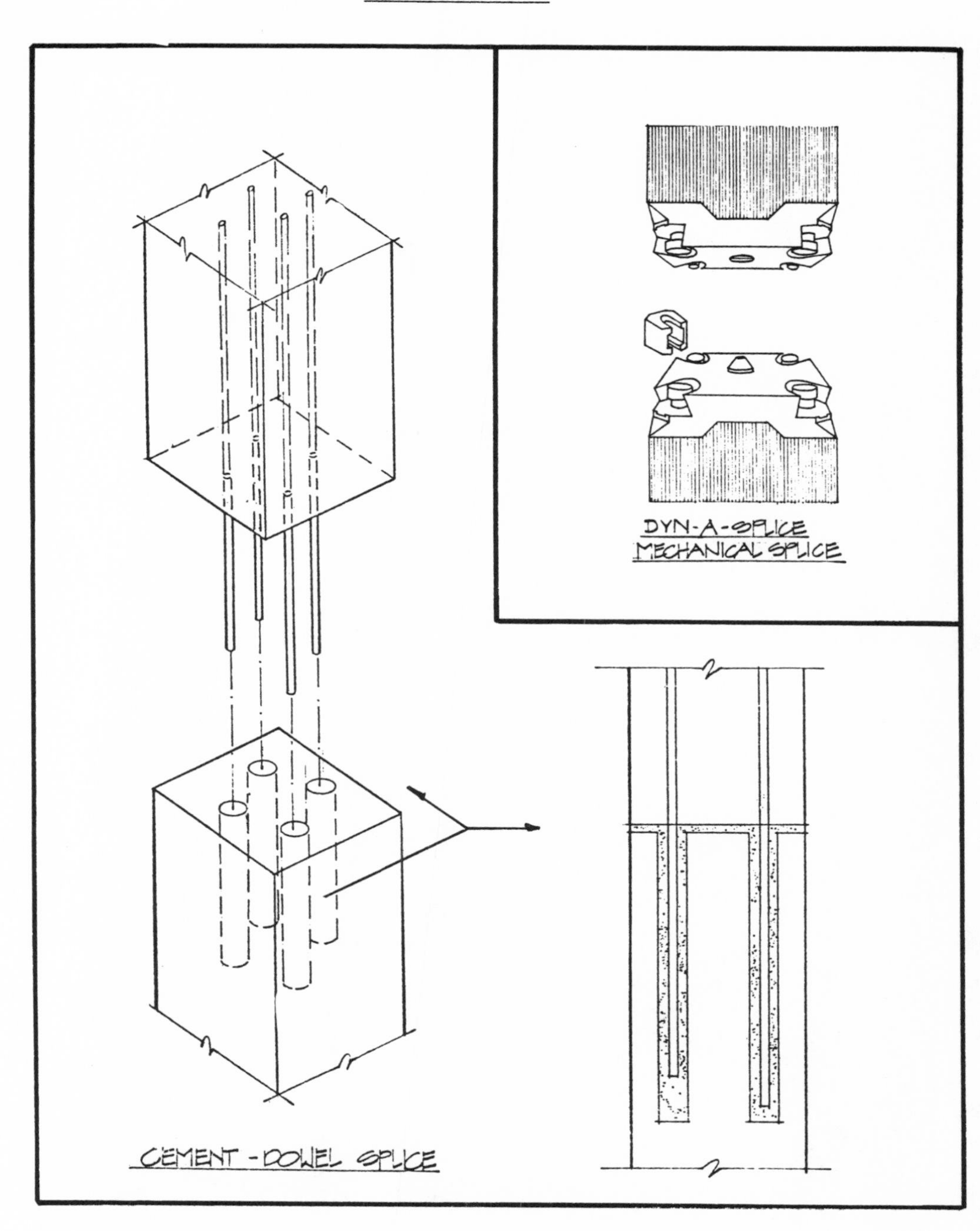

PRESTRESSED CONCRETE SHEET PILING

OYSTER POINT MARINA BREAKWATER - SO. SAN FRANCISCO

60 foot long prestressed concrete sheet pile units are shown here being slid into place in soft bay mud. The 14 inch thick units were shipped by barge from the fabricator's yard in Richmond. Installation equipment is also barge mounted.

Prestressed concrete sheet piling units are used to retain earth or other materials, such as in bulkhead walls for marine applications, or in special instances, as subterranean building foundation walls. They are also installed as breakwaters to protect harbors and marinas. In those applications where they support vertical loads of any appreciable magnitude, they must be tied back, or braced at sufficient intervals so that the bending stresses do not limit the axial load carrying capacity of the pile. Interaction diagrams for prestressed concrete sheet piling are constructed in a similar manner as for prestressed concrete bearing piles or columns. Various cross-sections have been used for sheet pile applications, but we shall only discuss the rectangular shape here as a standard product. The sections are normally made with a tongue and groove interlock to facilitate installation. Groove and groove configurations are also fabricated, where the area between the grooves is grouted above the waterline after installation. The area below the waterline usually transitions to the tongue and groove detail on one side, once again, for installation reasons. For more detailed practical information on sheet piles see the excellent text by Professor Ben C. Gerwick, Jr. entitled CONSTRUCTION OF PRESTRESSED CONCRETE STRUCTURES.

HUNTINGTON HARBOUR BULKHEAD WALL - HUNTINGTON BEACH

40 foot long prestressed concrete sheet pile units stockpiled in plant awaiting delivery.

AVAILABILITY

Solid rectangular units are fabricated in depths to 12". Depths greater than 12" are normally voided to reduce weight in handling. Widths are selected subject to a weight limitation per pile of 12 tons, unless a thorough investigation of on-site handling and installation limitations is performed. This limit, although admittedly somewhat arbitrary, assures that shipping will be optimized in that two pieces may be handled per load. Prestressed concrete sheet piles are fabricated by all of the manufacturers of prestressed concrete products. Some have universal prestressing lines with flat soffits specifically devoted to the production of rectangular sheet pile units. Others may temporarily convert their double tee or single tee beds into flat soffits by filling the stems with concrete.

OYSTER POINT MARINA BREAKWATER - SO. SAN FRANCISCO

View of prestressing bed with previous day's pour awaiting QC inspection and placement into storage. Note slight corner bevel to facilitate installation. Note also moveable side rails with keyway shape.

INSTALLATION CONSIDERATIONS

Sheet piling installation in the water usually employs a temporary wood or steel wale alignment system. Either wood or steel wales are supported on one end by the section of wall already driven, and on the open end by a temporary steel or concrete pile driven on line, and subsequently removed and "leapfrogged" ahead as wall installation proceeds. The wales are connected to the wall by large "C" clamps or with bolts through holes provided in the concrete units for this use, and also to hold up steel forming brackets used to construct the cast-in-place concrete coping. Driving or jetting proceeds with the tongue edge proceeding first in the direction of the driving. The bottoms of the piles are sloped so that the pile as it is driven is forced into the portion of the wall already installed, assuring plumbness. Without this, the units will tend to "walk," or experience a tendency for the bottom of the unit to splay out. The amount of slope depends upon the soil and method of installation. For units being jetted in sand, a full 45° bevel is usually provided. For units driven in silt or clay, only a slight corner bevel is required to keep the bottom edge in against the previously driven section. If hard driving is encountered, or in soft mud, too great a bevel could cause excessive shear key friction and impede driving. Given the manner in which the groove and tongue are related to the direction of driving, and the design of the bevelled bottom, the installation sequence of the wall must be known beforehand in order for the units to be properly designed.

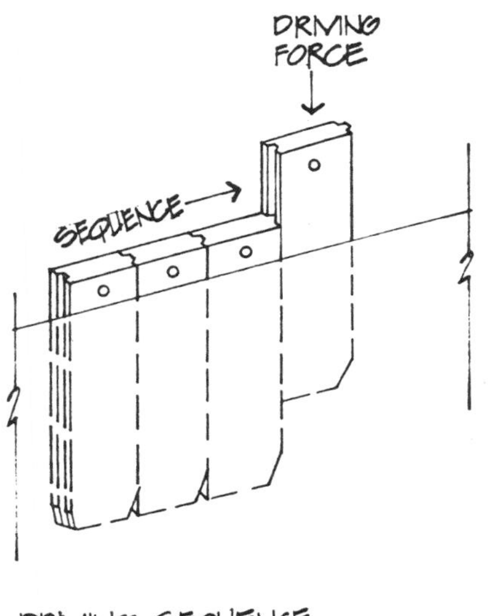

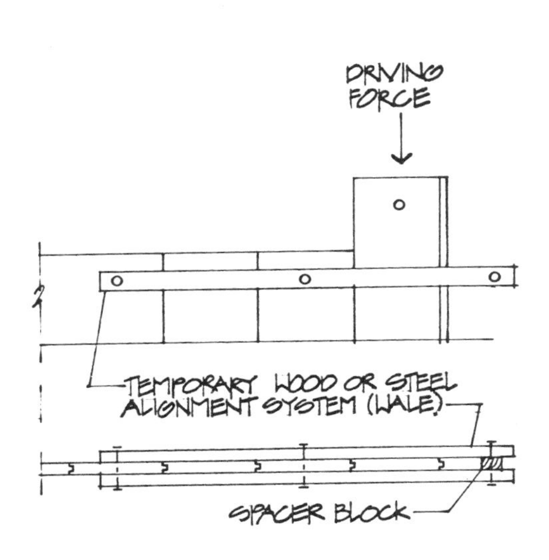

Jetting consists of using water pumped at high pressure through holes cast in the pile units to install the piles in granular soils. Jetting involves special techniques that are essential for efficient production in the field. The ducts cast in the sheet piles should be of sufficient size and number to accommodate an adequate volume of water in the pile. A minimum diameter of 2½" is recommended, with 4" preferred. 6 to 8% of the gross pile cross sectional area should be provided as jet tube opening. The bottom of the tube should be constricted to increase water velocity at discharge from the bottom of the pile, thereby increasing the movement of sand from the driving zone. Sufficient 2 or 3 stage pumps are required to develop the head and discharge essential for good efficient installation. Avoid the use of fragile PVC couplers at the top of the pile tube for providing a connection to the pump hoses. Sometimes a gang of pipes is connected to a manifold and placed over the pile to aid in jetting (external jetting). Sometimes in silty sands, the problem is not in getting the pile down but rather in keeping the pile from sinking too far before sufficient back flow occurs around the driven unit to develop side friction to maintain position. A cast-in-place concrete coping is usually poured to serve as an anchorage beam for tie back systems, and to dress-up any vertical irregularities of as-driven pile elevations. As was mentioned above, a 1½" diameter hole placed 15" down from the top of the pile serves to hold coping form brackets in position. The bracket will have a long vertical slot in the supporting angle to adjust to final installed pile elevations. Wall installation should be within an allowable alignment tolerance of ±2".

HUNTINGTON HARBOR BULKHEAD WALL

Twelve ton sheet pile unit being positioned for jetting. 85 ton crane on shore was used to hold the sheet piles during installation. Small barge mounted crane was used to hold auxiliary jetting equipment. Main jet pump was connected to hoses feeding water directly to pipes cast in the pile.

Another aspect of construction of sheet pile walls is the difficulty in maintaining exact horizontal control. For this reason, if the wall has changes in alignment, or closes upon itself, it is advisable to make up the corner units later based upon field measurements. The disadvantage of this is that additional plant and field set-up costs are involved, and time is lost in waiting for these special units. An alternative to the above is forming and pouring in place the corner or closure piece required for the wall at these points.

On the following pages are sketches depicting some typical conditions peculiar to sheet piling.

TYPICAL SHEET PILE UNITS

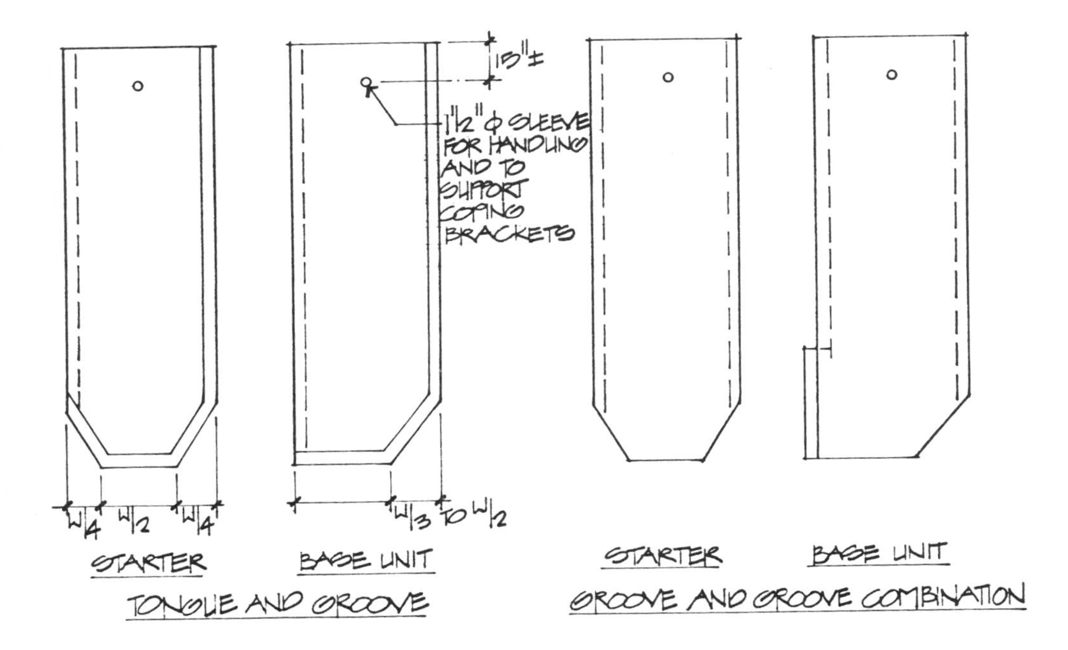

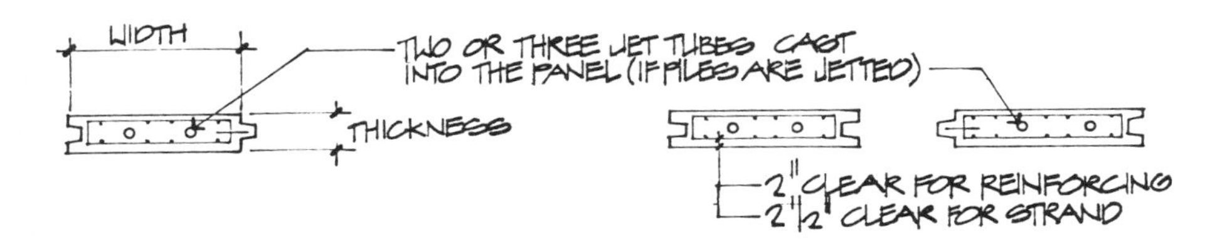

COPING

A cast-in-place concrete coping is usually formed after the wall installation is complete. This coping "dresses up" the top of the wall and hides minor variations in alignment. It also serves as a carrying beam to distribute tie back loads from wall active pressure.

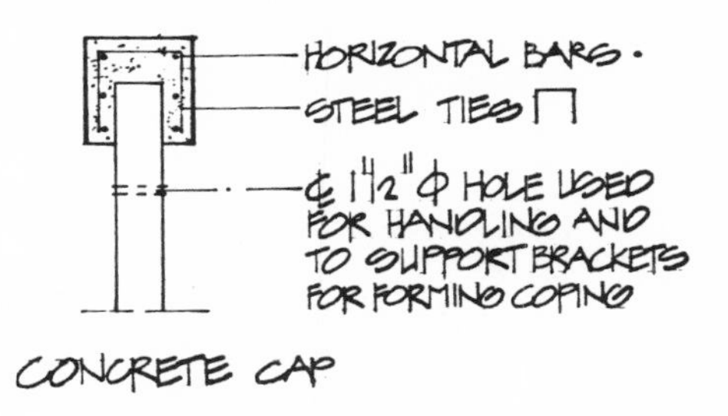

TYPICAL INSTALLATIONS

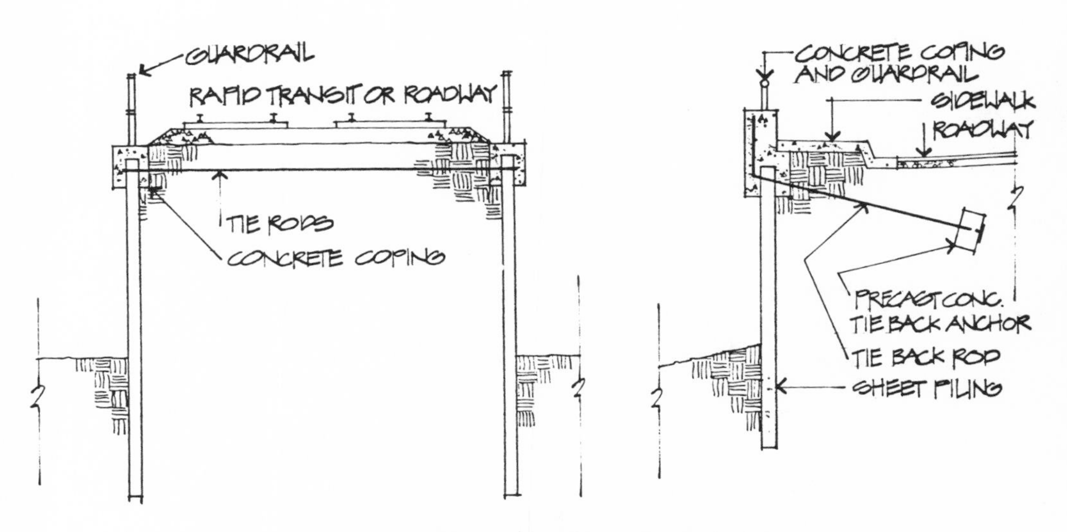

RAISED RAPID TRANSIT OR HIGHWAY BED

TYPICAL INSTALLATIONS

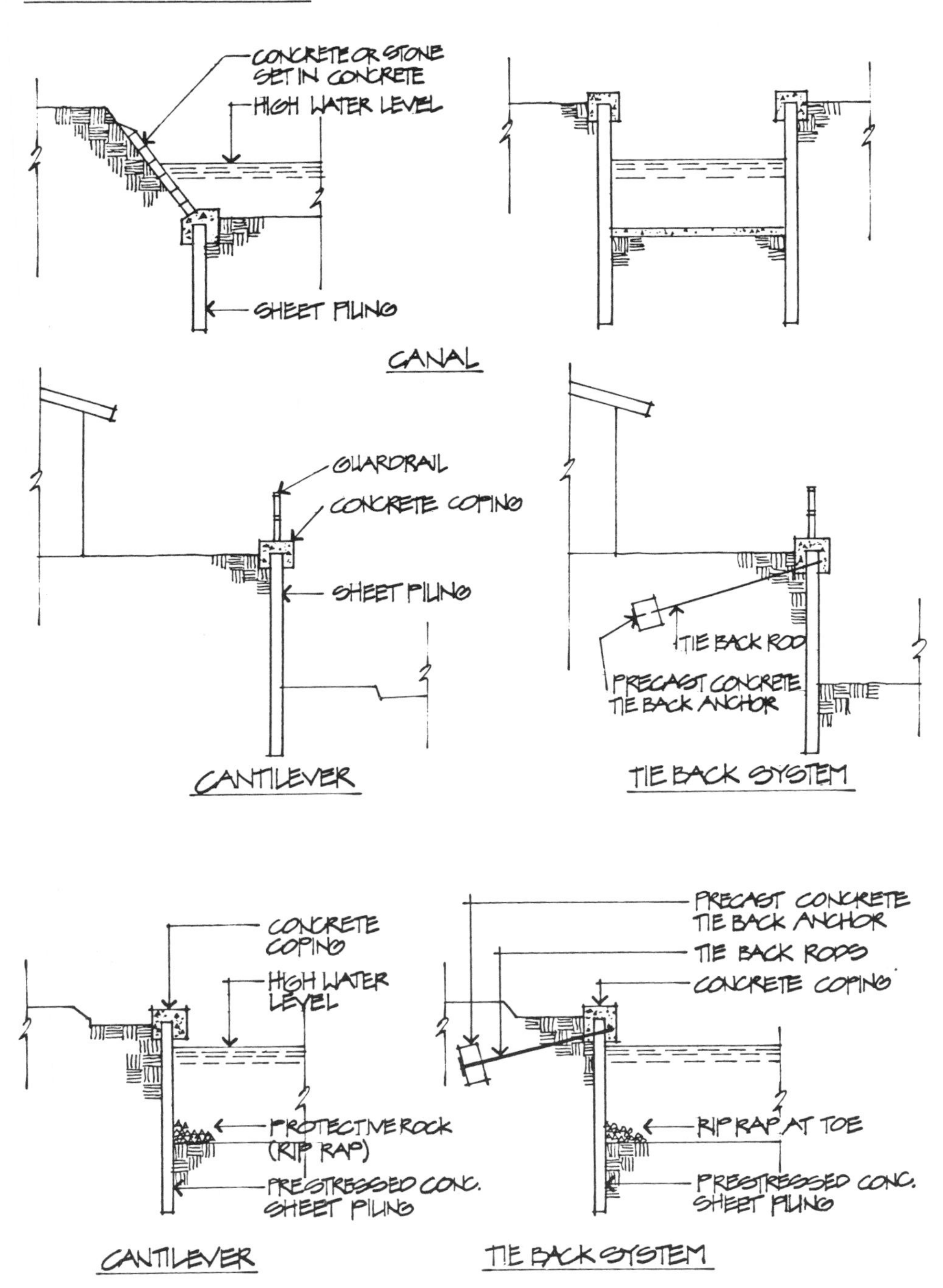

Sheet piling units are handled by lifting at cast in strand lift loops or manufactured inserts.

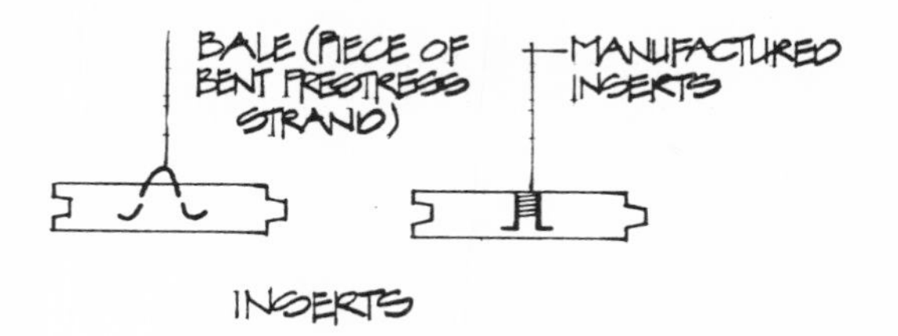

TYPICAL HANDLING CONDITIONS

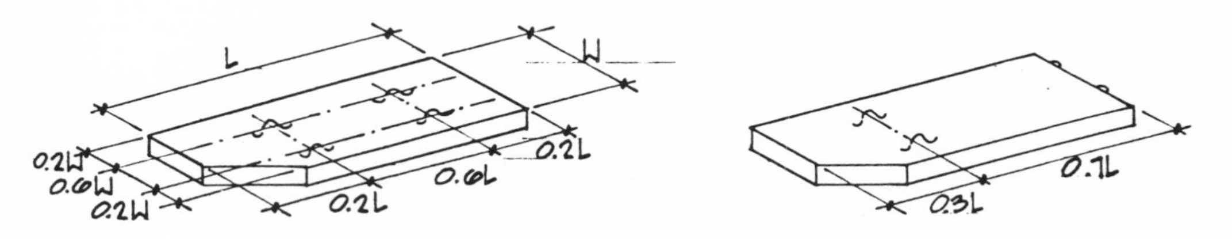

STRIPPING AND PLANT HANDLING (EQUAL NEGATIVE AND POSITIVE MOMENTS)

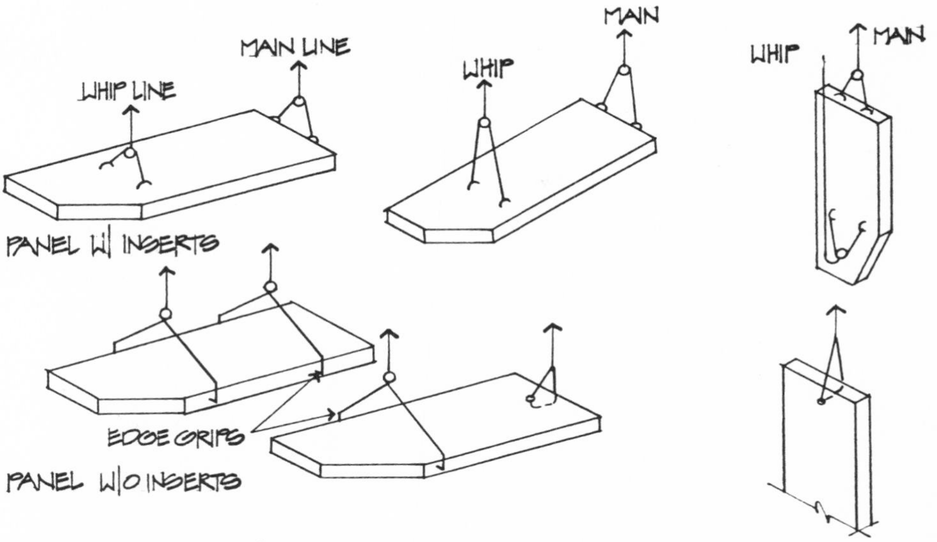

ERECTION HANDLING

HUNTINGTON HARBOUR
BULKHEAD WALL

12 ton sheet pile unit being "tripped up" into the vertical position from a flatbed trailer. Note hoses attached to jet pipes cast in pile, which will subsequently be hooked up to jet pumps located on an off-shore barge.

DESIGN CONSIDERATIONS

A thorough knowledge of the soil conditions at the site is essential in order for sheet piles to be correctly designed. Not only should the soils report indicate the active and passive soil pressures to be used in the design, but the report should also indicate the method of installation dictated by the soil present. If the soil is fine sand with a minimum of clay or silt, jetting is usually permitted. In addition the soils report will give the soil pressure diagram to be used in the design. If the piles are to be driven, then a minimum concentric prestress of 700 psi is indicated, and spiral or closed loop ties should be used for lateral reinforcing, with the spacing being very close at the ends, as for prestressed concrete bearing piles. If however, jetting is indicated, then the minimum prestress will be that required to satisfy bending and handling, and a certain amount of eccentricity is permitted, usually limited by that value producing an instantaneous camber of 1". The strands will be placed in the bottom of the member as cast, thereby facilitating handling, and also resulting in the smooth soffit side of the member being adjacent to the backfill material behind the wall after installation, to minimize soil friction and a consequent buildup of active soil pressure. The PCI Design Handbook on page 2-63 has preliminary design charts to assist the designer in selecting a pile thickness from a known value of service load moment. The balance of the sheet pile design proceeds in a similar manner as for bearing piles, in satisfying elastic design criteria for horizontal loading due to handling, vertical loading during installation, and in satisfying both elastic and ultimate design requirements for both bending and axial forces resulting from the in-situ condition. The following design example gives a practical demonstration of the principles discussed here.

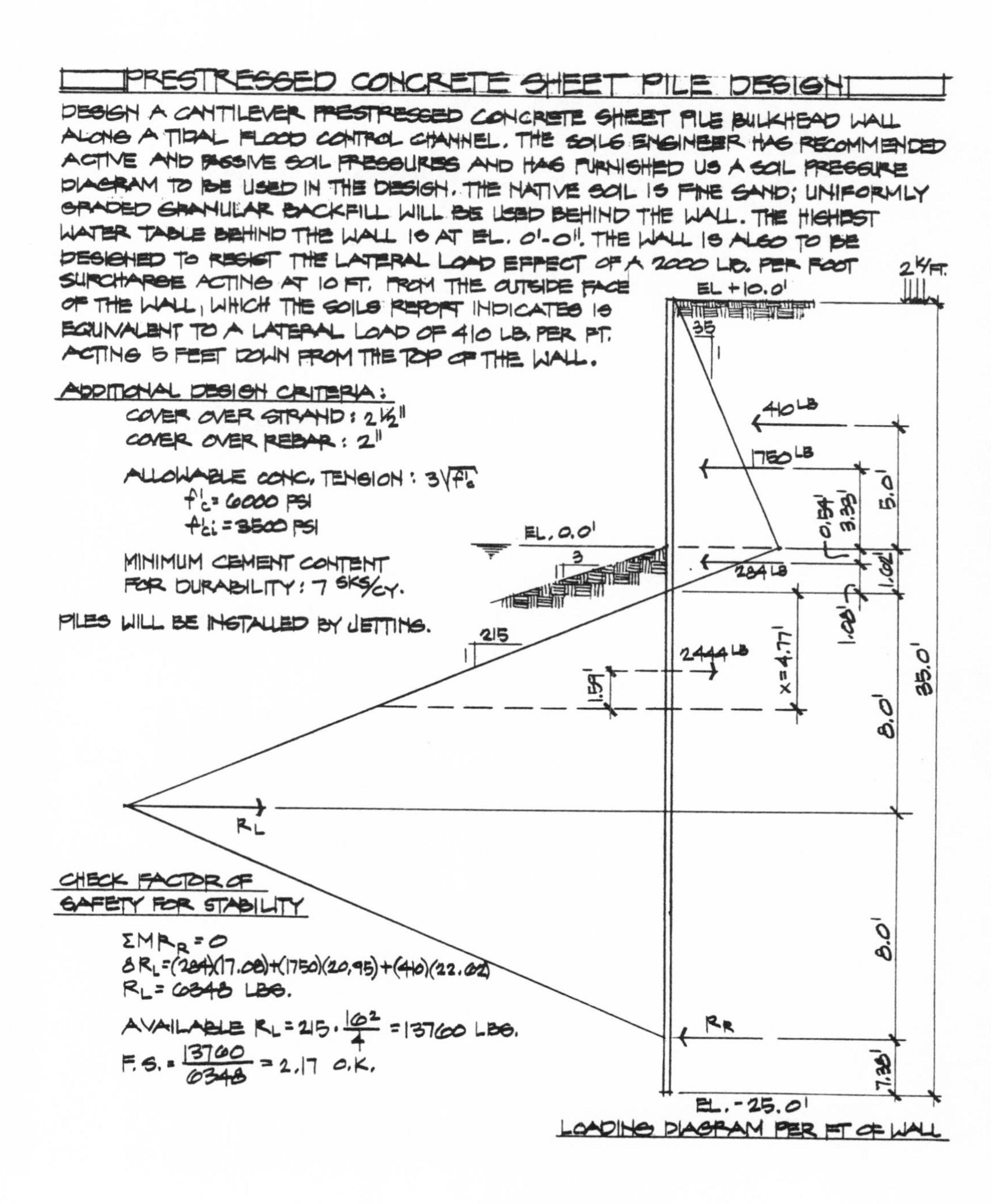

PRESTRESSED CONCRETE SHEET PILE DESIGN

DESIGN A CANTILEVER PRESTRESSED CONCRETE SHEET PILE BULKHEAD WALL ALONG A TIDAL FLOOD CONTROL CHANNEL. THE SOILS ENGINEER HAS RECOMMENDED ACTIVE AND PASSIVE SOIL PRESSURES AND HAS FURNISHED US A SOIL PRESSURE DIAGRAM TO BE USED IN THE DESIGN. THE NATIVE SOIL IS FINE SAND; UNIFORMLY GRADED GRANULAR BACKFILL WILL BE USED BEHIND THE WALL. THE HIGHEST WATER TABLE BEHIND THE WALL IS AT EL. 0'-0". THE WALL IS ALSO TO BE DESIGNED TO RESIST THE LATERAL LOAD EFFECT OF A 2000 LB. PER FOOT SURCHARGE ACTING AT 10 FT. FROM THE OUTSIDE FACE OF THE WALL, WHICH THE SOILS REPORT INDICATES IS EQUIVALENT TO A LATERAL LOAD OF 410 LB. PER FT. ACTING 5 FEET DOWN FROM THE TOP OF THE WALL.

ADDITIONAL DESIGN CRITERIA:

COVER OVER STRAND: 2½"
COVER OVER REBAR: 2"

ALLOWABLE CONC. TENSION: $3\sqrt{f'_c}$
$f'_c = 6000$ PSI
$f'_{ci} = 3500$ PSI

MINIMUM CEMENT CONTENT FOR DURABILITY: 7 SKS/CY.

PILES WILL BE INSTALLED BY JETTING.

CHECK FACTOR OF SAFETY FOR STABILITY

$\Sigma M R_R = 0$
$8 R_L = (284)(17.08) + (1750)(20.95) + (410)(22.02)$
$R_L = 6348$ LBS.

AVAILABLE $R_L = 215 \cdot \frac{16^2}{4} = 13760$ LBS.

$F.S. = \frac{13760}{6348} = 2.17$ O.K.

DETERMINE MAXIMUM BENDING MOMENT IN PILE:

PT. OF ZERO SHEAR

$$410+1750+284 = 215\frac{x^2}{2} \qquad \therefore x = 4.77'$$

$$M_{MAX} = (410)(11.39)+(1750)(9.72)+(284)(6.39)-(2444)(1.59) = \underline{19{,}609 \text{ LB-FT/FT}}$$

(2-63)* ∴ TRY 5'-0" WIDE × 10" THICK UNIT (11 T/EA.)

SECTION PROPERTIES

$A = 600 \text{ IN}^2 \quad WT = 0.63 \text{ K/FT}$

$I = 5000 \text{ IN}^4$

$S = 1000 \text{ IN}^3$

$d = 7.25'' \quad d' = 2.75''$

C.G.S.; 10"; 2.75"; 4.5"; 2.75"; 5'-0"

STRAND REQUIREMENT TO SATISFY SERVICE LOADS (ASSUME $e = 0.7''$)

$$F_f = \frac{\frac{M}{S} - f_t}{\frac{1}{A} + \frac{e}{S}} = \frac{\frac{19.61\times5\times12}{1000} - 3\sqrt{6000}}{\frac{1}{600} + \frac{0.7}{1000}} = \frac{1.176 - 0.232}{0.00237} = 398^K$$

(3-15) USING ½" φ - 270^K STRANDS, $F_f = 0.78 \times 28.9 = 22.5$ K/STRAND

$$N = \frac{398}{22.5} = 17.68 \quad \therefore \text{ USE 18 - ½" φ - 270}^K \text{ STRANDS}$$

TRY 12 IN. INSIDE FACE & 6 IN. OUTSIDE FACE

$$e = \frac{(12)(7.25)+(6)(2.75)}{18} - 5.0 = 0.75'' > e \text{ ASSUMED - O.K.}$$

CHECK COMPRESSION - FINAL CONDITION

$$f_c = \frac{(22.5)(18)}{600} + \frac{(22.5)(18)(0.75)}{1000} - \frac{(19.61)(5)(12)}{1000}$$

$$= -0.675 + 0.303 - 1.176 = -1.548 < 0.45\sqrt{f'_c} \quad \text{O.K.}$$

(3-14) CHECK ULTIMATE STRENGTH

(3-53)(3-5)

$$C\bar{\omega}_p = (1.13)\frac{(12)(.153)(270)}{(60)(7.25)(6)} = 0.215 \qquad f_{se} = 147 \text{ KSI}$$

FROM CHART $f_{ps}/f_{pu} = 0.95 \quad \therefore f_{ps} = (0.95)(270) = 256.5$ KSI

$$a/2 = \frac{A_{ps} f_{ps}}{1.7 f'_c b} = \frac{(1.837)(256.5)}{(1.7)(6)(60)} = 0.77''$$

$$M_u = \phi A_{ps} f_{ps}(d - a/2) = (0.9)(1.837)(256.5)(7.25-0.77) = \underline{2748^{K\text{-}IN}}$$

$$UM_H = 1.7 \times 5 \times 19.61 \times 12 = \underline{2000^{K\text{-}IN} < M_u \text{ O.K.}}$$

*◯ PAGE REFERENCE IN 2ND EDITION OF PCI DESIGN HANDBOOK

CHECK SHEAR

$UV = 1.7 \times 5 \times (0.410 + 1.750 + 0.284) = 20.8^K$

$v_u = \frac{20.8}{(0.85)(60)(0.8)(10)} = 0.051 \text{ KSI} < \frac{1}{2} v_c$

NO WEB REINFORCING REQUIRED.

(3-66) CAMBER AT TIME OF DRIVING

(3-27) $\Delta\uparrow = \frac{F_i e \ell^2}{8 E_{ci} I} \times C.F. = \frac{(468)(0.75)(35)^2(144)}{(8)(3600)(5000)} \times 1.80 = 0.77''$

O.K. FOR JETTED PILE

CHECK DEFLECTION AT TOP IN INSTALLED CONDITION

ASSUME FIXITY AT 10 FT. BELOW THE CHANNEL BOTTOM PER UBC-76, SEC. 2908(d).

$\Delta = \left[(2.05)(15)^2(60-15) + (8.75)(13.33)^2(60-13.33) + (1.42)(9.46)^2(60-9.46) - (17.4)(2.8)^2(60-2.8)\right]\left[\frac{1728}{(6)(4700)(5000)}\right]$

$\Delta = 1.13''$

2.05K, 8.75K, 1.42K, 17.4K, 5.0', 1.67', 3.87', 9.46', 2.8', 20.0'

(3-27) WITH CREEP FACTORS

$\Delta = (\Delta\uparrow - \Delta\downarrow) \times C.F. = (0.42)(2.45) - (1.13)(3.00)$

$= -2.36''$ (EXCESSIVE)

∴ USE THICKER AND NARROWER PILE; TRY 12" x 4'-0" SECTION

SECTION PROPERTIES

$A = 576 \text{ IN}^2$

$WT = 10.5\ ^T/_{EA.} = 0.6\ ^K/_{FT}$

$I = 6912 \text{ IN}^4$

$S = 1152 \text{ IN}^3$

$d = 9.25''$

$d' = 2.75''$

C.G.S., 12", 6.5", 2.75", 2.75", 4'-0"

RECALCULATED LOADS & MOMENTS

$M = 4 \times 19.61 = 78.4^{K\text{-}FT}$ OR $941^{K\text{-}IN}$

$UM = 1.7 \times 941 = 1600^{K\text{-}IN}$

$UV = \frac{4}{5} \times 20.8 = 16.64^K$

SERVICE LOAD STRAND REQUIREMENT (ASSUME $e = 0.6''$)

(3-14) $$F_f = \frac{\frac{941}{1152} - 0.232}{\frac{1}{576} + \frac{0.6}{1152}} = 259^K; \; N = \frac{259}{28.9 \times .78} = 11.5$$

USE 7-½" ⌀ STRANDS IN BOTTOM & 5 IN TOP ($e = 0.54''$)

$$f_t = -\frac{271}{576} - \frac{(271)(0.54)}{1152} + \frac{941}{1152} = -0.470 - 0.127 + 0.817 = +0.220 < 3\sqrt{f'_c} \quad \text{O.K.}$$

(3-53) CHECK ULTIMATE STRENGTH

(3-5) $$c\bar{\omega}_p = (1.13)\frac{(7)(.1531)(270)}{(48)(9.25)(6)} = 0.123 \quad \therefore f_{ps} = (0.96)(270) = 259.2 \text{ KSI}$$

$$a/2 = \frac{(1.072)(259.2)}{(1.7)(6)(48)} = 0.57''$$

$$M_u = (0.9)(1.072)(259.2)(9.25 - 0.57) = 2170^{K \cdot IN} > UM \quad \text{O.K.}$$

CHECK SHEAR

$$v_u = \frac{16.64}{(0.85)(48)(0.8)(12)} = 0.042 \text{ KSI} < \tfrac{1}{2} v_c \quad \therefore \text{NO WEB REINF. REQ'D.}$$

(3-66) CAMBER

(3-27) $$\Delta\uparrow = \frac{(312)(0.54)(35)^2(144)}{(8)(3600)(6912)} = 0.15'' \times 1.8 = 0.27'' \quad \text{O.K.}$$

DEFLECTION AT TOP IN INSTALLED CONDITION (SEE SHT. 33)

$$\Delta\downarrow = (4/5)\left(\frac{5000}{6912}\right)(1.13) = 0.65''$$

WITH CREEP

$$\Delta_{L.T.} = (0.15)(2.45) - (0.65)(3.0) = -1.58'' \quad \text{O.K.}$$

CHECK HANDLING

STRIPPING

TRY 2 PT PICK

IMPACT FACTOR = 1.5

0.2L | 0.6L | 0.2L

$M^+ = M^- = 0.0214\, wL^2$

$$M_s = 1.5 \times 0.0214 \times 0.6 \times \overline{35}^2 \times 12 = 283^{K-IN}$$

$$f_t^- = -\frac{312}{576} + \frac{(312)(0.54)}{1152} + \frac{283}{1152} = -0.542 + 0.146 + 0.245 = -0.151 \quad \text{O.K.}$$

SHIPPING ($I = 2.0$) TRY 2 PTS

$$f_t^- = -\frac{271}{576} + \frac{(271)(.54)}{1152} + \frac{283 \times 4/3}{1152} = -0.470 + 0.127 + 0.327 = -0.016 \quad \text{O.K.}$$

ERECTION - TRY LIFTING AT ENDS

$I = 1.25$

$M = (1.25)(0.125)(0.6)(35)^2(12) = 1378^{K\text{-}IN}$

$$f_t^+ = -\frac{271}{576} - \frac{(271)(0.54)}{1152} + \frac{1378}{1152}$$

$= -0.470 - 0.127 + 1.196 = +0.599$ N.G.

$$\frac{7.5\sqrt{f'_c}}{F.S.} = \frac{0.580}{1.65} = +0.352$$

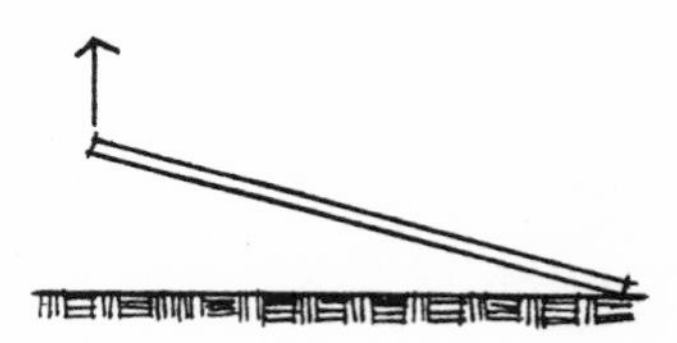

CHECK ERECTION AT BOTTOM STRIPPING POINT AND ERECTION LOOP

$M = (1.25)(0.0685)(0.6)(35)^2(12) = 755^{K\text{-}IN}$

$f_t^+ = -0.470 - 0.127 + \frac{755}{1152} = +0.058$ KSI O.K.

0.8L 0.2L

$M^+ = 0.0685\ wL^2$

HANDLING SUMMARY

STRIP STORE AND SHIP BY HANDLING AT LOOPS CAST IN 0.2 L PTS. ERECT AT LOWER STRIPPING LOOPS AND TOP ERECTION LOOPS.

DESIGN NOTE:

THE METHOD OF INSTALLATION DETERMINES THE DESIGN PHILOSOPHY USED IN THE REINFORCING OF THE PILE. IF THIS PILE WERE INSTALLED BY DRIVING TO BEARING, THEN THE PRESTRESSING WOULD BE CONCENTRIC, AND WOULD HAVE SPIRAL CONFINEMENT REINFORCING OR CLOSED LOOP STIRRUPS ARRANGED AS IN THE STANDARD BEARING PILES.

REINFORCING SUMMARY - 12" × 4'-0" P/S CONC. SHEET PILE

$f'_c = 6000$ PSI

$f'_{ci} = 3500$ PSI

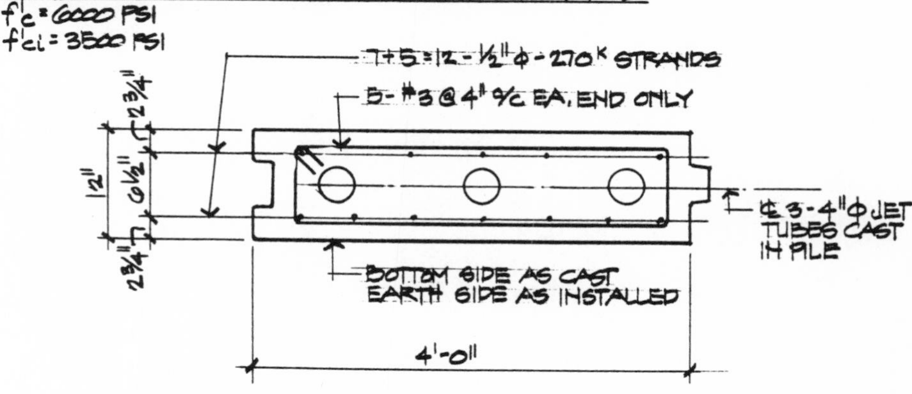

MISCELLANEOUS STRUCTURAL SHAPES

CANDLESTICK PARK - SOUTH SAN FRANCISCO

"L" shaped stadium slabs were furnished in 1961 for the original stadium construction, and again in 1973 for the conversion to allow the park to be used for football as well as baseball. The 28 foot long prestressed concrete units were cast with the vertical riser in the down position, and then rotated 180° prior to shipping.

In addition to the standard sections discussed in the previous component design chapters, two other elements have had prominence in the development of the plant cast precast concrete industry. These elements are the single tee and the stadium slab. While each has had extensive use, the lack of general marketing applications of each has prevented them from becoming truly standard sections.

SINGLE TEE

The single tee, originally referred to as the Lin Tee (after T. Y. Lin) was originally the standard long span deck element in both Eastern and Western United States. The single tee was originally used in parking structures, with the units set one next to the other with each having its own supporting column, or spread apart, supporting cast in place concrete or hollow core slab units spanning in between. Often times, the cast in place slab in-fill design solution was post tensioned transverse to the direction of tee span. The other principle use of the single tee was in achieving long spans over gymnasiums and pools in school construction. In California in recent years, the relaxing of fire requirements in this type of construction in building codes, coupled with the lack of awareness of specifiers and buyers of the other features and benefits of precast prestressed concrete construction have lead to a reduction in this area of product application. None the less, many manufacturers have single tee forms, and this shape can be used anywhere a double tee deck member would be specified. In fact, the wise specifier will allow the option of bidding the single tee as an alternate for double tee construction, thereby increasing competition with resulting economy.

FAB PARKING STRUCTURE - LOS ANGELES

Spread single tees span 63 feet to provide clear span parking in this seven level structure. Hollow core planks span between the tees transversely. Temporary shoring provided efficient use of the composite tee section.

In the erection stage, additional bracing is required for single tees. The flanges are usually blocked at installation to prevent overturning prior to pouring cast in place concrete topping. Often times, flange weld plates are provided so that the most recently placed unit can be immediately welded off to the adjacent structure, thereby rendering the tee stable. When using single tees in a spread apart design supporting hollow core planks, the tee flange requires temporary support, usually consisting of a series of pipe shores supporting a continuous header beneath the flangetip. Also, the plank should be supported on continuous hard-board or Korolath bearing strips placed back 2" from the edge of the flange to prevent localized spalling and bond failures at the flange tip.

Shown below is the "standard" single tee section made by several California prestressers. Note that the flange thickness, slope, and chamfer dimensions may vary slightly with individual manufacturers. It is also wise to check the availability of the deeper sections in the preliminary stages of project development.

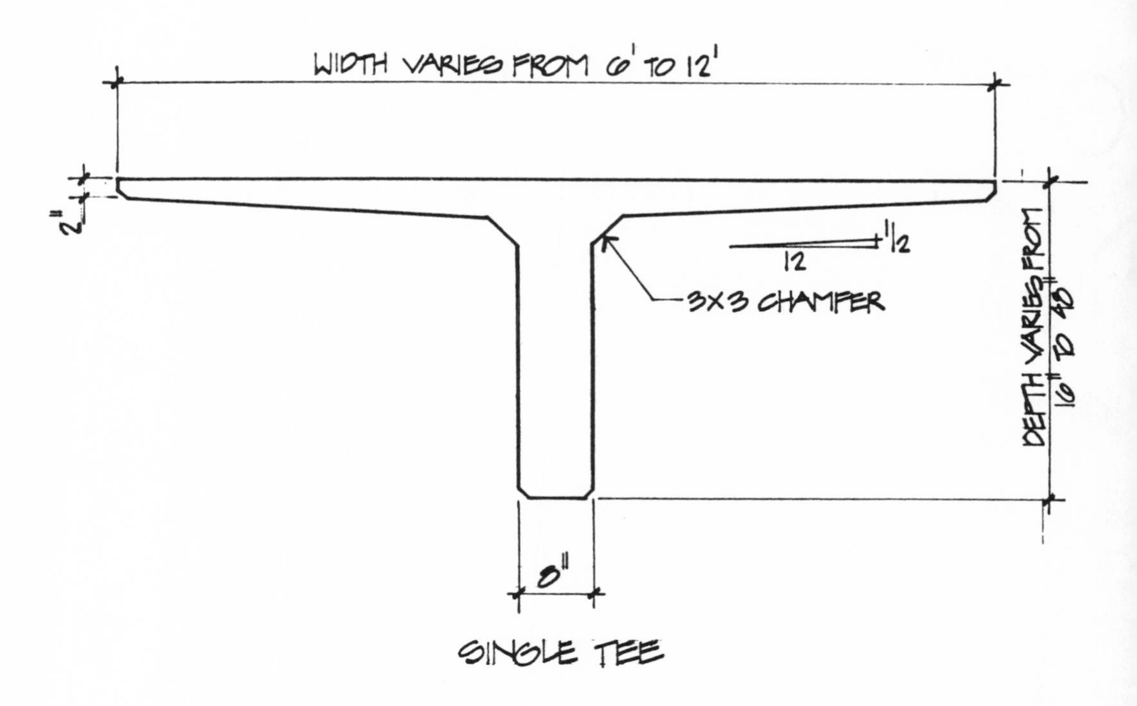

SINGLE TEE

On the following pages are shown preliminary design charts for 8' wide and 10' wide single tees.

8 WIDE SINGLE TEE LOAD TABLES

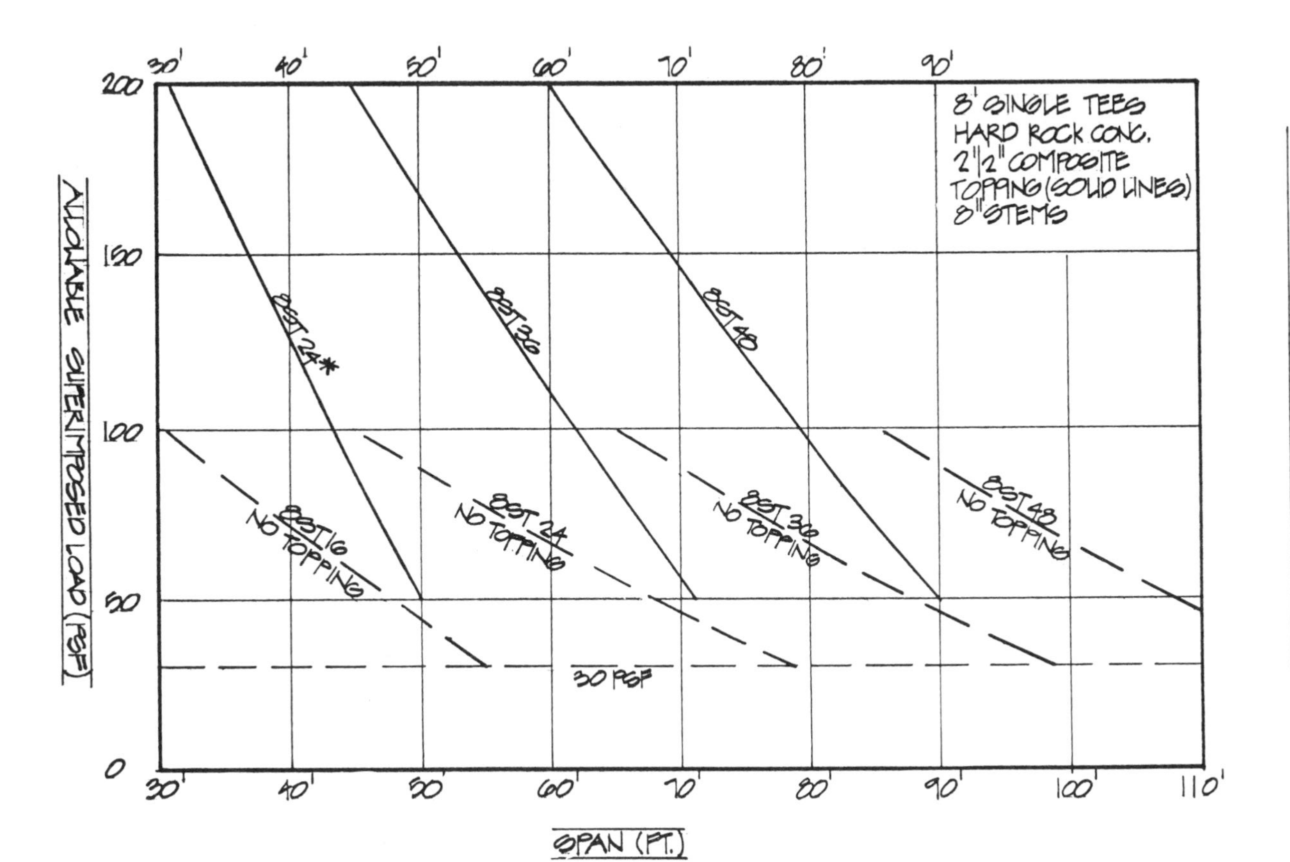

*8ST24 MEANS 8' WIDE SINGLE TEE 24" DEEP

10' WIDE SINGLE TEE LOAD TABLES

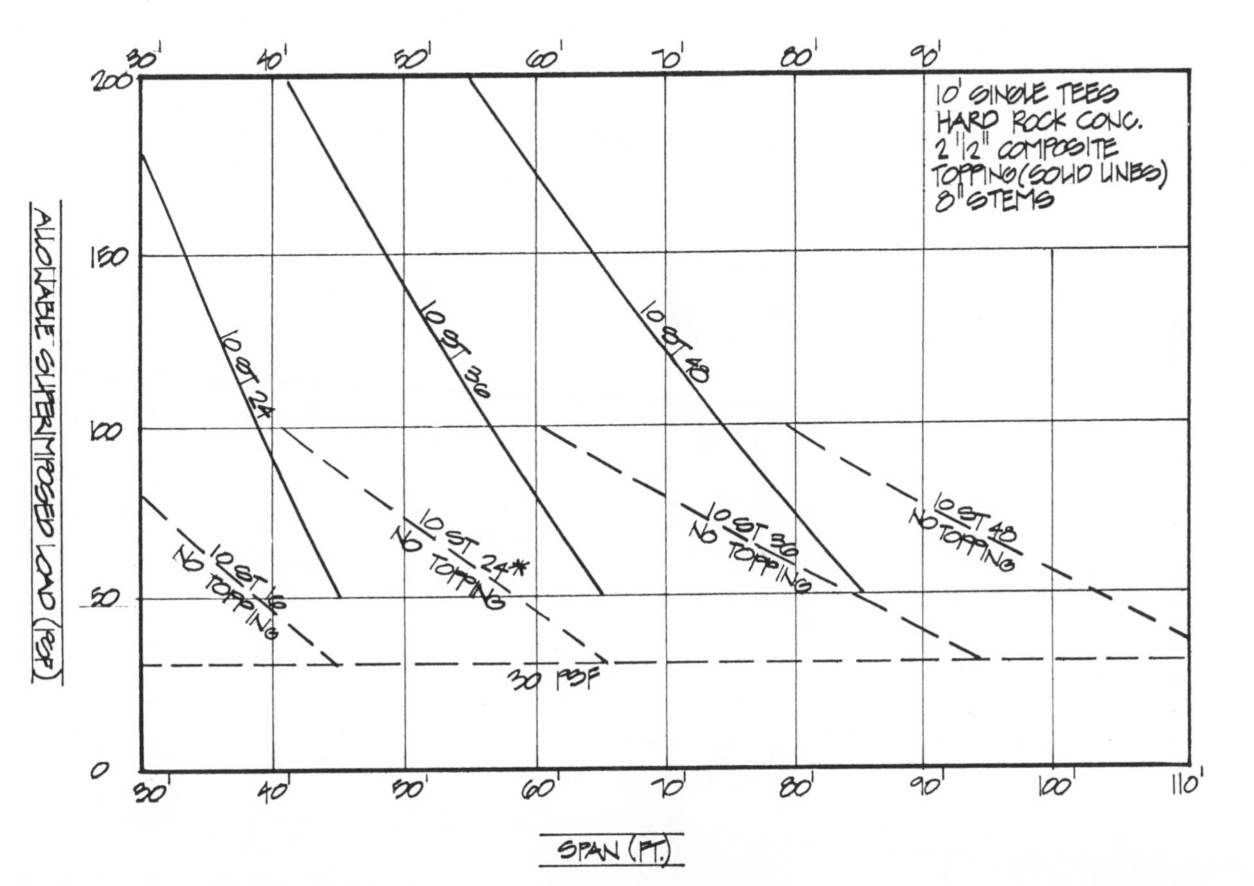

* 10 ST 24 MEANS 10' WIDE SINGLE TEE 24" DEEP

AMBASSADOR COLLEGE PHYSICAL EDUCATION FACILITY - PASADENA

Single tees form both the walls and roof of this multi building complex providing beauty as well as diversity of function.

STADIUM SLAB UNITS

In the early days of our industry, the use of prestressed concrete elements to provide an economical solution to stadium construction was obvious, and that advantage has not changed. The early prestressers made a shape that is generally not available today, namely the 4' wide double tee. This once standard unit was modified slightly to provide a slab unit varying from 10 to 16 inches in depth in a 3' width. This shallow slab solution consisted of a continuous wood seat unit placed directly on the top horizontal edge of the prestressed unit. Single tees were also modified to achieve a similar solution, with the overhanging flange constituting the seat. Later more sophisticated seating requirements for auditoriums coupled with higher angle seating to accommodate semi-enclosed or fully enclosed facilities dictated the selection of a deeper unit. The dimensions commonly selected for slab units now reflect this philosophy, coupled with dimensions which are multiples of optimum tread and riser dimensions. For large projects, casting repetition justifies the initial expense in setting up for multiple slab units 3 wide. The following sketches demonstrate the 3 most common design solutions for stadium slab units.

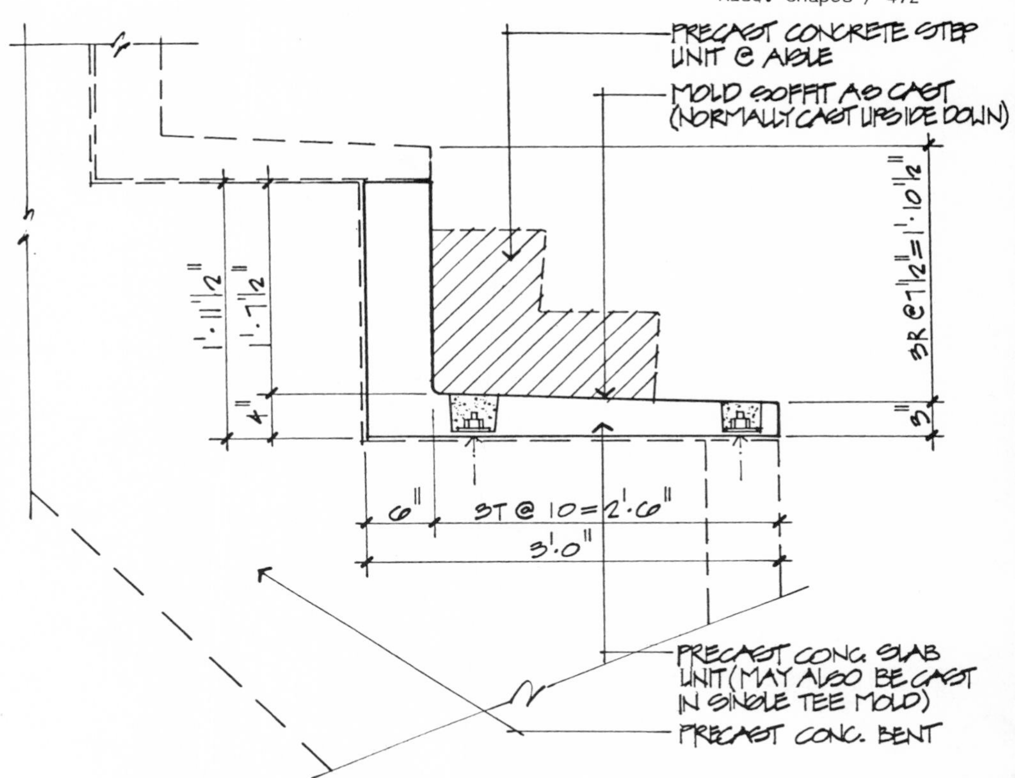

STADIUM SLAB UNIT

THE FORUM - INGLEWOOD

These single riser units were cast inverted in a single tee mold, and rotated 180 degrees prior to erection. The sawtooth cast-in-place concrete bents could have been economically done in precast concrete also

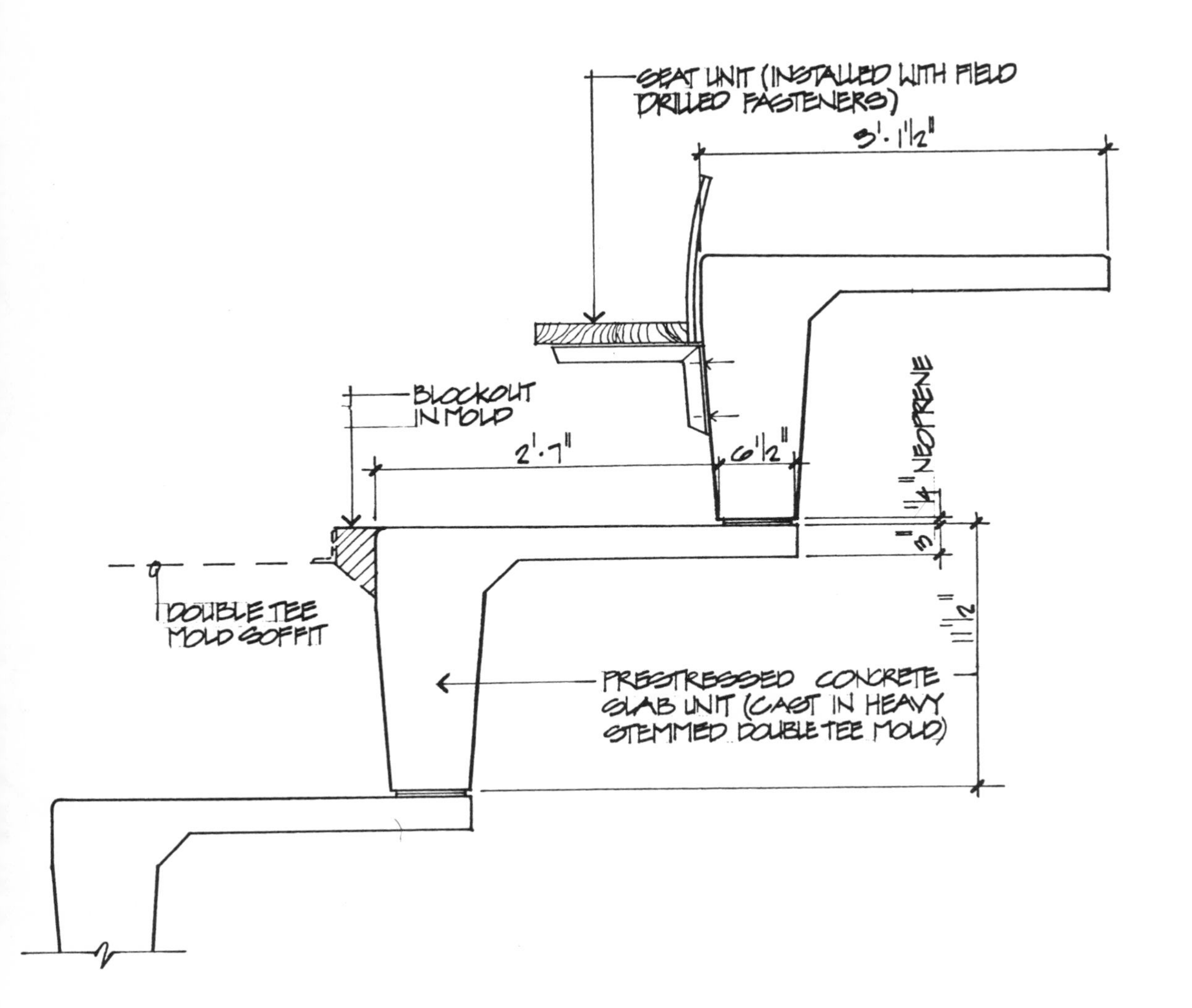

STADIUM SLAB UNIT
FABRICATED IN STD. DOUBLE TEE MOLD

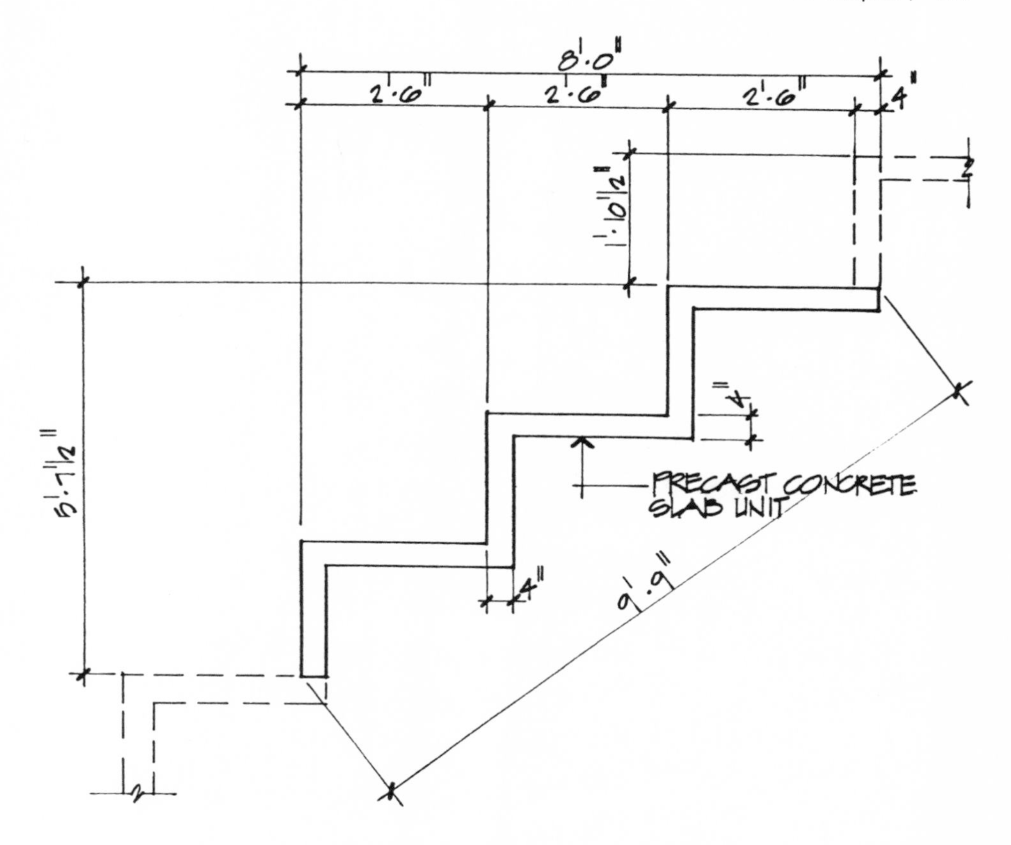

MULTIPLE STADIUM SLAB UNIT

CANDLESTICK PARK
SO. SAN FRANCISCO

Looking closely at the underside of these stadium slabs one can see evidence of hand finishing, indicating that the slabs were cast inverted. The smooth mold finish on the exposed surface has optimum weathering characteristics and appearance.

GUIDE SPECIFICATIONS

CALIFORNIA FIRST BANK BUILDING - SAN FRANCISCO

Careful consideration of desired finishes and required tolerances in drafting the specifications for this project, along with consultation with plant precast concrete manufacturing personnel in preliminary project planning, helped make this project a success. 110,000 square feet of architectural precast concrete facade in 1,576 pieces clad this 324 foot high building. The use of white cement and natural granite fine and coarse aggregates cast under factory controlled conditions made possible the uniform off-white color desired by the architect. Plant casting also made possible the close tolerances specified to assure that the complex interlocking column cover and spandrel units fit together.

The specifications are a very important part of the contract documents, along with the drawings, and the sub-contract agreement between the general contractor and the precast concrete manufacturer. The purpose of the specifications is to define the extent of the precast concrete contractor's responsibilities, and give clearly defined demarcations between his work and that to be performed by other trades. The specifications outline the work to be performed, the quality standards that are to be met, the materials to be used, methods of application, and the responsibilities for design, quality control, fabrication, handling in the plant, transportation, erection, cleaning, and satisfaction of contract requirements for final payment. The specifications are written for each specific project, and are tailor made for the scope and complexity of the precast and prestressed concrete components required on a specific project. Because of this, the specification writer must be thoroughly familiar with precast concrete materials, manufacturing processes, and erection and connection practices. Often times, the problems that occur on a project are a result of poorly written specifications that do not clearly define scope of work and extent of responsibilities of the various subcontractors that interface with the work of the precaster.

The guide specifications given in this chapter are to assist the designer or specification writer in preparing a detailed specification for a specific project. When defining specific requirements in the specification, the desired results, or performance criteria should be given without detailing the manufacturing or construction procedure. In this way, the end result is correctly assigned prime importance, rather than procedures which may vary with various precast concrete manufacturers.

Some of the items of principle importance in a properly written specification are:

QUALITY CONTROL

In plant quality control should be performed by the manufacturer's in house quality control department, as certified by either the Prestressed Concrete Institute's Plant Certification Program or ICBO certification. In the absence of effective in plant quality control, the specs should automatically provide for an outside testing agency to perform this function, if owner provided testing is not called for in Division 1.

MOCK-UP

The correct way to satisfy finish requirements for architectural precast concrete projects is by making provision for a mock-up, usually consisting of the first piece cast on a project, which the architect subsequently approves to set the standard for the project. Many misunderstandings have been created by basing the required finish on a 12" x 12" sample.

HANDLING

Handling information should be shown on the precaster's drawings. Handling points for stripping, storage, and shipping should be shown on the shop drawings. Erection pick points and final lifting procedures should be shown on the erection drawings. It is not sufficient to merely state that the precaster is responsible for the handling of the units in such a manner as to preclude cracking; this information should be indicated on the drawings for ready reference by shop, field, and inspection personnel.

BRACING PLAN

When the precast or prestressed components are not permanently tied into the structure prior to releasing from erection equipment, temporary bracing is required to maintain stability to resist construction, wind, or seismic forces. The method by which this bracing is to be performed is to be shown on a bracing plan, based upon an analysis by a civil engineer registered in the State of California. Once again, it is not sufficient to state that this is the precaster's responsibility; the above documentation is essential to assure the safety of other trades and personnel working on the site.

WATER CEMENT RATIO

A water cement ratio of 0.40 is recommended for both architectural precast concrete and prestressed concrete projects, to assure sufficient durability to stand the test of time. In addition, minimum cement content should be specified when there is evidence that the members will be exposed to corrosive atmospheres or sea water.

ERECTION ACCESS

The general contractor's responsibility to provide wide firmly compacted access roads and sufficient access all around and/or into the structure as required should be clearly defined. Heavy loads and long reaches with heavy precast elements require firm solid footing for crane outrigger supports.

CAMBER

The general contractor and interfacing subcontractors should be aware of camber and differential camber in estimating quantities of topping pours, setting screeds, correctly attaching partitions, etc.

TOLERANCES

Realistic tolerances for both production and other structural elements in the field should be spelled out to permit efficient installation of precast members. These required tolerances also affect detailing of materials that connect to the precast, such as windows and curtain wall systems.

PATCHING

In the normal course of handling and erection, spalls and chipping do occur, and patching is done in the field to correct these conditions. Proper bonding agents and admixtures should be indicated in the specifications to be used for this patching.

PREPAYMENT IN THE YARD

Precast concrete manufacturers spend large amounts of money for material and labor in producing and stockpiling prefabricated elements while work goes on at the site. If the members were site cast, progress payments would be made, so payment for in plant value is justified.

QUALITY CONTROL LABORATORY - BASALT ROCK CO., NAPA

Here we see concrete cylinders being broken early in the morning to ascertain if adequate release strength has been attained to strip yesterday's prestress bed pours. Precast Plant Quality Control Manager Bob Molinari performs the test while Plant Superintendent Bill Harlan looks on.

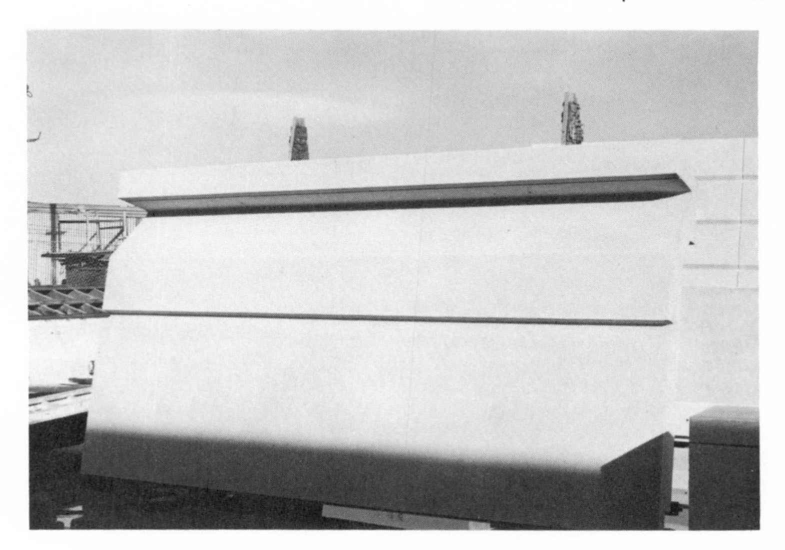

HASTINGS COLLEGE OF LAW - SAN FRANCISCO

The panel exterior finish is as the architect saw it in his dream because he had the wisdom to provide for a mock-up in the specifications for his project. Here we see the mock-up panel for the Hastings College of Law cladding in the fabricator's yard, approved by the architect, and setting the standard for production on the project. This unit was subsequently installed on the building.

GUIDE SPECIFICATIONS FOR ARCHITECTURAL PRECAST CONCRETE (PLANT CAST)
CSI DESIGNATION 03450

THIS DOCUMENT

This document provides a basis for specifying in-plant fabrication and field erection of architectural precast concrete with a variety of textures and finishes. It does not include field-fabricated precast concrete panels, precast structural concrete, nor does it include dampproofing, special coatings applied to the panels, caulking around the panels, or loose attaching hardware.

DRAWINGS AND SPECIFICATIONS

Drawings:

The Architects or Engineers plans will show locations and necessary sections and dimensions to define the size and shape of the architectural precast concrete. Indicate location of joints, both functional and aesthetic, and illustrate details between units. When more than one type of panel material or finish is used, indicate the location of each type on the drawings. Illustrate the details of corners of the structure and interfacing with other materials. Whether sizes and locations of steel reinforcement, and details and locations of typical and special connection items and inserts are shown may be determined by local practices. If reinforcement and connections are not detailed, identify the requirements for design and indicate load support points and space allowed for connections.

Specifications:

Describe the type and quality of the materials incorporated into the units, the design strength of the concrete, the finishes, and the tolerances for casting and erection. The methods and techniques required to achieve similar results will vary with individual precasters. Specifying the results desired without specifically defining manufacturing procedures will ensure a concise and accurate interpretation, and in turn encourage the best competitive bidding.

Coordination:

The responsibility for suppy of items to be placed on or in

the structure in order to receive the precast concrete units depends on the type of structure and varies with local practice. Clearly specify responsibility for supply and installation of hardware. When building frame is structural steel, erection hardware is normally supplied and installed as part of the structural steel. When building frame is cast-in-place concrete, hardware, if not predesigned or shown on drawings, is normally supplied by precast manufacturer and placed by the General Contractor to a hardware layout prepared by the precast supplier. Assurance that type and quantity of hardware items required to be cast into the precast units for other trades are specified and not duplicated, is of greater importance than the supplier. Specialty items, however, should be supplied from the trade requiring them. Verify that materials specified in the section on flashings are galvanically compatible with reglets or counterflashing receivers installed under this section. Check that concrete coatings, adhesives, and sealants specified in other sections are compatible with each other and with the form release agents or surfaces to which they are to be applied.

GUIDE SPECIFICATION

This Guide Specification is intended to be used as a basis for the development of an office master specification or in the preparation of specifications for a particular project. In either case, this Guide Specification must be edited to fit the conditions of use. Particular attention should be given to the deletion of inapplicable provisions. Include necessary items related to a particular project. Include appropriate requirements where blank spaces have been provided.

SECTION 03450

ARCHITECTURAL PRECAST CONCRETE (PLANT CAST)

Part 1 - General

1.01 DESCRIPTION

A. Related Work Specified Elsewhere

1. Concrete Reinforcement: Section 03200

(Delete when reinforcing steel is specified in this section. Architectural precast concrete reinforcing requirements differ from that used for cast-in-place reinforcing and should be specified in this section.)

2. Cast-in-Place Concrete: Section 03300

3. Precast Prestressed Concrete: Section 03420

4. Structural Metal Framing: Section 05100

5. Metal Fabrications: Section 05500

(miscellaneous iron, anchor bolts, or other anchorage devices required for installing precast concrete panels, but furnished and installed in cast-in-place concrete by others.)

6. Waterproofing: Section 07100

(for exposed face of panels)

7. Insulation: Section 07200

(for insulation field applied to precast panels. Insulation cast in precast panels during manu-

facture should be specified in this section.)

8. Flashing and Sheet Metal: Section 07600

   (counterflashing receivers and reglets, unless included in this section.)

9. Sealants and Caulking: Section 07951

10. Glazing Accessories: Section 08850

    (when reglets for use with structural glazing gaskets. Delete when specified in this section.)

11. Painting: Section 09900

    (field touch-up painting. Delete when specified in this section.)

B. Work Installed but Furnished by Others
(delete when furnished by precast manufacturer)

1. Counterflashing Receivers or Reglets, Section 07600

2. Inserts for Window Washing Equipment, Section 03252

C. Testing Agency Provided by Owner
(delete when testing agency is provided by precast manufacturer or contractor. Coordinate with Division 1)

1.02 QUALITY ASSURANCE

A. Acceptable Manufacturers: Minimum of _______ years production experience in architectural precast concrete work of quality and scope required on this project.

(Experience required is usually 2 to 5 years. The manufacture of architectural precast concrete requires a greater degree of craftsmanship than most other concrete products, and therefore requires some prequalification of the manufacturer. Plant certification, as provided in the PCI Plant Certification Program, is satisfactory evidence.)

1. Manufacturer must be able to show that he has experienced personnel, physical facilities,

established quality control procedures, and a management capability sufficient to produce the required units without causing delay to the project.

2. When requested by the Architect, the manufacturer shall submit written evidence of the above requirements.

** OR **

A. Acceptable Manufacturers:
(list a minimum of three)

B. Erector Qualifications: Regularly engaged for at least ______ years in erection of architectural precast concrete units similar to those required on this project.

(Experience required is usually 2 to 5 years.)

C. Qualifications of Welders and Tackers: In Accordance with AWS D 1.1

(Qualified within the past year. Delete when welding is not required.)

D. Testing: In general compliance with testing provisions of Prestressed Concrete Institute MNL-117, "Manual for Quality Control for Plants and Production of Architectural Precast Concrete Products."

E. Testing Agency:
(delete when provided by owner)

1. Not less than ______ years experience in performing concrete tests specified in this section.

(Usually 2 to 5 years experience is adequate)

2. Capable of performing testing in accordance with ASTM E 329.

3. Inspected by Cement and Concrete Reference Laboratory of the National Bureau of Standards.

F. Requirements of Regulatory Agencies: design, manufacture, and installation of architectural precast concrete shall be in accordance with the Uniform Building Code, 1979 edition.

** OR **

(list applicable code)

G. Allowable Tolerances

1. Length and width of precast units measured at face adjacent to mold:

   a. units 10 ft. or under: ±1/8 in.

   b. units 10 ft. to 20 ft.: +1/8 in. - 3/16 in.

   c. units 20 ft. or over: +1/8 in., - 3/16 in. - 1/16 in. per each 10 ft. over 20 ft.

   d. length and width of window openings in panels: ±1/8 in.

   e. all other tolerances as per PCI MNL 117

H. Source Quality Control

1. Quality control and inspection procedures to comply with applicable sections of PCI MNL 117.

2. Water absorption test on unit shall be conducted in accordance with ASTM C127 and ASTM C128 as applicable.

   (Water absorption test is an early indication of weather staining ((rather than durability)). Verify the water absorption of the proposed face mix, which for average exposures and based upon normal weight concrete ((150 lbs. per cubic foot)) should not exceed 5% to 6% by weight. As an improved weathering ((staining)) precaution, lower absorption between 3% to 4% ((by weight)) is feasible with some concrete mixes and consolidation methods. In order to establish comparable absorption figures for all materials the current trend is to specify absorption percentages by volumes. The stated limits for absorption would in volumetric terms correspond to 12% to 14% for average exposures and 8% to 10% for special conditions.)

I. Project Mock-up

1. After standard samples are accepted for color and texture, fabricate full-scale unit meeting design

requirements, stand up in plant, and request Architect's review and acceptance.

(Full-scale samples or inspection of the first production unit are sometimes required, especially when a new design concept or new manufacturing process or other unusual circumstance indicates that proper evaluation cannot otherwise be made. It is difficult to assess appearance from small samples.)

2. Mock-up to be standard of quality for architectural precast concrete work, when accepted by Architect.

(Use to determine range of acceptability with respect to color and texture variations, surface defects and over-all appearance.)

3. Incorporate mock-up into work in location reviewed by Architect after keeping unit in plant for the duration of production, thereby establishing the standard of acceptable quality.

** OR **

4. Mock-up is to be kept in the plant for the duration of production, thereby establishing the standard of acceptable quality. The mock-up is not to be incorporated into the finished project.

1.03 SUBMITTALS

A. Samples

(Number of samples and submittal procedures should be specified in Division 1.)

1. Submit samples representative of finished exposed face showing typical range of color and texture prior to commencement of manufacture.

** OR **

2. Submit samples to match color and finish of approved sample on file in Architect's office prior to commencement of manufacture.

(If the back face of a precast unit is to be exposed, samples of the workmanship, color, and texture of the backing should be shown as well as the facing.)

2. Sample size: approximately 12 in. x 12 in. and of appropriate thickness, representative of the proposed finished product.

B. Fabricator's Drawings

1. Shape drawings

   (Submitted for approval prior to mold fabrication.)

2. Production shop drawings containing:

   a. unit shapes (elevations and sections) and dimensions

   b. finishes

   c. reinforcing, joint, and connection details

   d. piece marks cross referenced to erection drawings

   e. lifting and erection inserts

   f. other items cast into panels

   g. relationship to adjacent material

   (Production show drawings are not normally submitted for approval, except in special cases where the Architect, Engineer, or Contractor agrees to assume responsibility.)

3. Erection drawings containing:

   a. connection details

   b. location and details of hardware attached to structure

   c. handling procedures and sequence of erection for special condtions; erection bracing plans are required when the precast concrete element is not permanently tied into the structure after disengagement from the crane or other erection equipment. This bracing plan shall be prepared by a civil engineer registered in the State of California

   (If sequence of erection is critical to the structural stability of the structure, it should be noted on the plans and specified)

   d. elevations showing locations of individual units by piece mark from shop drawings

4. Anchor bolt or insert plate location plans

(Normally prepared by Precaster and provided to Contractor for work by other trades)

5. Field work sheets

(Required when alterations to existing structures are required prior to erection of precast concrete panels. This work could consist of chipping, cutting, burning, drilling or setting expansion bolts, etc., by either the Precaster or other trades, as applicable and specified)

C. Test Reports: Submit on request, reports on materials, compressive strength tests on concrete and water absorption tests on units.

(Schedule of required tests, number of copies of test reports, and how distributed are included in Testing Laboratory Services, Section ________.)

D. Design Calculations: Prepare and submit on request structural design calculations for units in accordance with the PCI MNL 121 "Manual for Structural Design of Architectural Precast Concrete."

(Design and construction responsibilities are discussed in the Introduction of MNL 121.)

1.04 PRODUCT DELIVERY, STORAGE, AND HANDLING

A. Delivery and Handling

1. Handle and transport units in a position consistent with their shape and design in order to avoid excessive stresses or damage.

2. Lift or support units only at the points shown on the erection shop drawings. If required, show how each panel type is to be handled from stripping to erection.

3. Place nonstaining resilient spacers of even thickness between each unit.

4. Support units during shipment on nonstaining shock-absorbing material.

5. Do not place units on ground.

B. Storage at Project Site

1. Store units to protect from contact with soil,

staining, and from physical damage.

2. Store units, unless otherwise specified, with nonstaining, resilient supports located in same positions as when transported.

3. Store units on firm, level, and smooth surfaces.

4. Place stored units so that identification marks are discernible.

Part 2 - Products

2.01 MATERIALS

A. Concrete

1. Portland cement:

a. ASTM C 150, type__________, _________color.

(Type I - general use; Type II - moderate sulfate resisting; Type III - high early strength. Color: grey, white, buff, etc.)

b. for exposed surfaces use same brand, type and source of supply throughout.

(To minimize color variation. Specify source of supply when color shade is important.)

2. Air entraining agent: ASTM C 260.

(Delete if air entraining agent is not required.)

3. Water reducing, retarding, accelerating admixtures: ASTM C 494.

(Delete if water reducing, retarding or accelerating admixtures are not required. Calcium chloride or admixtures containing significant amount of calcium chloride should not be allowed.)

4. Coloring agent:

a. synthetic mineral oxide

b. harmless to concrete set and strength (consider effects upon concrete prior to final selection)

c. stable at high temperature

d. Sunlight and alkali-fast

(Investigate use of naturally colored fine aggregate in lieu of coloring agent.)

5. Aggregates:

a. provide fine and course aggregates for each type of exposed finish from a single source (pit or quarry) for entire job. They shall be clean, hard, strong, durable, and inert, free of staining or deleterious material.

(Base choice on visual inspection of concrete sample and on assessment of certified test reports. Use same type and source of supply to minimize color variation. Fine aggregate is not always from same source as coarse aggregate.)

b. hard rock aggregates for concrete shall conform to ASTM C 33

c. light weight aggregates for concrete shall conform to ASTM C 330

(Grading requirements for the above are generally vaived or modified.)

d. Material and color:

(Specify type of stone desired, such as crushed marble, quartz, limestone, granite, or locally available gravel. Some light weight aggregates, limestones, and marbles may not be acceptable as facing aggregates. Omit where sample is to be matched.)

e. maximum size and gradation __________

(State required sieve analysis. Omit where sample is to be matched.)

6. Water: free from deleterious matter that may interfere with the color, setting or strength of the concrete.

(Potable water is normally acceptable.)

B. Reinforcing Steel

1. Deformed steel bars:

   ASTM A 615

   (Use grade 40 for #3 and #4 bars; use grade 60 for #5 bars and larger.)

2. Welded wire fabric

   a. plain - ASTM A 185

   b. deformed - ASTM A 497

3. Fabricated steel bar or rod mats:

   ASTM A 184

4. Prestressing strand:

   ASTM A 416, Grade ____________

   (250 ksi or 270 ksi. Used in long or thin panels to control cracking from handling and service load conditions.

C. Cast-in-Anchors

(Loose attachment hardware specified under Miscellaneous Metals, Section 05500)

1. Materials

   a. carbon steel bars:

      ASTM A 306, Grade 65

      (for completely encased anchors)

   b. structural steel: ASTM A 36

      (for carbon steel connection assemblies)

   c. stainless steel: ASTM A 666, Type 304, Grade____

      (Grade A or Grade B, stainless steel attachments for use when resistance to staining merits extra cost)

   d. carbon steel plate: ASTM A 283, Grade ______

      (Grade A, B, C, or D)

e. malleable iron castings: ASTM A 47, Grade ____

(Grades 32510, 35018)

f. carbon steel castings: ASTM A 27, Grade 60-30

g. anchor bolts: ASTM A 307

** OR **

ASTM A 325

** ** **

(for steel bolts, nuts and washers)

h. welded headed studs: AWS D1.1, Part VI, Section 4

2. Finish

a. shop primer: FS TT-0-86, oil base paint, type 1

** OR **

a. shop primer: SSPC-Paint 14

** OR **

a. shop primer: Manufacturer's standard

** ** **

(for exposed carbon steel anchors)

b. hot-dip galvanized: ASTM A 153

(for exposed carbon steel anchors where corrosive environment justifies the additional cost. Field welding should not be permitted on galvanized element)

c. cadmium coating: ASTM A 165

(for threaded fasteners)

d. zinc rich coating: MIL-P-21035, self curing, one component, sacrificial organic coating

(for field spot painting)

D. Receivers for Flashing: 28 ga. formed ___________

(stainless steel, copper, zinc. Coordinate with flashing specification to avoid dissimilar metals. Delete when included in flashing and sheet metal section. Specify whether precaster or others furnish.)

** OR **

D. Receivers for Flashing: Polyvinyl chloride extrusions.

** ** **

E. Sandwich Panel Insulation _______________

(Specify type of insulation, such as foamed plastic ((polystyrene and polyurethane)), glasses ((foamed glass and fiberglass)), foamed or cellular light weight concretes, or light weight mineral aggregate concretes. Thickness of sandwich panel insulation governed by wall U-value requirements.)

F. Grout

(Indicate required strengths on contract drawings.)

1. Cement grout: Portland cement, sand, and water sufficient for placement and hydration.

2. Nonshrink grout: Premixed, packaged ferrous and non-ferrous aggregate shrink-resistant grout.

3. Epoxy-resin grout: Two-component mineral-filled epoxy-polysulfide, FS MMM-G-560___________, Type _______________, Grade C.

   (check with local suppliers to determine availability and types of epoxy-resin grouts.)

2.02 MIXES

A. Concrete Properties

(The back-up concrete and the surface finish concrete can be of one mix design, depending upon resultant finish, or the surface finish (facing mix) concrete can be separate from the back-up concrete. Clearly indicate specific requirements or allow manufacturer's option.)

1. Water-cement ratio: Maximum 40 lbs. of water to 100 lbs. of cement

   (Keep to a minimum consistent with strength and durability requirements and placement needs.)

2. Air entrainment: Amount produced by adding dosage of air entraining agent that will provide 19% ± 3% of entrained air in standard 1:4 sand mortar as tested according to ASTM C 185.

   (Gradation characteristics of most facing mix concrete will not allow use of a given percentage of air.)

** OR **

2. Air entrainment: Minimum 3%; maximum 6%, when tested in accordance with ASTM C 173 or ASTM C 231.

3. Coloring agent: Not more than 10% of cement weight.

   (Amount used should not have any detrimental effects on concrete qualities.)

4. 28-day compressive strength: Minimum of 5000 psi when tested by 6x12 or 4x8 in. cylinders.

** OR **

4. 28-day compressive strength: Minimum 6250 psi when tested on 4 in. cubes.

   (Vary strength to match requirements. Strength requirements for facing mixes and back-up mixes may differ. Also the strength at time of removal from the forms should be stated if critical to the engineering design of the units. The strength level of the concrete should be considered satisfactory if the average of each set of any three consecutive cylinder strength tests equals or exceeds the specified strength and no individual test falls below the specified value by more than 500 psi.)

** ** **

B. Facing Mix

(Delete if separate face mix is not used.)

1. Minimum thickness of face mix after consolidation shall be at least one in. or a minimum of 1-1/2 times the maximum size of aggregates used, whichever is larger.

2. Water-cement and cement-aggregate ratios of face and back-up mixes shall be similar.

(Similar behavior with respect to shrinkage is necessary in order to avoid undue bowing and warping.)

2.03 FABRICATION

A. Manufacturing procedures shall be in general compliance with PCI MNL-117

B. Finishes

(Finishing techniques used in individual plants may vary considerably from one part of the continent to another, and between individual plants. Many plants have developed specific techniques supported by skilled operators or special facilities.)

1. Exposed face to match approved sample or mock-up panel.

(It is preferable to match approved sample rather than attempting to specify method of exposure.)

** OR **

1. Smooth finish; as cast using flat, smooth, non-porous molds.

(Difficult to obtain satisfactory finish. The use of grey cement mixes in smooth exposed surfaces is not recommended.)

** OR **

1. Smooth finish: As cast using fluted, sculptured, board finish or textured form liners.

** OR **

1. Textured finish

a. achieve finish on face surface of precast concrete units by form liners applied to inside of forms

b. distress finish by breaking off portion of face of each flute (also known as split rib).

c. achieve uniformity of cleavage by alternately striking opposite sides of flute

** OR **

1. Exposed aggregate finish

a. apply even coat of retarder to face of mold

b. remove units from forms after concrete hardens

c. expose coarse aggregate by washing or brushing away surface mortar

d. expose aggregate to depth of ____________

(Finishes obtained vary from light etch to a depth of reveal of 1/2 in., but must relate to the size of aggregates. Matrix can be removed to a maximum depth of one-third the average diameter of coarse aggregate but not more than one-half the diameter of smallest sized coarse aggregate.)

** OR **

1. Exposed aggregate finish by acid etching

a. immerse unit in tank of acid solution

** OR **

a. treat surface of unit with brushes which have been immersed in acid solution.

(Use reasonably acid resistant aggregates, such as quartz or granite.)

b. protect hardware, connections and insulation from acid attack

** OR **

1. Exposed aggregate finish: Use power or hand tools to remove mortar and fracture aggregates at the surface of units (bushhammer).

(Use with softer aggregates, such as dolomite and marble.)

** OR **

1. Exposed aggregate finish:

   a. hand place large facing aggregate or brick, or cobblestones over form bottom

   b. produce mortar joints by keeping cast concrete 1/2 in. to 1 in. from face of unit

** OR **

1. Sandblasted finish: Sandblast away _______ of cement-sand matrix to expose aggregate face

   (Exposure of aggregate by sandblasting can vary from 1/16 in. or less to over 3/8 in. Remove matrix to a maximum depth of one-third the average diameter of coarse aggregate but not more than one-half the diameter of smallest sized coarse aggregate. Depth of sandblasting should be adjusted to suit the aggregate hardness and size.)

** OR **

1. Honed or polished finish

   (Honing and polishing of concrete are techniques which require highly skilled personnel. Use with aggregates, such as marble, onyx, and granite.)

   a. polish surface by continued mechanical abrasion with fine grit, followed by special treatment which includes filling of all surface holes and rubbing.

      (Delete if polished surface not desired.)

** OR **

1. Veneer faced finish

   a. cast concrete over ceramic tile, brick or stone placed in the bottom of the mold.

      (Full scale mock-up units with cut stone in actual production sizes, along with casting and curing of the units under realistic production conditions are essential for each new or major application or configuration of the cut stones. Bowing should be carefully measured over several weeks in the normal storage area and the final details of stone sizes and fastening determined to suit the observed behavior.)

b. connection of cut stone face material to concrete shall be by mechanical means

(Provide a complete bondbreaker between the cut stone face material and the concrete. Ceramic tile and brick are bonded to the concrete.)

** ** **

2. ____________ back surfaces of precast concrete units after striking surfaces flush to form finish lines.

((Smooth float finish), (smooth steel trowel), (light broom). (stipple finish). Use for exposed back surfaces of units.)

C. Cover

1. Provide at least 3/4 in. cover for reinforcing steel

(Increase cover requirements when units are exposed to corrosive environment or severe exposure conditions.)

2. Do not use metal chairs, with or without coating, in the finished face.

(For smooth cast facing, stainless steel chairs may be permitted.)

3. Provide embedded anchors, inserts, plates, angles, and other cast-in items with sufficient anchorage and embedment for design requirements.

D. Curing

(A wide variation exists in acceptable curing methods, ranging from no curing in some warm humid areas, to carefully controlled moisture-pressure-temperature-curing. Consult with local panel manufacturers to avoid unrealistic curing requirements.)

Cure precast concrete units until 2000 psi minimum concrete strength has been attained, or that strength required by design to allow the units to be stripped, before removing the units from the forms.

E. Panel Identification

1. Mark each precast panel to correspond to identification mark on shop drawings for panel location.

2. Mark each precast panel with date cast.

F. Acceptance: Architectural precast units which do not meet the color and texture range or the dimensional tolerances may be rejected at the option of the Architect, if they cannot be satisfactorily corrected.

2.04 CONCRETE TESTING

A. Make one compression test at 28 days for each day's production of each type of concrete.

(This test should be only a part of an in-plant quality control program.)

B. Specimens

1. Provide two test specimens for each compression test.

(One test specimen may be used to check the stripping strength.)

2. Obtain concrete for specimens from actual production batch.

3. 6 in. x 12 in. or 4 in. x 8 in. concrete test cylinder, ASTM C 31

** OR **

3. ___________________ sized concrete cube, ____________

(Specify size. Cube speciments are usually 4 in. units, but 2 in. or 6 in. units are sometimes required. Larger specimens give more accurate test results than smaller ones. Source: (molded individually), (sawed from slab).)

** ** **

4. Cure specimens using the same methods used for the precast concrete units until the units are stripped, then moist cure specimens until test.

C. Keep quality control records available for the Architect upon request for two years after final acceptance.

(These records should include mix designs, test reports, inspection reports, member identification numbers along with date cast, shipping records and erection reports.)

Part 3 - Execution

3.01 INSPECTION

A. Before erecting architectural precast concrete, the General Contractor shall verify that structure and anchorage inserts not within tolerances required to erect panels have been corrected.

B. Determine field conditions by actual measurements.

3.02 ERECTION

A. Clear, well-drained unloading areas and road access around and in the building (where appropriate) shall be provided and maintained by the General Contractor to a degree that the hauling and erection equipment for the architectural precast concrete products are able to operate under their own power.

B. Erect adequate barricades, warning lights or signs to safeguard traffic in the immediate area of hoisting and handling operations.

C. Set precast units level, plumb, square and true within the allowable tolerances. General Contractor shall be responsible for providing lines, center and grades in sufficient detail to allow installation.

D. Provide temporary supports and bracing as required to maintain position, stability and alignment as units are being permanently connected. For load bearing architectural units, the precast concrete manufacturer shall furnish sketches and calculations to the Architect indicating the method to be used to temporarily brace the wall panels prior to their incorporation into the building by dry packing joints, welding, and/or pouring and curing grouted connections, and pouring cast-in-place concrete topping. This bracing shall be designed to resist a temporary seismic load from all contributing dead loads of 0.2 g. Allowable stresses may be taken to 90% of ultimate values for this analysis. In addition, the requirements of Cal-OSHA, Article 29, shall be met. These calculations shall be performed by a Civil Engineer registered in the State of California.

E. Non-cumulative tolerances for location of precast units shall be in accordance with PCI MNL 117, Section 6.2.4.

1. Precast concrete panel joints: 5/8" nominal ± 1/4" (minimum joint size may increase due to clearances required to accommodate movement or "story" drift resulting from seismic movements in the building frame)

2. Face alignment: ± 1/4"

3. Jog in alignment: ± 1/4"

4. Joint taper 1/40 inch per foot, maximum of 1/4"

F. Set non-load bearing units dry without mortar, attaining specified joint dimension with lead, plastic or asbestos cement spacing shims.

(Shims should be near the back of the unit to prevent their causing spall on face of unit if shim is loaded. The selection of the width and depth of field-molded sealants, for the computed movement in a joint, should be based on the maximum allowable strain in the sealant.)

G. Finally anchor in place precast units by bolting, welding, completing dry packed joints, grouting sleeves and pockets, and/or pouring in-situ cast-in-place concrete joints as indicated on approved erection drawings.

(The erector shall protect units from damage caused by field welding or cutting operations and provide non-combustible shields as necessary during these operations. Precast units shall be fastened in place as indicated on the approved erection drawings.)

3.03 PATCHING

A. Mix and place patching mixture to match color and texture of surrounding concrete and to minimize shrinkage.

(Patching is normally accomplished prior to final cleaning and caulking. It is recommended that the precaster execute all repairs or approve the methods proposed for such repairs by other qualified personnel. The precaster should be compensated for repairs of any damage for which he is not responsible. Patching should be acceptable providing the structural adequacy of the product and the appearance is not impaired.)

B. Adhere large patch to hardened concrete with bonding agent.

(Bonding agent should not be used with small patches.)

3.04 CLEANING

(State whether erector or General Contractor responsible for cleaning.)

A. After installation: ________________ shall clean soiled precast concrete surfaces with detergent and water, using fiber brush and sponge, and rinse thoroughly with clean water.

** OR **

A. Clean precast concrete panels with __________________.

((Acid-free commercial cleaners), (steam cleaning), (water blasting), (sand blasting). Use sand blasting only for units with original sand blasted finish. Ensure that materials of other trades are protected when cleaning panels.)

** ** **

B. Use acid solution only to clean particularly stubborn stains after more conservative methods have been tried unsuccessfully.

C. Use extreme care to prevent damage to precast concrete surfaces and to adjacent materials.

D. Rinse thoroughly with clean water immediately after using cleaner.

3.05 PROTECTION

A. The erector shall be responsible for any chipping, spalling, cracking or other damage to the units after delivery to the job site. After installation is completed, any further damage shall be the responsibility of the General Contractor.

* END OF SECTION *

The following should be placed in the General Conditions Section of the project specifications and renumbered for that section:

A.1 Payments

A.1.1 Monthly progress payments equal to 90% of the in-plant value will be made to the manufacturer for all products fabricated and stocked in the manufacturer's plant prior to delivery. Progress payments shall not relieve the manufacturer from compliance with terms of his contract with the Buyer.

(Industry practice is to fabricate, in advance, products for each individual project in accordance with design requirements and dimensions for that project. Such products cannot normally be used on any other project. As monthly payment would be made for such products if they were

fabricated on the site, monthly progress payments for such material fabricated off site and stored for delivery and erection as scheduled are justified.)

A.1.2 Full payment for all products delivered and/or installed will be made within 35 days of completion of all work under the contract and acceptance by the Architect.

REFERENCE STANDARDS

The information provided below includes the reference standards used in the preparation of this document and is intended only as an aid in understanding the reference standards contained in the Guide Specification part of this document. Later editions of these reference standards may require revision to the Guide Specification text. Guidance on the use of reference standards is contained in CSI Manual of Practice.

INTERNATIONAL CONFERENCE OF BUILDING OFFICIALS (ICBO)

UBC-79 Uniform Building Code, 1979 Edition is the general building code enforced throughout California.

STRUCTURAL ENGINEERS ASSOCIATION OF CALIFORNIA (SEAOC)

Recommended Lateral Force Requirements and Commentary, Fourth Edition - 1975, often referred to as the "Blue Book."

AMERICAN CONCRETE INSTITUTE (ACI)

ACI318-77, Building Code Requirements for Reinforced Concrete, covers design and construction of buildings of reinforced concrete. It is written so that it may be incorporated verbatim or adopted by reference in a general building code.

AMERICAN SOCIETY FOR TESTING AND MATERIALS (ASTM)

ASTM A 27, Mild-to-Medium-Strength Carbon-Steel Castings for General Application, covers carbon-steel castings with tensile strength from 60,000 psi to 70,000 psi. Seven grades of steel castings are covered as follows: Grades N-1, N-2, U-60-30, 60-30, 64-35, 70-36, and 70-40. The numerals of the last five grades correspond to the significant numbers of the specified tensile strength and yield point (i.e., grade 60-30 requires a minimum tensile strength of 60,000 psi and a yield point of 30,000 psi). Grades N-1 and N-2 are not required to be mechanically tested.

ASTM A 36, Structural Steel, covers carbon steel shapes, plates, bars of structural quality for use in riveted, bolted or welded construction of bridges and buildings, and general construction purposes. Included are chemical requirements, tensile properties (that is, minimum tensile strength, minimum yield point and elongation), and bend test requirements.

ASTM A 47, Malleable Iron Castings, covers ferritic malleable irons used in general engineering castings for service at both normal and elevated temperatures. Grade 32510 castings have a minimum tensile strength of 50,000 psi. Grade 35018 castings have a minimum tensile strength of 53,000 psi.

ASTM A 82, Cold-Drawn Steel Wire for Concrete Reinforcement, covers cold-drawn steel wire to be used as such, or in fabricated form, for the reinforcement of concrete, in gages not less than 0.08 in. nor greater than 0.625 in. Included are the processes by which the steel may be made, the gage numbers and the equivalent diameter in inches, the tensile properties required, the bending properties, the permissible variation in gage, and finish required on the wire.

ASTM A 108, Standard Quality Cold Finished Carbon Steel Bars, covers cold finished steel bars used for the manufacture of headed welding studs, in grades 1010 through 1020 with a minimum tensile strength of 60,000 psi.

ASTM A 153, Zinc Coating (Hot-Dip) on Iron and Steel Hardware, covers zinc coatings applied by the hot-dipped process on iron and steel. Includes weights, distribution, adherence, testing and inspection of coatings, and fabrication considerations.

ASTM A 165, Electrodeposited Coatings of Cadmium on Steel, covers requirements for electro-plated cadmium coatings on steel articles. Three types of coatings are covered: Type NS, Type OS, and Type TS. The minimum thickness of cadmium coatings on significant surfaces of finished articles are 0.00050 in. for Type NS coating, 0.00030 in. for Type OS coatings, and 0.00015 in. for Type TS coating. The dimensional tolerance of most threaded articles normally does not permit the application of coating thickness much greater than Type TS.

ASTM A 167, Stainless and Heat-Resisting Chromium-Nickel Steel Plate, Sheet, and Strip, covers soft, corrosion-resisting chromium-nickel steel plate, sheet, and strip.

ASTM A 184, Fabricated Deformed Steel Bar Mats for Concrete Reinforcement, covers material in mat or sheet form fabricated from steel bars to be used for the reinforcement of concrete. These mats consist of two layers of bars assembled at right angles to each other and clipped or welded at all or some of the intersections to form a rectangular grid.

ASTM A 185, Welded Steel Wire Fabric for Concrete Reinforcement, covers cold-drawn steel wire (ASTM A 82), as drawn or galvanized fabricated into sheet or mesh by arranging a series of longitudinal and transverse wires accurately spaced at right angles to each other and welding them together at all points of intersection. Included are the minimum requirements for bending of the finished fabric, the minimum weld strength between transverse spaces and the wire gage differential permitted.

ASTM A 307, Carbon Steel Externally and Internally Threaded Standard Fasteners, ranging in diameter from 1/4" to 4", based upon a minimum tensile strength of 60,000 psi for grade A material.

ASTM A 325, High Strength Bolts for Structural Steel Joints, including suitable nuts, and plan hardened washers, in minimum tensile strengths from 105,000 psi to 120,000 psi.

ASTM A 416, Uncoated Seven-Wire Stress-Relieved Strand for Prestressed Concrete, covers two grades of seven-wire, uncoated stressed-relieved steel strand for use in pretensioned and post-tensioned prestressed concrete construction. Grade 250 and grade 270 have minimum ultimate strengths of 250,000 psi and 270,000 psi respectively, based on the nominal area of the strand.

ASTM A 421, Uncoated Stress Relieved Wire for Prestressed Concrete, covers two types of uncoated stress-relieved ground high-carbon steel wire commonly used in prestressed linear concrete construction. Type BA wire is used for applications in which cold-end deformation is used for anchoring purposes (button anchorage), and Type WA wire is used for applications in which the ends are anchored by wedges, and no cold-end deformation of the wire is involved (wedge anchorage).

ASTM A 496, Deformed Steel Wire for Concrete Reinforcement, covers cold-worked, deformed steel wire to be used as such, or in fabricated form, for the reinforcement of concrete in sizes having nominal cross-sectional areas not less than 0.01 $in.^2$ nor greater than 0.31 $in.^2$

ASTM A 497, Welded Deformed Steel Wire Fabric for Concrete Reinforcement, covers welded wire fabric made from cold-worked deformed wire, or a combination of deformed and nondeformed wires, to be used for the reinforcement of concrete. The wire conforms to ASTM A 496 either solely or in combination with ASTM A 82 Cold-Drawn Steel Wire. Longitudinal and transverse members are electrical-reinforced welded to provide an average weld shear strength of 8,000 lbs.

ASTM A 615, Deformed and Plain Billet-Steel Bars for Concrete Reinforcement, covers deformed and plain billet-steel concrete-reinforcement bars. A deformed bar is defined as a bar that is intended for use as reinforcement in reinforced concrete construction. The surface of the bar is provided with lugs or protrusions (hereinafter called deformations) which inhibit longitudinal movement of the bar relative to the concrete which surrounds the bar in such construction and conform to the provisions of this specification. Bars are of two minimum yield levels, namely, 40,000 psi and 60,000 psi, designated as Grade 40 and Grade 60, respectively.

ASTM A 616, Rail-Steel Deformed and Plain Bars for Concrete Reinforcement, covers deformed and plain rail steel concrete reinforcement bars. A deformed bar is defined as a bar that is intended for use as reinforcement in reinforced concrete construction. The surface of the bar is provided with lugs or protrusions (hereinafter called deformations) which inhibit longitudinal movement of the bars relative to the concrete which surrounds the bars in such construction and conform to the provisions of this specification. Bars are of two minimum yield levels, namely, 50,000 psi and 60,000 psi, designated as Grade 50 and Grade 60, respectively.

ASTM A 617, Axle-Steel Deformed and Plain Bars for Concrete Reinforcement, covers deformed and plain axle steel concrete reinforcement bars. A deformed bar is defined as a bar that is intended for use as reinforcement in reinforced concrete construction. The surface of the bar is provided with lugs or protrusions (hereinafter called deformations) which inhibit longitudinal movement of the bar relative to the concrete which surrounds the bar in such construction and conform to the provisions of this specification. Bars are to two minimum yield levels, namely, 40,000 psi and 60,000 psi, designated as Grade 40 and Grade 60, respectively.

ASTM A 663, Carbon Steel Bars Subject to Mechanical Property Requirements, covers carbon steel bars furnished in the as-rolled condition, subject to mechanical property requirements and intended for general constructional applications. Grade designations range from 45 to 80 in increments of 5, and indicate the tensile strength (i.e., the minimum tensile strength of grade 65 is 65,000 psi).

ASTM A 666, Austenitic Stainless Steel, Sheet, Strip, Plate, and Flat Bar for Structural Applications, covers six types of steels in 4 grades or strength levels. Of primary interest in architectural precast concrete application are Grade A (30,000 psi yield strength) and Grade B (40,000 psi yield strength).

ASTM C 31, Making and Curing Concrete Testing Specimens in the Field, covers procedures for making and curing test specimens from concrete being used in construction.

ASTM C 33, Concrete Aggregates, covers fine and coarse aggregates other than light-weight aggregate for use in concrete. Included are descriptions of fine aggregate and of coarse aggregate, grading requirements, limits for deleterious substances, the physical properties of both fine and coarse aggregates and a list of the methods of sampling and testing.

ASTM C 39, Test for Compressive Strength of Cylindrical Concrete Specimens, covers determination of compressive strength of cylindrical concrete specimens, such as molded cylinders and drilled cores.

ASTM C127, Specific Gravity and Absorption of Coarse Aggregates, covers the determination of bulk and apparent specific gravity and absorption of coarse aggregate. The absorption test determines the percentage by weight, of water absorbed by the aggregate specimen after being immersed at room temperature for 24 hours, compared with the specimen dried to constant weight at a temperature of 212° to 230°F and then cooled to room temperature.

ASTM C 128, Specific Gravity and Absorption of Fine Aggregate, covers the determination of bulk and apparent specific gravity and absorption of fine aggregate. These tests are similar to those required for coarse aggregate in ASTM C 127 but are modified as required to obtain accurate test results with smaller aggregate.

ASTM C 138, Standard Test Method for Unit Weight, Yield, and Air Content of Concrete, covers the determination of the weight per cubic foot of freshly mixed concrete and gives formulas for the calculation of the yield, cement content, and air content of the concrete.

ASTM C 150, Portland Cement, covers eight types of Portland cement, Type 1 through Type V and Type IA through Type IIIA. The "A" designation indicates an air-entraining cement. Ordinarily only Type I, Type II, and Type III cements are used for precast concrete units, with an air-entraining agent added at the mixer if air-entrained concrete is required.

ASTM C 171, Sheet Materials for Curing Concrete, covers materials in sheet form used for covering the surface of concrete to inhibit moisture loss during the curing period, and in the case of white reflective type materials, to also reduce the temperature rise in concrete exposed to radiation from the sun.

ASTM C 172, Standard Method for Sampling Fresh Concrete, gives the proper procedure for obtaining representative samples of fresh concrete as delivered to the molds on which tests are to be performed to determine compliance with quality control requirements of the precast prestressed concrete specification.

ASTM C 173, Standard Test Method for Air Content of Freshly Mixed Concrete by the Volumetric Method, which allows the determination of the air content of freshly mixed concrete containing any type of aggregate, whether it be dense, cellular, or lightweight.

ASTM C 185, Air Content of Hydraulic Cement Mortar, covers the determination of whether or not the hydraulic cement under test meets the air-entraining or non-air-entraining requirements of the applicable hydraulic cement specification for which the test is being made. The air content of concrete is influenced by many factors other than the potential of the cement for air entrainment.

ASTM C 231, Standard Test Method for Air Content of Freshly Mixed Concrete by the Pressure Method, which is based upon the observation of the change in volume of concrete with a change in pressure. This method is not applicable to lightweight aggregate concrete, or concrete made with aggregates of high porosity, for which ASTM C 173 should be used (Volumetric Method of Determining Air Content).

ASTM C 260, Air-Entraining Admixtures for Concrete, covers materials used as air-entraining admixtures to be added to concrete mixtures in the field. Included are the performance requirements of the material such as bleeding time of setting, compressive strength, flexural strength, resistance to freezing and thawing, and length change. Methods of taking samples for testing are also included.

ASTM C 309, Liquid Membrane-Forming Compounds for Curing Concrete, covers liquid membrane-forming compounds suitable for application on horizontal and vertical concrete surfaces to retard the loss of water during the early hardening period and, in the case of the white-pigmented compound, to also reduce the temperature rise in concrete exposed to radiation from the sun. The compounds covered by this specification are suitable for use as curing media for fresh concrete, and may also be used for further curing of concrete, and may also be used for curing.

ASTM C 330, Light-Weight Aggregates for Structural Concrete, covers light-weight aggregates intended for use in structural concrete in which prime considerations are lightness in weight and compressive strength of the concrete. The two general types of aggregates covered are those prepared by expanding, calcining, or sintering products such as blast furnace slag, clay diatomite, fly ash, shale or slate, and those prepared by processing natural materials such as pumice, scoria or tuff.

ASTM C 494, Chemical Admixtures for Concrete, covers material for use as chemical admixtures to be added to Portland cement concrete mixtures in the field for the purpose or purposes indicated for the five types as follows:

Type A - Water-reducing admixtures
Type B - Retarding admixtures
Type C - Accelerating admixtures
Type D - Water-reducing and retarding admixtures, and
Type E - Water-reducing and accelerating admixtures

ASTM E 329, Inspection and Testing Agencies for Concrete and Steel as Used in Construction, defines the duties and responsibilities and establishes minimum requirements for personnel and equipment of public and independent commercial materials inspection and testing agencies engaged in inspection and testing of concrete and steel as used in construction. It states the training and experience required by an agency manager, a supervising laboratory technician, and a supervising field technician, and lists the testing equipment required in the laboratory.

AMERICAN WELDING SOCIETY (AWS)

AWS D1.1, Structural Welding Code, covers types of welding and welding procedures approved for use without performing procedure qualification tests, and describes welding qualification procedures for welders, welding operators and tackers.

AWS D12.1, Reinforcing Steel Welding Code, covers requirements and procedures for welding reinforcing steel.

FEDERAL SPECIFICATIONS (FS)

FS TT-P-86G, Paint, Red-Lead-Base, Ready-Mix, covers four types of paint suitable for use on iron and steel surfaces:

Type I - Red Lead-linseed oil paint
Type II - Red lead-mixed pigment-alkyd varnish-linseed oil paint
Type III - Red lead-alkyd varnish paint
Type IV - Red lead-mixed pigment-phenolic varnish paint

Types I and II are intended for priming steel, either in the shop or field. Type I will bond to surfaces even with the presence of small amounts of corrosive materials found impractical to remove.

FS-MMM-G-650B, Grout, Adhesive, Epoxy Resin, Flexible, Filled, covers a two-component mineral filled thixo-tropic and flexible epoxy-resin-base grout. Type I is for use between 68°F and 104°F. Grade C is for use when application is by trowel or putty knife.

MILITARY SPECIFICATIONS (MIL)

SSPC-Paint 14-64T, No. 14 red lead, iron oxide and linseed oil primer, covers a slow drying primer for structural steel. It has excellent rust inhibitive characteristics, wetting ability, and resistance to weathering before finished coating. It may also be used for intermediate coats.

PRESTRESSED CONCRETE INSTITUTE (PCI)

MNL-117, Manual for Quality Control for Plants and Production of Architectural Precast Concrete Products, covers production guide lines for the manufacture of high quality plant cast architectural precast concrete products. It is not intended as a specification document, although many requirements of a specification have been outlined.

MNL-119, Architectural Precast Concrete Drafting Handbook, serves as the industry standard for the preparation of production shop drawings and erection drawings for architectural precast concrete products.

MNL-121, Manual for Structural Design of Architectural Precast Concrete, provides the designer with information to design the precast elements, the support systems for these elements, and the connections.

MNL-122, Architectural Precast Concrete, covers the complete design development of architectural precast concrete from conceptual design to detailing and specifying.

MNL-124, Design for Fire Resistance of Precast Prestressed Concrete, gives procedures for rational design for fire, including requirements for joinery and sealants at panel joints.

INSPECTION

During manufacture:

A. Check qualifications of testing agencies.

B. Where in-plant testing is to be performed, see that manufacturer's testing equipment is calibrated by testing agency personnel, and that manufacturer's testing personnel are qualified to perform work.

C. Check casting and after-casting tolerances. Inspectors should be provided with quality instruments for measuring tolerances.

D. Verify that production panels match texture and color of accepted project mock-up or panel samples.

E. Examine concrete test results.

Before installation:

A. Check precast concrete panel erector's experience and qualifications.

B. Check welder's qualifications.

C. Assure that precast concrete panels are stored on site with resilient and stain resistant spacers.

D. Verify that panel identification marks are easily discernible.

E. Check field dimensions affecting erection.

F. Check bracing plan for temporarily unstable components/ structures.

During installation:

A. Check erection tolerances.

B. See that units are clean and exposed metal spot painted.

C. Check that proper handling points are used as per erection drawings.

After installation:

A. Inspect repairs for accurate color and texture match of surrounding concrete.

B. Inspect panels after cleaning to see that they are properly prepared to receive caulking and damp-proofing.

GLOSSARY

Refer to MNL-117 for definitions of terms.

QUALITY CONTROL AT BASALT ROCK PLANT - NAPA

Strict quality control is essential to the production of prestressed concrete products. Basalt Rock Company's in house quality control program is periodically reviewed by the PCI plant certification program. Here we see inspector Tom Ruise rodding concrete cylinders as per ASTM C 31.

GUIDE SPECIFICATIONS FOR PRECAST PRESTRESSED CONCRETE (PLANT CAST)

CSI DESIGNATION 03420

THIS DOCUMENT

This document provides a basis for specifying in-plant fabrication and field erection of structural precast and prestressed concrete units, such as double tees, beams, hollow core plank, flat slabs, columns, prestressed structural wall elements, bearing piles and sheet piling. It does not cover architectural precast concrete panels, site cast precast and post tensioned concrete, or cast-in-place concrete topping and closure pours.

DRAWINGS AND SPECIFICATIONS

Drawings:

Indicate the number and size of precast prestressed sections on a layout plan. Show the magnitude and distribution of imposed loads. Indicate required reinforcing, prestressing, and connection material. Indicate the location and size of openings and supports, and the position of inserts, anchors, and connections. Include setting plans with details showing connection of precast prestressed sections to adjacent construction such as beam-to-column, beam-to-girder, and column base. Illustrate related items such as bearing pads, grout, cast-in-place concrete, and weld plates. For complicated or unusual situations with temporarily unstable elements provide an erection drawing indicating the bracing method to be employed.

Specifications:

State the required qualifications of the precast prestressed section manufacturer and erector. Define the materials to be used including concrete, tendons, and bearing pads. Require submission of shop drawings and erection procedures. List the allowable tolerances for fabrication and erection. State fire ratings, if required. Define in-plant quality assurance measures to be used.

Coordination:

Assure that specifications for framing and supporting members contain tolerances compatible with precast prestressed sections. When building foundations of interfacing structure contain cast-in hardware to facilitate connections to precast elements, provide hardware setting drawings to the general contractor, if not predesigned or shown on the contract drawings.

GUIDE SPECIFICATION

This Guide Specification is intended to be used as a basis for the development of an office master specification or in the preparation of specifications for a particular project. In either case, this Guide Specification must be edited to fit the conditions of use. Particular attention should be given to the deletion of inapplicable provisions. Include necessary items related to a particular project. Include appropriate requirements where blank spaces have been provided.

SECTION 03420

PRECAST PRESTRESSED CONCRETE (PLANT CAST)

Part 1 - General

1.01 DESCRIPTION

A. Related Work Specified Elsewhere

1. Concrete Formwork: Section 03100

2. Concrete Reinforcement: Section 03200

(Delete when reinforcing steel is specified in this section. Precast prestressed concrete reinforcing requirements differ from that used for cast-in-place reinforcing and should be specified in this section.)

3. Cast-in-Place Concrete: Section 03300

4. Site Cast Post Tensioned Concrete: Section 03435

5. Architectural Precast Concrete (Plant Cast): Section 03450

6. Unit Masonry: Section 04200

7. Metal Fabrications: Section 05500

(miscellaneous iron, anchor bolts, or other anchorage devices required for installing precast prestressed units but furnished and installed in cast-in-place concrete by others.)

8. Waterproofing: Section 07100

9. Insulation: Section 07200

   (for insulation field applied to precast prestressed components. Insulation cast in precast units during manufacture should be specified in this section.)

10. Flashing and Sheet Metal: Section 07600

   (Counterflashing receivers and reglets, unless included in this section.)

11. Sealants and Caulking: Section 07951

12. Painting: Section 09900

   (field touch-up painting. Delete when specified in this section.)

B. Work Installed but Furnished by Others

(Delete when furnished by precast manufacturer.)

1. Counterflashing Receivers or Reglets: Section 07600

2. Inserts for suspended ceiling systems: Section 09540

3. Elevator Guides: Section 14200

C. Testing Agency Provided by Owner

(Delete when testing agency is provided by precast manufacturer or contractor. Coordinate with Division 1)

1.02 QUALITY ASSURANCE

A. Acceptable Manufacturers: Minimum of _______ years production experience in precast prestressed concrete work of quality and scope required on this project.

(Experience required is usually 2 to 5 years. Plant certification, as provided in the PCI Plant Certification Program, is satisfactory evidence.)

1. Manufacturer must be able to show that he has experienced personnel, physical facilities, established quality control procedures, and a management capability sufficient to produce the required units without causing delay to the project.

2. When requested by the Architect, the manufacturer shall submit written evidence of the above requirements.

** OR **

A. Acceptable Manufacturers:

(List a minimum of three)

B. Erector Qualifications: Regularly engaged for at least ______ years in erection of precast prestressed concrete units similar to those required on this project.

(Experience required is usually 2 to 5 years.)

C. Qualifications of Welders and Tackers: In Accordance with AWS D 1.1

(Qualified within the past year. Delete when welding is not required.)

D. Testing: In general compliance with testing provisions of Prestressed Concrete Institute MNL 116, "Manual for Quality Control for Plants and Production of precast prestressed concrete products."

E. Testing Agency:

(delete when provided by owner)

1. Not less than ______ years experience in performing concrete tests specified in this section.

   (Usually 2 to 5 years experience is adequate)

2. Capable of performing testing in accordance with ASTM E 329.

3. Inspected by Cement and Concrete Reference Laboratory of the National Bureau of Standards.

F. Requirements of Regulatory Agencies: design, manufacture, and installation of precast prestressed concrete shall be in accordance with the Uniform Building Code, 1979 edition.

** OR **

(list applicable code)

G. Allowable Tolerances shall be in accordance with PCI MNL 116

H. Source Quality Control

1. Quality control and inspection procedures to comply with applicable sections of PCI MNL 116.

1.03 SUBMITTALS

A. Fabricators Drawings

1. Production Shop Drawings Containing:

a. Elevation view of each member

b. Sections and details to indicate quantities and position of reinforcing steel, anchors, inserts, connections, accessories, joints, and openings

c. Lifing and erection inserts

d. Dimensions and finishes

e. Reinforcing and prestress strands

f. Tensioning and detensioning sequences and schedules

g. Estimated cambers at release and at erection

h. Chamfers and radii of corners

i. Welds

j. handling methods to be used for stripping, storage, and shipping

k. piece marks cross referenced to erection drawings

l. other items cast into the prestressed units

m. openings

B. Erection drawings containing:

a. Plans and elevations, as required locating and defining all material furnished by the manufacturer, by piece mark.

b. Sections and details showing connections, cast-in items, and their relation to the structure and the prestressed units

c. Erection sequences and field handling requirements; erection bracing plans are required when the precast or prestressed element is not permanently tied into the structure after disengagement from the crane or other erection equipment. This bracing plan shall be prepared by a civil engineer registered in the State of California

d. Description of all loose erection hardware, usually listed on bills of material.

3. Anchor Bolt or Insert Plate Location Plans

(Normally prepared by precaster and provided to contractor for work by other trades)

4. Field work sheets

(Required when alterations to existing structures are required prior to erection of prestressed concrete units. This work could consist of chipping, cutting, burning, drilling or setting expansion bolts, etc., by either the Precaster or other trades, as applicable and specified)

C. Test Reports: Submit reports on materials, compressive strength tests on concrete and water absorption tests on units.

(Schedule of required tests, number of copies of test reports, and how distributed are included in Testing Laboratory Services, Section__________.)

D. Design Calculations: Prepare component design calculations as required to satisfy loading induced by handling and erection forces as well as final dead and live loads. Design calculations of products not defined on the contract drawings shall be performed by a registered civil engineer in the State of California experienced in the design of precast prestressed concrete.

E. Permissible Design Deviations

1. Design deviations will be permitted only after the architect/engineer's written approval of the manufacturer's proposed design, supported by

complete design calculations and drawings.

2. Design deviations shall provide an installation equivalent to the basic intent without incurring additional cost to the owner.

(The design and architectural drawings normally will be prepared using a local precast prestressed concrete manufacturer's design data and load tables. Dimensional changes which would not materially affect architectural and structural properties or details usually are permissible.

Most precast prestressed concrete is cast in continuous steel forms; therefore connection devices on the formed surfaces must be contained within the member since penetration of the form is impractical.

Camber will generally occur in prestressed concrete members having eccentricity of the stressing force. If camber considerations are important, check with your local prestressed concrete manufacturer to secure estimates of the amount of camber and of camber movement with time and temperature change.

Architectural details must recognize the existence of this camber and camber movement in connection with:

(1) Closures to interior non-load bearing partitions.

(2) Closures parallel to prestressed concrete members (whether masonry, windows, curtain walls or others) must be properly detailed for appearance.

(3) Floor slabs receiving cast-in-place topping. The elevation of top of floor and amount of concrete topping must allow for camber of prestressed concrete members.

Design cambers less than obtained under normal design practices are possible but this usually requires the addition of strands or non-prestressed steel reinforcement and price should be checked with the local manufacturer.

As the exact cross section of precast prestressed

members may vary somewhat from producer to producer, permissible deviations in member shape from that shown on the contract drawings might enable more manufacturers to quote on the project. Manufacturing procedures also vary between plants and permissible modifications to connection details, inserts, etc., will allow the manufacturer to use devices he can best adapt to his manufacturing procedure.

Be sure that loads shown on the contract drawings are easily interpreted. For instance, on members which are to receive concrete topping, be sure to state whether all superimposed dead and alive loads on precast prestressed members do or do not include the weight of the concrete topping.

It is best to list the live load, superimposed dead load, topping weight, and weight of the member, all as separate loads. Where there are two different live loads (e.g., roof level of a parking structure) indicate how they are to be combined.)

1.04 PRODUCT DELIVERY, STORAGE, AND HANDLING

A. Delivery and Handling

1. Handle and transport units in a position consistent with their shape and design in order to avoid excessive stresses or damage.

2. Lift or support units only at the points shown on the erection shop drawings. If required, show how the units are to be handled from stripping to erection.

3. Place nonstaining resilient spacers of even thickness between each unit.

4. Support units during shipment on nonstaining shock-absorbing material.

5. Do not place units on ground without dunnage.

6. Use lifting slings or spreader bars to keep angle between lifted member and cable greater than 45°.

7. Handle and store members to protect from dirt and damage.

8. Place members in storage so that identification marks are discernible.

B. Storage at Project Site

1. Store units to protect from contact with soil, staining, and from physical damage.

2. Store units, unless otherwise specified, with nonstaining, resilient supports located in same positions as when transported.

3. Separate stacked members by dunnage across full width of each bearing point.

<u>Part 2 - Products</u>

2.01 MATERIALS

A. Concrete

1. Portland cement:

a. ASTM C 150, type __________

(Type I - general uses; Type II - moderate sulfate resisting; Type III - high early strength.)

b. for exposed surfaces use same brand, type, and source of supply throughout.

(To minimize color variation. Specify source of supply when color shade is important.)

2. Air entraining agent: ASTM C 260.

(Delete if air entraining agent is not required.)

3. Water reducing, retarding, accelerating admixtures: ASTM C 494.

(Delete if water reducing, retarding or accelerating admixtures are not required. Calcium chloride or admixtures containing significant amount of calcium chloride should not be allowed.)

4. Aggregates

a. hard rock aggregates for concrete shall conform to ASTM C 33

b. lightweight aggregates for concrete shall conform to ASTM C 330

(Grading requirements for the above are generally waived or modified)

c. material

(Specify type of stone desired, such as limestone, granite, or locally available gravel.)

d. maximum size and gradation ______

(State required sieve analysis.)

6. Water: Free from foreign materials in amounts harmful to concrete and embedded steel.

(Potable water is normally acceptable.)

B. Reinforcing Steel

1. Deformed steel bars:

ASTM A 615

(Use grade 40 for #3 and #4 bars; use grade 60 for #5 bars and larger. When welding of bars is required, use all grade 40 bars, or establish weldability in accordance with AWS D12.1.)

2. Welded wire fabric:

a. plain - ASTM A 185

b. deformed - ASTM A 497

3. Wire:

Cold drawn steel: ASTM A 82

4. Fabricated steel bar or rod mats:

ASTM A 184

C. Prestressing strand:

ASTM A 416, Grade _______. Certified test reports of ultimate strength and typical stress strain curves shall be furnished by the steel manufacturer.

(250 ksi or 270 ksi.)

D. Cast-in-Anchors and Inserts

(Loose attachment hardware specified under Miscellaneous Metals, Section 05500.)

1. Materials

a. carbon steel bars:

ASTM A 306, Grade 65

(for completely encased anchors)

b. structural steel: ASTM A 36

(for carbon steel connection assemblies)

c. stainless steel: ASTM A 666, Type 304, Grade __________

(Grade A or Grade B, stainless steel attachments for use when resistance to staining merits extra cost)

d. carbon steel plate: ASTM A 283, Grade _______

(Grade A, B, C, or D)

e. malleable iron castings: ASTM A 47, Grade __________

(Grades 32510, 35018)

f. carbon steel castings: ASTM A 27, Grade 60-30

g. anchor bolts: ASTM A 307

** OR **

ASTM A 325

(for steel bolts, nuts and washers)

h. welded headed studs: AWS D1.1, Part VI, Section 4; ASTM A 108

i. deformed bar anchors: ASTM A 496

2. Finish

a. shop primer: FS TT-0-86, oil base paint, type 1

** OR **

a. shop primer: SSPC-Paint 14

** OR **

a. shop primer: Manufacturer's standard

(for exposed carbon steel anchors)

b. hot-dip galvanized: ASTM A 153

(for exposed carbon steel anchors where corrosive environment justifies the additional cost. Field welding should not be permitted on galvanized element)

c. cadmium coating: ASTM A 165

(for threaded fasteners)

d. zinc rich coating: MIL-P-21035, self curing, one component, sacrificial organic coating

(for field spot painting)

D. Receivers for Flashing: 28 ga. formed ____________

(stainless steel, copper, zinc. Coordinate with flashing specification to avoid dissimilar metals. Delete when included in flashing and sheet metal section. Specify whether precaster or others furnish.)

** OR **

D. Receivers for Flashing: Polyvinyl chloride extrusions.

E. Rigid Insulation ____________

(Specify type of insulation, such as foamed plastic ((polystyrene and polyurethane)), glasses ((foamed glass and fiberglass)), foamed or cellular light weight concretes, or light weight mineral aggregate concretes.)

F. Grout

(Indicate required strengths on contract drawings)

1. Cement grout: Portland cement, sand, and water sufficient for placement and hydration.

2. Nonshrink grout: Premixed, packaged ferrous and non-ferrous aggregate shrink-resistant grout.

3. Epoxy-resin grout: Two-component mineral-filled epoxy-polysulfide, FS MMM-G-560________________, Type ________________, Grade C.

   (check with local suppliers to determine availability and types of epoxy-resin grouts.)

4. Neat cement grout: Cement and water paste, such used for grouting pile anchor dowels.

G. Bearing Pads

1. Elastomeric: conforming to Section 51-1.12H of the CALSTRANS standard specifications.

   (Pads specified have a strength of 2500 psi. For many applications, commercial grade pads are adequate and are more economical, but strengths vary and should be determined in advance by the specifier.)

2. Hardboard or Korolath strips - 1/8" thick

   (acceptable for hollow core plank and solids slabs)

3. Tetraflouroethylene (TFE) reinforced with glass fibers and applied to stainless steel or structural steel plates. May also be bonded to neoprene.

   (ASTM D2116 applies only to basic TFE resin molding and extrusion material in powder or pellet form. Physical and mechanical properties must be specified by naming manufacturer or other methods.)

H. Curing Materials: liquid membrane forming compound, ASTM C 309

* * OR * *

H. Curing Materials: Sheet Materials, ASTM C 171

2.02 CONCRETE MIXES

A. Mixing Procedures: same as for cast-in-place concrete, Section 03300.

B. Concrete Properties:

1. Water-cement ratio: Maximum 40 lbs of water to 100 lbs. of cement.

(keep to a minimum consistent with strength, durability requirements and placement needs. Concrete to be exposed to walt water or other corrosive environments should contain at least 7½ sacks of cement per cubic yard, in addition to the above water cement ratio and increased cover requirements)

2. Air entrainment: Amount produced by adding dosage of air entraining agent that will provide 19% ± 3% of entrained air in standard 1:4 sand mortar as tested according to ASTM C 185.

* * OR * *

2. Air entrainment: minimum 3%; maximum 6%, when tested in accordance with ASTM C 173 or ASTM C 231.

3. 28 Day Compressive Strength: minimum of _______psi when tested by 6 x 12 or 4 x 8 in. cylinders.

   (6000 psi readily achievable with hard rock coarse aggregates or 5000 psi with sand-lightweight coarse aggregate. When testing 4 inch cubes, required strengths shall be 25% higher than the above for correct correlation to cylinder strength)

4. Strength at release of prestress: minimum of ____________psi.

   (maximum strength achievable is 4200 psi 15 hours from time of placement, with 3500 psi being a common standard release strength)

5. Use of Calcium chloride, chloride ions or other salts is not permitted.

2.03 FABRICATION

A. Manufacturing procedures shall be in general compliance with PCI MNL-116.

B. Formwork

1. Construct molds to withstand tensioning and detensioning operations.

2. Construct molds to maintain units within specified tolerances with radii or chamfers at corners.

3. Securely attach anchorage devices to molds in locations not interfering with proper positioning

of strand, reinforcement, or placing of concrete.

C. Stressing Procedures

1. All prestressing steel shall be accurately placed and firmly tied in position during the pouring operations. Tolerances shall be those specified on the drawings. In absence of such notes covering any particular section, the maximum deviation of the c.g.s. of the total force shall not exceed 1/4 inch or 3% of the specified distance of the c.g.s. measured from the bottom fiber, whichever is greater.

2. Prestressing strands. Strands shall not be allowed to become pitted with rust. Light coating of rust is permissible and short of visible pitting will not be cause for rejection. Form oil shall not be permitted to coat the strands.

3. Pretensioning. May be by either single strand or multiple strand jacking. Where multiple jacking is used, all strands shall be brought to a reasonably uniform preliminary tension. Final stress shall be measured by a calibrated pressure gauge of adequate capacity and not less than 5" diameter dial. Total elongation shall be checked regularly and the total stress as measured by gauge and elongation shall not differ more than 5%.

4. Deflected strands. Methods of deflecting, type of strand holding devices, and deflection points shall be approved by the Engineer or Architect.

5. Safety. All operations will be performed in a safe manner and in strict compliance with all applicable state and local regulations.

6. Detensioning. Transfer of prestress shall not be made until the concrete strength has reach 3,000 psi for strands 3/8" and under, or 3,500 for 7/16" strands and over. Should structural requirements for unusual conditions indicate higher or lower concrete strength at transfer, they will be noted in the structural drawings. Transfer of prestress may be accomplished by gradual release of the tensioning jacks or by burning of strands. Where the burning method is used, the sequence of cutting strands shall be such as to prevent severe unbalance of the loading. Prior to transfer of prestress, forms shall be loosened, or removed if necessary, to allow free movement of the casting.

D. Concrete Work

1. Conveying. Concrete shall be conveyed from the mixer to the place of final deposit by methods which will prevent separation, segregation, or loss of materials.

2. Vibrating. All concrete shall be consolidated in the form by approved high frequency vibration. This vibration may be either internal or external or a combination of both. Where external vibration is used, forms must be of a design adequate to withstand such external vibrating without distortion or failure.

3. Curing. Curing by any of the following methods shall be maintained until release strength has been reached:

   Steam or hot water curing at maximum temperature not exceeding 165°F. After placement of the concrete, members shall be held for a minimum pre-steaming period of not less than 3 hours.

   Sealing the exposed surfaces with an approved sealing compound or membrane.

   Hot air combined with moisture.

   Continuous water curing.

4. Finishing. Rock pockets and chipped corners shall be patched and cleaned by accepted methods. No rock pocket or other void reaching into the prestressing steel shall be patched without permission of the Engineer or Architect.

   Structural members will be accepted with form finish appearance but with any leakage fins removed.

   Where members are used architecturally, they are so designated and specific finishing requirements are as follows:

   (Concrete of the consistency normally used for prestressed concrete is difficult to place without some surface flaws. This is particularly true for air bubbles trapped against flat or sloping upper surfaces of forms. Air or water bubbles resulting in "bug holes" have no significance structurally and are objectionable only if a perfectly smooth surface is required.)

E. Finishes

1. Standard Underside: Resulting from casting against approved forms using good industry practice in cleaning of forms, design of concrete mix, placing and curing. Small surface holes caused by air bubbles, normal color variations, normal form joint marks, and minor chips and spalls will be tolerated, but no major or unsightly imperfections, honeycomb, or other defects will be permitted.

2. Standard Top: Roughened as a result of vibrating screed or raking to produce a surface roughness of an amplitude of 1/4", to assure horizontal shear transfer to cast in place concrete topping.

3. Smooth Top: For applications such as untopped floor or roof elements or sections used as wall panels, additional hand finishing and trowelling shall be employed to produce a smooth finish.

4. Float Finish

5. Broom Finish

6. Exposed Vertical Ends: Strands shall be recessed and the ends of the member will receive sacked finish.

(Other formed finishes which may be specified are:

Commercial Finish. Concrete may be produced in forms that impart a texture to the concrete (e.g., plywood or lumber). Fins and large protrusions shall be removed and large holes shall be filled. All faces shall have true, well-defined surfaces. Any exposed ragged edges shall be corrected by rubbing or grinding.

Architectural Grade B Finish. All air pockets and holes over 1/4 in. in diameter shall be filled with a sand-cement paste. All form offsets or fins over 1/8 in. shall be ground smooth.

Architectural Grade A Finish. In addition to the requirements for Architectural Grade B Finish, all exposed surfaces shall be coated with a neat cement paste using an acceptable float. After thin pastecoat has dried, the surface shall be rubbed vigorously with burlap to remove loose particles. These requirements are not applicable to extruded products using zero-slump concrete in their process.)

7. Special Finish: If required, listed as follows:

(Special finishes, if required, should be described in this section of the specifications and noted on the contract drawings, pointing out which members require special finish. Such finishes will involve additional cost and consultation with the manufacturer is recommended. A sample of such finishes should be made available for review prior to bidding.)

F. Openings: Primarily on thin sections, the manufacturer shall provide for those openings 10 in. round or square or larger as shown on the structural drawings. Other openings shall be located and field drilled or cut by the trade requiring them after the precast prestressed products have been erected. Openings shall be approved by Architect/Engineer before drilling or cutting.

(This paragraph requires other trades to field drill holes needed for their work, and such trades should be alerted to this requirement through proper notation in their sections of the specifications. Some manufacturers prefer to install openings smaller than 10 in. which is acceptable if their locations are properly identified on the structural drawings.)

G. Fasteners: The manufacturer shall cast in structural inserts, bolts, and plates as detailed or required by the contract drawings.

(Exclude this requirement from extruded sections)

H. Product Identification: Mark each unit with correct piece mark, project or contract number, and date cast.

I. Handling: Remove units from molds, store, and ship in accordance with the information given on the production shop drawings.

J. Acceptance: Units which do not meet dimensional tolerance standards, or quality control requirements pertaining to concrete strength, reinforcing or prestressing shall be rejected; units judged deficient in finish quality may be corrected at the option of the Architect.

2.04 CONCRETE TESTING

A. Make one compression test at 28 days for each day's production of each class of concrete.

(This test should be only a part of an in-plant quality control program.)

B. Specimens

1. Provide a minimum of 4 test specimens for each compression test.

   (One test specimen may be used to check the stripping strength.)

2. Obtain concrete for specimens from actual production batch.

3. 6 in. x 12 in. or 4 in. x 8 in. concrete test cylinder, ASTM C 31

** OR **

3. ______________ sized concrete cube, ______________.

   (Specify size. Cube specimens are usually 4 in. units, but 2 in. or 6 in. units are sometimes required. Larger specimens give more accurate test results than smaller ones. Source: molded individually, sawed from slab.)

4. Cure specimens using the same methods used for the precast concrete units until the units are stripped, then moist cure specimens until test.

C. Slump Tests. Shall be in accordance with ASTM test designation C-143. Periodically, slump tests shall be made of individual samples taken at approximately one-quarter and three-quarter points of an individual load to check the uniformity of the mixing operation. The Kelly Ball test may be used if proper correlation is made.

D. Tests for Air Content. If air entraining agents are used, tests shall be made at the same time that specimens for compression tests are made. The air content of the concrete shall be determined preferably by direct measurement, using the pressure method in accordance with ASTM test designation C-231 or gravimetrically, in accordance with ASTM test designation C-138.

E. Modulus of Elasticity. When camber and/or deflection is critical, the Engineer may request tests to determine the modulus of elasticity of the concrete mix to be used in the fabrication. When such tests are requested they will be paid for by the Owner.

F. Test Results

1. If the measured slump or air content falls outside

the limits specified, a check test shall be made. In the event of a second failure, the load of concrete represented shall be rejected.

2. To conform to the requirements of these specifications, the average of any three consecutive strength tests of the laboratory-cured specimens representing each class of concrete shall be equal to or greater than the specified strength, fc, and not more than 10 per cent of the strength tests shall have values less than the specified strength.

This is not applicable to the cylinders and tests made to determine transfer strength.

(With uniform procedures, individual strength tests are meaningless and can only be evaluated statistically. Permissibility of a small percentage of tests below the specified level recognizes the unavoidable variability of strength measurements. Conformance to these requirements assures that the concrete is being produced and controlled in such a way that it should be capable of developing adequate strength in the work. Evaluation of strength tests is discussed in ACI 214-65.)

G. Keep quality control records available for the Architect upon request for two years after final acceptance.

(These records should include mix designs, test reports, inspection reports, member identification numbers along with date cast, shipping records and erection reports.)

Part 3 - Execution

3.01 INSPECTION

A. Before erecting precast prestressed concrete members, the General Contractor shall verify that structure and anchorage inserts not within tolerances required to erect these units have been corrected.

B. Determine field conditions by actual measurements.

3.02 ERECTION

A. Clear, well-drained unloading areas and road access around and in the building (where appropriate) shall be provided and maintained by the General Contractor to a degree that the hauling and erection equipment for the prestressed precast concrete products are able to operate under their own power.

B. Erect adequate barricades, warning lights or signs to safeguard traffic in the immediate area of hoisting and handling operations.

C. Set precast units level, plumb, square, and true within the allowable tolerances. General Contractor shall be responsible for providing lines and grades in sufficient detail to allow installation.

D. In addition, the General Contractor shall also be responsible for:

1. Providing true, level bearing surfaces on all field placed bearing walls and other field placed supporting members.

2. Placement and accurate alignment of anchor bolts, plates or dowels in column footings, grade beams and other field placed supporting members.

3. All shoring required for composite beams and slabs.

(Construction tolerances for cast-in-place concrete, masonry, etc., should be specified in those sections of the specifications.)

E. Provide temporary supports and bracing as required to maintain position, stability, and alignment as units are being permanently connected. In addition, for those elements not able to be permanently tied into the structure prior to disengagement from the erection equipment, a bracing plan shall be prepared by a registered civil engineer in the employ of the precast concrete manufacturer.

This bracing shall be designed to resist a temporary seismic load from all contributing dead loads of 0.2 g. Allowable stresses may be taken to 90% of ultimate values for this analysis. In addition, the requirements of Cal-OSHA, Article 29, shall be met.

F. Erection Tolerances: Individual pieces are considered plumb, level, and aligned if the error does not exceed 1:500 excluding structural deformations caused by loads.

G. Delivery, Storage and Handling

1. Precast concrete members shall be lifted and supported during manufacturing, stockpiling, transporting and erection operations only at the lifting or supporting points, or both, as shown on the erection and shop drawings, and with approved lifting devices. All lifting

devices shall have a minimum safety factor of 4.

2. Transportation, site handling and erection shall be performed with acceptable equipment and methods, and by qualified personnel.

H. Storage at Jobsite

1. Store all units off ground.

2. Place stored units so that identification marks are discernible.

3. Separate stacked members by dunnage across full width of each bearing point.

4. Stack so that lifting devices are accessible and undamaged.

5. Do not use upper member of stacked tier as storage area for shorter member or heavy equipment.

I. Installation: Installation of precast prestressed concrete shall be performed by the manufacturer or a competent erector. Members shall be lifted by means of suitable lifting devices at points provided by the manufacturer.

J. Alignment: Members shall be properly aligned and leveled as required by the approved erection drawings. Variations between adjacent members shall be reasonably leveled out by jacking, loading, or any other feasible method as recommended by the manufacturer and acceptable to the Architect/Engineer.

K. Perform welding according to AWS D1.1 and AWS D12.1

L. Remove lifting devices and grout surfaces flush with concrete.

(Delete for deck members that receive cast-in-place concrete topping)

M. Refinish damaged surfaces to match adjacent areas.

3.03 ATTACHMENTS

A. Subject to approval of the Architect/Engineer, precast prestressed products may be drilled or "shot" provided no contact is made with the prestressing steel. Should spalling occur, it shall be repaired by the trade doing the drilling or the shooting.

3.04 CLEANING

(State whether erector or General Contractor responsible for cleaning.)

A. After Installation: _______________ shall clean soiled precast concrete surfaces with detergent and water, using fiber brush and sponge, and rinse thoroughly with clean water.

* * OR * *

A. Clean precast concrete surfaces with __________________.

(Acid-free commercial cleaners, steam cleaning, water blasting, sand blasting. Use sand blasting only for units with original sand blasted finish.)

B. Use acid solution only to clean particularly stubborn stains after more conservative methods have been tried unsuccessfully.

C. Use extreme care to prevent damage to precast concrete surfaces and to adjacent materials.

D. Rinse thoroughly with clean water immediately after using cleaner.

3.05 PROTECTION

A. The erector shall be responsible for any chipping, spalling, cracking or other damage to the units after delivery to the job site. After installation is completed, any further damage shall be the responsibility of the General Contractor.

* END OF SECTION *

The following should be placed in the General Conditions Section of the project specifications and renumbered for that section:

A.1 Payments

A.1.1 Monthly progress payments equal to 90% of the in-plant value will be made to the manufacturer for all products fabricated and stocked in the manufacturer's plant prior to delivery. Progress payments shall not relieve the manufacturer from compliance with terms of his contract with the Buyer.

(Industry practice is to fabricate, in advance,

products for each individual project in accordance with design requirements and dimensions for that project. Such products cannot normally be used on any other project. As monthly payment would be made for such products if they were fabricated on the site, monthly progress payments for such material fabricated off site and stored for delivery and erection as scheduled are justified.)

A.1.2 Full payment for all products delivered and/or installed will be made within 35 days of completion of all work under the contract and acceptance by the Architect.

REFERENCE STANDARDS

The information provided below includes the reference standards used in the preparation of this document and is intended only as an aid in understanding the reference standards contained in the Guide Specification part of this document. Later editions of these reference standards may require revision to the Guide Specification text. Guidance on the use of reference standards is contained in CSI Manual of Practice.

INTERNATIONAL CONGRESS OF BUILDING OFFICIALS (ICBO)

UBC-79, Uniform Building Code, 1979 Edition is the general building code enforced throughout California.

STRUCTURAL ENGINEERS ASSOCIATION OF CALIFORNIA (SEAOC)

Recommended Lateral Force Requirements and Commentary, Fourth Edition - 1975, often referred to as the "Blue Book."

AMERICAN CONCRETE INSTITUTE (ACI)

ACI 318-77, Building Code Requirements for Reinforced Concrete, covers design and construction of buildings of reinforced concrete. It is written so that it may be incorporated verbatim or adopted by reference in a general building code.

AMERICAN SOCIETY FOR TESTING AND MATERIALS (ASTM)

ASTM A 27, Mild-to-Medium-strength Carbon-Steel Castings for General Application, covers carbon-steel castings with tensile strength from 60,000 psi to 70,000 psi. Seven grades of steel castings are covered as follows: Grades N-1, N-2, U-60-30, 60-30, 64-35, 70-36, and 70-40. The numerals of the last five grades correspond to the significant numbers of the specified tensile strength and yield point (i.e., grade 60-30 requires a minimum tensile strength of 60,000 psi and a yield point of 30,000 psi. Grades N-1 and N-2 are not required to be mechanically tested.

ASTM A 36, Structural Steel, covers carbon steel shapes, plates, bars of structural quality for use in riveted, bolted or welded construction of bridges and buildings, and general construction purposes. Included are chemical requirements, tensile properties (that is, minimum tensile strength, minimum yield point and elongation), and bend test requirements.

ASTM A 47, Malleable Iron Castings, covers ferritic malleable irons used in general engineering castings for service at both normal and elevated temperatures. Grade 32510 castings have a minimum tensile strength of 50,000 psi. Grade 35018 castings have a minimum tensile strength of 53,000 psi.

ASTM A 82, Cold-Drawn Steel Wire for Concrete Reinforcement, covers cold-drawn steel wire to be used as such, or in fabricated form, for the reinforcement of concrete, in gages not less than 0.08 in. nor greater than 0.625 in. Included are the processes by which the steel may be made, the gage numbers, and the equivalent diameter in inches, the tensile properties required, the bending properties, the permissible variation in gage, and finish required on the wire.

ASTM A 108, Standard Quality Cold-Finished Carbon Steel Bars, covers cold finished steel bars used for the manufacture of headed welding studs, in grades 1010 through 1020 with a minimum tensile strength of 60,000 psi.

ASTM A 153, Zinc Coating (Hot-Dip) on Iron and Steel Hardware, covers zinc coatings applied by the hot-dipped process on iron and steel. Includes weights, distribution, adherence, testing and inspection of coatings, and fabrication considerations.

ASTM A 165, Electrodeposited Coatings of Cadmium on Steel, covers requirements for electro-plated cadmium coatings on steel articles. Three types of coatings are covered: Type NS, Type OS, and Type TS. The minimum thickness of cadmium coatings on significant surfaces of finished articles are 0.00050 in. for Type NS coating, 0.00030 in. for Type OS coating, and 0.00015 in. for Type TS coating. The dimensional tolerance of most threaded articles normally does not permit the application of coating thickness much greater than Type TS.

ASTM A 167, Stainless and Heat-Resisting Chromium-Nickel Steel Plate, Sheet, and Strip, covers soft, corrosion-resisting chromium-nickel steel plate, sheet, and strip.

ASTM A 184, Fabricated Deformed Steel Bar Mats for Concrete Reinforcement, covers material in mat or sheet form fabricated from steel bars to be used for the reinforcement of concrete. These mats consist of two layers of bars assembled at right angles to each other and clipped or welded at all or some of the intersections to form a rectangular grid.

ASTM A 185, Welded Steel Wire Fabric for Concrete Reinforcement, covers cold-drawn steel wire (ASTM A 82), as drawn or galvanized fabricated into sheet or mesh by arranging a series of longitudinal and transverse wires accurately spaced at right angles to each other and welding them together at all points of intersection. Included are the minimum requirements for bending of the finished fabric, the minimum weld strength between transverse spaces and the wire gage differential permitted.

ASTM A 307, Carbon Steel Externally and Internally Threaded Standard Fasteners ranging in diameter from 1/4" to 4", based upon a minimum tensile strength of 60,000 psi for Grade A material.

ASTM A 325, High Strength Bolts for Structural Steel Joints, including suitable nuts and plain hardened washers in minimum tensile strengths from 105,000 psi to 120,000 psi.

ASTM A 416, Uncoated Seven-Wire Stress-Relieved Strand for Prestressed Concrete, covers two grades of seven-wire, uncoated, stressed-relieved steel strand for use in pretensioned and post-tensioned prestressed concrete construction. Grade 250 and grade 270 have minimum ultimate strengths of 250,000 psi and 270,000 psie respectively, based on the nominal area of the strand.

ASTM A 421, Uncoated Stress-Relieved Wire for Prestressed Concrete, covers two types of uncoated stress-relieved ground high-carbon steel wire commonly used in prestressed linear concrete construction. Type BA wire is used for applications in which cold-end deformation is used for anchoring purposes (button anchorage) and Type WA wire is used for applications in which the ends are anchored by wedges, and no cold-end deformation of the wire is involved (wedge anchorage).

ASTM A 496, Deformed Steel Wire for Concrete Reinforcement, covers cold-worked, deformed steel wire to be used as such, or in fabricated form, for the reinforcement of concrete in sizes having nominal cross-sectional areas not less than 0.01 in.^{2} nor greater than 0.31 in.^{2}.

ASTM A 497, Welded Deformed Steel Wire Fabric for Concrete Reinforcement, covers welded wire fabric made from cold-worked deformed wire, or a combination of deformed and nondeformed wires, to be used for the reinforcement of concrete. The wire conforms to ASTM A 496 either solely or in combination with ASTM A 82 Cold-Drawn Steel Wire. Longitudinal and transverse members are electrical-reinforced welded to provide an average weld shear strength of 8,000 lbs.

ASTM A 615, Deformed and Plain Billet-Steel Bars for Concrete Reinforcement, covers deformed and plain billet-steel concrete-reinforcement bars. A deformed bar is defined as a bar that is intended for use as reinforcement in reinforced concrete construction. The surface of the bar is provided with lugs or protrusions (hereinafter called deformations) which inhibit longitudinal movement of the bar relative to the concrete which surrounds the bar in such construction and conform to the provisions of this specification. Bars are of two minimum yield levels, namely, 40,000 psi and 60,000 psi, designated as Grade 40 and Grade 60, respectively.

ASTM A 616, Rail-Steel Deformed and Plain Bars for Concrete Reinforcement, covers deformed and plain rail steel concrete reinforcement bars. A deformed bar is defined as a bar that is intended for use as reinforcement in reinforced concrete construction. The surface of the bar is provided with lugs or protrusions (hereinafter called deformations) which inhibit longitudinal movement of the bars relative to the concrete which surrounds the bars in such construction and conform to the provisions of this specification. Bars are of two minimum yield levels, namely, 50,000 psi and 60,000 psi, designated as Grade 50 and Grade 60, respectively.

ASTM A 617, Axle-Steel Deformed and Plain Bars for Concrete Reinforcement, covers deformed and plain axle steel concrete reinforcement bars. A deformed bar is defined as a bar that is intended for use as reinforcement in reinforced concrete construction. The surface of the bar is provided with lugs or protrusions (hereinafter called deformations) which inhibit longitudinal movement of the bar relative to the concrete which surrounds the bar in such construction and conform to the provisions of this specification. Bars are of two minimum yield levels, namely, 40,000 psi and 60,000 psi, designated as Grade 40 and Grade 60, respectively.

ASTM A 648, Hard Drawn Steel Wire for Prestressing, covers 3 classes of uncoated, high-strength hard drawn steel wire for use in the manufacture of prestressed concrete pipe. This material may also be used for spiral reinforcement for prestressed concrete columns and prestressed concrete piling.

ASTM A 663, Merchant Quality Hot-Rolled Carbon Steel Bars Subject to Mechanical Property Requirements, covers carbon steel bars furnished in the as-rolled condition, subject to mechanical property requirements and intended for general constructional applications. Grade designations range from 45 to 80 in increments of 5, and indicate the tensile strength (i.e., the minimum tensile strength of grade 65 is 65,000 psi).

ASTM A 666, Austenitic Stainless Steel, Sheet, Strip, Plate and Flat Bar for Structural Applications, covers six types of steels in 4 grades or strength levels. Of primary interest in precast concrete application are Grade A (30,000 psi yield strength) and Grade B (40,000 psi yield strength).

ASTM C 31, Making and Curing Concrete Testing Specimens in the Field, covers procedures for making and curing test specimens from concrete being used in construction.

ASTM C 33, Concrete Aggregates, covers fine and coarse aggregates other than light-weight aggregate for use in concrete. Included are descriptions of fine aggregate and of coarse aggregate, grading requirements, limits for deleterious substances, the physical properties of both fine and coarse aggregates and a list of the methods of sampling and testing.

ASTM C 39, Test for Compressive Strength of Cylindrical Concrete Specimens, covers determination of compressive strength of cylindrical concrete specimens, such as molded cylinders and drilled cores.

ASTM C 127, Specific Gravity and Absorption of Coarse Aggregates, covers the determination of bulk and apparent specific gravity and absorption of coarse aggregates. The absorption test determines the percentage by weight, of water absorbed by the aggregate specimen after being immersed at room temperature for 24 hours, compared with the specimen dried to constant weight at a temperature of 212° to 230°F and then cooled to room temperature.

ASTM C 128, Specific Gravity and Absorption of Fine Aggregates, covers the determination of bulk and apparent specific gravity and absorption of fine aggregate. The tests are similar to those required for coarse aggregate in ASTM C 127 but are modified as required to obtain accurate test results with smaller aggregate.

ASTM C 138, Standard Test Method for Unit Weight, Yield, and Air Content of Concrete, covers the determination of the weight per cubic foot of freshly mixed concrete, and gives formulas for the calculation of the yield, cement content, and the air content of the concrete.

ASTM C 143, Standard Test Method for Slump of Portland Cement Concrete, covers the traditional method used for determination of slump, both in the plant and in the laboratory.

ASTM C 150, Portland Cement, covers eight types of Portland cement, Type 1 through Type V and Type 1A through Type IIIA. The "A" designation indicates an air-entraining cement. Ordinarily only Type I, Type II, and Type III cements are used for precast concrete units, with an air-entraining agent added at the mixer if air-entrained concrete is required.

ASTM C 171, Sheet Materials for Curing Concrete, covers materials in sheet form used for covering the surface of concrete to inhibit moisture loss during the curing period, and in the case of white reflective type materials, to also reduce the temperature rise in concrete exposed to radiation from the sun.

ASTM C 172, Standard Test Method for Sampling Fresh Concrete, gives the proper procedure for obtaining representative samples of fresh concrete as delivered to the molds on which tests are to be performed to determine compliance with quality control requirements of the precast prestressed concrete specification.

ASTM C 173, Standard Test Method for Air Content of Freshly Mixed Concrete by the Volumetric Method, which allows the determination of the air content of freshly mixed concrete containing any type of aggregate, whether it be dense, cellular, or lightweight.

ASTM C 185, Air Content of Hydraulic Cement Mortar, covers the determination of whether or not the hydraulic cement under test meets teh air-entraining or non-air-entraining requirements of the applicable hydraulic cement specification for which the test is being made. The air content of concrete is influenced by many factors other than the potential of the cement for air entrainment.

ASTM C 231, Standard Test Method for Air Content of Freshly Mixed Concrete by the Pressure Method, which is based upon the observation of the change in volume of concrete with a change in pressure. This method is not applicable to lightweight aggregate concrete, or concrete made with aggregates of high porosity, for which ASTM C 173 should be used (Volumetric Method for Determining Air Content).

ASTM C 260, Air-Entraining Admixtures for Concrete, covers materials used as air-entraining admixtures to be added to concrete mixtures in the field. Included are the performance requirements of the material such as bleeding time of setting, compressive strength, flexural strength, resistance to freezing and thawing, and length change. Methods of taking samples for testing are also included.

ASTM C 309, Liquid Membrane-Forming Compounds for Curing Concrete, covers liquid membrane-forming compounds suitable for application on horizontal and vertical concrete surfaces to retard the loss of water during the early hardening period and, in the case of the white-pigmented compound, to also reduce the temperature rise in concrete exposed to radiation from the sun. The compounds covered by this specification are suitable for use as curing media for fresh concrete, and may also be used for further curing of concrete, and may also be used for curing.

ASTM C 330, Light-Weight Aggregates for Structural Concrete, covers light-weight aggregates intended for use in structural concrete in which prime considerations are lightness in weight and compressive strength of the concrete. The two general types of aggregates covered are those prepared by expanding, calcining, or sintering products such as blast furnace slag, clay diatomite, fly ash, shale or slate, and those prepared by processing natural materials such as pumice, scoria or tuff.

ASTM C 494, Chemical Admixtures for Concrete, covers material for use as chemical admixtures to be added to Portland cement concrete mixtures in the field for the purpose or purposes indicated for the five types as follows:

Type A - Water-reducing admixtures
Type B - Retarding admixtures
Type C - Accelerating admixtures
Type D - Water-reducing and retarding admixtures, and
Type E - Water-reducing and accelerating admixtures

ASTM D 2000, Standard Classification System for Rubber Products, which tabulates properties of elastomeric materials which may be used as bearing pads for precast prestressed concrete elements. This classification system is based upon the premise that all rubber and vulcanized elastomeric products can be arranged into characteristic material designations. This classification system provides guidance to the engineer in selection of practical elastomers and to provide a method for specifying these materials by use of a simple "line call-out" designation; which indicates the degree of heat resistance, the resistance of the elastomer to swelling in oil, durometer hardness, minimum tensile strength, grade number, and other suffix designations which further identify performance characteristics of the elastomer.

ASTM D2116, FEP - Fluorocarbon Molding and Extrusion Materials, covers the manufacture of this material used for low friction slide bearings.

ASTM E 329, Inspection and Testing Agencies for Concrete and Steel as Used in Construction, defines the duties and responsibilities and establishes minimum requirements for personnel and equipment of public and independent commercial materials inspection and testing agencies engaged in inspection and testing of concrete and steel as used in construction. It states the training and experience required by an agency manager, a supervising laboratory technician, and a supervising field technician, and lists the testing equipment required in the laboratory.

AMERICAN WELDING SOCIETY (AWS)

AWS D1.1, Structural Welding Code, covers types of welding and welding procedures approved for use without performing procedure qualification tests, and describes welding qualification procedures for welders, welding operators and tackers.

AWS D12.1, Reinforcing Steel Welding Code, covers requirements and procedures for welding reinforcing steel.

FEDERAL SPECIFICATIONS (FS)

FS TT-P-86G, Paint, Red-Lead Base, Ready-Mix, covers four types of paint suitable for use on iron and steel surfaces:

Type I - Red Lead - linseed oil paint
Type II - Red Lead - mixed pigment-alkyd varnish-linseed oil paint
Type III - Red Lead - alkyd varnish paint
Type IV - Red Lead - mixed pigment-phenolic varnish paint

Types I and II are intended for priming steel, either in the shop or field. Type I will bond to surfaces even with the presence of small amounts of corrosive materials found impractical to remove.

FS-MMM-G-650B, Grout, Adhesive, Epoxy Resin, Flexible, Filled, covers a two-component mineral filled thixo-tropic and flexible epoxy-resin-base grout. Type I is for use between 68°F and 104°F. Grade C is for use when application is by trowel or putty knife.

MILITARY SPECIFICATIONS (MIL)

MIL-P-21035 (AMD 1), Paint, High Zinc Dust Content, Galvanizing, Repair, covers high zinc dust paint originally intended for use in regalvanizing welds in steel. Paint meeting this standard must have 94% pigment by weight of nonvolatile content, and pigment must be 97% zinc by analysis.

STRUCTURAL STEEL PAINTING COUNSEL (SSPC)

SSPC-Paint 14-64T, No. 14 red lead, iron oxide and linseed oil primer, covers a slow drying primer for structural steel. It has excellent rust inhibitive characteristics, wetting ability, and resistance to weathering before finished coating. It may also be used for intermediate coats.

PRESTRESSED CONCRETE INSTITUTE (PCI)

MNL-116, Manual for Quality Control for Plants and Production of Precast Prestressed Concrete Products, covers production guidelines for the manufacture of high quality structural plant cast precast and prestressed concrete products. It is not intended as a specification document, although many requirements have been outlined.

MNL-120, PCI Design Handbook (Second Edition), provides the designer with information to design precast and prestressed concrete components, their connections, and to a limited degree total structures. Some of the building design concepts are limited by the state of the art of building construction in seismic zones as permitted by the Uniform Building Code, 1979 Edition.

MNL-124, Design for Fire Resistance of Precast Prestressed Concrete, gives procedures for rational design for fire, including requirements for joinery of prestressed and precast concrete components.

INSPECTION

During Manufacture:

A. Check qualifications of testing agencies.

B. Where in-plant testing is to be performed, see that manufacturer's testing equipment is calibrated by testing agency personnel, and that manufacturer's testing personnel are qualified to perform work.

C. Check casting and after-casting tolerances. Inspectors should be provided with quality instruments for measuring tolerances.

D. Examine mill certificates of prestressing strand and reinforcing.

E. Examine concrete test results.

Before Installation:

A. Check prestressed concrete erector's experience and qualifications.

B. Check welder's qualifications.

C. Assure that precast prestressed concrete members are stored on site using proper dunnage.

D. Verify that component identification marks are easily discernible.

E. Check field dimensions affecting erection.

F. Check bracing plan for temporarily unstable components/ structures.

During Installation:

A. Check erection tolerances.

B. See that units are clean and exposed metal spot painted.

C. Check that proper handling points are used as per erection drawings.

After Installation.

A. Inspect repairs for adequate match to surrounding concrete.

B. Inspect units after cleaning to see that they are properly prepared to receive caulking and dampproofing, if applicable.

GLOSSARY

Refer to MNL-116 for definitions of terms.

- NOTES -

- NOTES -

- NOTES -

- NOTES -

- NOTES -

- NOTES -